Lecture Notes in Computer Science 9340

Commenced Publication in 1973
Founding and Former Series Editors:
Gerhard Goos, Juris Hartmanis, and Jan van Leeuwen

Editorial Board

More information about this series at http://www.springer.com/series/7407

Marijn Heule · Sean Weaver (Eds.)

Theory and Applications of Satisfiability Testing – SAT 2015

18th International Conference
Austin, TX, USA, September 24–27, 2015
Proceedings

 Springer

Editors
Marijn Heule
University of Texas
Austin, TX
USA

Sean Weaver
Trusted Systems Research Group
Fort Meade, MD
USA

ISSN 0302-9743 ISSN 1611-3349 (electronic)
Lecture Notes in Computer Science
ISBN 978-3-319-24317-7 ISBN 978-3-319-24318-4 (eBook)
DOI 10.1007/978-3-319-24318-4

Library of Congress Control Number: 2015948854

LNCS Sublibrary: SL1 –Theoretical Computer Science and General Issues

Springer Cham Heidelberg New York Dordrecht London
© Springer International Publishing Switzerland 2015

Printed on acid-free paper

Springer International Publishing AG Switzerland is part of Springer Science+Business Media
(www.springer.com)

Preface

This volume contains the papers presented at the 18th International Conference on Theory and Applications of Satisfiability Testing (SAT 2015), held during September 24–27, 2015 in Austin, Texas, USA. SAT 2015 was colocated with Formal Methods in Computer-Aided Design (FMCAD 2015) and was hosted by the University of Texas at Austin.

The International Conference on Theory and Applications of Satisfiability Testing (SAT) is the primary annual meeting for researchers focusing on the theory and applications of the propositional satisfiability problem, broadly construed: Besides plain propositional satisfiability, it includes Boolean optimization (including MaxSAT and Pseudo-Boolean (PB), constraints), Quantified Boolean Formulas (QBF), Satisfiability Modulo Theories (SMT), and Constraint Programming (CP) for problems with clear connections to propositional reasoning. Many hard combinatorial problems can be tackled using SAT-based techniques, including problems that arise in formal verification, artificial intelligence, operations research, biology, cryptology, data mining, machine learning, mathematics, etc. Indeed, the theoretical and practical advances in SAT research over the past 20 years have contributed to making SAT technology an indispensable tool in various domains.

SAT 2015 welcomed scientific contributions addressing different aspects of SAT, including (but not restricted to) theoretical advances (including exact algorithms, proof complexity, and other complexity issues), practical search algorithms, knowledge compilation, implementation-level details of SAT solvers and SAT-based systems, problem encodings and reformulations, applications, as well as case studies and reports on insightful findings based on rigorous experimentation.

A total of 70 papers were submitted to SAT 2015, distributed into 44 regular papers (up to 15 pages excluding references), 17 short papers (up to eight pages excluding references), and nine tool papers (up to six pages excluding references). In contrast to recent SAT conferences, no paper submission was found to be out of scope for the conference. All 70 submissions were assigned for review to at least four Program Committee members and their selected external reviewers. Continuing the procedure initiated in SAT 2012, the review process included an author-response period, during which the authors of submitted papers were given the opportunity to respond to the initial reviews for their submissions. For reaching final decisions, a Program Committee discussion period followed the author-response period. This year, external reviewers supporting the Program Committee were also invited to participate directly in the discussions for the papers they reviewed. In the end, the Program Committee decided to accept 21 regular papers, two short papers, and seven tool papers. Two short papers were downgraded to tool papers.

In addition to presentations on the accepted papers, the scientific program of SAT 2015 included three invited talks:

- Dimitris Achlioptas (University of California Santa Cruz, USA)
  Random Formulas are Irrelevant, Right?
- Anna Slobodova (Centaur Technology, USA)
  Pragmatic Approach to Formal Verification
- Aaron Tomb (Galois, Inc., USA)
  Applying Satisfiability to the Analysis of Cryptography

SAT 2015 hosted various affiliated events, including two workshops on September 23:

- Sixth International Workshop on Pragmatics of SAT (PoS 2015)
  Organizers: Daniel Le Berre and Allen Van Gelder;
- Third International Workshop on Quantified Boolean Formulas (QBF 2015)
  Organizers: Florian Lonsing and Martina Seidl;

  and three competitions and system evaluations:

- SAT Race 2015
  Organizers: Tomas Balyo, Carsten Sinz, and Markus Iser;
- Max-SAT Evaluation 2015
  Organizers: Josep Argelich, Chu-Min Li, Felip Manya, and Jordi Planes;
- Pseudo-Boolean Evaluation 2015
  Organizers: Norbert Manthey and Peter Steinke

We would like to thank everyone who contributed to making SAT 2015 a success. First and foremost we would like to thank the members of the Program Committee and the additional external reviewers for their careful and thorough work, without which it would not have been possible for us to put together such an outstanding conference program. We also wish to thank all the authors who submitted their work for our consideration. We thank the SAT Association chair Armin Biere, vice chair John Franco, and treasurer Hans Kleine Büning for their help and advice in organizational matters. We wish to thank the workshop chair Albert Oliveras. The EasyChair conference system provided invaluable assistance in coordinating the submission and review process, as well as in the assembly of these proceedings. We also thank the local organization team for their efforts with practical aspects of local organization.

Finally, we gratefully thank the University of Texas at Austin, the SAT Association, the Artificial Intelligence journal, CyberPoint, Galois, Inc., Intel, and Microsoft Research for financial and organizational support for SAT 2015.

July 2015   Marijn Heule
                                                              Sean Weaver

Organization

Program Committee

Fahiem Bacchus	University of Toronto, Canada
Olaf Beyersdorff	University of Leeds, UK
Armin Biere	Johannes Kepler University, Austria
Leonardo De Moura	Microsoft Research, USA
Uwe Egly	Vienna University of Technology, Austria
John Franco	University of Cincinnati, USA
Enrico Giunchiglia	DIST - University of Genova, Italy
Youssef Hamadi	Microsoft Research, UK
Marijn Heule	The University of Texas at Austin, USA
Holger Hoos	University of British Columbia, Canada
Alexander Ivrii	IBM, Israel
Jie-Hong Roland Jiang	National Taiwan University, Taiwan
Matti Järvisalo	University of Helsinki, Finland
Oliver Kullmann	Swansea University, UK
Daniel Le Berre	CNRS - Université d'Artois, France
Ines Lynce	INESC-ID/IST, University of Lisbon, Portugal
Sharad Malik	Princeton University, USA
Panagiotis Manolios	Northeastern University, USA
Norbert Manthey	TU Dresden, Germany
Joao Marques-Silva	University College Dublin, Ireland
Alexander Nadel	Intel, Israel
Nina Narodytska	Samsung Research America, USA
Jakob Nordström	KTH Royal Institute of Technology, Sweden
Albert Oliveras	Technical University of Catalonia, Spain
Karem Sakallah	University of Michigan, USA
Roberto Sebastiani	DISI, University of Trento, Italy
Martina Seidl	Johannes Kepler University Linz, Austria
Bart Selman	Cornell University, USA
Laurent Simon	Labri, Bordeaux Institute of Technology, France
Carsten Sinz	Karlsruhe Institute of Technology, Germany
Stefan Szeider	Vienna University of Technology, Austria
Sean Weaver	Department of Defense, USA
Xishun Zhao	Institute of Logic and Cognition, Sun Yat-Sen University, China

Additional Reviewers

Abío, Ignasi
Aleksandrowicz, Gadi
Arbel, Eli
Audemard, Gilles
Balabanov, Valeriy
Balyo, Tomáš
Bapst, Victor
Bayless, Sam
Ben-Haim, Yael
Berkholz, Christoph
Bingham, Jesse
Bova, Simone
Cao, Weiwei
Chamarthi, Harsh Raju
Creignou, Nadia
Diller, Martin
Elffers, Jan
Fawcett, Chris
Fröhlich, Andreas
Ganian, Robert
de Haan, Ronald
Ignatiev, Alexey
Jain, Mitesh
Janota, Mikolas
Kim, Eun Jung
Kotthoff, Lars
Krakovski, Roi
Lagniez, Jean-Marie
Lauria, Massimo
Lee, Nian-Ze
Li, Chu-Min
Lonca, Emmanuel

Lonsing, Florian
López-Ibáñez, Manuel
Manquinho, Vasco
Marek, Victor
Martins, Ruben
Mencía, Carlos
Miksa, Mladen
Morgado, Antonio
Nevo, Ziv
Nieuwenhuis, Robert
Oetsch, Johannes
Oikarinen, Emilia
Ordyniak, Sebastian
Previti, Alessandro
Ray, Sayak
de Rezende, Susanna F.
Ryvchin, Vadim
Shen, Yuping
Slivovsky, Friedrich
Steinke, Peter
Strichman, Ofer
Subramanyan, Pramod
Trentin, Patrick
Tu, Kuan-Hua
Van den Broeck, Guy
Vinyals, Marc
Vizel, Yakir
Wang, Hung-En
Wetzler, Nathan
Widl, Magdalena
Yamada, Akihisa
Yue, Weiya

Pragmatic Approach to Formal Verification

Anna Slobodova

Centaur Technology, Taipet, Taiwan
anna@centtech.com

After more than two decades of hard work by researchers in academia and industry, formal methods have been accepted as a viable part of the hardware design and validation process. We now have a better understanding of what is a cost-effective use of formal methods, and companies even set aside some resources for the further development of formal tools. Spreading formal methods into industrial scale software verification broadened the user population and increased motivation for development of such tools.

Centaur Technology is one of the companies that adopted formal verification (FV) as a part of their production flow. Our company designs Intel compatible x86-64 microprocessors. It does it with a relatively small team. To assure the quality of the design, a lot of effort is spent in the process of validation. Since new additions to x86-64 architecture widened the data on which instructions are performed, classic simulation provides even less coverage with respect to all possible inputs to the system than a few years ago. The FV team at Centaur Technology was created as a reaction to this trend as well as the fact that the capacity of formal tools has reached a level where they can be successful even on industrial scale designs. There are also more publicly available off-shelf formal point tools (SAT, SMT, Model-checkers, etc.) that can be incorporated into more complex validation framework. A pilot project [1] that discovered a corner case bug in floating-point arithmetic was convincing enough to justify investing in a small FV team.

Our verification framework is built on top of the ACL2 theorem prover [2]. There are many decision procedures built in the logic of ACL2 and proved correct within this logic. For example, packages exist, defined inside the logic, for Binary Decision Diagrams (BDD) [3] and for And-Inverter Graph (AIG) manipulation [4]. It has a symbolic simulator called GL [5] that can automate the proof of theorems over finite domains. GL can be combined with word-level symbolic simulation of a hardware model to relate that model to its specification.

While we strive for rigorous analysis, our verification approach is very pragmatic. We connected some "trusted" tools to ACL2, for instance various Satisfiability solvers (e.g., Glucose, Penelope, Lingeling, Riss3G), or the ABC model/equivalence checking tool [6]. The results from these tools are tagged as "trusted" (unverified) by ACL2. However, for tools that provide a proof trace, in some cases, we can verify the correctness of those results within ACL2. We prefer this approach to blindly trusting the tools. However, when we exhaust approaches that have verifiable results, we are willing to use unverified methods as well.

Our team has worked on a variety of design and verification problems [7, 8] including microcode verification [9] and transistor-level validation. Our main focus remains the verification of Register-transfer-level (RTL) designs written in Verilog or System Verilog. These designs are translated to a formal model with well defined semantics in ACL2 which allows for a rigorous analysis. The formal model is represented using our specialized word-level expression language called SVEX. An SVEX model can be symbolically executed and compared to the symbolically evaluated specification using GL. This process includes a word-level rewriting before it is bit-blasted into AIG (and later into Conjunctive Normal Form – CNF), or BDD representation. A SAT solver or BDD package is used to check equivalences. It is worthwhile to mention that the translation steps from SVEX to AIGs and from AIGs to CNFs or BDDs are proved correct – in fact, the whole process creates an ACL2 theorem, where the only unverified portion is typically due to the use of a SAT solver. Such theorems usually prove properties of microoperations as executed by the RTL design of the Centaur microprocessor. This is similar to the Symbolic Trajectory Evaluation based approach taken by the Intel FV team [10]. However, the main difference is that our approach allows for the richer reasoning provided by ACL2 including proofs about specifications, composition of GL theorems, etc.

The specifications of the microoperations used for the verification of RTL can be used to define operational semantics of the machine that executes microcode. This allows for a seamless transition between two different verification domains – hardware and microcode. We also use these microoperation specifications as the basis of a formal executable model of a subset of the x86 ISA that can be validated by running against Intel, AMD or Centaur's existing microprocessors, running a few hundred thousand tests per second. This increases our confidence in our specifications and clarifies inconsistencies in the ISA manual [11]. Overall, our framework offers a more holistic approach to microprocesor validation.

In this talk we give you a top-down overview of our verification methodology. We point out problems that can be solved using automatic methods, and describe the aspects of our work that cannot be easily automated. We hope for feedback that may help us to be more efficient in use of your tools (SAT solvers in particular). We also invite anybody who would like to collaborate on improving the AIG to CNF encoding or SVEX to AIG encoding.

References

1. Hunt Jr, W.A., Swords, S.: Centaur technology media unit verification. In: Bouajjani, A., Maler, O. (eds.) CAV 2009. LNCS, vol. 5643, pp. 353–367. Springer, Heidelberg (2009)
2. Kaufmann, M., Moore, J.S.: ACL2 home page. http://www.cs.utexas.edu/users/moore/acl2
3. Swords, S., Hunt Jr, W.A.: A mechanically verified AIG-to-BDD conversion algorithm. In: Kaufmann, M., Paulson, L.C. (eds.) ITP 2010. LNCS, vol. 6172, pp. 435–449. Springer, Heidelberg (2010)
4. Davis, J., Swords, S.: Verified aig algorithms in acl2. In: Gamboa, R., Davis, J. (eds.) Proceedings International Workshop on the ACL2 Theorem Prover and its Applications, Laramie, Wyoming, USA, May 30–31, 2013. Volume 114 of Electronic Proceedings in Theoretical Computer Science. Open Publishing Association, pp. 95–110 (2013)

5. Swords, S.: A verified framework for symbolic execution in the ACL2 theorem prover. Ph.D. thesis, Department of Computer Sciences, The University of Texas at Austin (2010)
6. Brayton, R., Mishchenko, A.: ABC: an academic industrial-strength verification tool. In: Touili, T., Cook, B., Jackson, P. (eds.) CAV 2010. LNCS, vol. 6174, pp. 24–40. Springer, Heidelberg (2010)
7. Hunt, Jr., W.A., Swords, S., Davis, J., Slobodova, A.: Use of formal verification at centaur technology. In: Hardin, D.S. (ed.) Design and Verification of Microprocessor Systems for High-Assurance Applications, pp. 65–88. Springer (2010)
8. Slobodova, A., Davis, J., Swords, S., Hunt, Jr., W.A.: A flexible formal verification framework for industrial scale validation. In: Proceedings of Formal Methods and Models for Codesign (MEMOCODE), pp. 89–97. IEEE (2011)
9. Davis, J., Slobodova, A., Swords, S.: Microcode verification – another piece of the microprocessor verification puzzle. In: Klein, G., Gamboa, R. (eds.) ITP 2014. LNCS, vol. 8558, pp. 1–16. Springer, Heidelberg (2014)
10. Kaivola, R., et al.: Replacing testing with formal verification in intel core i7 processor execution engine validation. In: Bouajjani, A., Maler, O. (eds.) CAV 2009. LNCS, vol. 5643, pp. 414–429. Springer, Heidelberg (2009)
11. Intel: Intel 64 and IA-32 Architectures Software Developer's Manual. Order Number: 325462–053US. (January 2015). http://www.intel.com/content/dam/www/public/us/en/documents/manuals/64-ia-32-architectures-software-developer-manual-325462.pdf. Accessed March 2015

Applying Satisfiability to the Analysis of Cryptography

Aaron Tomb

Galois, Inc., Portland, USA

Cryptographic algorithms and satisfiability (SAT) solvers are intrinsically well-matched. Most cryptographic algorithms are presented in terms of a bounded number of operations on finite collections of bits: either to shuffle bits around in unpredictable ways (as typical in block ciphers or hash functions), or to perform algebraic operations on modular integers represented as vectors of bits (as in public key cryptography). This property allows many cryptographic algorithms, as well as statements about properties of those algorithms, to be represented as potentially large but purely propositional expressions.

Once cryptographic algorithms are represented as boolean terms, analyzing them with SAT solvers is straightforward. Any query that can be described using purely existential or purely universal quantifiers over the free variables in these terms can, in theory, be decided. Such queries include checking functions for injectivity (important for key expansion [5] or random number generation [1]); finding collisions in hash functions [6]; comparing two alternative implementations of an algorithm for equality [2]; simplifying differential cryptanalysis [4]; and simplifying side channel attacks [7], among others.

Many of these queries, though theoretically decidable, are not solvable in practice. Finding hash collisions with modern secure hash functions, for instance, is intractable. However, even in these cases, SAT solvers can help measure the relative security of different algorithms. Reduced-round versions of widely-used hashing or encryption algorithms are frequently analyzable by modern solvers (discovering secret keys or hash collisions), and the relative difficulty of analysis between alternative algorithms can be a useful comparison factor [3].

In addition to pure satisfiability queries, some other interesting problems can be stated with alternating quantifiers around purely propositional bodies. Such problems include synthesis of cryptographic implementation code from a specification. While these problems are less thoroughly researched, they are within the theoretical domain of emerging tools, such as quantified boolean formula (QBF) solvers and the exists-forall extensions in some satisfiability modulo theories (SMT) solvers.

This talk will walk through some of the properties of cryptographic code that are within the reach of existing solvers, as well as other properties that are currently infeasible but that may be solvable with future tools. It will also describe some specific tools that can be useful for applying SAT solvers to the analysis of cryptographic algorithms.

References

1. Dörre, F., Klebanov, V.: Pseudo-random number generator verification: a case study. In: 7th Working Conference on Verified Software: Theories, Tools, and Experiments (2015)
2. Erkök, L., Matthews, J.: Pragmatic equivalence and safety checking in Cryptol. In: Proceedings of the 3rd Workshop on Programming Languages Meets Program Verification, PLPV 2009, pp. 73–82. ACM, New York (2008)
3. Homsirikamol, E., Morawiecki, P., Rogawski, M., Srebrny, M.: Security margin evaluation of SHA-3 contest finalists through SAT-based attacks. In: Cortesi, A., Chaki, N., Saeed, K., Wierzchoń, S. (eds.) CISIM 2012. LNCS, vol. 7564, pp. 56–67. Springer, Heidelberg (2012)
4. Kölbl, S., Leander, G., Tiessen, T.: Observations on the SIMON block cipher family. In: Gennaro, R., Robshaw, M. (eds.) Advances in Cryptology — CRYPTO 2015. LNCS, vol. 9215, pp. 161–185. Springer, Berlin Heidelberg (2015)
5. Lafitte, F., Markowitch, O., Van Heule, D.: SAT based analysis of LTE stream cipher ZUC. In: Proceedings of the 6th International Conference on Security of Information and Networks, SIN 2013, pp. 110–116. ACM, New York (2013)
6. Mironov, I., Zhang, L.: Applications of SAT solvers to cryptanalysis of hash functions. In: Biere, A., Gomes, C.P. (eds.) SAT 2006. LNCS, vol. 4121, pp. 102–115. Springer, Heidelberg (2006)
7. Renauld, M., Standaert, F.-X.: Algebraic side-channel attacks. In: Bao, F., Yung, M., Lin, D., Jing, J. (eds.) Inscrypt 2009. LNCS, vol. 6151, pp. 393–410. Springer, Heidelberg (2010)

Random Formulas are Irrelevant, Right?

Dimitris Achlioptas

Department of Computer Science
University of California Santa Cruz, Santa Cruz, USA

Let $F_k(n, m)$ denote a Boolean formula in Conjunctive Normal Form (CNF) with m clauses over n variables, whose clauses are chosen uniformly, independently and without replacement among all $2^k \binom{n}{k}$ non-trivial clauses of length k, i.e., clauses with k distinct, non-complementary literals. Say that a sequence of random events $\mathcal{E}_n$ occurs with high probability (w.h.p.) if $\lim_{n \to \infty} \Pr[\mathcal{E}_n] = 1$.

Franco and Paull pioneered the analysis of random k-CNF formulas in [9], where they noted that $F_k(n, m)$ is w.h.p. unsatisfiable if $m = rn$ and $r \geq 2^k \ln 2$. Chao and Franco [4] complemented this by proving that if $r < 2^k/k$, then UNIT CLAUSE PROPAGATION (UCP) alone finds a satisfying truth assignment w.h.p., thus establishing $m = \Theta(n)$ as the most interesting range for random k-SAT.

Chvátal and Szemerédi [5] proved that random k-CNF formulas w.h.p. have exponential resolution complexity, implying that if F is a random k-CNF formula with $r \geq 2^k \ln 2$, then w.h.p. *every* DPLL-type algorithm needs exponential time to prove its unsatisfiability. The works of Mitchell, Selman, and Levesque [12] and of Kirkpatrick and Selman [10] gave birth to the *Satisfiability Threshold Conjecture*: for every $k \geq 3$, the probability of satisfiability exhibits a 0/1 law around a critical density r_k. The conjecture was made particularly attractive by the apparent maximization of algorithmic hardness around r_k, attracting attention in computer science, mathematics, and, most fruitfully, statistical physics.

Non-rigorous but mathematically sophisticated methods of statistical physics predicted that while for low densities the set of satisfying assignments forms a single giant cluster, at the critical density $r \sim (2^k/k) \ln k$ it shatters into exponentially many tiny clusters, each of which is far apart from all others. Moreover, it was predicted that every path connecting satisfying assignments in different clusters must pass through assignments that violate $\Omega(n)$ constraints, and that the majority of variables inside each cluster are frozen, i.e., take the same value in all solutions in the cluster; thus getting even a single frozen variable wrong requires changing $\Omega(m)$ variables and going over a huge energy barrier to correct the error. This picture suggested an *algorithmic barrier* and fit perfectly with the fact that efforts to improve upon the UCP lower bound of $2^k/k$ by analyzing more sophisticated algorithms only improved the leading constant.

To overcome the algorithmic barrier, Achlioptas and Moore [3] introduced the second moment to the study of random k-CNF formulas. A long sequence of subsequent refining works culminated very recently in the seminal work of Ding, Sly, and Sun [8] determining the satisfiability threshold *exactly* for $k \geq k_0$, showing that it scales as $2^k \ln 2 - (1 + \ln 2)/2 + o_k(1)$, a mere additive constant below the trivial upper bound $2^k \ln 2$ (unfortunately $k_0 \sim 10^6$). While [8] proceeds along a different path than the original physics calculations, the latter offered indispensable "clues" on how to appropriately refine the second method. Notably, the second moment method offers no guidance whatsoever on how to *find* a satisfying assignment. In [6], by analyzing a new, but still relatively simple, algorithm Coja-Oghlan matched the shattering threshold of $(2^k/k) \ln k$ which remains to date the greatest density for which algorithms provably find solutions.

Over the course of the last fifteen years, a large fraction of the physics picture has been established rigorously. Specifically, the shattering threshold was established for all $k \geq 8$ by Achlioptas and Coja-Oghlan [1], while Achlioptas, Coja-Oghlan and Ricci-Tersenghi [2] proved that for $r \geq (4/5 + o_k(1))2^k \ln k$ the majority of variables are frozen in *every* cluster for all $k \geq 9$.

In [11], Mézard, Parisi, and Zecchina proposed a new satisfiability algorithm called Survey Propagation (SP) which performs extremely well of random 3-CNF formulas. Unfortunately, physics-style calculations of Montanari, Ricci-Tersenghi and Semerjian [13] showed that a close relative of SP, namely Belief Propagation, fails to overcome the algorithmic barrier as k is increased. Coja-Oghlan [7] made this a rigorous result, by an argument that strongly suggests that SP also fails to find satisfying assignments past the algorithmic barrier as k is increased.

References

1. Achlioptas, D., Coja-Oghlan, A.: Algorithmic barriers from phase transitions. In: Proceedings FOCS 2008, pp. 793–802 (2008)
2. Achlioptas, D., Coja-Oghlan, A., Ricci-Tersenghi, F.: On the solution-space geometry of random constraint satisfaction problems. Random Struct. Algorithms **38**(3), 251–268 (2011)
3. Achlioptas, D., Moore, C.: Random k-SAT: two moments suffice to cross a sharp threshold. SIAM J. Comput. **36**(3), 740–762 (2006)
4. Chao, M.-T., Franco, J.: Probabilistic analysis of a generalization of the unit-clause literal selection heuristics for the k-satisfiability problem. Inform. Sci. **51**(3), 289–314 (1990)
5. Chvátal, V., Szemerédi, E.: Many hard examples for resolution. J. Assoc. Comput. Mach. **35** (4), 759–768 (1988)
6. Coja-Oghlan, A.: A better algorithm for random k-sat. In: Albers, S., Marchetti-Spaccamela, A., Matias, Y., Nikoletseas, S., Thomas, W. (eds.) ICALP 2009, Part 1. LNCS, vol. 5555, pp. 292–303 (2009)
7. Coja-Oghlan, A.: On belief propagation guided decimation for random k-sat. In: Proceedings of SODA 2011, pp. 957–966 (2011)
8. Ding, J., Sly, A., Sun, N.: Proof of the satisfiability conjecture for large k. In: Proceedings of STOC 2015, pp. 59–68. ACM (2015)
9. Franco, J., Paull, M.: Probabilistic analysis of the Davis-Putnam procedure for solving the satisfiability problem. Discrete Appl. Math. **5**(1), 77–87 (1983)
10. Kirkpatrick, S., Selman, B.: Critical behavior in the satisfiability of random Boolean expressions. Science **264**(5163), 1297–1301 (1994)
11. Mézard, M., Parisi, G., Ricardo, Z.: Analytic and algorithmic solution of random satisfiability problems. Science **297**, 812–815 (2002)

12. Mitchell, D.G., Selman, B., Levesque, H.J.: Hard and easy distributions of sat problems. In: AAAI, pp. 459–465 (1992)
13. Montanari, A., Ricci-Tersenghi, F., Semerjian, G.: Solving constraint satisfaction problems through belief propagation-guided decimation. *CoRR*, abs/0709.1667

Contents

CCAnr: A Configuration Checking Based Local Search Solver for Non-random Satisfiability

Shaowei Cai[1]([✉]), Chuan Luo[2], and Kaile Su[3,4]

[1] State Key Laboratory of Computer Science, Institute of Software, Chinese Academy of Sciences, Beijing, China
shaoweicai.cs@gmail.com
[2] School of EECS, Peking University, Beijing, China
chuanluosaber@gmail.com
[3] Department of Computer Science, Jinan University, Guangzhou, China
[4] IIIS, Griffith University, Nathan, Australia
k.su@griffith.edu.au

Abstract. This paper presents a stochastic local search (SLS) solver for SAT named CCAnr, which is based on the configuration checking strategy and has good performance on non-random SAT instances. CCAnr switches between two modes: it flips a variable according to the CCA (configuration checking with aspiration) heuristic if any; otherwise, it flips a variable in a random unsatisfied clause (which we refer to as the focused local search mode). The main novelty of CCAnr lies on the greedy heuristic in the focused local search mode, which contributes significantly to its good performance on structured instances. Previous two-mode SLS algorithms usually utilize diversifying heuristics such as *age* or randomized strategies to pick a variable from the unsatisfied clause. Our experiments on combinatorial and application benchmarks from SAT Competition 2014 show that CCAnr has better performance than other state-of-the-art SLS solvers on structured instances, and its performance can be further improved by using a preprocessor CP3. Our results suggest that a greedy heuristic in the focused local search mode might be helpful to improve SLS solvers for solving structured SAT instances.

1 Introduction

The Satisfiability problem (SAT) is a prototypical NP-complete problem of importance in both theory and applications. Two popular approaches for solving SAT are conflict driven clause learning (CDCL) and stochastic local search (SLS). SLS algorithms for SAT perform a local search in the space of truth assignments by starting with a complete assignment, and then repeatedly flipping the truth value of a variable until a satisfying assignment has been found or some limits (usually the time limit) have been reached. The function for choosing the variable to be flipped is usually denoted as *pickVar*.

SLS algorithms for SAT mainly fall into two types: focused local search (FLS) and two-mode SLS. Focused local search (as called in [15,17]) always picks the flip variable from an unsatisfied clause [13,19]; two-mode SLS [1,5,10,16,18]

© Springer International Publishing Switzerland 2015
M. Heule and S. Weaver (Eds.): SAT 2015, LNCS 9340, pp. 1–8, 2015.
DOI: 10.1007/978-3-319-24318-4_1

switches between "global local search" (where the flip variable is chosen from a candidate set filtered from the set of all variables) and focused local search, usually depending on whether a local optimum is reached. Also, there is a significant line of research concerns about weighting techniques [8,14,20,21], which are usually utilized in two-mode SLS algorithms.

SLS is well known as the most effective approach for solving random satisfiable instances, and SLS solvers are often evaluated on random k-SAT benchmarks. For structured instances, SLS solvers have been considered not effective as complete solvers (particularly CDCL ones) for a long time. Nevertheless, recent progress shows promising results of SLS solvers, particularly two-mode SLS solvers, on crafted satisfiable instances. Modern two-mode SLS solvers are competitive and complementary with complete solvers in solving crafted instances. For example, in the crafted SAT track of SAT Competition 2009, the SLS solver Sattime solved 109 instances while the best CDCL solved 93 instances [11]; in SAT Competition 2013, CCAnr solved 21 crafted SAT instances that the best complete solver glucose (v2.3) failed to solve, while glucose solved 53 instances that CCAnr failed to solve in the same track (Hard-combinatorial SAT track).[1] Indeed, the top three solvers in the Hard-combinatorial SAT track of SAT Competition 2014, namely Sparrow2riss [2], CCAnr+glucose [3] and SGseq [9] are all hybrid solvers combining an SLS solver and a complete solver.

In this paper, we present the CCAnr solver, which is a two-mode SLS solver designed for solving structured instances. Existing two-mode SLS solvers for SAT, including state-of-the-art ones such as Sparrow [1], Sattime [11] and CCASat [5], employ greedy heuristics in the global local search mode with the aim of decreasing the number of unsatisfied clauses, and employ diversifying heuristics in the focused local search mode with the aim of better exploring the search space and avoiding local optima. However, CCAnr employs a greedy heuristic (i.e., picking the one with the greatest *score*) in the focused local search mode, which significantly contributes to its good performance on structured instances. Our experiments comparing different versions of CCAnr with various heuristics in the focused local search mode show that, for solving structured instances, it could be helpful to incorporate greedy heuristics in the focused local search mode.

The good performance of CCAnr as compared to other SLS solvers is also confirmed by the results in the SAT Competitions 2013 and 2014. In the Hard-combinatorial SAT track of SAT Competition 2013, CCAnr solves more instances than other SLS solvers except Sparrow+CP3, which solves only 2 more instances than CCAnr. In SAT Competition 2014, CCAnr+glucose is ranked second in the same track, solving only 1 less instance than Sparrow2riss. However, we note that both Sparrow+CP3 and Sparrow2riss utilize a powerful preprocessor CP3 [12] while CCAnr only performs unit propagation before local search. In this sense, CCAnr can be considered as the best pure SLS solver in this track. In this paper, we also combine CCAnr with the preprocessor CP3, and the resulting solver solves more instances in those benchmarks, yielding further improvement over other SLS solvers.

---

[1] http://satcompetition.org/edacc/SATCompetition2013/experiment/20/

2   Solver Description

CCAnr is built on top of the Swcca solver [4], and keeps the main technique there, namely the configuration checking (CC) strategy with an aspiration mechanism. In this section, we first introduce the CCA heuristic and the Swcca algorithm, and then describe the CCAnr algorithm by specifying the difference from Swcca. CCAnr is open-source and the code is available for download at http://lcs.ios. ac.cn/~caisw/SAT.html. Readers interested in the implementation may like to refer to the codes for more details.

2.1   The CCA Heuristic and Swcca

Originally proposed in [6], the configuration checking (CC) strategy aims at avoiding cycling (i.e., revisiting the already visited assignments too early) in local search. CC has proved effective in local search for SAT and has been widely used in recent SLS solvers. In the context of SAT, the idea of CC is to forbid flipping a variable if its *configuration* has not been changed after its last flip, where the *configuration* of a variable typically refers to truth values of all its neighbouring variables [4,5]. The CCA heuristic combines the CC strategy with an aspiration mechanism, which allows to flip a variable forbidden by CC if it has a significant *score*, recalling that a variable's *score* is the change on the number (or the total weight) of satisfied clause produced by its flip.

It is easy to see from the above discussions that, there are two kinds of variables of importance in the CCA heuristic, which are defined as follows.

- A *configuration changed decreasing* (CCD) variable is a decreasing variable (with positive *score*) whose configuration has been changed (i.e., at least one of its neighbouring variables has been flipped) since its last flip.
- A *significant decreasing* (SD) variable is a variable with $score(x) > g$, where g is a positive sufficient large integer, and in this work g is set to the averaged clause weight (over all clauses) $\overline{w}$.

A two-mode SLS algorithms based on the CCA heuristic works as follows. In the global local search mode, it prefers to pick a CCD variable with the greatest *score* to flip. If there are no CCD variables, an SD variable with the greatest *score* is selected there is one. If there are neither CCD variables nor SD variables, the algorithm updates the clause weights and switches to the the focused local search mode, where a variable in a random unsatisfied clause is picked to flip.

Based on the CCA heuristic, we have developed the Swcca algorithm [4], which has good performance on random 3-SAT instances and crafted instances. The pickVar function of Swcca is shown in Algorithm 1.

2.2   The CCAnr Algorithm

CCAnr is an improved version of Swcca for structured instances. Starting with a randomly generated complete assignment, CCAnr iteratively flips a variable until a satisfying assignment has been found or the given time limit has been reached. There are two differences between CCAnr and Swcca algorithms.

Algorithm 1. *pickVar* function in Swcca

---

1 **if** *CCD variables exist* **then return** a CCD variable with the greatest *score*, breaking ties in favor of the oldest one;
2 **if** *SD variables exist* **then return** an SD variable with the greatest *score*, breaking ties in favor of the oldest one;
3 increases clause weights of all unsatisfied clauses by one;
4 **if** $\overline{w} > \gamma$ **then** for each clause c_i, $w(c_i) := \lfloor \rho \cdot w(c_i) \rfloor + \lfloor (1 - \rho)\overline{w} \rfloor$;
5 pick a random unsatisfied clause c;
6 **return** the oldest variable in c;

---

1. the smoothing formula (line 4 in Algorithm 1) is generalized as $w(c_i) := \lfloor \rho \cdot w(c_i) \rfloor + \lfloor q \cdot \overline{w} \rfloor$.
2. in the focused local search mode, CCAnr picks the variable with the greatest *score* from an unsatisfied clause, breaking ties by favoring the oldest one (replace line 6 in Algorithm 1).

As will be shown in the experiment parts, the first modification, which indeed is just about parameter setting, has little impact on the performance of CCAnr. The second modification (i.e., the greedy heuristic in focused local search mode) makes the essential contributions to the good performance of CCAnr, and renders CCAnr much more effective than the original solver Swcca on structured instances, and also outperforms (although sometimes slightly) other state-of-the-art SLS solvers for structured instances.

2.3   Implementation Details

CCAnr is implemented in C++, and compiled by g++ with the 'O3' optimization option. There are three parameters in CCAnr: the threshold parameter γ, and two factor parameters ρ and q, all of which belong to the clause weighting scheme. The parameters are set as follows: $\gamma = 300$; $\rho = 0.3$; q is set to 0 if $r \leq 15$, and 0.7 otherwise (r is the clause-to-variable ratio of the instance). This setting is adopted in CCAnr in SAT Competition 2013 and CCAnr+glucose in SAT Competition 2014. After the competition, we found that actually setting q to 0.7 (which equals $1 - \rho$, as the same setting in Swcca) for all instances has very close performance.

3   Experimental Results

According to results of SAT Competitions 2013 and 2014, CPSparrow, CCAnr and Sattime are the currently the best three SLS solvers for structured SAT instances. However, CCAnr itself did not participate in SAT Competition 2014; instead, the hybrid solver CCAnr+glucose did. We carried out experiments to compare CCAnr with CPSparrow (version sc14) and Sattime2014r as well as Sattime2013 (the version used in SGseq [9] in Hard-combinatorial track of SAT Competition 2014).

We also develop a solver called CCAnr+CP3 which employs the CP3 preprocessor before calling CCAnr (as CPSparrow does), and include it in the experiments. Sattime already has a sophisticated preprocessing procedure in it.

Our experiments are conducted on Hard-combinatorial SAT benchmark (150 instances) which is also known as the crafted benchmark, and application SAT benchmark (150 instances) in SAT Competition 2014. The experiments are carried out on a machine under GNU/Linux, using 2 cores of Intel Core i7 2.4 GHz and 7.8 GByte RAM. Each solver was executed one time on each instance, as in competitions.

The results are summarized in Table 1, where the first row presents the results on the Hard-combinatorial SAT benchmark, and the second row presents the results on the application SAT benchmark. CCAnr+CP3 gives better performance than other solvers on both benchmarks. We also observe that even without CP3, CCAnr solves more instances than the two competitors CPSparrow and Sattime2014r in this experiment, although indeed the performance of CCAnr and CPSparrow is indistinguishable. In particular, CCAnr solves 13 instances in the Hard-combinatorial SAT benchmark which were not solved by CDCL solvers in SAT Competition 2014.

Table 1. Comparative results on the hard-combinatorial SAT and the application SAT benchmarks from SAT Competition 2014.

Benchmark	CCAnr+CP3		CCAnr		CPSparrow		Sattime2014r		Sattime2013	
	#solv.	par10 time	#solv.	par10 time	#solv.	par10 time	#solv.	par10 time	#solv.	par10 time
SC14_HC_SAT	**60**	**30115**	56	31427	55	31727	46	34784	43	35864
SC14_APP_SAT	**35**	**38440**	29	40449	28	40773	25	41822	23	42539

To demonstrate the importance of the greedy heuristic in the focused random mode in CCAnr, we also compare it with three alternatives. The alternatives are modified from from CCAnr as follows:

- CCAnr_fq (short for CCAnr with fixed q) is the CCAnr solver with a fixed setting of parameter q in the smoothing formula: $q = 0.7$ (i.e., $q = 1 - \rho$) for all instances (in this case, the clause weighting is the same with Swcca).
- CCAnr_rand picks a random variable in the selected unsatisfied clause in the focused local search mode.
- CCAnr_age picks the oldest variable in the selected unsatisfied clause in the focused local search mode. CCAnr_age corresponds to the Swcca algorithm, but use the parameter setting of SWT scheme in CCAnr, so that we can see the performance improvement due to the greedy heuristic.

The comparative results of CCAnr and its alternatives are reported in Table 2. The performance of CCAnr_fq is quite close to that of CCAnr, which indicates the conditional setting rule for parameter q in the smoothing formula (used in

Table 2. Comparative results of CCAnr and its alternatives on the hard-combinatorial SAT and the application SAT benchmarks from SAT Competition 2014.

Benchmark	CCAnr		CCAnr_fq		CCAnr_rand		CCAnr_age	
	#solv.	par10 time	#solv.	par10 time	#solv.	par10 time	#solv.	par10 time
SC14_HC_SAT	56	31427	54	32113	47	34514	41	36484
SC14_APP_SAT	29	40449	28	40755	25	41753	22	42709

CCAnr version in competitions) is not a main factor of the good performance of CCAnr. The performance of CCAnr is considerably better than that of CCAnr_age and CCAnr_rand. Noting that the only difference between CCAnr and these two alternative solvers is that, CCAnr uses a greedy heuristic in the focused local search mode while CCAnr_age and CCAnr_rand employ diversifying heuristics. This indicates that using greedy heuristics rather than diversifying ones in the focused local search mode might be helpful to improve SLS-based SAT solvers on structured benchmarks.

CCAnr has also been discovered to be very competitive with other SLS-based SAT solvers on solving Satisfiability Modulo Theories (SMT) instances, although it can not compete with the SLS-based SMT solver called BV-SLS which works on the theory level representation [7]. The experiments in [7] were conducted with two benchmarks of bit-vector formulas namely QF_BV and SAGE2. The QF_BV benchmark can be found in the SMT-LIB and is also part of the SMT Competition. The SAGE2 benchmark consists of problems generated as part of the SAGE project at Microsoft, describing some testcases for automated whitebox fuzz testing. The number of solved instances are reported in Table 3, which is taken from [7]. Z3 is a state-of-the-art SMT solver. As well as Z3 and BV-SLS, the experiments include state-of-the-art SLS-based SAT solvers or those have good performance on certain types of structured SAT instances.

Table 3. Number of solved instances in bit-vector SMT benchmarks

	QF_BV (7498 instances)	SAGE2 (8017 instances)
CCAnr	**5409**	64
CCASat	4461	8
probSAT	3816	10
Sparrow	3806	12
VW2	2954	4
PAWS	3331	**143**
YalSAT	3756	142
Z3 (Default)	7173	5821
BV-SLS	6172	3719

Acknowledgments. This work is supported by China National 973 Program 2014CB340301, National Natural Science Foundation of China 61370156 and 61472369. We would like to thank the anonymous referees for their helpful comments.

References

1. Balint, A., Fröhlich, A.: Improving stochastic local search for sat with a new probability distribution. In: Strichman, O., Szeider, S. (eds.) SAT 2010. LNCS, vol. 6175, pp. 10–15. Springer, Heidelberg (2010)
2. Balint, A., Manthey, N.: SparrowToRiss. In: Proc. of SAT Competition 2014: Solver and Benchmark Descriptions, p. 77 (2014)
3. Cai, S., Luo, C., Su, K.: CCAnr+glucose in SAT competition 2014. In: Proc. of SAT Competition 2014: Solver and Benchmark Descriptions, p. 17 (2014)
4. Cai, S., Su, K.: Configuration checking with aspiration in local search for SAT. In: Proceedings of the Twenty-Sixth AAAI Conference on Artificial Intelligence, AAAI 2012, pp. 334–340 (2012)
5. Cai, S., Su, K.: Local search for Boolean Satisfiability with configuration checking and subscore. Artif. Intell. **204**, 75–98 (2013)
6. Cai, S., Su, K., Sattar, A.: Local search with edge weighting and configuration checking heuristics for minimum vertex cover. Artif. Intell. **175**(9–10), 1672–1696 (2011)
7. Fröhlich, A., Biere, A., Wintersteiger, C.M., Hamadi, Y.: Stochastic local search for satisfiability modulo theories. In: Proceedings of the Twenty-Ninth AAAI Conference on Artificial Intelligence, AAAI 2015, pp. 1136–1143 (2015)
8. Hutter, F., Tompkins, D.A.D., H. Hoos, H.: Scaling and probabilistic smoothing: efficient dynamic local search for SAT. In: Van Hentenryck, P. (ed.) CP 2002. LNCS, vol. 2470, pp. 233–248. Springer, Heidelberg (2002)
9. Li, C.M., Habet, D.: Description of RSeq2014. In: Proc. of SAT Competition 2014: Solver and Benchmark Descriptions, p. 72 (2014)
10. Li, C.-M., Huang, W.Q.: Diversification and determinism in local search for satisfiability. In: Bacchus, F., Walsh, T. (eds.) SAT 2005. LNCS, vol. 3569, pp. 158–172. Springer, Heidelberg (2005)
11. Li, C.M., Li, Y.: Satisfying versus falsifying in local search for satisfiability. In: Cimatti, A., Sebastiani, R. (eds.) SAT 2012. LNCS, vol. 7317, pp. 477–478. Springer, Heidelberg (2012)
12. Manthey, N.: Coprocessor 2.0 – a flexible CNF simplifier. In: Cimatti, A., Sebastiani, R. (eds.) SAT 2012. LNCS, vol. 7317, pp. 436–441. Springer, Heidelberg (2012)
13. McAllester, D.A., Selman, B., Kautz, H.A.: Evidence for invariants in local search. In: Proceedings of the 14th National Conference on Artificial Intelligence, AAAI 1997, pp. 321–326 (1997)
14. Morris, P.: The breakout method for escaping from local minima. In: Proceedings of the 11th National Conference on Artificial Intelligence, AAAI 1993, pp. 40–45 (1993)
15. Papadimitriou, C.H.: On selecting a satisfying truth assignment. In: Proceedings of the 32nd Annual Symposium on Foundations of Computer Science, FOCS 1991, pp. 163–169 (1991)
16. Pham, D.N., Gretton, C.: gNovelty+. In: Solver Description of SAT Competition 2007 (2007)

17. Seitz, S., Alava, M., Orponen, P.: Focused local search for random 3-satisfiability. J. Stat. Mech. (2005). P06006
18. Selman, B., Kautz, H.A.: Domain-independent extensions to gsat: solving large structured satisfiability problems. In: Proceedings of the 13th International Joint Conference on Artificial Intelligence, IJCAI 1993, pp. 290–295 (1993)
19. Selman, B., Kautz, H.A., Cohen, B.: Noise strategies for improving local search. In: Proceedings of the 12th National Conference on Artificial Intelligence, AAAI 1994, pp. 337–343 (1994)
20. Thornton, J., Pham, D.N., Bain, S., Jr., V.F.: Additive versus multiplicative clause weighting for SAT. In: Proceedings of the 19th National Conference on Artificial Intelligence, AAAI 2004, pp. 191–196 (2004)
21. Wu, Z., Wah, B.W.: An efficient global-search strategy in discrete lagrangian methods for solving hard satisfiability problems. In: Proceedings of the 17th National Conference on Artificial Intelligence, AAAI 2000, pp. 310–315 (2000)

PBLib – A Library for Encoding
Pseudo-Boolean Constraints into CNF

Tobias Philipp and Peter Steinke[✉]

Knowledge Representation and Reasoning Group,
Technische Universität Dresden, 01062 Dresden, Germany
peter.steinke@tu-dresden.de

Abstract. PBLib is an easy-to-use and efficient library, written in C++, for translating pseudo-Boolean (PB) constraints into CNF. We have implemented fifteen different encodings of PB constraints. Our aim is to use efficient encodings, in terms of formula size and whether unit propagation maintains generalized arc consistency. Moreover, PBLib normalizes PB constraints and automatically uses a suitable encoder for the translation. We also support incremental strengthening for optimization problems, where the tighter bound is realized with few additional clauses, as well as conditions for PB constraints.

1 Introduction

Many applications such as hardware verification and model checking benefit from the impressive developments in the area of SAT solving by translating high level descriptions into propositional formulas in conjunctive normal form (CNF) [24,30]. Pseudo-Boolean (PB) constraints are expressions of the form $\sum_{i=1}^{n} w_i \cdot x_i \lhd k$ and require that the weighted sum over the literals x_i is $\lhd$-related with k. They frequently occur in scheduling, planning, and translations of problems from languages like CSP, ASP or integer programming. Moreover, optimization problems like MaxSAT, minimal unsatisfiable core extraction, maximal satisfying subformulas, and PB optimization itself, rely on good translations from PB constraints into CNF [4,5,18,21,22]. However, there is no straightforward translation into CNF [6,11,15,16,25].

In this paper, we present *PBLib*, an easy-to-use and efficient library, written in C++, and distributed under the MIT license[1]. The library contains *fifteen different encodings* for PB constraints, which differ in the number of clauses, auxiliary variables and further properties. For instance, *generalized arc consistency* (GAC), a notion developed in the area of constraint programming [25], allows to cut off the search space as soon as possible. Therefore, maintaining generalized arc consistency by unit propagation is an important property of encodings. A weaker property than GAC is that *unit propagation detects inconsistent assignments*. The size of an encoding is another performance criteria, since SAT solvers often perform better when the number of clauses is small [8,19,29]. Additionally,

---

[1] available at http://tools.computational-logic.org/content/pblib.php

© Springer International Publishing Switzerland 2015
M. Heule and S. Weaver (Eds.): SAT 2015, LNCS 9340, pp. 9–16, 2015.
DOI: 10.1007/978-3-319-24318-4_2

PBLib performs *constraint normalization*, and supports *incremental strengthening* as well as *conditionals*. Experiments have shown that PBLib performs better than minisat+ [10].

The rest of this paper is structured as follows: We formally introduce the concept of encodings and generalized arc consistency in Sect. 2. Afterwards, we describe the concepts in PBLib and present code examples in Sect. 3. In Sect. 4, we give an overview of the tools included in PBLib. Then, we evaluate different encodings and compare PBLib with minisat+ in Sect. 5. Finally, we conclude in Sect. 6.

2   Pseudo-Boolean Constraints and Encodings

We assume that the reader is familiar with the concepts in propositional logic. *Pseudo-Boolean (PB) constraints* are expressions of the form $\sum_{i=1}^{n} w_i \cdot x_i \lhd k$, where x_i are literals, $w_i \in \mathbb{Z}$ are the *associated weights* for the literals x_i for every $i \in \{1, \ldots, n\}$, $k \in \mathbb{Z}$, and $\lhd \in \{=, \leq, \geq\}$ is the *comparator*. A *cardinality constraint* is a PB constraint, where all weights are equal to 1. Depending on the comparator, we call a cardinality constraint an *at-most-one, at-least-one* or *exactly-one* constraint, if $k = 1$.

Formally, the formula F *encodes* the original formula G iff *1)* F entails G, and *2)* for every model I of the formula G there is a model I' of F such that $I(x) = I'(x)$ for every variable x occurring in G. The first condition states that every model of the encoding is a model of the original formula. The second condition states that every model of the original formula can be transformed to a model of the encoding by modifying the interpretation of the auxiliary variables.

We consider the following two structural properties: generalized arc consistency (GAC) and inconsistency detection. Both are important inference rules in constraint programming and can significantly reduce the search space [25]. As in [23], we describe the notions of GAC and inconsistency detection in terms of the entailment relation. An *assignment* J is a consistent set of literals. We say that J is *consistent* w.r.t. a constraint C iff the formula $(\bigwedge_{x \in J} x) \wedge C$ is satisfiable. Otherwise, J is *inconsistent* w.r.t. C. An encoding *detects inconsistencies by unit propagation* if unit propagation in the encoding derives the empty clause, if the assignment J is inconsistent w.r.t. C. Informally, an assignment is GAC, if the assignment contains all entailed literals. Formally, a consistent assignment J is *GAC w.r.t. the constraint* C iff for every variable y occurring in C, $y \in J$ whenever $(\bigwedge_{x \in J} x) \wedge C \models y$, and $\neg y \in J$, whenever $(\bigwedge_{x \in J} x) \wedge C \models \neg y$. *Unit propagation maintains GAC*, if unit propagation transforms a consistent assignment to generalized arc consistent assignment.

3   Description of the PBLib

Table 1 presents the encodings offered by PBLib. Unit propagation in the offered encodings, except sorting and adder networks, detects inconsistent assignment and maintains generalized arc consistency. We also offer a variant of the watchdog

Table 1. A catalog of encodings offered by PBLib, categorized into different fragments of PB constraints.

at-most-one	cardinality	pseudo-Boolean
sequential counter[1] [27]	BDD[2] [10,14]	BDD [10,14]
bimander [13]	cardinality networks [1]	adder networks [10]
commander [17]	adder networks [10]	watchdog [23]
k-product [7]		sorting networks [10]
binary [2]		binary merge [20]
pairwise		sequential weight counter[3] [12]
nested		

[1] similar to BDD, latter and regular encoding,[2] similar to sequential counter
[3] similar to BDD but useful for incremental encoding

and binary merge encoding for which unit propagation detects only inconsistent assignments, with the advantage of fewer clauses.

3.1 Components of the PBLib

PB constraints. In the PBLib, a PB constraint $\sum_{i=1}^{n} w_i \cdot x_i \lhd k$ is specified with a list of weighted literals, a comparator and an integer k, where every 64 bit integer is accepted as weight and as k. The comparator can be either less equal, greater equal, or a combination of both. Hence it is possible to specify a single constraint like $\sum_{i=1}^{n} w_i \cdot x_i \leq k_1 \land \sum_{i=1}^{n} w_i \cdot x_i \geq k_2$. Note that GAC and inconsistency detection refers to single PB constraint using $\leq$ or $\geq$ as comparator.

PreEncoder. The *PreEncoder* normalizes PB constraints such that the following holds: 1. $n > 0$, 2. $1 \leq w_i \leq k$ for every $i \in \{1, \ldots, n\}$, 3. no literal in a constraint occurs twice, and 4. the comparator is either less equal or both: less equal and greater equal. Moreover, it detects trivial constraints such as units and tautologies, directly encodes them, and applies some simplifications such as removing unnecessary comparators.

ClauseDatabase. As container for the clauses in a formula a *ClauseDatabase* is used. The PBLib contains different instances of ClauseDatabases such as a *VectorClauseDatabase* that stores each clause in a vector, and a *SATSolver-ClauseDatabase* that stores each clause in a minisat-like [9] SAT solver. The ClauseDatabase is a simple interface, requiring only an implementation for the addition method for single clauses. This makes it easy to integrate PBLib in projects. Moreover, every ClauseDatabase can process minisat+ like Boolean circuits [10] by translating them into clauses.

AuxVarManager. For handling auxiliary variables, PBLib uses an auxiliary variable manager, called *AuxVarManager*. Initialized with a fresh variable, AuxVarManager returns the next free variable upon request. It is possible to reset already used auxiliary variables as well as marking individual variables as fresh variables. Hence the AuxVarManager helps to keep the set of used variables tight.

Encoder. The PBLib contains 15 different encoders, where each produces different clause sets. Some encoders are only applicable for specific subsets of PB constraint, e.g. at-most-one or cardinality constraints. In the framework of the PBLib, it is easy to extend the set of encoders with new encodings.

IncrementalData. It is required to use the class *IncPBConstraint* to represent PB constraints that supports incrementally strengthening. After the initial encoding of such a constraint, the *IncPBConstraint* stores *IncrementalData* internally that allows to restrict the constraint with a tighter bound. This allows the implementation of an easy to handle SAT-based linear optimization algorithm.

Conditionals. PB and incremental PB constraints can be augmented with *conditions*, i.e. finite conjunctions of literals. This is achieved by adding the complementary literals to all activation clauses of the encoding. For example, we can express the following constraint with a single constraint in PBLib:

$$(x_5 \wedge \overline{x_6}) \rightarrow (-3 \leq -7x_1 + 5\overline{x_2} + 9\overline{x_3} - 3\overline{x_{10}} + 7x_{10} \leq 8)$$

PB2CNF. The *PB2CNF* class handles constructed PB constraints: It normalizes the constraint, classifies it, and chooses a suitable encoding depending on the kind and size of the constraint. Produced clauses are stored in the given ClauseDatabase and auxiliary variables are managed by the AuxVarManager.

3.2 Example

We demonstrate how to encode the constraint $3\overline{x_1} - 2\overline{x_2} + 7x_3 \geq -4$ in the following example. First, we reserve space for two vectors containing the decision literals and their associated weights, and for the resulting formula, which is a vector of vectors of literals. Moreover, we specify the first free variable in firstAuxVar. Finally, we call the method encodeGeq that encodes the constraint and stores the result in formula.

```
#include "PB2CNF.h"
int main() {
  PBLib::PB2CNF pb2cnf;
  vector< int64_t > weights = {3, -2, 7};
  vector< int32_t > literals = {-1, -2, 3};
  vector< vector< int32_t > > formula;
  int32_t firstAuxVar = 4;
  int64_t k = -4;
  pb2cnf.encodeGeq(weights, literals, k, formula, firstAuxVar);
}
```

You can also add a less equal and a greater equal comparator, as well as incremental constraints. For the latter one, we need the generic formula container ClauseDatabase and an instance of AuxVarManager. Moreover, we have to use the configurations class of the PBLib. In the following example, the constraint $-5 \leq -7x_1 + 5\overline{x_2} + 9\overline{x_3} - 3\overline{x_{10}} + 7x_{10} \leq 100$ is encoded:

```
using namespace PBLib;
PBConfig config = make_shared< PBConfigClass >();
VectorClauseDatabase formula(config);
PB2CNF pb2cnf(config);
AuxVarManager auxvars(11);
vector< WeightedLit > literals =
  {WeightedLit(1, -7), WeightedLit(-2, 5), WeightedLit(-3, 9),
   WeightedLit(-10, -3), WeightedLit(10, 7)};
IncPBConstraint constraint(literals, BOTH, 100, -5);
pb2cnf.encodeIncInital(constraint, formula, auxvars);
```

We can increase the bounds:

```
    constraint.encodeNewGeq(3, formula, auxvars);
    constraint.encodeNewLeq(8, formula, auxvars);
```

The constraint $-3 \leq -7x_1 + 5\overline{x_2} + 9\overline{x_3} - 3\overline{x_{10}} + 7x_{10} \leq 8$ is encoded with the code above in combination with the formula encoded with encodeIncInital.

4   Included Tools

The PBLib includes the following programs: *pbencoder*, *pbsolver* and a *fuzzer*. *pbencoder* takes as input a list of PB constraints in the OPB format [26] and encodes them into CNF. The result is printed on the standard output. *pbsolver* solves a OPB instance by translating the PB constraints and afterwards solving the resulting CNF formula with a back-end SAT solver such as *minisat 2.2* [28]. For optimization instances, pbsolver iteratively encodes upper bounds until the optimum is reached. The program *fuzzer* randomly generates PB constraints, and afterwards encode them with different configurations. This program helps to find bugs in new or customized implementations.

5   Empirical Evaluation and Related Work

We evaluated *pbsolver* on all *new* instances in the PB competition 2012, in total 2782 instances[2]. Note that the most recent PB competition was held in the year 2012. Besides various encodings inside the PBLib, we compared the performance of *minisat+* [10] on this benchmark. The evaluation was performed on a PC cluster with Intel E5-2690 CPUs (2.90 GHz) and 2 GB RAM.

minisat+ follows the same approach for solving PB constraints: It translates them into CNF and applies an iterative solving strategy for optimization problems. In contrast to PBLib, minisat+ uses only three encodings: BDDs, sorting networks and adder networks. BDDs in minisat+ are encoded with three clauses per BDD node instead of only two as in the PBLib and presented in [14]. For a fair comparison, we used minisat 2.2 as back-end SAT solver in pbsolver and in minisat+. Figure 1 shows the results of the evaluation. Adder and sorting

---

[2] available at http://www.cril.univ-artois.fr/PB12/

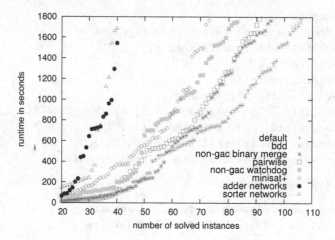

Fig. 1. A comparison of different encodings in the PBLib and minisat+ using 2782 new instances in the PB competition 2012

networks have the least number of clauses, but the worst runtime, because in both encodings unit propagation neither detects inconsistent assignments nor maintains generalized arc consistency. We observe that the plot for the default configuration and the use of BDDs for general PB constraint are nearly the same. This can be explained since PBLib decides for the BDD encoding due to the low number of clauses. In contrast to minisat+, PBLib in pbsolver solves significantly more instances in the timeout of 1800 seconds. This has two reasons: First, PBLib uses a better BDD encoding. Second, minisat+ does not distinguish between at-most-one, at-most-k and PB constraint. Therefore, all constraints, but clauses, are handled in the same way. Instead, PBLib detects these special cases and chooses a more appropriate encoding.

The Java library Boolvar/PB [3] is also related to the PBLib: It uses basically the same encodings as minisat+, but sorting networks have been replaced by the watchdog encoding.

6   Conclusion

In this paper, we presented PBLib, an efficient and easy-to-use library for encoding PB constraints into clause sets. It normalizes PB constraints before encoding them, and can automatically choose between fifteen different encodings that vary in size and propagation properties. Moreover, PBLib supports incremental strengthening and conditional PB constraints. Experiments have shown that our library outperforms minisat+ in the recent benchmark of the PB evaluation. It is distributed under the MIT license.

In future, we plan to implement more encodings, and to increase confidence in the tools by mechanically verifying them.

References

1. Abío, I., Nieuwenhuis, R., Oliveras, A., Rodríguez-Carbonell, E.: A parametric approach for smaller and better encodings of cardinality constraints. In: Schulte, C. (ed.) CP 2013. LNCS, vol. 8124, pp. 80–96. Springer, Heidelberg (2013)
2. Frisch, A.M., Peugniez, T.J., Doggett, A.J., Nightingale, P.W.: Solving non-Boolean satisfiability problems with stochastic local search: A comparison of encodings. Journal of Automated Reasoning **35** (2005)
3. Bailleux, O.: Boolvar/PB 1.0 (2012). http://boolvar.sourceforge.net/
4. Belov, A., Janota, M., Lynce, I., Marques-Silva, J.: On computing minimal equivalent subformulas. In: Milano, M. (ed.) CP 2012. LNCS, vol. 7514, pp. 158–174. Springer, Heidelberg (2012)
5. Boros, E., Hammer, P.L.: Pseudo-Boolean optimization. Discrete Applied Mathematics **123**(1–3), 155–225 (2002)
6. Brewka, G., Eiter, T., Truszczyński, M.: Answer set programming at a glance. Communications of the ACM **54**(12), 92–103 (2011)
7. Chen, J.: A new SAT encoding of the at-most-one constraint. In: ModRef 2010 (2010)
8. Eén, N., Biere, A.: Effective preprocessing in SAT through variable and clause elimination. In: Bacchus, F., Walsh, T. (eds.) SAT 2005. LNCS, vol. 3569, pp. 61–75. Springer, Heidelberg (2005)
9. Eén, N., Sörensson, N.: An extensible SAT-solver. In: Giunchiglia, E., Tacchella, A. (eds.) SAT 2003. LNCS, vol. 2919, pp. 502–518. Springer, Heidelberg (2004)
10. Een, N., Sörensson, N.: Translating pseudo-Boolean constraints into SAT. Journal on Satisfiability, Boolean Modeling and Computation **2**, 1–26 (2006)
11. Großmann, P., Hölldobler, S., Manthey, N., Nachtigall, K., Opitz, J., Steinke, P.: Solving periodic event scheduling problems with SAT. In: Jiang, H., Ding, W., Ali, M., Wu, X. (eds.) IEA/AIE 2012. LNCS, vol. 7345, pp. 166–175. Springer, Heidelberg (2012)
12. Hölldobler, S., Manthey, N., Steinke, P.: A compact encoding of pseudo-boolean constraints into SAT. In: Glimm, B., Krüger, A. (eds.) KI 2012. LNCS, vol. 7526, pp. 107–118. Springer, Heidelberg (2012)
13. Hölldobler, S., Nguyen, V.H.: On SAT-encodings of the at-most-one constraint. In: ModRef 2013 (2013)
14. Abío, I., Nieuwenhuis, R., Oliveras, A., Rodríguez-Carbonell, E.: BDDs for pseudo-boolean constraints – revisited. In: Sakallah, K.A., Simon, L. (eds.) SAT 2011. LNCS, vol. 6695, pp. 61–75. Springer, Heidelberg (2011)
15. Kautz, H., Selman, B.: Planning as satisfiability. In: Neumann, B. (ed.) ECAI 1992, pp. 359–363. John Wiley & Sons Inc., New York (1992)
16. Kautz, H., Selman, B.: Pushing the envelope: planning, propositional logic, and stochastic search. In: AAAI 1996, pp. 1194–1201. MIT Press (1996)
17. Klieber, W., Kwon, G.: Efficient CNF encoding for selecting 1 from n objects. In: CFV 2007 (2007)
18. Liffiton, M.H., Sakallah, K.A.: Algorithms for computing minimal unsatisfiable subsets of constraints. Journal of Automated Reasoning **40**(1), 1–33 (2008)
19. Manthey, N., Heule, M.J.H., Biere, A.: Automated reencoding of boolean formulas. In: Biere, A., Nahir, A., Vos, T. (eds.) HVC. LNCS, vol. 7857, pp. 102–117. Springer, Heidelberg (2013)

20. Manthey, N., Philipp, T., Steinke, P.: A more compact translation of pseudo-boolean constraints into CNF such that generalized arc consistency is maintained. In: Lutz, C., Thielscher, M. (eds.) KI 2014. LNCS, vol. 8736, pp. 123–134. Springer, Heidelberg (2014)

21. Manthey, N., Steinke, P.: npSolver - a SAT based solver for optimization problems. In: POS 2012 (2012)

22. Nadel, A.: Boosting minimal unsatisfiable core extraction. In: Bloem, R., Sharygina, N. (eds.) FMCAD 2010, pp. 121–128 (2010)

23. Bailleux, O., Boufkhad, Y., Roussel, O.: New encodings of pseudo-boolean constraints into CNF. In: Kullmann, O. (ed.) SAT 2009. LNCS, vol. 5584, pp. 181–194. Springer, Heidelberg (2009)

24. Plaisted, D.A., Greenbaum, S.: A structure-preserving clause form translation. Journal of Symbolic Computation 2(3), 293–304 (1986)

25. Rossi, F., Beek, P.V., Walsh, T.: Handbook of Constraint Programming. Elsevier Science Inc., New York (2006)

26. Roussel, O., Manquinho, V.: Input/output format and solver requirements for the competitions of pseudo-Boolean solvers (2012)

27. Sinz, C.: Towards an optimal CNF encoding of boolean cardinality constraints. In: van Beek, P. (ed.) CP 2005. LNCS, vol. 3709, pp. 827–831. Springer, Heidelberg (2005)

28. Sorensson, N., Een, N.: MiniSat - a SAT solver with conflict-clause minimization. In: Bacchus, F., Walsh, T. (eds.) SAT 2005. LNCS. Springer, Heidelberg (2005)

29. Subbarayan, S., Pradhan, D.K.: NiVER: non-increasing variable elimination resolution for preprocessing SAT instances. In: H. Hoos, H., Mitchell, D.G. (eds.) SAT 2004. LNCS, vol. 3542, pp. 276–291. Springer, Heidelberg (2005)

30. Tseitin, G.S.: On the complexity of derivation in propositional calculus. In: Siekmann, J.H., Wrightson, G. (eds.) Automation of Reasoning. Symbolic Computation, pp. 466–483. Springer, Heidelberg (1983)

Speeding up MUS Extraction
with Preprocessing and Chunking

Valeriy Balabanov[1] and Alexander Ivrii[2]([⊠])

[1] National Taiwan University, Taipei, Taiwan
balabasik@gmail.com
[2] IBM Research, Haifa, Israel
alexi@il.ibm.com

Abstract. In this paper we present several improvements to extraction
of a minimal unsatisfiable subformula (MUS) of a Boolean formula. As
our first contribution, we describe model rotation on preprocessed formulas and show that preprocessing significantly improves model rotation.
We find very convenient to adopt the framework of labeled CNF formulas
and we present our algorithms in this more general framework. We use
the assumption-based approach for computing MUSes due to its simplicity and the ability to use any SAT-solver as the back-end. However, this
comes with a price: it is well-known that the assumption-based approach
performs significantly worse than the resolution-based approach. This
leads to our second contribution, we show how to bridge the gap between
the two approaches using "chunking". An extensive experimental evaluation shows that our method significantly outperforms state-of-the-art
solutions in the context of group MUS extraction.

1 Introduction

The problem of computing a minimal explanation of unsatisfiability of a Boolean
formula has attracted a lot of recent research. Given an unsatisfiable Boolean
formula F in conjunctive normal form (CNF), computing an MUS of F is the
problem of finding a *minimal unsatisfiable subset* of its clauses (that is, an unsatisfiable subset of clauses such that removal of any clause renders it satisfiable).
Further, in many applications the formula F comes partitioned into several
disjoint groups (subsets), and computing a *group MUS* (GMUS) (also called
high-level MUS) is the problem of finding a minimal unsatisfiable subset of these
groups [1,2]. Of course, MUS is a special case of GMUS, where each group consists of a single clause. A further generalization of GMUS, called *labeled MUS*
(LMUS), was recently introduced in [3–5] and corresponds to the case where the
groups do not need to be disjoint. This name comes from an alternative description of the problem, in which each clause in F has an associated set of *labels*, and
the problem consists of finding a minimal subset L of labels with the property
that the subset of clauses of F with all their labels in L is unsatisfiable.

All state-of-the-art MUS extraction techniques make use of Boolean satisfiability (SAT) solvers, and the key observation from [3] is that by working with

© Springer International Publishing Switzerland 2015
M. Heule and S. Weaver (Eds.): SAT 2015, LNCS 9340, pp. 17–32, 2015.
DOI: 10.1007/978-3-319-24318-4_3

labeled CNF formulas (LCNF), one can easily enable various preprocessing techniques for MUS extraction – a task which is not simple otherwise. For example, the problem of finding an MUS of the original formula can be translated to a certain LMUS computation on the preprocessed formula, but not in general to an MUS or a GMUS computation. Further, [3] experimentally demonstrates the practical importance of preprocessing on standard group MUS benchmarks.

As our first contribution, we amplify the importance of preprocessing for MUS and GMUS computations. To this end, we adopt the LCNF framework and adapt the recursive model rotation (RMR) [6] from MUS computations to this more general context. We denote the new technique by LRMR (label-based recursive model rotation). We emphasize that LRMR is a general technique and is applicable to preprocessed formulas for MUS and GMUS computations. We experimentally show that LRMR on preprocessed formulas finds significantly more necessary labels than LRMR on original formulas, which directly translates into a reduction in the number of SAT queries required for the computation. Moreover, we show that the benefit of preprocessing for model rotation continuously increases as one increases the preprocessing effort.

Efficiently implementing preprocessing in an MUS extractor leads to a dilemma. Let us recall that there are two standard approaches, both based on incremental SAT solving. In the *assumption-based* approach (as implemented in MUSER2 [7]) one creates a fresh activation literal for each clause (group or label) and uses the common interface of a SAT solver (such as MINISAT [8]) to solve a formula under assumptions. Setting some of the activation literals to TRUE and some to FALSE enables to activate and deactivate groups of clauses. In the *resolution-based* approach (as implemented in HaifaMUC [9]) one uses the ability of certain SAT solvers to produce a resolution proof in the case that a problem is unsatisfiable. There is in fact a hybrid approach implemented in MINISAT$_{abb}$ [10], which technically uses assumptions but modifies the SAT solver to organize the assumptions in a form of a partial resolution graph.

The dilemma is as follows. On the one hand, using the pure assumption-based approach has the ability to use any SAT solver for the back-end, and allows for a significantly simpler implementation, which is indirectly corroborated by the fact that neither HaifaMUC nor MINISAT$_{abb}$ fully support preprocessing in the context of GMUS extraction. On the other hand, it is well-known that assumption-based approaches are slower than resolution-based approaches. There are several reasons for this, see [10,11]. First, in incremental SAT solving with *many* assumptions, learned clauses become significantly larger, and this leads to an increased memory consumption, more cache misses, and a significantly increased average number of literals that need to be considered in BCP and in conflict clause analysis. More importantly, modern SAT solvers use a crucial optimization described in [12] that allows one to minimize learned conflict clauses by using additional resolutions with existing clauses. However, this additional minimization is only applied when the resulting conflict clause is a strict sub-clause of the initial conflict clause: in other words, conflict clauses become shorter (and more useful) at the expense of bringing more clauses into the proof. Having many activation literals to a large extent prevents this minimization, which makes the learned conflict clauses significantly less useful, and which in turn significantly degrades the performance of an assumption-based MUS extractor. An especially tantalizing illustration of this phenomenon can be witnessed by running MUSER2 on its own minimal cores (either from regular MUS or GMUS benchmarks) and noticing that in many cases the time required to solve the minimized formula is *an order of magnitude larger* than the time required to solve the original formula. We note that neither HaifaMUC nor MINISAT$_{abb}$ have the above problem, and their performance is only improved when running on the fully minimized formulas. In fact, MINISAT$_{abb}$ solves the problem by using *assumption aware clause minimization* ([10], Section 2.5), in which additional resolutions are allowed as long as only the set of non-activation literals is decreased. In this sense MINISAT$_{abb}$ performs exactly as a resolution-based approach. It is also interesting to note that HaifaMUC uses the same observation in the opposite way and in some cases prevents performing additional resolutions if this involves new groups to be pulled into the proof ([11], Optimization B).

As our second contribution, we show how a single LMUS computation can be reduced to a sequence of simpler LMUS computations by splitting the set of labels into *chunks* and by iteratively analyzing these chunks. This approach is closely related to the approach of computing a minimal equivalent subformula (MES) by an incremental reduction into GMUS (see [13], Section 3.4). In the assumption-based approach, the size of each chunk represents a trade-off. Considering smaller chunks makes the overall approach less incremental and degrades the quality of label-set refinement. However, it decreases the number of activation literals introduced in each SAT query and in particular allows for a better conflict clause minimization (all the clauses without an activation literal become usable as additional resolvents). We experimentally show that when the formula is minimal or close to minimal (as for example obtained using *trimming*, see for example [14]), chunking essentially bridges the gap between the assumption and the resolution based approaches.

In summary, this paper presents two interdependent contributions. First, we amplify the role of preprocessing for computing MUSes and GMUSes, and we suggest to use the assumption-based approach as it does not require any modification to the SAT-solver and leads to a significantly simpler and less error-prone implementation. However this introduces a problem that the conflict clauses discovered by the SAT solver become less useful, and we solve this problem (to a large extent) using chunking. We experimentally show that the combined algorithm significantly outperforms state-of-the-art techniques when computing group MUSes. We speculate that an implementation of preprocessing and label-based model rotation in the resolution-based approach would improve the performance even further.

The rest of the paper is organized as follows. Section 2 describes preliminaries, including the LCNF framework from [3]. In Section 3 we describe the algorithms for LMUS extraction and label-based model rotation, and in Section 4 we describe the algorithm for chunking. Section 5 contains experimental results. Section 6 concludes the paper.

2   Preliminaries

We briefly recall some basic notions of satisfiability solving. A *literal* is either a Boolean variable or its negation. A *clause* is a disjunction of literals and a CNF is a conjunction of clauses. When convenient, we view clauses and CNFs as sets.

Given a CNF formula $\mathcal{F}$, we denote the set of variables that occur in $\mathcal{F}$ by $Var(\mathcal{F})$. An *assignment* τ for $\mathcal{F}$ is a map $\tau : Var(\mathcal{F}) \to \{0, 1\}$. Assignments are extended to clauses and formulas according to the semantics of classical propositional logic. By $Unsat(\mathcal{F}, \tau)$ we denote the set of clauses of $\mathcal{F}$ that are falsified by τ. If $\tau(\mathcal{F}) = 1$, then τ is a *model* of $\mathcal{F}$. If a formula $\mathcal{F}$ has (resp. does not have) a model, then $\mathcal{F}$ is *satisfiable* (resp. *unsatisfiable*). By SAT and UNSAT we denote the set of all satisfiable (resp. unsatisfiable) CNF formulas.

Next, we recall several notions of minimal unsatisfiability actively studied in the literature.

Definition 1 (Minimal Unsatisfiable Subformula, MUS). *Given an unsatisfiable CNF formula* $\mathcal{F} = C_1 \cup \cdots \cup C_n$, *an MUS of* $\mathcal{F}$ *is a subset* $\mathcal{F}' = C_{i_1} \cup \cdots \cup C_{i_k}$ *of* $\mathcal{F}$ *such that* $\mathcal{F}' \in$ UNSAT *and, for every* $1 \le j \le k$, $\mathcal{F}' \setminus C_{i_j} \in$ SAT.

Definition 2 (Group Minimal Unsatisfiable Subformula, GMUS [1,2]). *Given an explicitly partitioned unsatisfiable CNF formula* $\mathcal{F} = \mathcal{G}_0 \cup \cdots \cup \mathcal{G}_n$, *a group oriented MUS of* $\mathcal{F}$ *is a subset* $\mathcal{F}' = \mathcal{G}_0 \cup \mathcal{G}_{i_1} \cup \cdots \cup \mathcal{G}_{i_k}$ *of* $\mathcal{F}$ *such that* $\mathcal{F}' \in$ UNSAT *and, for every* $1 \le j \le k$, $\mathcal{F}' \setminus \mathcal{G}_{i_j} \in$ SAT.

A further extension of Definition 2 to the case where the groups do not need to be disjoint was recently proposed in [3,5]. Let L be a non-empty set of clause labels. A *labeled CNF (LCNF)* is a tuple $\langle \mathcal{F}, \lambda \rangle$, where $\mathcal{F}$ is a CNF formula and $\lambda : \mathcal{F} \to 2^L$ is a labeling function. In this way, with each clause C we associate its set of labels $\lambda(C)$, and a labeled CNF formula can be also viewed as a conjunction (or a set) of labeled clauses $\{\langle C, \lambda(C) \rangle\}$. We also denote a labeled clause $\langle C, \lambda(C) \rangle$ as $C_{\lambda(C)}$, and somewhat abusing notation, we also use the symbol $\mathcal{F} = \{C_{\lambda(C)}\}$ to represent a labeled CNF formula.

Given an LCNF formula $\mathcal{F} = \{\langle C, \lambda(C) \rangle\}$, we denote the set of labels that occur in $\mathcal{F}$ by $Label(\mathcal{F})$. Let $K \subseteq Label(\mathcal{F})$. We define $\mathcal{F}^K \subseteq \mathcal{F}$ as a subset of labeled clauses of $\mathcal{F}$ with at least one label in K: $\mathcal{F}^K = \{\langle C, \lambda(C) \rangle \mid \lambda(C) \cap K \neq \emptyset\}$. We define the *induced* formula $\mathcal{F}|_K$ as a subset of labeled clauses of $\mathcal{F}$ with all their labels in K: $\mathcal{F}|_K = \{\langle C, \lambda(C) \rangle \mid \lambda(C) \subseteq K\}$. Finally, we define a *projection* of $\mathcal{F}$ onto $K \subseteq L$ by restricting the labels of labeled clauses in $\mathcal{F}$ to K: $\mathtt{Project}(\mathcal{F}, K) = \{\langle C, \lambda(C) \cap K \rangle\}$.

Satisfiability of an LCNF formula is defined in terms of satisfiability of its clauses, and we reuse the notation SAT (resp. UNSAT) for the set of satisfiable (resp. unsatisfiable) LCNF formulas.

Definition 3 (Label Minimal Unsatisfiable Subformula, LMUS [3,5]). *Given an LCNF formula* $\mathcal{F}$, *a set of labels* $K \subseteq \lambda(\mathcal{F})$ *is a label minimal unsatisfiable subset (LMUS) of* $\mathcal{F}$ *if* $\mathcal{F}|_K \in$ UNSAT *and,* $\forall K' \subsetneq K$, $\mathcal{F}|_{K'} \in$ SAT.

We emphasize that LMUS is defined in terms of labels of the labeled formula. Note that LMUS can be viewed as an extension of GMUS (and hence also of MUS) by associating a label for each of the groups. Given an LCNF formula $\mathcal{F}$ and a label $l \in Label(\mathcal{F})$, we say that l is *necessary* for $\mathcal{F}$ if $\mathcal{F}$ is UNSAT and $\mathcal{F} \setminus \mathcal{F}^{\{l\}}$ is SAT. Next, we recall from [3] the two key benefits of the LMUS framework.

Labeled Preprocessing Techniques. Using the LMUS framework allows one to easily apply the (labeled versions of) common preprocessing techniques, including subsumption, self-subsumption, and variable-elimination. As one example, *labeled subsumption* asserts that whenever $\langle C_1, \lambda_1 \rangle$ and $\langle C_2, \lambda_2 \rangle$ are two labeled clauses of $\mathcal{F}$ with $C_1 \subseteq C_2$ and $\lambda_1 \subseteq \lambda_2$, then $\langle C_2, \lambda_2 \rangle$ can be removed. As another example, given two labeled clauses $\langle x \vee C_1, \lambda_1 \rangle$ and $\langle \neg x \vee C_2, \lambda_2 \rangle$, their *labeled resolvent* is defined as $\langle C_1 \cup C_2, \lambda_1 \cup \lambda_2 \rangle$, and *labeled variable elimination* allows replacing all labeled clauses involving a variable or its negation by all labeled resolvents on this variable. We refer to [3] for additional details.

LMUS Implementation in the Assumption-Based Framework. Given an LCNF formula $\mathcal{F}$, we can create a fresh variable p_l for every label $l \in Label(\mathcal{F})$, and replace every labeled clause $\langle C, \lambda \rangle$ by a regular clause $(C \vee \bigvee_{l \in \lambda(C)} p_l)$. In other words, each label of every clause now appears as an additional literal in that clause. Moreover, the labeled preprocessing techniques correspond to the classical preprocessing techniques, as long as variable elimination is disallowed to eliminate variables p_l (note that this functionality of *freezing* variables is already supported in most modern SAT solvers). In other words, supporting LMUS computations is virtually free in the assumption-based approach.

3 LMUS Extraction

Algorithm 1 presents a high-level algorithm for computing an LMUS of an unsatisfiable LCNF. Without the optimizations on lines 2 and 15, it can be viewed as a direct adaptation of the hybrid algorithm for computing an MUS (see [15], Algorithm 3) to the more general framework of LCNFs.

Let us first present a quick overview of Algorithm 1 omitting many important details. It accepts an unsatisfiable LCNF formula $\mathcal{F}$ as input and maintains the following data structures: the set of necessary labels L (initially empty), the *working set of labels* L_w (initialized to $Label(\mathcal{F})$), and the *working formula* $\mathcal{F}_w$ (initialized to $\mathcal{F}$). On each iteration of the while-loop (line 6) we choose a label l from L_w, and invoke a SAT solver (line 8) on the formula $\mathcal{F}_w \setminus \mathcal{F}_w^{\{l\}}$ obtained from $\mathcal{F}_w$ by removing all clauses labeled with l. We use the following standard notation: SAT returns a triple $(st, \tau, \mathcal{U})$, where if the formula is satisfiable, then $st = true$ and τ is a satisfying assignment, and if the formula is unsatisfiable, then $\tau = false$ and $\mathcal{U}$ is an unsatisfiable core. If $\mathcal{F}_w \setminus \mathcal{F}_w^{\{l\}}$ is unsatisfiable (lines 9–11), then l is not necessary, in which case the clauses of $\mathcal{F}_w^{\{l\}}$ are permanently removed from the working formula $\mathcal{F}_w$, and l is removed from the working set

---

Algorithm 1. LHYB($\mathcal{F}$) – Hybrid LMUS Extraction

Input : $\mathcal{F}$ — a (trimmed) unsatisfiable LCNF formula
Output: LMUS L of $\mathcal{F}$

```
1  begin
2  |  LSimplify(F)                                    // Main preprocessing
3  |  L ← ∅                                           // LMUS under-approximation
4  |  Lw ← Label(F)                                   // Working set of labels
5  |  Fw ← F                                          // Working formula
6  |  while Lw ≠ 0 do
7  |  |  l ← PickLabel(Lw)
8  |  |  (st,τ,U) = SAT(Fw \ Fw^{l})
9  |  |  if st=false then
10 |  |  |  Lw ← Lw ∩ Label(U)                        // Refinement
11 |  |  |  Fw ← Fw|L∪Lw
12 |  |  else
13 |  |  |  LModelRotation(Fw, L, τ)                  // Rotation
14 |  |  |  Lw ← Lw \ L
15 |  |  if condition then
16 |  |  |  LSimplify(Fw)                             // Additional preprocessing
17 |  |  |  Lw ← Label(Fw)
18 |  return L
```

---

of labels L_w. Otherwise (lines 12–14), l is necessary, in which case l is added to L and removed from L_w.

Next we describe in detail the important optimizations required for this algorithm to be practically efficient.

Label-Set Refinement. *Label-set refinement* ([15,16]) is a technique that takes advantage of the capability of modern SAT solvers to produce unsatisfiable cores (in the resolution-based approach) or return conflicting assumptions (in the assumption-based approach). Consider the case that the formula $\mathcal{F}_w \setminus \mathcal{F}_w^{\{l\}}$ is unsatisfiable and let $\mathcal{U}$ denote an unsatisfiable core. The key observation is that (in addition to l) any label l', with the property that no clause marked by l' appears in $\mathcal{U}$, can be removed from L. In practice this is a crucial optimization technique that allows one to remove multiple unnecessary labels in a single SAT solver call.

Main Preprocessing. The main novelty of the LMUS computation is the ability to apply *preprocessing* techniques to simplify the formula [3]. In Algorithm 1, the main preprocessing is accomplished by a call to LSimplify at the start of the algorithm (line 2). Following [3], we use labeled clause elimination, labeled variable elimination, labeled subsumption (including labeled unit propagation), and labeled self-subsumption. As described previously, in the assumption-based framework each label has its own activation variable, and all of these techniques are already supported by MINISAT (if we disallow variable elimination to eliminate activation variables of the labels).

There are several key benefits of preprocessing. First, the formula generally becomes easier to solve, leading to a decreased runtime of an average SAT solver query (see line 8). Second, labeled variable elimination can drop many (unnecessary) labels. We illustrate this on an example. Suppose that $\mathcal{F} = \{(p \vee q)_{\{l_1\}}, (\neg p \vee r)_{\{l_2\}}, (\neg q \vee \neg r)_{\{l_3\}}, (\neg q \vee s)_{\{l_4\}}, (\neg r)_{\{l_5\}}, (\neg s)_{\{l_6\}}\}$. Suppose that variable elimination first eliminates variable p, replacing the clauses $(p \vee q)_{\{l_1\}}, (\neg p \vee r)_{\{l_2\}}$ by their labeled resolvent clause $(q \vee r)_{\{l_1, l_2\}}$, and then further eliminates variable q resulting in the LCNF $\{(r \vee s)_{\{l_1, l_2, l_4\}}, (\neg r)_{\{l_5\}}, (\neg s)_{\{l_6\}}\}$. Note that there are no more clauses labeled by l_3 and hence l_3 is unnecessary. This example also demonstrates that even if in the original LCNF each clause has at most one label (as the result of a conversion from an MUS or a GMUS problem), after preprocessing each clause will in general have several labels. We will show later that this makes preprocessing extremely beneficial for model rotation.

Let us recall that MINISAT has two integer-valued parameters that control the influence of variable elimination on a more fine-grained level: *cl-lim* that prohibits eliminating a variable if it produces a resolvent clause of a larger length, and *grow* that prohibits eliminating a variable if the number of clauses in the CNF increases by more than this value. We have experimentally found that for GMUS extraction it is highly beneficial to increase these parameters beyond their default values: in our implementation we set *cl-lim* to 200 and *grow* to 40.

Additional Preprocessing. The labeled preprocessing techniques can also be naturally integrated into the main while-loop of the algorithm (lines 15–17). We note that as the formula $\mathcal{F}_w$ gets progressively simplified throughout the computation, this has value even after the main preprocessing. However, to keep the presented experimental data as clean as possible, we do not use this technique in the experiments.

Label-based Model Rotation. *Label-based recursive model rotation* (LRMR) can be viewed as both a generalization of the recursive model rotation for regular MUSes [6] and variable-based model rotation for variable-based MUSes [17]. Let us first make the following observation.

Proposition 1. *Let $\mathcal{F}$ be a labeled CNF formula, let τ be a complete assignment to $Var(\mathcal{F})$, let $Unsat(\mathcal{F}, \tau)$ denote the subset of clauses of $\mathcal{F}$ that are falsified by τ, and let $L = \bigcap_{C \in Unsat(\mathcal{F}, \tau)} Label(C)$. Then any label in L is necessary for $\mathcal{F}$.*

Proof. Take any $l \in \bigcap_{C \in Unsat(\mathcal{F}, \tau)} Label(C)$. By this assumption, $Unsat(\mathcal{F}, \tau) \subseteq \mathcal{F}^{\{l\}}$, and hence $\mathcal{F} \setminus \mathcal{F}^{\{l\}} \subseteq \mathcal{F} \setminus Unsat(\mathcal{F}, \tau) \in$ SAT. Thus l is necessary. $\qquad\square$

Following this proposition, let us call an assignment τ a *witness for necessity of a label l* in $\mathcal{F}$ (or simply a *witness for l*) if it satisfies the condition $l \in \bigcap_{C \in Unsat(\mathcal{F}, \tau)} Label(C)$.

Clearly, when the main SAT query $\mathtt{SAT}(\mathcal{F}_w \setminus \mathcal{F}_w^{\{l\}})$ in Algorithm 1 is satisfiable and τ is the assignment returned by the SAT solver, τ represents a witness for l in $\mathcal{F}_w$. However, it is possible that there are other labels shared among all the clauses of $Unsat(\mathcal{F}_w, \tau)$ – and these labels can be immediately declared as necessary for $\mathcal{F}_w$, without additional SAT calls. A special case of interest is when $Unsat(\mathcal{F}_w, \tau)$ consists of a single clause – in this case all the other labels of that clause are necessary. As an example, consider the LCNF $\mathcal{F} = \{(r \vee s)_{\{l_1, l_2, l_4\}}, (\neg r)_{\{l_5\}}, (\neg s)_{\{l_6\}}\}$ from before, and consider the satisfying assignment τ to $\mathcal{F} \setminus \mathcal{F}^{\{l_1\}}$ given by $\tau = \{r = 0, s = 0\}$. This same assignment is a witness for each of the labels l_1, l_2 and l_4.

Model rotation exploits the idea of starting with an assignment τ that is a witness for some label l, and obtaining an assignment τ' that is a witness for some other label l' by flipping the value of one of the variables. For example, flipping the value of r in the example above leads to a witness $\tau' = \{r = 1, s = 0\}$ for l_5, while flipping the value of s leads to a witness $\tau'' = \{r = 0, s = 1\}$ of l_6.

A natural question is which variables should be considered for flipping. Let us assume that τ is a witness for l, and suppose that τ' is obtained by flipping the value of a variable $x \in Var(\mathcal{F})$. If $x \notin Var(Unsat(\mathcal{F}, \tau))$ then $Unsat(\mathcal{F}, \tau') \supseteq Unsat(\mathcal{F}, \tau)$ – and so at best τ' is a witness for the same labels as τ. On the other hand, even if the variable x is shared among all of the clauses of $Unsat(\mathcal{F}, \tau)$, there is no guarantee that flipping x will result in another witness. As an example, consider $\mathcal{F} = \{(x)_{\{l_1\}}, (\neg x)_{\{l_2\}}, (\neg x)_{\{l_3\}}\}$ and an assignment $\tau = \{x = 0\}$ that is a witness for l_1. The assignment $\tau' = \{x = 1\}$ is not a witness for either l_2 or l_3. In addition, it is also possible that x is not shared among all of the clauses of $\mathcal{F}$ but flipping the value of x does result in another witness. Consider $\mathcal{F} = \{(z \vee y)_{\{l_1, l_2\}}, (x \vee z)_{\{l_1, l_3\}}, (\neg y \vee \neg x)_{\{l_2\}}, (\neg z)_{\{l_3\}}, (\neg x \vee z)_{\{l_2\}}\}$ and a witness $\tau = \{x = 0, y = 0, z = 0\}$ for l_1. In this case $Unsat(\mathcal{F}, \tau) = \{(z \vee y), (z \vee x)\}$ and x is not a shared variable, yet $\tau' = \{x = 1, y = 0, z = 0\}$ is a witness for l_2. Back to the general case, when τ' happens to also be a witness, then it can be analyzed just in the same way as τ, leading to a recursive process of detection of necessary labels and construction of witnesses, which we refer to as *label-based recursive model rotation* (LRMR).

Algorithm 2 presents the algorithm for LRMR. Note that the new necessary label l found by the SAT call in Algorithm 1 is always added to the set L as a result of LRMR. Motivated by the examples above, we consider the set of all variables present in at least one clause of $Unsat(\mathcal{F}, \tau)$ as candidates for rotation. For each such variable v, we create a new assignment by flipping the value of v in τ and then compute the set of labels witnessed by τ'. The purpose of the if-statement on line 6 of Algorithm 2 is to prevent the algorithm from re-detecting labels that are already known to be necessary and to bound the number of recursive calls to `LModelRotation`. In practice, LRMR is a light-weight technique for detection of necessary labels and represents a crucial optimization that allows one to detect multiple necessary labels in a single SAT solver call.

In our implementation we actually consider the *eager* version of LRMR [11] which allows revisiting necessary labels detected by the previous invocation of

---

Algorithm 2. LModelRotation($\mathcal{F}, L, \tau$) – Label-based Model Rotation

---

    Input   : $\mathcal{F}$ — an unsatisfiable labelled CNF formula

               : L – a set of necessary labels for $\mathcal{F}$

               : τ – a model of $\mathcal{F} \setminus \mathcal{F}^{\{l\}}$ for some $l \in Label(\mathcal{F})$

    Effect : L contains l and possibly additional necessary labels of $\mathcal{F}$

1 **begin**

2      $L \leftarrow L \cup \{l\}$ // l is a new necessary label

3   **foreach** $v \in Var(Unsat(\mathcal{F}, \tau))$ **do**

4          $\tau' \leftarrow \tau|_{\neg v}$ // flip v in τ

5   **foreach** $l \in \bigcap_{C \in Unsat(\mathcal{F}, \tau')} Label(C)$ **do**

6   **if** $l \notin L$ **then**

7   LModelRotation($\mathcal{F}, L, \tau'$)   // recurse

LModelRotation from Algorithm 1 – as it is very likely that starting with a different initial model would lead to discovering a different set of new necessary labels.

4 Chunking

The idea of splitting an MUS computation into *chunks* has already appeared in the context of computing minimal equivalent subformulas using an incremental reduction to GMUS [13]. In this section we adapt this idea to the context of LMUS computation.

Let $\mathcal{F}$ be an unsatisfiable LCNF formula and suppose that the set $L = Label(\mathcal{F})$ of labels of $\mathcal{F}$ is partitioned into two chunks: $L = L_1 \cup L_2$. We first concentrate on minimizing the L_1-labels required for unsatisfiability. To this end, we consider the formula $\mathcal{F}_1 = \text{Project}(\mathcal{F}, L_1)$, obtained by removing all the non-L_1 labels from the clauses of $\mathcal{F}$, and apply the LMUS computation to $\mathcal{F}_1$. Let $K_1 \subseteq L_1$ represent the result of this computation. It follows that the formula $\mathcal{F} \setminus \mathcal{F}^{L_1 \setminus K_1}$ (obtained from $\mathcal{F}$ by removing all the clauses with a label in $L_1 \setminus K_1$) remains unsatisfiable, and that no additional label in K_1 can be removed. Next, starting from the formula $\mathcal{F} \setminus \mathcal{F}^{L_1 \setminus K_1}$, we concentrate on minimizing the L_2-labels required for unsatisfiability. As before, we consider the formula $\mathcal{F}_2 = \text{Project}(\mathcal{F} \setminus \mathcal{F}^{L_1 \setminus K_1}, L_2)$ obtained by removing all the non-L_2 labels. Let $K_2 \subseteq L_2$ represent the result of the LMUS computation applied to $\mathcal{F}_2$. Again, it follows that the formula $\mathcal{F} \setminus \mathcal{F}^{(L_1 \setminus K_1) \cup (L_2 \setminus K_2)}$ remains unsatisfiable, and that no additional label in K_2 can be removed. In other words, $K_1 \cup K_2$ is an LMUS of the original formula $\mathcal{F}$.

Algorithm 3 is based on the iterative and slightly optimized application of the observation above. The input to the algorithm is an unsatisfiable LCNF formula $\mathcal{F}$ and a disjoint partitioning of its set of labels into chunks. As the first step, we simplify $\mathcal{F}$ using preprocessing (see Section 3 for details). The algorithm maintains the *working formula* $\mathcal{F}_w$. Consider an iteration of the for-loop. First, we create an auxiliary LCNF $\mathcal{G}_j$ by removing from each clause of $\mathcal{F}_w$ all the

---

Algorithm 3. LHYB($\mathcal{F}, \{L_1, \ldots, L_n\}$) – LMUS Extraction with chunks

---

Input : $\mathcal{F}$ — a (trimmed) unsatisfiable LCNF formula
      : $L = \{L_1, \ldots, L_n\}$ – a partition of $Label(\mathcal{F})$ into n chunks
Output : LMUS L of $\mathcal{F}$

```
1 begin
2   LSimplify(F)                                    // Main preprocessing
3   Fw ← F                                          // Working formula
4   for j ← 1 to n do
5       Gj = Project(Fw, Lj)                        // Leave only Lj labels
6       Kj ← LHYB(Gj)                               // Call Algorithm 1
7       Fw ← Fw \ Fw^{Lj\Kj}    // Remove clauses with Lj \ Kj labels
8       Fw ← Project(Fw, Lj+1 ∪ ··· ∪ Ln)          // Remove Kj labels
9   return K1 ∪ ··· ∪ Kn
```

---

non-L_j labels (line 5). Next, we apply Algorithm 1 to compute an LMUS K_j of $\mathcal{G}_j$ (line 6), after which all the clauses with a label in $L_j \setminus K_j$ are permanently removed from $\mathcal{F}_w$ (line 7). Finally, we remove all the K_j labels from every clause of $\mathcal{F}_w$ (and in some sense finalize them).

Proposition 2. *The set $K_1 \cup \cdots \cup K_n$ computed by Algorithm 3 is an LMUS of $\mathcal{F}$.*

We illustrate the execution of Algorithm 3 on a simple example. Let $\mathcal{F} = \{(p \vee r)_{\{l_1\}}, (q \vee r)_{\{l_2\}}, (\neg p \vee r)_{\{l_3\}}, (\neg q \vee r)_{\{l_4\}}, (\neg r)_{\{l_5\}}\}$, $L_1 = \{l_1, l_2\}$, $L_2 = \{l_3, l_4\}$ and $L_3 = \{l_5\}$. Suppose that no preprocessing is applied. We set $\mathcal{F}_w = \mathcal{F}$ (line 3). On the first iteration of the for-loop we create the formula $\mathcal{G}_1 = \{(p \vee r)_{\{l_1\}}, (q \vee r)_{\{l_2\}}, (\neg p \vee r)_{\{\}}, (\neg q \vee r)_{\{\}}, (\neg r)_{\{\}}\}$ (line 5). Suppose that the LMUS computation applied to $\mathcal{G}_1$ results in $K_1 = \{l_2\}$ (line 6). This allows to remove from $\mathcal{F}_w$ all the clauses with an l_1-label, resulting in $\mathcal{F}_w = \{(q \vee r)_{\{l_2\}}, (\neg p \vee r)_{\{l_3\}}, (\neg q \vee r)_{\{l_4\}}, (\neg r)_{\{l_5\}}\}$ (line 7). Furthermore, we remove the l_2-label from every clause, resulting in $\mathcal{F}_w = \{(q \vee r)_{\{\}}, (\neg p \vee r)_{\{l_3\}}, (\neg q \vee r)_{\{l_4\}}, (\neg r)_{\{l_5\}}\}$ (line 8). On the second iteration of the for-loop, we create the formula $\mathcal{G}_2 = \{(q \vee r)_{\{\}}, (\neg p \vee r)_{\{l_3\}}, (\neg q \vee r)_{\{l_4\}}, (\neg r)_{\{\}}\}$ (line 5), with $K_2 = \{l_4\}$ (line 6), and we update $\mathcal{F}_w$ to $\{(q \vee r)_{\{\}}, (\neg q \vee r)_{\{\}}, (\neg r)_{\{l_5\}}\}$ (lines $7 - 8$). On the final iteration of the for-loop we detect that $K_3 = \{l_5\}$, and the algorithm terminates with the LMUS $K_1 \cup K_2 \cup K_3 = \{l_2, l_4, l_5\}$.

In the assumption-based approach, the size of each chunk represents an important trade-off. On the one hand, using chunks of smaller size significantly reduces the number of assumptions passed to the SAT solver, and as a result significantly improves the runtime of each SAT query. As we show in the experiments, using chunks (of reasonable size) leads to a dramatic performance speed-up on formulas that are already minimal or close to minimal. On the other hand, each invocation of Algorithm 1 from Algorithm 3 requires to create a fresh instance of a SAT solver, making the overall approach less incremental. Even more importantly, the two crucial optimizations used in Algorithm 1, namely refinement and rotation, become *local* to each chunk. As we show in the experiments, rotation is not a problem, but the lack of global refinement leads to larger formulas and consequently to a significant performance slow-down when the initial LCNF formula contains a lot of redundancy. To summarize: *chunking works extremely well after trimming, but not otherwise.*

We finish this section with a few additional implementation details on the integration of Algorithm 1 in Algorithm 3. As the main preprocessing is now performed in Algorithm 3, we completely disable all preprocessing in Algorithm 1. Second, we use the satisfying assignments discovered by Algorithm 1 for rotation on the global formula, thus also discovering necessary labels in other chunks.

5 Experiments

5.1 Overall GMUS Evaluation

We have implemented Algorithm 3 (assumption-based approach with preprocessing and chunking) on top of the SAT solver MINISAT. We call our implementation IBMUC[1]. For all practical purposes IBMUC should be viewed as a variant of MUSER2.

The default options for IBMUC were chosen as follows: the main preprocessing in Algorithm 3 is set to the most aggressive mode (MINISAT's options "-grow=40" and "-cl-lim=200"); the size of each chunk is set to 4000; all preprocessing in Algorithm 1 is turned off; groups were evaluated in the original order (sorted by increasing index).

We focus on GMUS extraction, and we consider the following state-of-the-art tools for comparison: MUSER2[7], MINISAT$_{abb}$[10], HaifaMUC[9,11]. It should be noted that MUSER2 represents a pure assumption-based approach, HaifaMUC represents a pure resolution-based approach, while MINISAT$_{abb}$ represents a hybrid approach (or more precisely, a resolution-based approach implemented via assumptions). For fair comparison, we select MINISAT as the default back-end SAT solver in each of these tools. Neither MINISAT$_{abb}$ nor HaifaMUC support preprocessing for GMUS extraction, and we enable preprocessing in MUSER2 (using the "-minisats" option).

For our experiments we use all 197 group MUS benchmarks from the MUS track of the 2011 SAT Competition[5] under various initial trimming scenarios. More precisely, we have modified HaifaMUC to perform iterative trimming of the original formula, and we consider the GMUS formulas obtained after 1, 5, and 10 rounds of trimming (we refer to these as T1, T5 and T10 respectively). In addition, we consider original formulas (ORIG) and fully minimized formulas (CORE) obtained by running IBMUC on T10 formulas (the choice of IBMUC is

---

[1] The tool was developed during the internship of the first author at IBM. Due to legal issues, we cannot make the source code publicly available.
[2] http://logos.ucd.ie/wiki/doku.php?id=muser
[3] http://fmv.jku.at/musaddlit
[4] We thank Vadim Ryvchin for providing us the version used in [11].
[5] http://www.satcompetition.org/2011

irrelevant since all of the tools produce the same or very similar results in most cases). The statistics for GMUS formulas considered are presented in Table 1, where "Trim Ratio" represents the ratio of the number of groups in trimmed formulas and those in CORE, and "Trim Time" represents the trimming time. However, these numbers should only be considered as a proof of concept, since we believe that an extension of DMUSer [14] to GMUSes would provide both better trim ratios and better runtimes.

Table 1. Summary for GMUS formulas considered

	ORIG	T1	T5	T10	CORE
Trim Ratio	7.6	1.9	1.3	1.2	1
Trim Time	0	726	1764	2686	NA

Trim ratio: ratio of the number of groups in trimmed formulas and those in CORE;
Trim time: trimming time in seconds (NA for CORE).

All of the experiments are performed on a 2.0 GHz Linux-based machine with an Intel Xeon E7540 processor and 32 GB of RAM. The time-limit for each run is set to 2000 seconds.

Table 2. Summary for GMUS extraction under different trimming strength

	ORIG	T1	T5	T10	CORE
HaifaMUC	**11548**	10759	8969	8775	6166
MINISAT$_{abb}$	19649 (1)	**9202**	7503	6295	5577
MUSER2	12032	9375 (1)	7501	8111 (1)	74345 (14)
IBMUC	43917 (12)	12446 (2)	**5143**	**4619**	**2752**

Total runtimes in seconds and number of timeouts (in parentheses).

The cumulative runtimes for each of the tools HaifaMUC (HM), MINISAT$_{abb}$ (MA), MUSER2 (M2) and IBMUC (IB) are presented in Table 2 and displayed in Figure 1. In addition, for IBMUC we explicitly show the time spent on satisfiable SAT calls (SAT), unsatisfiable SAT calls (UNSAT), and preprocessing and rotation (OTHER). Several comments are in order. First, chunking significantly deteriorates performance on original formulas (or more generally on formulas with many redundancies) as it limits refinement to be local for each chunk, which in turn significantly increases the total time for UNSAT queries. On the other hand, chunking is crucial in the assumption-based approach when considering fully minimized formulas (or more generally formulas with very few redundancies), justifying our statements from the introduction. The two resolution-based approaches HaifaMUC and MINISAT$_{abb}$ are more stable: their runtimes are consistently decreased with additional trimming.

We can also see that IBMUC clearly outperforms other tools on T5, T10 and CORE formulas, with the corresponding speed-ups of at least 1.45x, 1.36x and 2x respectively. A more detailed analysis shows that out of 197 instances, the combinatorial problem "4pipe.gcnf" is particularly hard for all the assumption-based solvers; if it were excluded from the evaluation, then the corresponding

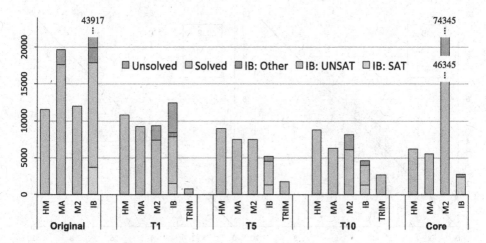

Fig. 1. Detailed time statistics corresponding to Table 2. The columns for IB at ORIG and for M2 at CORE are truncated.

speed-ups would rise to 1.9x, 2x and 3x respectively. By also taking the trimming time into account, we can see that medium trimming (T5) benefits all MUS extractors, while the more aggressive trimming (T10) does not pay off the additional effort spent for its computation (although as mentioned previously this could change if our trimming procedure were improved). Thus we choose T5 and CORE as the most interesting scenarios, since the latter could be important in the context of GMUS validation. The scatter plots in Figure 2 show a detailed comparison of GMUS extraction times of HaifaMUC (on X-axis) versus IBMUC, MUSER2, MINISAT$_{abb}$ (on Y-axis). For better visualization, the plots follow a logarithmic scale, and the instances with solving times greater than 400 seconds are truncated. We can see that in both scenarios HaifaMUC and MINISAT$_{abb}$ perform very similarly, while IBMUC is a clear winner.

5.2 Parameter Impact

In Figure 3 we present a detailed analysis of the impact of preprocessing and chunking on IBMUC's performance on T5 formulas. We have excluded the benchmark "4pipe.gcnf" from the evaluation (since its running time by far dominates all the other runtimes). For the left two plots, we vary the initial preprocessing effort, with the data points representing (left to right): no preprocessing, MINISAT's default preprocessing (grow=0, cl-lim=20), more aggressive (grow=6, cl-lim=30), even more aggressive (grow=10, cl-lim=50), and (our default) most aggressive (grow=40, cl-lim=200). For the right two plots, we vary the number of chunks (or more precisely the number of groups in each chunk), with the data points representing (left to right): a single chunk, 6000 groups per chunk, 5000 groups per chunk, (our default) 4000 groups per chunk, 3000 groups per chunk. In the top plots we present the cumulative number of satisfiable solver queries (SAT) and the corresponding time, and in the bottom two plots we present the

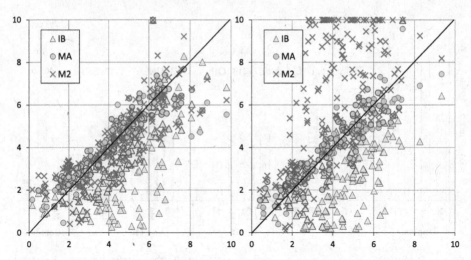

Fig. 2. Detailed solver-to-solver comparison on T5 (left) and CORE (right) formulas, with `HaifaMUC` on X-axis, and `MINISAT`$_{abb}$, `MUSER2` and `IBMUC` on Y-axis.

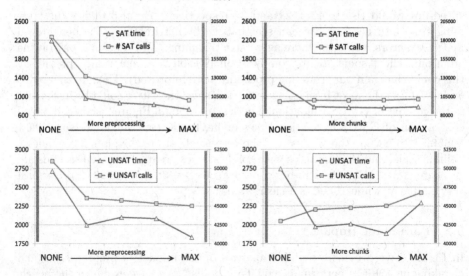

Fig. 3. Influence of preprocessing and chunking on T5 formulas. Left: preprocessing. Right: chunking. Top: number of SAT calls and SAT time. Bottom: number of UNSAT calls and UNSAT time.

cumulative number of unsatisfiable solver queries (UNSAT) and the corresponding time. By computing the ratio between the total time and the number of queries, we can estimate an average time for SAT and for UNSAT queries for each data point.

From the top-left plot we can see that preprocessing is extremely effective at decreasing the total number of SAT queries required. This effect is due solely to improved model rotation. We can also note that preprocessing makes the average SAT query slightly faster, however this effect is not significant. From the bottom-left plot we see that preprocessing is also effective in reducing the total number of UNSAT queries. We speculate that this is due to the refinement capability of preprocessing: even after trimming many unnecessary labels get removed when applying variable elimination, subsumption and self-subsumption. In addition, we note that preprocessing also has a clear beneficial effect on the average UNSAT query time.

From the top-right plot we can see that chunking does not have any effect on the number of rotations found (and hence on the number of SAT calls), however chunking helps to reduce the average time for a SAT query and hence the total SAT time. Finally, the bottom-right plot shows that chunking in general degrades the performance of core refinement (the number of UNSAT calls increases as there are more chunks) and justifies our claim that the size of each chunk represents an interesting trade-off (with an empirically best value being 4000).

We have performed many additional experiments. We have experimented with the order of label removal – no significant effect on T5 formulas. We have replaced MINISAT by Glucose [18] as the SAT solver back-end, which resulted in additional significant performance improvements – the time required to solve "4pipe.gcnf" was reduced from 1900s to 20s, and the time required to solve the remaining formulas was reduced from 3243s to 2400s. We emphasize that absolutely no modification to Glucose was required, justifying the wider applicability of the assumption-based approach. We have also conducted experiments on T10 and CORE formulas – the conclusions are exactly the same as for T5 formulas (up to a slightly different best empirical value for the chunk size).

We have also conducted preliminary experiments on standard MUS benchmarks and a preliminary conclusion is that by applying preprocessing too aggressively (for example, with grow=40) helps rotation but hurts solving. In other words, we have witnessed a significant reduction in the number of SAT solver queries (just as for GMUSes), but each individual solver query becomes significantly more expensive (contrary to GMUSes). One promising solution, which we leave to further work, is to consider different formulas for solving and for rotation (each obtained from the original formula by applying a suitable amount of preprocessing).

6   Conclusions

In this paper we have further extended the ideas from [3] on using preprocessing for MUS computations, and in particular presented an algorithm for recursive model rotation on preprocessed (labeled) formulas and showed that it has a clear benefit for GMUS computation. In addition, we have shown that the main weakness of the assumption-based approach for MUS computation can be easily addressed using chunking. In conclusion, *a careful choice of the back-end SAT solver, trimming, preprocessing, chunking and recursive rotation leads to the best overall approach.*

References

1. Liffiton, M.H., Sakallah, K.A.: Algorithms for computing minimal unsatisfiable subsets of constraints. J. Autom. Reasoning **40**(1), 1–33 (2008)
2. Nadel, A.: Boosting minimal unsatisfiable core extraction. In: Proceedings of 10th International Conference on Formal Methods in Computer-Aided Design, FMCAD 2010, Lugano, Switzerland, October 20–23, pp. 221–229 (2010)
3. Belov, A., Järvisalo, M., Marques-Silva, J.: Formula preprocessing in MUS extraction. In: Piterman, N., Smolka, S.A. (eds.) TACAS 2013 (ETAPS 2013). LNCS, vol. 7795, pp. 108–123. Springer, Heidelberg (2013)
4. Belov, A., Morgado, A., Marques-Silva, J.: SAT-based preprocessing for MaxSAT. In: McMillan, K., Middeldorp, A., Voronkov, A. (eds.) LPAR-19 2013. LNCS, vol. 8312, pp. 96–111. Springer, Heidelberg (2013)
5. Belov, A., Marques-Silva, J.: Generalizing redundancy in propositional logic: Foundations and hitting sets duality. CoRR, abs/1207.1257 (2012)
6. Belov, A., Marques-Silva, J.: Accelerating MUS extraction with recursive model rotation. In: International Conference on Formal Methods in Computer-Aided Design, FMCAD 2011, Austin, TX, USA, October 30 - November 02, 2011, pp. 37–40 (2011)
7. Belov, A., Marques-Silva, J.: Muser2: An efficient MUS extractor. JSAT **8**(3/4), 123–128 (2012)
8. Eén, N., Sörensson, N.: An extensible SAT-solver. In: Giunchiglia, E., Tacchella, A. (eds.) SAT 2003. LNCS, vol. 2919, pp. 502–518. Springer, Heidelberg (2004)
9. Ryvchin, V., Strichman, O.: Faster extraction of high-level minimal unsatisfiable cores. In: Sakallah, K.A., Simon, L. (eds.) SAT 2011. LNCS, vol. 6695, pp. 174–187. Springer, Heidelberg (2011)
10. Lagniez, J.-M., Biere, A.: Factoring out assumptions to speed up MUS extraction. In: Järvisalo, M., Van Gelder, A. (eds.) SAT 2013. LNCS, vol. 7962, pp. 276–292. Springer, Heidelberg (2013)
11. Nadel, A., Ryvchin, V., Strichman, O.: Accelerated deletion-based extraction of minimal unsatisfiable cores. JSAT **9**, 27–51 (2014)
12. Sörensson, N., Biere, A.: Minimizing learned clauses. In: Kullmann, O. (ed.) SAT 2009. LNCS, vol. 5584, pp. 237–243. Springer, Heidelberg (2009)
13. Belov, A., Janota, M., Lynce, I., Marques-Silva, J.: On computing minimal equivalent subformulas. In: Milano, M. (ed.) CP 2012. LNCS, vol. 7514, pp. 158–174. Springer, Heidelberg (2012)
14. Belov, A., Heule, M.J.H., Marques-Silva, J.: MUS extraction using clausal proofs. In: Sinz, C., Egly, U. (eds.) SAT 2014. LNCS, vol. 8561, pp. 48–57. Springer, Heidelberg (2014)
15. Marques-Silva, J., Lynce, I.: On improving MUS extraction algorithms. In: Sakallah, K.A., Simon, L. (eds.) SAT 2011. LNCS, vol. 6695, pp. 159–173. Springer, Heidelberg (2011)
16. Dershowitz, N., Hanna, Z., Nadel, A.: A scalable algorithm for minimal unsatisfiable core extraction. In: Biere, A., Gomes, C.P. (eds.) SAT 2006. LNCS, vol. 4121, pp. 36–41. Springer, Heidelberg (2006)
17. Belov, A., Ivrii, A., Matsliah, A., Marques-Silva, J.: On efficient computation of variable MUSes. In: Cimatti, A., Sebastiani, R. (eds.) SAT 2012. LNCS, vol. 7317, pp. 298–311. Springer, Heidelberg (2012)
18. Audemard, G., Simon, L.: Predicting learnt clauses quality in modern SAT solvers. In: IJCAI 2009, Proceedings of the 21st International Joint Conference on Artificial Intelligence, Pasadena, California, USA, July 11–17, 2009, pp. 399–404 (2009)

Improved Algorithms for Sparse MAX-SAT and MAX-k-CSP

Ruiwen Chen$^{(\boxtimes)}$ and Rahul Santhanam$^{(\boxtimes)}$

School of Informatics, University of Edinburgh, Edinburgh, UK
{rchen2,rsanthan}@inf.ed.ac.uk

Abstract. We give improved deterministic algorithms solving sparse instances of MAX-SAT and MAX-k-CSP. For instances with n variables and cn clauses (constraints), we give algorithms running in time poly$(n) \cdot 2^{n(1-\mu)}$ for

- $\mu = \Omega(\frac{1}{c})$ and polynomial space solving MAX-SAT and MAX-k-SAT,
- $\mu = \Omega(\frac{1}{\sqrt{c}})$ and exponential space solving MAX-SAT and MAX-k-SAT,
- $\mu = \Omega(\frac{1}{ck^2})$ and polynomial space solving MAX-k-CSP,
- $\mu = \Omega(\frac{1}{\sqrt{ck^3}})$ and exponential space solving MAX-k-CSP.

The previous MAX-SAT algorithms have savings $\mu = \Omega(\frac{1}{c^2 \log^2 c})$ for running in polynomial space [15] and $\mu = \Omega(\frac{1}{c \log c})$ for exponential space [5]. We also give an algorithm with improved savings for satisfiability of depth-2 threshold circuits with cn wires.

Keywords: Satisfiability algorithm · MAX-SAT · MAX-k-CSP

1 Introduction

The maximum satisfiability problem (MAX-SAT) is to find an assignment that maximizes the number of satisfied clauses in a CNF formula. MAX-k-SAT is the special case where all clauses have at most k literals. For instances with n variables and $m = cn$ clauses, a trivial brute-force search solves MAX-SAT in time $O(mn2^n)$. We are interested in better algorithms running in time $\tilde{O}(2^{n(1-\mu)})$ for $\mu > 0$; we will call μ the *savings* over exhaustive search, and we use $\tilde{O}()$ to ignore polynomial factors. To the best of our knowledge, the best savings is $\mu = \Omega(\frac{1}{c \log c})$ obtained by Dantsin and Wolpert [5] for an exponential-space algorithm. For polynomial space algorithms, the best savings is $\mu = \Omega(\frac{1}{c^2 \log^2 c})$ shown by Sakai, Seto, and Tamaki [15] recently.

R. Chen—Supported by the European Research Council under the European Union's Seventh Framework Programme (FP7/2007-2013)/ ERC Grant Agreement no. 615075.

R. Santhanam—Supported by the European Research Council under the European Union's Seventh Framework Programme (FP7/2007-2013)/ ERC Grant Agreement no. 615075.

© Springer International Publishing Switzerland 2015
M. Heule and S. Weaver (Eds.): SAT 2015, LNCS 9340, pp. 33–45, 2015.
DOI: 10.1007/978-3-319-24318-4_4

The algorithm of Sakai, Seto, and Tamaki [15] is based on concentrated shrinkage under restrictions, which was used by Santhanam [16] for solving the satisfiability problem on de Morgan formulas. Santhanam [16] observed that, by greedily restricting the most frequent variables in a formula until p fraction of the variables are left, the formula size shrinks with high probability by a factor of p^Γ for $\Gamma \geqslant 1.5$. We will call Γ the *shrinkage exponent* for de Morgan formulas with respect to greedy restrictions. The satisfiability algorithm [16] recursively restricts $n - \Omega(n)$ variables, and then gets nontrivial savings since almost all restricted formulas have size much smaller than the number of variables left. Sakai, Seto, and Tamaki [15] showed that a similar shrinkage property holds for MAX-k-SAT instances, which leads to an algorithm with savings $\mu = \Omega(\frac{1}{c^2 k^2})$ for instances with cn clauses. For solving MAX-SAT, they applied Schuler's width reduction [1,17] to reduce MAX-SAT to MAX-k-SAT for $k = O(\log n)$; their final MAX-SAT algorithm [15] has savings $\mu = \Omega(\frac{1}{c^2 \log^2 c})$.

In this work, we improve the savings in [15] further to the following.

Theorem 1. *There is a polynomial-space algorithm solving MAX-SAT instances with n variables and cn clauses in time $\tilde{O}(2^{n(1-\mu)})$ for $\mu = \Omega(\frac{1}{c})$.*

Our algorithm is based on an improvement of concentrated shrinkage under greedy restrictions. We define a *measure* on MAX-SAT instances, which takes into account the numbers of clauses of different widths. We show that by a greedy restriction of all but p fraction of the variables, the measure shrinks with high probability by a factor of p^Γ for $\Gamma \geqslant 2$. This improved shrinkage exponent allows us to get better savings in the algorithm for the maximization problem. Furthermore, since the measure does not depend on the clause width, we do not need Schuler's width reduction which was used by [15], and our algorithm does not differentiate between MAX-SAT and MAX-k-SAT.

We further improve the savings when the algorithm is allowed to run in exponential space. Here we use Williams' algorithm [19] for MAX-2-SAT as a black-box, and improve the shrinkage exponent to $\Gamma \geqslant 3$ by defining a different measure on MAX-SAT instances. This improved shrinkage exponent again implies better savings in the algorithm.

Theorem 2. *There is an exponential-space algorithm solving MAX-SAT instances with n variables and cn clauses in time $\tilde{O}(2^{n(1-\mu)})$, for $\mu = \Omega(\frac{1}{\sqrt{c}})$.*

This improves the previous best-known result with savings $\mu = \Omega(\frac{1}{c \log c})$ by Dantsin and Wolpert [5] for solving MAX-SAT in exponential space.

Our approach is quite generic; we also apply it to solve sparse MAX-k-CSP. Specifically, we give a measure for MAX-k-CSP instances, and show that the measure shrinks nontrivially with probability 1 under greedy restrictions. This allows us to give the following algorithms.

Theorem 3. *For MAX-k-CSP instances with n variables and cn constraints, there is a polynomial-space algorithm running in time $\tilde{O}(2^{n(1-\mu)})$ with savings $\mu = \Omega(\frac{1}{ck^2})$, and an exponential-space algorithm with savings $\mu = \Omega(\frac{1}{\sqrt{c}k^3})$.*

All our algorithms extend to *counting* the number of optimal assignments for the *weighted* version of the problem, where each clause/constraint is given a weight, and the goal is to maximize the total weight of satisfied clauses/constraints.

We also consider depth-2 threshold circuits, which can be viewed as a generalization of MAX-SAT. Impagliazzo, Paturi, and Schneider [10] recently gave a satisfiability algorithm with savings $\mu = \frac{1}{c^{O(c^2)}}$ for depth-2 threshold circuits of cn wires. Using our shrinkage approach, we improve the savings slightly to $\frac{1}{c^{O(c)}}$, with a much simpler analysis.

1.1 Related Work

Exact algorithms for sparse MAX-SAT and MAX-k-SAT have been well studied. For MAX-SAT instances with n variables and $m = cn$ clauses, the best savings of polynomial-space algorithms was $\Omega(\frac{1}{c^2 \log^2 c})$ [15] (and $\Omega(\frac{1}{c \log^3 c})$ for a randomized algorithm), improving a previous result $\Omega(\frac{1}{2^{O(c)}})$ [13]. The best savings of exponential-space algorithms was $\Omega(\frac{1}{c \log c})$ [5]. In this work, we improve the savings for both polynomial-space and exponential-space algorithms.

There are also algorithms with running time expressed as $\tilde{O}(2^{\delta m})$ for a constant $\delta < 1$, where m is the number of clauses/constraints. For example, the best such algorithms achieved $\delta \leqslant 0.4057$ for MAX-SAT [2], $\delta \leqslant 0.1583$ for MAX-2-SAT [7], and $\delta \leqslant 0.1901$ for MAX-2-CSP [7]. However, such algorithms are not better than exhaustive search for $m > n/\delta$.

For general (non-sparse) instances, the only non-trivial exact algorithm is Williams' algorithm for MAX-2-CSP (and MAX-2-SAT), which runs in time $\tilde{O}(2^{n\omega/3})$ for $\omega < 2.376$ and in exponential space. It is open whether we have non-trivial MAX-2-CSP algorithms running in polynomial space, or generalize Williams' algorithm for MAX-3-CSP or MAX-3-SAT. In this work, we will use Williams' algorithm as a blackbox for improving algorithms for sparse MAX-k-CSP and MAX-SAT.

The shrinkage approach to satisfiability algorithms was initiated by Santhanam [16] for de Morgan formulas. The algorithm was improved later [3,4,12] by improving the shrinkage exponents with respect to certain greedy restrictions. In particular, the improvement in [4] follows from a measuring technique of [9,14] for de Morgan formulas.

The measuring and shrinkage technique we use in this work is also related to the "measure and conquer" approach [6], which was used to give improved exact algorithms for graph problems such as maximum independent set. The main difference is that, the usual "measure and conquer" approach reduces the measure additively in each recursive step, whereas the shrinkage approach reduces the measure by a multiplicative factor (depending on the shrinkage exponent), and moreover the reduction only occurs with high probability in the latter case.

1.2 Organization of the Paper

We give preliminaries in Section 2. Since our MAX-SAT algorithms require more involved analysis than MAX-k-CSP, we first present our MAX-k-CSP algorithms in Section 3, and then MAX-SAT algorithms in Section 4. In Section 5, we apply a similar approach to improve satisfiability algorithms for depth-2 threshold circuits.

2 Preliminaries

2.1 MAXSAT, MAX-k-SAT, MAX-k-CSP

Let $x_1, \ldots, x_n$ be boolean variables. A *literal* is either a variable or its negation. A *clause* is a disjunction of literals; a *k-clause* is a clause on k literals. The *MAX-SAT* problem is to find, given a collection of clauses, an assignment to the variables maximizing the number of satisfied clauses (we call such an assignment *optimal*). MAX-k-SAT is the special case of MAX-SAT where all clauses have at most k literals. The *weighted MAX-SAT* problem generalizes MAX-SAT by associating with each clause an integer weight, and the goal is to find an assignment maximizing the total weight of satisfied clauses.

A *k-constraint* is a boolean function on k variables. The *(weighted) MAX-k-CSP* problem generalizes (weighted) MAX-k-SAT by allowing arbitrary constraints rather than disjunctions of literals. We will also consider the problems of counting the number of optimal assignments for the above optimization problems.

2.2 Concentration Bounds

A sequence of random variables $X_0, X_1, \ldots, X_n$ is a *supermartingale* with respect to a sequence of random variables $R_1, \ldots, R_n$ if $\mathbf{E}[X_i \mid R_{i-1}, \ldots, R_1] \leqslant X_{i-1}$, for $1 \leqslant i \leqslant n$. We need the following variant of Azuma's inequality.

Lemma 1 ([3]). *Let $\{X_i\}_{i=0}^n$ be a supermartingale with respect to $\{R_i\}_{i=1}^n$. Let $Y_i = X_i - X_{i-1}$. If, for every $1 \leqslant i \leqslant n$, the random variable Y_i (conditioned on $R_{i-1}, \ldots, R_1$) assumes two values each with probability $1/2$, and there exists a constant $c_i \geqslant 0$ such that $Y_i \leqslant c_i$, then, for any λ, we have*

$$\mathbf{Pr}[X_n - X_0 \geqslant \lambda] \leqslant \exp\left(-\frac{\lambda^2}{2\sum_{i=1}^n c_i^2}\right).$$

3 MAX-k-CSP

3.1 Known Algorithms for MAX-2-CSP

Williams [19] gave an algorithm with constant savings solving general (non-sparse, weighted) MAX-2-CSP and MAX-2-SAT. In fact, Williams's algorithm also counts the number of optimal assignments.

Theorem 4 ([11,19]). *For MAX-2-CSP instances with n variables and m constraints where each constraint has a weight at most W, there is an algorithm which counts the number of optimal assignments in time $O(\mu(nmW) \cdot 2^{n\omega/3})$, where $\omega < 2.376$ and $\mu(b) = b \log b \log \log b$.*

Note that, Williams' algorithm requires exponential space; it is not known whether there are polynomial-space algorithms with constant savings.

For sparse instances of MAX-2-CSP with cn constraints, several known algorithms [8,18] have savings of the form $\mu = \Omega(\frac{1}{c})$. In the following, we show one simple algorithm with savings $\Omega(\frac{1}{c})$. The algorithm is based on a greedy restriction of the most frequent variables appearing in 2-constraints. Although the hidden constant in the savings $\Omega(\frac{1}{c})$ is not the best, we wish to use it as a warm-up for our later algorithms.

Lemma 2. *There is a polynomial-space algorithm solving MAX-2-CSP with n variables and cn 2-constraints in time $\tilde{O}(2^{n(1-\Omega(\frac{1}{c}))})$.*

Proof. Given a MAX-2-CSP instance with cn 2-constraints, we assume each 2-constraint is over two distinct variables. Let x be a variable which appears the maximum number of times in the 2-constraints. This means x appears in at least $2c$ of 2-constraints. We make two branches by fixing $x = 0$ in one branch and $x = 1$ in the other. In each branch, the number of remaining 2-constraints is at most $cn - 2c = cn(1 - 2/n) \leqslant cn(1 - 1/n)^2$. Then recursively restrict the most frequent variables one at a time. After $n - pn$ steps, for $p = 1/(2c)$, there will be pn variables left unfixed, but the number of remaining 2-constraints will be at most $cn(1-1/n)^2 \cdots (1-1/(pn+1))^2 = cn \cdot p^2 = pn/2$. Then after at most $pn/2$ more steps (restricting the most frequent variable in each step), all 2-constraints will be eliminated, and we get a MAX-1-CSP instance (on $pn/2$ variables).

We can maintain the number of satisfied constraints along each recursive branch, and, at the end of each branch, solve MAX-1-CSP by setting each variable to a value which satisfies at least as many constraints as the other. The total number of branches is at most $2^{n-pn/2}$. Therefore, the running time is $\tilde{O}(2^{n-pn/2}) = \tilde{O}(2^{n(1-\frac{1}{4c})})$, and the algorithm uses polynomial space. $\quad\square$

Note that, this algorithm can be extended to solve weighted MAX-2-CSP and also count the number of optimal assignments, by maintaining necessary information along the recursive branches. If W is the maximum weight of the constraints, the running time will be $\tilde{O}(2^{n(1-\frac{1}{4c})} \cdot \log W)$.

3.2 A Polynomial-space Algorithm for MAX-k-CSP

We first extend the algorithm in Lemma 2 to solve MAX-k-CSP. We introduce a *measure* on the instances, and use greedy restrictions such that the measure (and thus, the size of the instance) reduces non-trivially.

Let F be a MAX-k-CSP instance on n variables. For each i-constraint C in F, we define $\sigma(C) = \sigma_i \equiv i(i-1)$. Let $\sigma(F) = \sum_{C \in F} \sigma(C)$. Consider a restriction ρ where we randomly pick a variable and fix it. For an i-constraint C,

$$\mathbf{E}_\rho[\sigma(C|_\rho)] \leqslant \frac{n-i}{n}\sigma_i + \frac{i}{n}\sigma_{i-1} = \sigma_i \cdot \left[1 - \frac{i}{n}\left(1 - \frac{\sigma_{i-1}}{\sigma_i}\right)\right] = \sigma_i \cdot \left(1 - \frac{2}{n}\right).$$

We then have $\mathbf{E}_\rho[\sigma(F|_\rho)] \leqslant \sigma(F)(1 - 2/n) \leqslant \sigma(F)(1 - 1/n)^2$. By averaging, we can deterministically find one variable (in polynomial time) such that, after fixing it to either 0 or 1, $\sigma(F)$ reduces by a factor of $(1-1/n)^2$. If we repeat this recursively until pn variables left, the measure on the restricted instance will be at most $\sigma(F)p^2$. Our algorithm follows from this non-trivial shrinkage.

Theorem 5. *For MAX-k-CSP instances on n variables with cn constraints, there is an algorithm running in $\tilde{O}(2^{n(1-\Omega(\frac{1}{ck^2}))})$ time and polynomial space.*

Proof. Let F be an instance with cn constraints; then $\sigma(F) \leqslant cn\sigma_k = cnk(k-1)$. The algorithm recursively restricts one variable at a time. At the i-th step, restrict a variable x such that the measure reduces by a factor of $(1-1/(n-i+1))^2$ for both restrictions $x = 0$ and $x = 1$. After $n - pn$ steps, for $p = \frac{1}{ck(k-1)}$, the restricted instance F' has $\sigma(F') \leqslant \sigma(F)p^2 \leqslant pn$. Note that, this holds for all recursive branches.

Suppose the number of i-constraints left in F' is b_i; then since $\sigma(F') = \sum_{i=2}^k b_i i(i - 1) \leqslant pn$, we have $\sum_{i=2}^k b_i(i - 1) \leqslant pn/2$. Therefore, after $pn/2$ recursive steps (fix all but one variable in each i-constraint), all remaining constraints have width 1, and we can solve MAX-1-CSP in polynomial time. The recursion tree has at most $2^{n-pn/2}$ branches. The total running time is at most $\text{poly}(n) \cdot 2^{n(1-p/2)}$, and the algorithm uses polynomial space. $\qquad\square$

3.3   An Exponential-Space Algorithm for MAX-k-CSP

We next give an algorithm with improved running time but using exponential space. The algorithm reduces MAX-k-CSP to MAX-2-CSP via greedy restrictions as before, and then solves MAX-2-CSP using Williams' algorithm [19]. Using a different measure on the instances, we can improve the shrinkage exponent, and get better savings in the running time.

Let F be an instance with n variables and cn constraints. For each i-constraint C, we change the measure to $\sigma(C) = \sigma_i \equiv i(i - 1)(i - 2)$. Then $\sigma(F) \leqslant cn \cdot k(k - 1)(k - 2)$. Under a restriction ρ which randomly fixes one variable, we have, for $i \geqslant 3$,

$$\mathbf{E}[\sigma(C|_\rho)] \leqslant \sigma_i \cdot \left[1 - \frac{i}{n}\left(1 - \frac{\sigma_{i-1}}{\sigma_i}\right)\right] = \sigma_i \cdot \left(1 - \frac{3}{n}\right) \leqslant \sigma_i \cdot \left(1 - \frac{1}{n}\right)^3.$$

Then $\mathbf{E}[\sigma(F|_\rho)] \leqslant \sigma(F)\left(1 - \frac{1}{n}\right)^3$. By averaging, we can deterministically find a variable such that $\sigma(F)$ shrinks by $\left(1 - \frac{1}{n}\right)^3$.

Theorem 6. *For MAX-k-CSP instances with n variables and cn constraints, there is an algorithm running in time $\tilde{O}(2^{n(1-\mu)})$ for $\mu = \Omega(\frac{1}{\sqrt{ck^3}})$.*

Proof. We recursively restrict variables one by one. At the i-th step, restrict a variable x such that the measure reduces by $(1 - 1/(n - i + 1))^3$. After $n - pn$ steps for $p = \sqrt{\frac{1}{ck(k-1)(k-2)}}$, let F' be the restricted instance; we have $\sigma(F') \leqslant \sigma(F) \cdot p^3 \leqslant pn$. Then we can further restrict $pn/6$ variables such that all remaining constraints have width at most 2 (restrict all but two variables in each constraint). We get MAX-2-CSP instances for each branch and solve by Williams' algorithm in Theorem 4. The running time is at most

$$\text{poly}(n) \cdot 2^{n-pn+pn/6} \cdot 2^{(pn-pn/6)\cdot 2.376/3} = \text{poly}(n) \cdot 2^{n(1-\Omega(\frac{1}{\sqrt{ck^3}}))}.$$

$\square$

Note that, the algorithm uses exponential space as required by Williams' algorithm. In general, this shrinkage approach gives a reduction from sparse instances of large width to dense instances of small width.

Theorem 7. *If, for some r, MAX-r-CSP is solvable in time $2^{n(1-\delta)}$, where δ is independent of the number of clauses, then, for all $k > r$, MAX-k-CSP instances with cn constraints are solvable in time $\tilde{O}(2^{n(1-\mu)})$ for $\mu = \Omega(\frac{\delta}{c^{1/r}k^{1+1/r}})$.*

The proof is essentially the same as in Theorem 6, by changing $\sigma_i = i(i - 1)\cdots(i - r)$. We omit the proof here.

We also note that, the algorithm in Theorem 6 can be generalized to count optimal assignments for even weighted instances. As required by Williams' algorithm in Theorem 4, for maximum constraint weight W, the running time increases by a factor of $\text{poly}(W)$. This is in contrast with the polynomial-space algorithms in Lemma 2 and Theorem 5, where the running time increases by a factor of $O(\log W)$.

4   MAX-SAT and MAX-k-SAT

The algorithms for MAX-k-CSP also apply to the special case MAX-k-SAT. However, we can still improve the savings by eliminating the dependency on the clause width, and also generalize the algorithms to solve MAX-SAT. Here we still use the greedy restriction approach, but need a more involved analysis on the shrinkage of the instance size.

4.1   A Polynomial-Space Algorithm for MAX-SAT

Let F be a MAX-SAT instance on n variables and cn clauses. We associate with each i-clause a measure σ_i. Let C be an i-clause, for $i \geq 2$. Let ρ be a restriction which randomly picks and fixes one variable. Then C becomes an $(i - 1)$-clause or a constant each with probability $i/2n$. Thus,

$$\mathbf{E}[\sigma(C|_\rho)] \leq \frac{n-i}{n}\sigma_i + \frac{i}{2n}\sigma_{i-1} = \sigma_i \cdot \left[1 - \frac{i}{n}\left(1 - \frac{\sigma_{i-1}}{2\sigma_i}\right)\right].$$

We can choose

$$\sigma_1 = 0, \quad \sigma_2 = 1, \text{ and } \sigma_i = 2, \quad i \geq 3.$$

It is easy to check that, for all $i \geq 2$, $\mathbf{E}[\sigma(C|_\rho)] \leq \sigma_i(1-2/n) \leq \sigma_i(1-1/n)^2$. Then we have $\mathbf{E}[\sigma(F|_\rho)] \leq \sigma(F)(1-1/n)^2$. By averaging, we can deterministically find a variable x such that $[\sigma(F|_{x=1}) + \sigma(F|_{x=0})]/2 \leqslant \sigma(F)(1 - 1/n)^2$. Note that, this only bounds the average of $\sigma(F|_{x=1})$ and $\sigma(F|_{x=0})$.

The MAX-SAT algorithm then follows by restricting variables recursively. Although we only have shrinkage on average in each step, we can argue that, shrinkage happens with high probability over the whole process, using a similar approach as in [3,16].

Theorem 8 (Theorem 1 Restated). *There is a polynomial-space algorithm solving MAX-SAT instances with n variables and cn clauses in time $\tilde{O}(2^{n(1-\mu)})$ for $\mu = \Omega(\frac{1}{c})$.*

Proof. Let F be an instance with n variables and cn clauses, and thus $\sigma(F) \leqslant 2cn$. The algorithm recursively restricts one variable at a time.

Let $F_0 := F$, and F_i be the restricted instance after the i-th step. At the i-th step, we find a variable x in F_{i-1} such that, by randomly fixing x to 0 or 1, $\mathbf{E}[\sigma(F_i)] \leqslant \sigma(F_{i-1})(1 - \frac{1}{n-i+1})^2$.

Define

$$Z_i = \log \sigma(F_i) - \log \sigma(F_{i-1}) - 2 \log \left(1 - \frac{1}{n-i+1} \right).$$

By Jensen's inequality, conditioned on the random bits assigned to the first $i - 1$ variables, $\mathbf{E}[Z_i] \leqslant 0$. We also have $Z_i \leqslant c_i := -2 \log \left(1 - \frac{1}{n-i+1} \right)$, since $\sigma(F_i) \leqslant \sigma(F_{i-1})$.

Then $\{\sum_{j=1}^{i} Z_j\}$ is a supermartingale with respect to the random bits assigned to restricted variables. By Lemma 1, for any λ,

$$\mathbf{Pr}\left[\sum_{j=1}^{i} Z_j \geqslant \lambda \right] \leqslant \exp\left(-\frac{\lambda^2}{2\sum_{j=1}^{i} c_j^2} \right).$$

The left-hand side is

$$\mathbf{Pr}\left[\log \sigma(F_i) - \log \sigma(F) - 2\log\left(\tfrac{n-i}{n}\right) \geqslant \lambda \right] = \mathbf{Pr}\left[\sigma(F_i) \geqslant e^\lambda \sigma(F)\left(\tfrac{n-i}{n}\right)^2 \right].$$

For each $1 \leqslant j \leqslant i$, by $\log(1+x) \leqslant x$, we have $c_j = 2\log(1 + \frac{1}{n-j}) \leqslant \frac{2}{n-j}$. Thus, $\sum_{j=1}^{i} c_j^2 \leqslant \frac{4}{n-i-1}$ since

$$\sum_{j=1}^{i} \left(\frac{1}{n-j} \right)^2 \leqslant \sum_{j=1}^{i} \left(\frac{1}{n-j-1} - \frac{1}{n-j} \right) \leqslant \frac{1}{n-i-1}.$$

For $i = n - pn$, $\lambda = \ln 2$, and $pn > 20$, we get

$$\mathbf{Pr}\left[\sigma(F_i) \geqslant 4cnp^2 \right] \leqslant \mathbf{Pr}\left[\sigma(F_i) \geqslant 2\sigma(F)p^2 \right] \leqslant e^{-\lambda^2(pn-1)/8} < 2^{-pn/20}.$$

Choose $p = \delta/4c$, for δ to be fixed later. We have that, after restricting $n - pn$ variables, with probability at least $1 - 2^{-pn/20}$, there are at most $4cnp^2 = \delta pn$ remaining clauses of width at least 2.

Claim. There is a polynomial-space algorithm running in time $\tilde{O}(2^{n/2})$ which solves MAX-SAT for instances with n variables and δn clauses of width at least 2, for $\delta \leqslant 0.1$.

Proof. We recursively restrict arbitrary variables that appear in clauses of width at least 2, and stop when all remaining clauses have width 1; then solve MAX-1-SAT easily. Let $m = \delta n$ be the number of clauses of width at least 2. The recursion tree has size bounded by the recurrence $T(n, m) \leqslant T(n - 1, m) + T(n - 1, m - 1)$. This is at most $\binom{n}{\delta n} \leqslant \left(\frac{e}{\delta}\right)^{\delta n} < 2^{n/2}$, where the first inequality follows from $\binom{n}{k} \leqslant \left(\frac{ne}{k}\right)^k$. □

Choose $\delta = 0.1$. By the above claim, the running time for branches left with at most δpn clauses of width at least 2 is $2^{n-pn} \cdot 2^{pn/2} = 2^{n-pn/2}$. For the other branches, we use brute-force search; the running time is at most $2^n \cdot 2^{-pn/20}$. Therefore, the total running time is bounded by $\text{poly}(n) \cdot 2^{n(1-\Omega(1/c))}$. □

4.2 An Exponential-Space Algorithm for MAX-SAT

To improve the savings in the running time, we can improve the shrinkage exponent by reducing the instances to MAX-2-SAT, and apply Williams' algorithm [19].

We let $\sigma_1 = \sigma_2 = 0$, $\sigma_3 = 1$, $\sigma_4 = 2$, and $\sigma_i = 3$, for $i \geqslant 5$. It is easy to see that, under one step of random restriction ρ, for an i-clause C where $i \geqslant 3$,

$$\sigma(C|_\rho) \leqslant \sigma_i \cdot \left[1 - \frac{i}{n}\left(1 - \frac{\sigma_{i-1}}{2\sigma_i}\right)\right] \leqslant \sigma_i \cdot \left(1 - \frac{3}{n}\right).$$

Thus, for an instance F,

$$\sigma(F|_\rho) \leqslant \sigma(F) \cdot \left(1 - \frac{3}{n}\right) \leqslant \sigma(F) \cdot \left(1 - \frac{1}{n}\right)^3.$$

Theorem 9 (Theorem 2 restated). *For MAX-SAT instances on n variables with cn constraints, there is an algorithm running in time $\tilde{O}(2^{n(1-\mu)})$ for $\mu = \Omega(\frac{1}{\sqrt{c}})$.*

The proof is similar to the proof of Theorem 1. We can greedily restrict $n - pn$ variables. Then with high probability (at least $1 - 2^{-pn/20}$), there are at most $4cnp^3$ clauses of width at least 3; we solve such restricted instances following the claim below for $p = \sqrt{\delta/4c}$ and $\delta = 0.01$. Otherwise, we use brute-force search. The total running time is bounded by $\tilde{O}(2^{n(1-\Omega(p))})$. We omit the complete proof.

Claim. MAX-SAT instances on n variables and δn clauses of width larger than 2, for $\delta \leqslant 0.01$, are solvable in time $\tilde{O}(2^{0.9n})$ and exponential space.

Proof. We recursively restrict variables appearing in clauses of width larger than 2; when all clauses have width at most 2, we use Williams' algorithm in Theorem 4 for MAX-2-SAT. The total running time is bounded by $\binom{n}{\delta n} \cdot 2^{2.376n/3} \leqslant \left(\frac{e}{\delta}\right)^{\delta n} \cdot 2^{2.376n/3} < 2^{0.9n}$.   □

Again, we have the following reduction from sparse instances of large width to dense instances of small width.

Theorem 10. *If for some r, MAX-r-SAT is solvable in time $2^{n(1-\delta)}$, where δ is independent of the number of clauses, then MAX-SAT for instances with cn clauses is solvable in time $\tilde{O}(2^{n(1-\mu)})$ for $\mu = \Omega(\frac{\delta}{c^{1/r}})$.*

Sakai et al. [15] uses Schuler's width reduction to reduce MAX-SAT instances with cn clauses to MAX-k-SAT instances for $k = O(\log c)$, and then solve MAX-k-SAT in time $O(2^{n(1-\mu)})$ for $\mu = \Omega(\frac{1}{c^2 k^2})$; their final algorithm [15] for MAX-SAT runs in time $O(2^{n(1-\mu)})$ for $\mu = \Omega(\frac{1}{c^2 \log^2 c})$. Our result avoids Schuler's width reduction and improves the running time.

Our MAX-SAT algorithms can be extended to solve the weighted version and also the counting problem. For instances with clause weight at most W, the running time of the polynomial-space algorithm (Theorem 1) increases by a factor of $\log(W)$, and the running time of the exponential-space algorithm (Theorem 2) increases by a factor of $\text{poly}(W)$.

5   Sparse Depth-2 Threshold Circuits

A *threshold circuit* is a boolean circuit where all internal gates are threshold gates; a *threshold gate* on k inputs computes a function $\text{sign}(\sum_{i=1}^{k} w_i x_i + w_0)$ where w_i's are integer weights. A depth-2 threshold circuit on n inputs has one output threshold gate at the top, a layer of threshold gates in the middle, and n input gates at the bottom. Obviously, a MAX-SAT instance with m clauses is a special case of depth-2 threshold circuits with $m + 1$ threshold gates.

Impagliazzo et al. [10] showed a nontrivial satisfiability algorithm for depth-2 threshold circuits with linear number of wires (a *wire* is an edge in the underlying graph of the circuit). For depth-2 threshold circuits with cn wires, the algorithm runs in time $\tilde{O}(2^{(1-\mu)n})$ for $\mu = \frac{1}{c^{O(c^2)}}$. They first give an algorithm for the special case where there are few threshold gates, and then applied restrictions to eliminate threshold gates non-trivially. Using our shrinkage approach, we can improve the parameters in gate elimination, which implies better savings of the algorithm.

Lemma 3. *Given a depth-2 threshold circuits with n variables and cn wires, there is a set U of at least pn variables for $p = \frac{1}{c^{O(c)}}$ such that the number of threshold gates depending on at least two variables in U is at most δpn, for any constant $\delta < 1$. Furthermore, U can be constructed in polynomial time.*

This lemma follows directly from the following claim. Impagliazzo et al. [10] showed this result for $p = \frac{1}{c^{O(c^2)}}$ using a more dedicated analysis.

Claim. Let $\mathcal{S}$ be a collection of subsets of $[n]$ such that $\sum_{A \in \mathcal{S}} |A| \leqslant cn$. Then there is a subset $U \subseteq [n]$ of size at least pn for $p = \frac{1}{c^{O(c)}}$ such that there are at most δpn subsets in $\mathcal{S}$ containing at least two elements in U, for any constant $\delta < 1$. Furthermore, U can be constructed in polynomial time.

Proof. Define $\sigma(\mathcal{S}) = \sum_{A \in \mathcal{S}} |A| - |\mathcal{S}|$. At the beginning, let $U = [n]$ and obtain $\mathcal{T}$ from $\mathcal{S}$ by eliminating singletons; note that $\sigma(\mathcal{T}) = \sigma(\mathcal{S})$. We will greedily eliminate elements from U and subsets in $\mathcal{T}$ such that $\sigma(\mathcal{T})$ reduces non-trivially. We maintain $\mathcal{T}$ as the collection of subsets each containing at least two elements from U.

At each step, check whether $|\mathcal{T}| \leqslant \delta |U|$. If so, then we are done and return U.

Otherwise, it holds that $|\mathcal{T}| > \delta |U|$; we will greedily eliminate one element and continue the process. Suppose that $\sum_{A \in \mathcal{T}} |A| = c'|U|$ for $c' \leqslant c$. Then we can easily find some $x \in U$ which appears in at least c' subsets in $\mathcal{T}$. Eliminate x from U and from all subsets in $\mathcal{T}$; we also remove any singletons from $\mathcal{T}$. Let $U' = U \setminus \{x\}$, and denote by $\mathcal{T}'$ the new collection. Then

$$\sum_{A \in \mathcal{T}'} |A| \leqslant c'|U| - c' = c'(|U| - 1) \leqslant c|U'|,$$

and since $\sigma(\mathcal{T}) = c'|U| - |\mathcal{T}| < (c' - \delta)|U|$,

$$\sigma(\mathcal{T}') \leqslant \sigma(\mathcal{T}) - c' < \sigma(\mathcal{T})\left(1 - \frac{c'}{(c'-\delta)|U|}\right) \leqslant \sigma(\mathcal{T})\left(1 - \frac{1}{|U|}\right)^{c/(c-\delta)}.$$

Fix $p' = (2(c-\delta))^{-(c-\delta)/\delta} = c^{-O(c)}$. If $|\mathcal{T}| \leqslant \delta |U|$ holds at the i-th step for some $i \leqslant n - p'n$, then we have returned U at the i-th step; the claim holds for $p = |U|/n = c^{-O(c)}$.

If the process continues for $n - p'n$ steps ($|\mathcal{T}| > \delta |U|$ holds at each step), then the collection $\mathcal{T}$ after $n - p'n$ steps has $\sigma(\mathcal{T}) \leqslant (cn - \delta n)p'^{c/(c-\delta)} \leqslant p'n/2$. Note that we still have $p'n$ elements in U. Then, for each $A \in \mathcal{T}$, eliminate all but one arbitrary element of A from U and all subsets in $\mathcal{T}$; we also remove any singletons from $\mathcal{T}$. Since $\sigma(\mathcal{T}) \leqslant p'n/2$, we can eliminate at most $p'n/2$ elements in total. Finally, there are at least $p'n/2$ elements left in U, but $\mathcal{T}$ becomes empty; that is $|\mathcal{T}| = 0 \leqslant \delta |U|$. We return U; the claim holds for $p \geqslant p'/2 = c^{-O(c)}$. $\square$

We need the following algorithm [10] for the special case where there are few threshold gates.

Lemma 4 ([10]). *For depth-2 threshold circuits with n variables and δn threshold gates for $\delta < 0.099$, there is a satisfiability algorithm running in time $\tilde{O}(2^{(1-\mu)n})$ for a constant $\mu > 0$.*

We then get the following improved algorithm by combining greedy restrictions in Lemma 3 with the algorithm of Lemma 4.

Theorem 11. *There is a satisfiability algorithm for depth-2 circuits with n variables and cn wires running in time $\tilde{O}(2^{n(1-\mu')})$ for $\mu' = \frac{1}{c^{O(c)}}$.*

Proof. Given a depth-2 threshold circuit with n variables and cn wires, we fix $\delta < 0.099$, and find a subset U of pn variables for $p = \frac{1}{c^{O(c)}}$ as in Lemma 3. We restrict all variables not in U. By enumerating assignments to variables not in U, we get 2^{n-pn} branches. Each branch gives a restricted circuit on pn variables and δpn threshold gates; then apply the algorithm of Lemma 4 for each branch, which takes time $\tilde{O}(2^{(1-\mu)pn})$ for a constant μ.

The total running time is bounded by $\tilde{O}(2^{n-pn} \cdot 2^{(1-\mu)pn}) = \tilde{O}(2^{n-n/c^{O(c)}})$. □

6 Open Questions

A major open question is to improve the savings to $\Omega(1/\operatorname{polylog}(c))$ for solving MAX-SAT/MAX-k-SAT on instances of cn clauses. For depth-2 threshold circuits with cn wires, it would be interesting to improve the savings to $\Omega(1/\operatorname{poly}(c))$, or give an algorithm for circuits with cn gates, instead of wires. It is challenging to get constant savings for solving non-sparse MAX-k-SAT, for any $k \geqslant 3$.

References

1. Calabro, C., Impagliazzo, R., Paturi, R.: The complexity of satisfiability of small depth circuits. In: Chen, J., Fomin, F.V. (eds.) IWPEC 2009. LNCS, vol. 5917, pp. 75–85. Springer, Heidelberg (2009)
2. Chen, J., Kanj, I.: Improved exact algorithms for max-sat. Discrete Applied Mathematics **142**(1–3), 17–27 (2004)
3. Chen, R., Kabanets, V., Kolokolova, A., Shaltiel, R., Zuckerman, D.: Mining circuit lower bound proofs for meta-algorithms. In: Proceedings of the 29th Annual IEEE Conference on Computational Complexity, CCC 2014 (2014)
4. Chen, R., Kabanets, V., Saurabh, N.: An improved deterministic #SAT algorithm for small de morgan formulas. In: Csuhaj-Varjú, E., Dietzfelbinger, M., Ésik, Z. (eds.) MFCS 2014, Part II. LNCS, vol. 8635, pp. 165–176. Springer, Heidelberg (2014)
5. Dantsin, E., Wolpert, A.: MAX-SAT for formulas with constant clause density can be solved faster than in $\mathcal{O}(2^n)$ time. In: Biere, A., Gomes, C.P. (eds.) SAT 2006. LNCS, vol. 4121, pp. 266–276. Springer, Heidelberg (2006)
6. Fomin, F., Grandoni, F., Kratsch, D.: A measure & conquer approach for the analysis of exact algorithms. J. ACM **56**(5), 25:1–25:32 (2009)
7. Gaspers, S., Sorkin, G.: A universally fastest algorithm for max 2-sat, max 2-csp, and everything in between. J. Comput. Syst. Sci. **78**(1), 305–335 (2012)
8. Golovnev, A., Kutzkov, K.: New exact algorithms for the 2-constraint satisfaction problem. Theor. Comput. Sci. **526**, 18–27 (2014)
9. Impagliazzo, R., Nisan, N.: The effect of random restrictions on formula size. Random Structures and Algorithms **4**(2), 121–134 (1993)

10. Impagliazzo, R., Paturi, R., Schneider, S.: A satisfiability algorithm for sparse depth two threshold circuits. In: 54th Annual IEEE Symposium on Foundations of Computer Science, FOCS 2013, Berkeley, CA, USA, October 26–29, 2013, pp. 479–488 (2013)
11. Koivisto, M.: Optimal 2-constraint satisfaction via sum-product algorithms. Inf. Process. Lett. **98**(1), 24–28 (2006)
12. Komargodski, I., Raz, R., Tal, A.: Improved average-case lower bounds for demorgan formula size. In: Proceedings of the Fifty-Fourth Annual IEEE Symposium on Foundations of Computer Science, pp. 588–597 (2013)
13. Kulikov, A.S., Kutzkov, K.: New bounds for MAX-SAT by clause learning. In: Diekert, V., Volkov, M.V., Voronkov, A. (eds.) CSR 2007. LNCS, vol. 4649, pp. 194–204. Springer, Heidelberg (2007)
14. Paterson, M., Zwick, U.: Shrinkage of de Morgan formulae under restriction. Random Structures and Algorithms **4**(2), 135–150 (1993)
15. Sakai, T., Seto, K., Tamaki, S.: Solving sparse instances of Max SAT via width reduction and greedy restriction. In: Sinz, C., Egly, U. (eds.) SAT 2014. LNCS, vol. 8561, pp. 32–47. Springer, Heidelberg (2014)
16. Santhanam, R.: Fighting perebor: new and improved algorithms for formula and qbf satisfiability. In: Proceedings of the Fifty-First Annual IEEE Symposium on Foundations of Computer Science, pp. 183–192 (2010)
17. Schuler, R.: An algorithm for the satisfiability problem of formulas in conjunctive normal form. J. Algorithms **54**(1), 40–44 (2005)
18. Scott, A., Sorkin, G.: Linear-programming design and analysis of fast algorithms for max 2-csp. Discret. Optim. **4**(3–4), 260–287 (2007)
19. Williams, R.: A new algorithm for optimal 2-constraint satisfaction and its implications. Theor. Comput. Sci. **348**(2–3), 357–365 (2005)

Laissez-Faire Caching for Parallel #SAT Solving

Jan Burchard[✉], Tobias Schubert, and Bernd Becker

Albert-Ludwigs-University Freiburg, Georges-Köhler-Allee 051,
79110 Freiburg, Germany
{burchard,schubert,becker}@informatik.uni-freiburg.de

Abstract. The problem of counting the number of satisfying assignments of a propositional formula (#SAT) can be considered to be the big brother of the well known SAT problem. However, the higher computational complexity and a lack of fast solvers currently limit its usability for real world problems.

Similar to SAT, utilizing the parallel computation power of modern CPUs could greatly increase the solving speed in the realm of #SAT. However, in comparison to SAT there is an additional obstacle for the parallelization of #SAT that is caused by the usage of conflict learning together with the #SAT specific techniques of component caching and sub-formula decomposition. The combination can result in an incorrect final result being computed due to incorrect values in the formula cache. This problem is easily resolvable in a sequential solver with a depth-first node order but requires additional care and handling in a parallel one. In this paper we introduce laissez-faire caching which allows for an arbitrary node computation order in both a sequential and parallel solver while ensuring a correct final result. Additionally, we apply this new caching approach to build *countAntom*, the world's first parallel #SAT-solver.

Our experimental results clearly show that *countAntom* achieves considerable speedups through the parallel computation while maintaining correct results on a large variety of benchmarks coming from different real-world applications. Moreover, our analysis indicates that laissez-faire caching only adds a small computational overhead.

1 Introduction

The problem of counting the number of satisfying assignments of a propositional formula (#SAT) can be directly derived from the simpler Boolean satisfiability problem (SAT) which has become one of the important work horses in modern computer science, especially in the domains of artificial intelligence, planning, model checking, and hardware test.

Its #P-completeness [1] however makes #SAT a much harder problem which is reflected in the comparatively small size of currently solvable formulas.

Any real world applicability, though, highly depends on the possibility of solving larger formulas. Applications for #SAT could then be found wherever problems are encoded as propositional formulas. Especially, every SAT instance is a possible #SAT instance: If a formula encodes a problem such that it is

© Springer International Publishing Switzerland 2015
M. Heule and S. Weaver (Eds.): SAT 2015, LNCS 9340, pp. 46–61, 2015.
DOI: 10.1007/978-3-319-24318-4_5

satisfiable if the event X can occur, the number of satisfying assignments often represents the likelihood of X to occur.

One possibility of increasing the speed of the computation is to harness to parallel computation power of modern CPUs. This approach has shown great speedups for the SAT problem [2–4] but has, thus far, not been extended to #SAT.

Modern #SAT solvers contain several improvements like component caching and sub-formula decomposition which are generally not present in a SAT solver. The combination of these techniques with conflict learning can cause the final result to be computed incorrectly due to incorrect values in the cache [5]. In current sequential #SAT solvers like *cachet* [5] or *sharpSat* [6] this problem can easily be resolved because the decision tree is traversed in a depth-first manner. Since such a node order cannot be guaranteed in a parallel solver this simple solution is not applicable anymore.

In this paper we introduce *laissez-faire caching* as a solution to incorrect cache values which is independent of the node order. It can therefore be used in a parallel solver as well as in a sequential one where a changed node order might increase the solving speed too. Utilizing this new caching scheme we created *countAntom*, the first parallel #SAT solver which is competitive or even superior to [6] when used as a sequential solver and moreover gives large speedups in parallel mode.

The remainder of this paper is structured as follows: Sections 2 and 3 give a short overview of the mathematical notation and of modern #SAT solving techniques, respectively. Next, Section 4 introduces laissez-faire caching which is followed by an explanation of the general layout of *countAntom* in Section 5. In Section 6 we present our experimental results. Section 7 concludes with a short summary and outlook onto future work.

2   Basic Notations

A propositional formula is built up of variables $(v_1, v_2, \ldots)$ which are linked by operators $(\neg, \wedge, \vee, \oplus, \ldots)$. In the context of (#)SAT, formulas are usually given in conjunctive normal form (CNF). A formula in CNF is composed of clauses connected by conjunctions ($\wedge$). Each clause consists of literals connected with disjunctions ($\vee$). A literal is a variable or a negated variable (v or $\neg v$). For simplicity, a set notation is used: A clause $v_1 \vee \neg v_2 \vee v_3$ is represented as a set of literals $\{v_1, \neg v_2, v_3\}$ and a formula $c_1 \wedge c_2 \wedge c_3$ as a set of clauses $\{c_1, c_2, c_3\}$.

A (partial) variable assignment π assigns (some) variables a truth value (*true* or *false*). Based on a variable assignment, a clause or the whole formula can be evaluated as either *true* or *false* (e.g., for a clause c and an assignment π we write $c|_\pi = false$, iff c evaluates to *false* for π). If the clause or formula is evaluated as *true*, it is called *satisfied*, if it is evaluated as *false*, it is called *unsatisfied*. To satisfy a formula in CNF all clauses have to be satisfied. To satisfy a clause it is sufficient that one of its literals evaluates as *true*.

A variable that is not assigned is *free*. The concept of free variables is extended to literals: A literal is free if its variable is free.

A clause with free literals which is not satisfied (yet) is called *open*. An open clause that has exactly one free literal l is *unit*. Unit clauses imply a variable assignment which makes l evaluate as *true* because each clause has to be satisfied in order to satisfy the entire formula. Given a partial variable assignment π the *residual* formula consists of the remaining open clauses of the original formula without the assigned literals.

The model count of a formula φ (abbreviated as $mc(\varphi)$) is the number of different variable assignments π with $\varphi|_\pi = true$.

Given two clauses c_1 and c_2 with a shared variable v occurring positively in the one and negatively in the other, the clauses can be combined into a new clause through resolution. The resolvent $c_1 \otimes c_2$ is computed by combining the literals of c_1 and c_2 and removing all occurrences of v: $c_1 \otimes c_2 = (c_1 \cup c_2) \setminus \{v, \neg v\}$. Adding the resolvent of two clauses to a formula results in an equivalent formula and thus the model count does not change, either.

3   #SAT Solving

There are different approaches to #SAT solving which either compute exact or approximative results (with or without any guarantees). This paper presents an exact #SAT solver which computes the number of satisfying assignments by counting the number of satisfied branches in the decision tree.

Even though approximative solvers outperform their exact counterparts on some formula classes they can also be considerably slower on others. Thus a combined approach of running both an approximative and exact solver in parallel could offer the highest solving speed even when the result does not have to be correct [7]. Hence further improving the performance of exact search based #SAT solvers still remains an important objective.

Like most exact #SAT solvers *countAntom* is a modified DPLL [8] based SAT solver which is extended with various techniques that are usually too costly in the SAT context but provide speedups for #SAT. Of greatest interest and gain are sub-formula decomposition [9] and formula caching [10]. Additionally common SAT solver improvements like preprocessing, a fast deduction implementation, an advanced decision heuristic and the learning of conflict clauses are used.

3.1   Conflict Learning

Whenever the current variable assignment is in conflict (e.g., because a variable has to be assigned to *true* and *false* at the same time to satisfy the formula) the solver creates a conflict clause. This conflict clause attempts to capture the reason behind the current conflict (which might only be a subset of the current variable assignment) to avoid repeating the same assignment pattern again. The conflict clause is generated by resolving clauses of the current formula and is then added as a new clause.

3.2   Sub-Formula Decomposition

During the computation the residual formula might become divisible into multiple sets of clauses with no variable occurring in more than one set. Each set corresponds to a sub-formula which can be solved separately. The final model count in one branch of the decision tree is then computed by multiplying the results from all sub-formulas.

When splitting a formula with $n + m$ variables into two sub-formulas with n and m variables, the worst case computation complexity is reduced from $\mathcal{O}(2^{n+m})$ to $\mathcal{O}(2^n + 2^m)$. To increase the chance of a successful split operation, learned conflict clauses are not considered for this operation because they might add additional dependencies between otherwise disjoint sub-formulas which were not present in the original formula.

For graphical representation the concept of the decision tree in which each node represents a decision variable and each outgoing edge of a node corresponds to an assignment of that variable to either *true* or *false* is extended into that of a component tree (see Figure 1 for an example). When a node is split into sub-formulas they are represented as sibling nodes each of which is computed separately.

3.3   Formula Caching

Nodes with the same residual formula might occur multiple times in different locations in the component tree. Formula caching allows the solver to re-use previously computed results in case the same residual formula is encountered again. In combination with sub-formula decomposition, the cache stores the value of every computed node in the component tree.

3.4   Incorrect Results in the Cache

When combining conflict learning with sub-formula decomposition and caching, it might happen that an incorrect value is stored in the cache [5]. This rare event can only occur when the residual formula is split into two or more sub-formulas and one of these sub-formulas is unsatisfiable. Since one sub-formula is unsatisfiable the whole branch of the component tree (and hence the residual formula) is unsatisfiable. Through the influence of learned conflict clauses, which prune the search space, the results for the remaining sibling sub-formulas might be incorrect because some satisfying assignments are erroneously ignored.

As an example consider the formula φ presented in [5]:

$$\varphi = \big\{ \underbrace{\{p_0, p_1, \neg a_1\}}_{c_1}, \underbrace{\{p_0, \neg p_2, a_2\}}_{c_2}, \underbrace{\{a_1, a_2, a_3\}}_{c_3}, \underbrace{\{\neg p_1, b_1\}}_{c_4}, \underbrace{\{\neg b_1, b_2\}}_{c_5}, \underbrace{\{p_2, \neg b_2\}}_{c_6} \big\}$$

(1)

The clause $c_x = \{p_0, \neg a_1, a_2\}$ can be inferred from φ through resolution using c_1, c_4, c_5, c_6 and c_2. Therefore, c_x could have been learned and added to the formula as a conflict clause. When working with the variable assignment

$p_0 = false$, $p_1 = true$, $p_2 = false$ the formula can be split into two sub-formulas $\varphi_1 = \{\{a_1, a_2, a_3\}\}$ and $\varphi_2 = \{\{b_1\}, \{\neg b_1, b_2\}, \{\neg b_2\}\}$. Clearly, φ_2 is unsatisfiable and thus is the entire branch of φ. There are 7 satisfying assignments for φ_1. However, taking the learned clause c_x into account a different result is computed: $mc(\varphi_1 \wedge c_x) = 5$. The cache only stores the results for sub-formulas of the original formula (disregarding any learned clauses like c_x) to increase the likeliness of a cache hit. Normally (without an unsatisfied sibling) this is valid because adding learned conflict clauses results in an equivalent formula and they can therefore be used or ignored at will. However, in case of an unsatisfied sibling $mc(\varphi_1) = 7 \neq mc(\varphi_1 \wedge c_x) = 5$. Thus storing $mc(\varphi) = 5$ in the cache is clearly incorrect and might result in an erroneous final result.

Without caching this does not cause any problem because the possibly incorrect result is multiplied by the model count of the unsatisfied sub-formula which is 0. Through caching, however, the incorrect value might be used in a satisfied branch of the tree causing an incorrect final model count (see Figure 1).

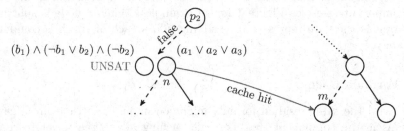

Fig. 1. A component tree with an unsatisfied node. Utilizing the cached result of any satisfied sibling of that node can result in an erroneous final model count. In this example node m uses the cached result of node n which is potentially incorrect.

4   Laissez-Faire Caching

Solving the problem of incorrect results in the cache in a sequential solver is comparatively simple. Assuming the solver traverses the component tree in a depth-first manner, removing possibly incorrect results from the cache as soon as an unsatisfiable sub-formula is encountered suffices. Through the depth-first node order any possibly incorrect result definitely is not used in a satisfiable branch of the component tree before it is deleted from the cache. Current sequential #SAT solvers like *cachet* use this method.

When such a node order cannot be guaranteed, for example in a parallel solver, this simple solution is insufficient to always provide correct results. A straightforward solution would be controlling the cache access through rules. These rules could ensure that results from the cache can only be used when the access is similar to a depth-first tree traversal. In a parallel solver such rules would degrade the cache performance because the node order by design is not depth-first. Therefore, many results in the cache could not be used by the threads. For example splitting the computation into two threads on a decision with each thread proceeding one of the resulting branches in a depth-first manner would not allow the threads to use any cached results created by the other thread.

Laissez-faire caching, our solution to this problem offers a more refined access strategy. It allows for access to all nodes stored in the cache regardless of their relation to the current node. Only in the rare event of a possibly incorrect value being stored in the cache, additional computations are required. The approach is based on the following three principles:

1. Use every result available in the cache without any limitations.
2. When a node n uses the result of the node m from the cache a new *dependency* between the nodes is created.
3. If a node that is stored in the cache becomes invalid it informs all nodes that depend on its value to initialize a re-computation to correct potentially incorrect intermediate results.

Clearly, normal cache access is as fast as it is without any modifications since there are no access rules. On a cache hit a new dependency has to be created which requires a constant amount of additional computation steps. Only in the event of a re-computation additional work has to be performed.

4.1 Node Re-Computation

When a node depends on an incorrect cache result it has to be recomputed. However, it is not clear whether the result of a node m with an unsatisfiable sibling is actually correct or not in the first place. For many nodes with unsatisfied siblings the correct value is computed. To verify if m's result was computed correctly the solver would have to re-compute it without any learned conflict clauses. This is not feasible because it might require a large amount of additional computation steps.

Instead, every possibly incorrect result is assumed to be incorrect. Of the nodes n_i which depend upon this result only one node n needs to be re-computed; the remaining dependencies can use the result of this re-computed node because they all represent the same sub-formula. Since this approach allows for the usage of all available conflict clauses during the re-computation of n it is far superior to validating every possible incorrect node m without conflict clauses.

When n has to be recomputed it also needs to inform its parent node p since the model count of p depends on a possibly incorrect value and may thus be incorrect as well. Should p be finished and stored in the cache, it has to be removed from the cache, the nodes depending on p's value have to be informed and p's parent has to be notified of the incorrect child in turn.

Thus removing a single node m from the cache can result in a cascade of invalidations. However, given the right re-computation order only a single node needs to be fully computed by the solver: As discussed above, only one dependency n of m needs to be recomputed. All other dependent nodes can then get an updated value from the cache which in turn allows their parent nodes to store a new result in the cache. This allows all nodes which depend on the parents to retrieve updated values from the cache. The update sequence continues for all invalidated nodes with n being the only node that needed a full re-computation.

4.2   Memory Consumption and Cleanup

A drawback of laissez-faire caching is that every node n that is a cache hit potentially needs to be recomputed because the node it depended on is removed from the cache again. To recompute n, the information of the sub-formula it represents as well as information about its parent node need to be available. Additionally, all nodes that depended in some manner on the value of n (i.e., all nodes on the path from n to the root of the component tree) might also become invalid when n is invalidated and must also be re-computable.

Therefore, almost the entire component tree has to be kept in memory during the computation. Given the size of common formulas, this requirement quickly becomes a space issue even with today's large memory modules. To reduce the memory consumption, the stored component tree is pruned periodically. During this operation all finished nodes are removed. This deletes both nodes that are stored in the cache and nodes that are cache hits.

Erasing nodes from the cache is in line with the observation that the usability of cache entries quickly decays over time [5]. Hence, the performance decrease of such a non-selective pruning is expected to be minimal compared to only removing a subset of nodes.

When removing nodes their dependencies have to be taken into account. To this end the concept of dependencies is extended from the original "node n uses node m's value from the cache" to a more general "node n in some way depends on node m". All dependencies of a node are moved to the node's parent on its deletion (see Figure 2). Should at any later point in time a node become invalidated, a larger part of the component tree might need to be re-computed because the exact node which requires re-computation has been removed.

As an example, assume that the node n_1 in Figure 2 is unsatisfiable. Hence the node n_2 and all of its siblings are possibly incorrect and invalidated. In the original component tree this would trigger the re-computation of the nodes n_4 and n_5. In the pruned component tree the entire negative branch of n_3 has to be recomputed because the solver has lost all knowledge of the nodes in this branch which creates a substantial overhead. The real extent of this overhead is analyzed with the experimental results in Section 6.3.

5   *countAntom* Solver Structure

Our parallel #SAT solver *countAntom* consists of independent solver threads each of which works very similar to a sequential #SAT solver. The implementation is based on the *antom* SAT solver [11] and the boost threading library [12] which provides simple to use parallelization with sufficient control over each thread.

The component tree is represented through one object for each node. To ensure thread-safety, each node provides its own mutex, guaranteeing that no

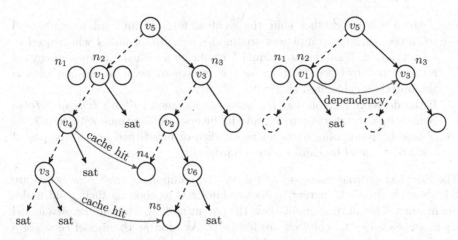

Fig. 2. A component tree before and after a memory cleanup operation. The cache hit dependencies are moved to the parents of the removed nodes and are joined into a single general dependency.

two threads can ever work on the same node at the same time. Deadlock freedom is provided through a special node locking order: When a thread attempts to lock a node n it must either already hold the lock for n's parent or it must be the first lock that the thread obtains. Thus, once a thread obtains its first lock for a node n all subsequent locking will occur in the subtree with root n. Locks are released in reverse order of acquisition.

Additionally, all access to shared structures like the cache is restricted through appropriate mutexes to ensure thread safety.

5.1   Node Computation

The node computation within the threads is similar to those of a sequential #SAT solver limited to a single step at a time:

1. Get the next component tree node n to be computed.
2. Check if a result for n is stored in the cache. In case of a cache hit add a new dependency and inform n's parent that a child node is finished. If there was a cache hit the computation of n is concluded, otherwise continue.
3. Backtrack from the last computed node m to n. Unlike the backtracking operation in a SAT solver n is not always a node on the path from the root of the component tree to m (see Figure 3). Thus, our backtrack operation first removes assignments up to the first common ancestor a of m and n and then re-assigns the required variables on the path from a to n.
4. Choose a decision variable and assign it to either *true* or *false*. As decision heuristic VSADS [13] is used.
5. Calculate the implications of the previous assignment and perform a conflict analysis if there is a conflict. In case of a conflict the branch is marked as unsatisfiable.

6. If there was no conflict split the residual formula into sub-formulas and store the resulting component tree nodes in a list of nodes which need to be computed. Should the residual formula be satisfied the split operation returns an empty set. In this case the number of satisfying assignments in the branch is calculated as $2^{\#free\ variables}$.
7. If the decision variable has not yet been assigned to both *true* and *false* remove its current assignment, add its inverse and continue with step 5.
8. Should both outgoing branches be satisfied or unsatisfied inform the parent node that one of its child nodes is finished.

The calculations thus move up and down the component tree: New nodes are added as children and parent nodes are informed as soon as their child nodes are finished. This in turn might allow the parent node to compute its own model count (when all of its children are finished). As soon as the model count of a satisfiable node is computed it is added to the cache. If a node is unsatisfiable the results of all its sibling nodes are not required anymore because the branch will remain unsatisfiable and the sibling's result cannot be used by any other node. Therefore these nodes are aborted and their values removed from the cache if they were already finished.

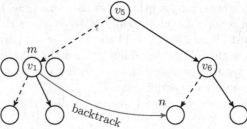

Fig. 3. During the backtrack from m to n the solver removes the assignment $v_5 = false$ and adds the assignments $v_5 = true$ and $v_6 = false$.

To simplify the parallel computation each node stores its current state (see Figure 4). The possible states consist of those found in a classic #SAT solver (the states *new, cache hit, waiting for child results, aborted* and *finished* represent the normal life-cycle of a node), and some additional ones which are required because of the removal of nodes during a memory cleanup operation (*incomplete* and *marked for deletion*).

Should a cache entry become invalid all its dependencies have to be recomputed and are, thus, returned to the *new* state. Additionally, their parents have to *wait for child results*, again if they were previously finished.

When nodes are removed from the component tree to reduce the memory consumption they are first *marked for deletion* and then deleted. The parents of deleted nodes become *incomplete*. If an *incomplete* node has an incorrect child or depended on an incorrect node it has to be completely recomputed because not all information on its children is available.

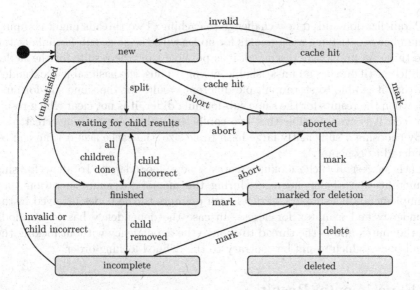

Fig. 4. The state diagram for a node in *countAntom* with caching and the option to remove nodes from the component tree.

5.2   Node Order

The order of computation plays a vital role for two reasons: Firstly, depending on which node is computed first it is possible that conflict clauses are learned earlier or the cache is utilized more efficiently. Secondly, each thread has to backtrack to the next node it has to compute. The further the backtracking distance, the lower is the overall performance of the solver.

Our implementation aims to keep the backtracking distance as small as possible while ensuring that each thread can always perform computations on a node. It does so by allowing each thread to keep a local list of nodes to be computed next. This list is implemented as a stack which results in a depth-first component tree traversal. Additionally, a global, shared node list is maintained by all threads to which nodes are added if its size is below a defined threshold. When a thread's local list becomes empty it takes a new node from the global list.

The local node lists ensure that the backtracking distance is kept to a minimum (similar to that of a sequential solver) while the global list is used for work sharing among the threads.

5.3   Caching

The cache is implemented as a hash table and shared among all threads. For each node a hash value is computed through the formula it represents. Hash conflicts are resolved through a linked conflict list for each bucket. Each cache access follows the laissez-faire approach.

Parallelization adds a new challenge to caching: Two threads might compute the number of satisfying assignments for nodes with the same sub-formula at the same time. As previously discussed, it is possible that the results for the nodes are different (if one has an unsatisfiable sibling). Thus, the hash table is extended such that it is able to store multiple different results for the same sub-formula. Even when the results for the same sub-formula differ, it is not clear which result is correct. It is even possible that two conflicting values are both incorrect.

By choosing a sufficiently large hash table size, the cache access time can be considered as constant.

Each successful cache lookup creates a new dependency. To allow for simple modification of dependencies during the memory cleanup operation they are implemented as simple C-structs with pointers to the nodes involved in the dependency and a mutex for access. In case the dependency has to be modified, the mutex allows the thread to change the dependency without locking the second node (which would be contrary to the defined locking order).

6   Experimental Results

We tested laissez-faire caching in combination with our parallel #SAT solver *countAntom*. Our solver is compared to *cachet* [5] and *sharpSAT* [6] which to the authors' knowledge are the fastest exact #SAT solvers available.

The benchmarks system uses two octa-core Intel Xeon E5-2650v2 2.60 GHz CPU with 64 GB of memory. If not stated otherwise the solver's memory limit is set to 4 GB. All times are given in seconds. The timeout (X) is set to 1 h. An entry of ? marks an internal solver error (e.g., a segmentation fault) due to which the solver did not finish the computation.

6.1   Benchmarks

In addition to the benchmarks presented in previous #SAT solver publications (which are labeled as *Classic*) several additional real-life #SAT formulas were selected:

The *Fault Injection* benchmarks (referred to as *FI easy* and *FI hard*) encode a fault injection through clock manipulation into a cryptographic circuit as described in [14]. The satisfiability probability of a formula is the likelihood of a successful injection when encrypting random data.

The *Output Probability* benchmarks contain the cone of influence of the output of a circuit and force that output to be *true*. Thus, the number of satisfying assignments corresponds to the probability that the output is *true* on a random input assignment. The circuits originate from the ITC 99 [15], ISCAS 89 [16] and ISCAS 85 [17] industrial benchmark sets.

Table 1 gives an overview of the selected formulas together with the single threaded runtime of the solvers.

Although all benchmarks are structured formulas, the solvers exhibit a large performance difference between the *Classic* and our real life benchmarks. This

Table 1. Runtime comparison among *cachet*, *sharpSAT* and *countAntom* (with a single solver thread).

	Formula	#variables	#clauses	Model count	*cachet*	*sharpSAT*	*countAntom*
Classic	2bitadd_10	590	1422	0	?	**29.80**	138.73
	4blocks	758	47820	55097	?	**15.47**	16.67
	bmc-ibm-2	2810	11683	$\approx 1.333 \cdot 10^{19}$	0.05	**0.04**	0.16
	bmc-ibm-3	14930	72106	$\approx 2.472 \cdot 10^{19}$	21.89	**2.27**	21.13
	bmc-ibm-4	28161	139716	$\approx 9.729 \cdot 10^{79}$	3598.67	**3.66**	199.64
	bmc-ibm-5	9396	41207	$\approx 2.458 \cdot 10^{171}$	195.72	14.15	**8.03**
	logistics.b	843	7301	$\approx 4.526 \cdot 10^{23}$	?	**0.88**	2.84
	logistics.c	1141	10719	$\approx 3.980 \cdot 10^{24}$	?	**36.51**	127.71
FI easy	easy_28	1244	4694	3791872	28.79	28.98	**4.89**
	easy_29	1201	4520	12684800	487.39	338.75	**40.68**
	easy_30	1177	4426	28830208	?	632.60	**58.95**
	easy_31	1123	4207	33171456	430.35	208.81	**2.91**
FI hard	hard_30	4705	17689	$\approx 1.771 \cdot 10^{32}$	X	X	**159.97**
	hard_31	4489	16816	$\approx 2.326 \cdot 10^{34}$	X	X	**65.83**
	hard_32	4245	15844	$\approx 8.514 \cdot 10^{34}$	X	X	**45.86**
	hard_33	4157	15484	$\approx 8.514 \cdot 10^{34}$	X	X	**28.50**
Output Prob	b14c_04	805	2194	$\approx 2.230 \cdot 10^{43}$	**219.14**	460.42	391.65
	b22c_79	511	1317	$\approx 8.308 \cdot 10^{34}$	574.75	118.14	**72.51**
	c0499_10	130	352	$\approx 1.099 \cdot 10^{12}$	3598.60	X	**1.73**
	c1908_08	291	884	4563402752	?	304.90	**71.99**
	cs38417_1411	379	900	$\approx 3.796 \cdot 10^{29}$	3598.54	2686.16	**146.69**
	cs38417_291	328	748	$\approx 2.293 \cdot 10^{28}$	1133.63	375.47	**52.67**

can be attributed to solver specific optimizations and the different origins of the formulas. The decision heuristic of *countAntom* was slightly modified for the *Classic* formulas to increase the performance. Overall, *countAntom* outperforms both *cachet* and *sharpSAT* on most of our real life benchmarks even with only a single thread.

6.2 Parallel Performance

Since only *countAntom* provides a parallel solving mode we analyze only its results in the remainder of this section. To measure the performance of our parallel implementation we compare the solving speed with a single thread to that with n threads. The quotient of these values is called the *speedup*.

We determined the runtime with 1 to 16 threads for each of the benchmark formulas with five repetitions. The arithmetic means of the results are shown in Table 2. The average speedup for each formula group is visualized in Figure 5.

Clearly, parallelization greatly increases the solving speed across all formula groups. The average speedup for all groups increases from 1 to 8 threads, reaching around 3 with 4 threads and between 3.5 and 5.5 with 8 threads.

Adding a 9th thread decreases the performance across all groups. With up to 8 threads the computation can be performed by a single processor. If the system's

Table 2. The arithmetic mean of the runtime of *countAntom* across 5 repetitions with 1 to 16 threads.

	Formula	1	2	3	4	5	6	7	8	9	...	14	15	16
							Number of threads							
Classic	2bitadd_10	140.0	32.8	30.3	47.2	22.0	21.9	21.8	25.8	33.2	...	17.6	19.6	**16.6**
	4blocks	16.8	12.4	10.3	9.73	9.08	8.91	8.21	8.10	8.17	...	7.78	**7.52**	7.56
	bmc-ibm-2	0.08	**0.07**	**0.07**	**0.07**	**0.07**	**0.07**	**0.07**	0.08	0.08	...	0.08	0.08	0.08
	bmc-ibm-3	21.3	14.4	10.3	10.3	10.5	9.75	10.4	9.04	8.80	...	**8.54**	9.03	9.44
	bmc-ibm-4	203.6	116.9	74.1	70.3	80.1	41.3	49.8	51.3	51.0	...	15.0	29.5	**14.9**
	bmc-ibm-5	7.97	4.42	3.39	2.86	2.63	2.05	1.90	1.66	1.91	...	1.48	1.79	**1.42**
	logistics.b	2.75	1.42	0.99	0.94	0.79	0.71	0.70	**0.66**	0.72	...	0.84	0.90	0.83
	logistics.c	130.1	33.6	27.0	19.3	15.9	14.5	12.6	13.0	11.6	...	11.7	11.4	**10.4**
FI Easy	easy_28	4.79	2.33	1.75	1.39	1.09	0.95	0.83	**0.75**	0.84	...	1.20	**0.75**	0.86
	easy_29	41.6	23.2	16.9	12.8	11.4	10.2	8.92	8.05	9.60	...	10.0	8.95	**7.54**
	easy_30	59.7	32.8	25.6	22.1	17.4	15.3	13.6	12.4	13.3	...	11.9	11.3	10.3
	easy_31	2.82	1.60	1.09	0.92	0.78	0.68	0.63	**0.59**	0.68	...	0.97	0.81	0.66
FI Hard	hard_30	164.4	87.9	64.2	64.3	66.8	58.2	40.6	34.9	32.6	...	26.8	25.2	**21.8**
	hard_31	67.7	34.6	26.1	22.0	22.6	18.5	16.8	14.8	16.0	...	13.9	12.4	12.1
	hard_32	48.0	24.5	21.0	17.1	14.9	12.7	12.1	11.0	12.3	...	10.7	9.60	9.16
	hard_33	29.5	15.1	11.0	10.3	8.41	8.42	6.95	7.15	9.44	...	8.07	7.36	6.05
Output Prob	b14c_04	392.8	284.1	216.8	211.2	184.9	175.7	162.0	161.1	161.5	...	154.9	150.7	145.6
	b22c_79	74.1	40.3	29.9	24.3	20.4	17.7	16.2	14.6	15.4	...	14.3	13.1	**12.8**
	c0499_10	1.63	0.81	0.91	0.73	**0.64**	0.74	0.73	0.73	0.79	...	1.55	1.44	1.88
	c1908_08	70.6	39.3	30.7	27.1	23.1	19.8	17.8	16.5	18.5	...	19.3	18.3	**15.8**
	cs38417_1411	137.7	84.5	67.4	52.3	46.4	40.9	36.2	32.7	35.0	...	33.7	31.3	**28.8**
	cs38417_291	49.6	29.6	24.2	19.4	**16.6**	19.1	18.2	16.7	19.3	...	22.2	20.5	18.6

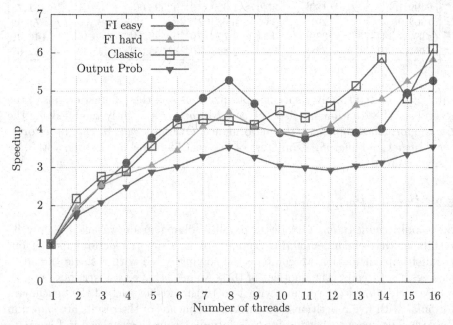

Fig. 5. The average speedup for the different benchmark groups with 1 to 16 threads.

second processor is added the speedup degrades due to the less tight coupling of the two CPUs and additional shared memory access overhead. Thus, the lower performance is due to the hardware structure of the benchmark system.

With even more threads the speedup increases again but at a much slower pace. For most formulas the maximum speedup is achieved when utilizing all 16 available cores.

6.3   Impact of Laissez-Faire Caching

Compared to the caching techniques used by sequential #SAT solvers, laissez-faire caching in combination with the parallelization adds computational overhead: During the calculations all nodes have to be kept in the memory. Furthermore each cache hits creates a new dependency. These dependencies require special attention during cleanup operations.

To evaluate the overhead, we compare the runtime of *countAntom* with and without laissez-faire caching in Table 3. To that end the dependencies are disabled and the cache is used as it would be in a sequential solver. Therefore, should a node be removed from the cache, nodes that used its results will not be informed nor invalidated. Of course this does not ensure the correctness of the final result and is only useful in this evaluation.

Table 3. The arithmetic mean across 30 repetitions for the runtime and number of decisions of *countAntom* using 16 threads with and without adding laissez-faire caching dependencies.

		Without dependencies		With dependencies		
	Formula	Time	Decisions	Time	Decisions	Time dif.
Classic	2bitadd_10	18.81	445697.87	21.49	472866.50	+14.24%
	4blocks	7.46	33846.07	7.55	35581.60	+1.24%
	bmc-ibm-2	0.08	369.60	0.08	430.90	+1.68%
	bmc-ibm-3	8.53	8904.63	8.48	8955.20	-0.56%
	bmc-ibm-4	17.76	516041.60	21.88	520252.13	+23.18%
	bmc-ibm-5	1.43	38053.63	1.56	42288.77	+8.57%
	logistics.b	0.68	28246.40	0.76	27475.37	+11.40%
	logistics.c	9.56	530838.50	10.93	526091.10	+14.37%
FI easy	easy_28	0.67	36781.73	0.81	37818.23	+21.24%
	easy_29	6.37	584328.57	7.32	582706.80	+15.00%
	easy_30	8.46	679935.27	10.12	685038.97	+19.63%
	easy_31	0.68	48725.23	0.65	48882.03	-5.71%
FI easy	hard_30	21.71	1963895.83	24.39	1961081.03	+12.33%
	hard_31	10.24	919613.30	11.99	920852.03	+17.14%
	hard_32	7.79	685190.73	9.06	683700.13	+16.41%
	hard_33	5.88	530489.23	6.95	525454.10	+18.14%
Output Prob	b14c_04	125.93	11450672.97	144.38	11575403.60	+14.65%
	b22c_79	9.82	687743.87	12.18	689614.17	+24.05%
	c0499_10	1.36	169984.70	1.72	192928.97	+26.44%
	c1908_08	13.16	1548375.23	16.04	1565834.00	+21.96%
	cs38417_1411	22.50	2083893.87	29.02	2090450.30	+28.99%
	cs38417_291	14.51	1240944.37	19.35	1201234.33	+33.34%

The comparison clearly shows that dependencies have a negative impact both in terms of computation time and number of decisions. However, the average increase in computation time is relatively minor at 15.35 %. Thus, the number of required re-computations due to incorrect results in the cache appears to be low and the overall impact of laissez-faire caching acceptable. There are two outliers in the experiments: The formulas *bmc-ibm-3* and *easy_31* are solved faster *with* dependencies than without. This can, however, be attributed to general variations in solving speed due to the nondeterministic behavior of the parallel computation and the fast solving speed and is thus not analyzed further.

The overhead through the explicit representation of the component tree through one object for each node is inherent in our implementation and cannot be disabled for testing purposes.

7   Conclusion

This paper introduced two new contributions to the realm of #SAT solving: Firstly, *laissez-faire caching* offers a new caching technique that provides a solution to the problem of incorrect cache values which is independent from the node computation order. Thus, it can be used in any #SAT solver be it sequential or parallel that requires a not depth-first tree traversal. Secondly, *countAntom* is the world's first parallel #SAT solver.

Our experiments clearly show the viability of parallelization for #SAT: The speedup of *countAntom* with multiple cores is almost linear on many formulas and on average between 3.5 and 5.5 when using 8 threads on a octa-core CPU. Additionally, a speedup was observed for every single tested formula. Adding more threads which are executed on multiple different CPUs yields only a minor improvement. The overhead created by adding dependencies between nodes and recomputing nodes which depend on possibly incorrect values in the cache appears to be small at around 15 % on average.

Future improvements could focus on three areas: Firstly, the memory consumption of the current implementation of laissez-faire caching is rather large. This leads to many time consuming memory cleanup operations. Optimizing the size of each component tree node or switching to an implicit representation of nodes would therefore speedup the solving process. This is especially relevant for large formulas which already have a higher memory consumption

Secondly, as with any solver there are many parameters for different heuristics which are set manually. The goal in this area is to develop a heuristic for this parameter selection (for example by choosing different initial parameters for each thread and using the parameter set of the best one after a specific time). This would increase the performance especially for unknown formulas.

Thirdly, expanding the solver to work on systems without shared memory (for example on a computation grid) could allow it to tackle even larger formulas. One possibility is a hierarchical approach with many multi-threaded nodes sharing information and jointly solving the problem. This would require an additional communication layer to guide the computation.

Overall *countAntom* is the first step into the direction of faster #SAT solving and shows that #SAT can be effectively and efficiently solved in parallel.

References

1. Garey, M.R., Johnson, D.S.: Computers and Intractability; A Guide to the Theory of NP-Completeness. W.H. Freeman & Co., New York (1990)
2. Schubert, T., Lewis, M., Becker, B.: PaMiraXT: Parallel SAT solving with threads and message passing. Journal on Satisfiability, Boolean Modeling and Computation **6**, 203–222 (2009)
3. Biere, A.: Lingeling, plingeling and treengeling entering the sat competition 2013. In: Proceedings of SAT Competition 2013. vol. B-2013-1. University of Helsinki, Department of Computer Science Series of Publications (2013)
4. Hamadi, Y., Jabbour, S., Sais, L.: Manysat: a parallel SAT solver. Journal on Satisfiability, Boolean Modeling and Computation **6**(4), 245–262 (2009)
5. Sang, T., Bacchus, F., Beame, P., Kautz, H.A., Pitassi, T.: Combining component caching and clause learning for effective model counting. In: SAT 2004 - The Seventh International Conference on Theory and Applications of Satisfiability Testing 10–13 May 2004 Vancouver BC Canada Online Proceedings (2004)
6. Thurley, M.: sharpSAT – counting models with advanced component caching and implicit BCP. In: Biere, A., Gomes, C.P. (eds.) SAT 2006. LNCS, vol. 4121, pp. 424–429. Springer, Heidelberg (2006)
7. Meel, K.S.: Sampling techniques for boolean satisfiability. CoRR abs/1404.6682 (2014)
8. Davis, M., Logemann, G., Loveland, D.: A machine program for theorem-proving. Commun. ACM **5**(7), 394–397 (1962)
9. Bayardo, R.J., Pehoushek, J.D.: Counting models using connected components. In: AAAI National Conference, pp. 157–162 (2000)
10. Bacchus, F., Dalmao, S., Pitassi, T.: Algorithms and complexity results for #sat and bayesian inference. In: 44th annual IEEE Symposium on Foundations of Computer Science (FOCS), pp. 340–351 (2004)
11. Schubert, T., Lewis, M., Becker, B.: Antom - solver description. SAT Race (2010)
12. Kempf, B.: The boost.threads library. C/C++ Users Journal (2002)
13. Sang, T., Beame, P., Kautz, H.: Heuristics for fast exact model counting. In: Bacchus, F., Walsh, T. (eds.) SAT 2005. LNCS, vol. 3569, pp. 226–240. Springer, Heidelberg (2005)
14. Sauer, M., Burchard, J., Schubert, T., Polian, I., Becker, B.: Waveform-guided fault injection by clock manipulation. In: TRUDEVICE: First Workshop on Trustworthy Manufacturing and Utilization of Secure Devices (2013)
15. Corno, F., Reorda, M.S., Squillero, G.: RT-level ITC'99 benchmarks and first ATPG results. IEEE Des. Test **17**(3), 44–53 (2000)
16. Brglez, F., Bryan, D., Kozminski, K.: Combinational profiles of sequential benchmark circuits. In: IEEE International Symposium on Circuits and Systems, vol. 3, pp. 1929–1934 (1989)
17. Brglez, F., Fujiwara, H.: A neutral netlist of 10 combinational benchmark circuits and a target translator in fortran. In: Proceedings of IEEE Int'l Symposium Circuits and Systems (ISCAS 1985), pp. 677–692. IEEE Press, Piscataway (1985)

SATGraf: Visualizing the Evolution of SAT Formula Structure in Solvers

Zack Newsham[✉], William Lindsay, Vijay Ganesh, Jia Hui Liang,
Sebastian Fischmeister, and Krzysztof Czarnecki

University of Waterloo, Waterloo, Canada
znewsham@uwaterloo.ca

Abstract. In this paper, we present SATGraf, a tool for visualizing the evolution of the structure of a Boolean SAT formula in real time as it is being processed by a conflict-driven clause-learning (CDCL) solver. The tool is parametric, allowing the user to define the structure to be visualized. In particular, the tool can visualize the community structure of real-world Boolean satisfiability (SAT) instances and their evolution during solving. Such visualizations have been the inspiration for several hypotheses about the connection between community structure and the running time of CDCL SAT solvers, some which we have already empirically verified. SATGraf has enabled us in making the following empirical observations regarding CDCL solvers: First, we observe that the Variable State Independent Decaying Sum (VSIDS) branching heuristic consistently chooses variables with a high number of inter-community edges, i.e., high-centrality bridge variables. Second, we observe that the VSIDS branching heuristic and hence the CDCL search procedure is highly focused, i.e., VSIDS disproportionately picks variables from a few communities in the community-structure of input SAT formulas.

1 Introduction

Conflict-driven clause-learning (CDCL) SAT solvers have witnessed dramatic improvements in their efficiency over the last 20 years, and consequently have become drivers of progress in many areas of computer science such as formal verification [1,2]. There is general agreement that these solvers somehow exploit structure inherent in industrial instances. In order to understand what this structure is and the mechanism by which CDCL solvers exploit it, we need visualization/evolution tools that can help us formalize and visually check our hypotheses that can subsequently be verified using the scientific method.

In order to enable researchers to improve their intuitions of how CDCL solvers work, better understand the structure of industrial instances, and visualize in real-time how CDCL solvers exploit said structure, we built the SATGraf visualization/evolution tool. SATGraf takes as input a Boolean formula, and outputs a rendering of its variable-incidence graph (VIG) as well as showing how the structure evolves in real-time while being solved by a SAT solver. SATGraf is parametric, i.e., it can be configured to display any structure discoverable in a

© Springer International Publishing Switzerland 2015
M. Heule and S. Weaver (Eds.): SAT 2015, LNCS 9340, pp. 62–70, 2015.
DOI: 10.1007/978-3-319-24318-4_6

SAT formula. SATGraf enables researchers to formalize and visually check their hypotheses about the behavior of SAT solvers.

SATGraf has been invaluable to us in formulating and visually checking many hypotheses about CDCL SAT solvers that we proposed, which we were able to subsequently verify empirically. For example, in our paper on community structure and their impact on SAT solver performance [3] we provide empirical evidence that community structure correlates more strongly with the running time of CDCL solvers than traditional hypotheses such number of clauses, variables and their ratio. We used SATGraf to visually check that many classes of easy-to-solve industrial instances have "good" community structure. Another hypothesis that SATGraf helped us verify is that the VSIDS branching heuristic disproportionately favors high-centrality bridge variables, i.e., those that belong to clauses that lie between communities in the community structure of SAT instances.

Background: While SATGraf is able to display any user-defined structure, we focus here on community structure. The idea of decomposing graphs into *natural communities* arose in the study of complex networks. Modularity is a measure of the quality of the community structure of a graph and ranges from 0 to 1, where 0 is a poor community structure and 1 a strong community structure. Informally, we say a graph has poor community structure (modularity close to 0) if there are more inter-community edges than intra-community edges. Conversely, if the graph has more intra-community edges than inter-community edges, this correlates with good community structure (modularity close to 1). Modularity is often used in optimization methods for detecting community structure in networks. The precise definition and its calculation can be found in [4]. Many algorithms [5,6] have been proposed to solve the problem of finding an optimal community structure of a graph, the most well-known among them being the one from Girvan and Newman [5]. We refer the reader to these papers [5–7] for complete descriptions of community detection algorithms.

Contributions: We make the following contributions in this paper.[1]

The SATGraf Tool: We present SATGraf, a tool that enables researchers to visualize the community structure of a SAT instance and see its evolution while solving by a real world CDCL solver.

VSIDS & High-centrality Bridge Variables: Using SATGraf we observed that the VSIDS branching heuristic disproportionately picks high-centrality bridge variables in the community structure of input instances during the entire run of the solver.

Focused Search by CDCL Solvers: Using SATGraf we observed that the VSIDS branching heuristic disproportionately picks variables from a few communities in the community structure of SAT instances during the entire run of the solver.

---

[1] All code and data can be obtained from the SATGraf formula visualization/evolution tool website: http://satbench.uwaterloo.ca/satgraf/index

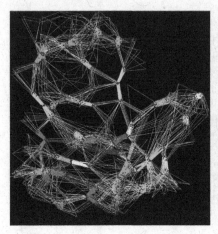
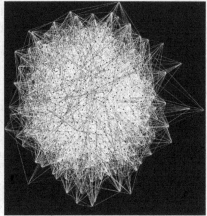

(a) Industrial instance: *aes_16_10_ keyfind_3*

(b) Random instance: *unif-k3-r4.267- v421-c1796-S4839562527790587617*

Fig. 1. Community structure of instances from the SAT 2013 Competition

Unnecessary Backtracking Steps: Using SATGraf we observed that back-tracking resets decisions and propagations unrelated to the current conflict.

2   How **SATGraf** Works

SATGraf is implemented in phases as described below:

Phase 1: First, SATGraf converts an input Boolean formula (in DIMACS format) into its corresponding graph. Currently the only format we consider for this is the variable-incidence graph, however other implementations such as the clause-incidence graph are possible.

Phase 2: Second, SATGraf computes structure metrics as defined by the user. Currently the user may choose from either the Clauset-Newman-Moore (CNM) algorithm [5] or the online (OL) community algorithm [6], however it is possible for the user to specify their own additional algorithms.

Phase 3: Third, SATGraf uses a user-specified layout algorithm to render the graph while maintaining the structure detected in phase 2. Currently the user may choose from either a modified version of layout algorithm by Kamada and Kawai (KK) [8] or the Fruch-Reingold (FR) algorithm [9]. Other "fast" layout options include a grid or circle solution, where communities are treated as separate graphs and use either the KK or FR layout algorithms, these communities are then placed on a grid or circle pattern. While these options do not display the structure as clearly, they scale better.

Phase 4: Finally, users of SATGraf can replay different stages of the evolution. While doing this they may also hide communities, edges or variables, that are not of interest to obtain a clearer view of those that are. To this end the user may also zoom in on specific communities within the graphical representation, and choose whether to hide, or colour variables that have been assigned values at any point during the evolution. The user may also choose to export the entire graphical evolution as a GIF file for later analysis, though this does create large files.

The modular design of SATGraf allows for easy integration of any other structure metric or layout algorithm for either of these categories. Figure 1 shows the graph generated by SATGraf for two instances from the SAT 2013 competition [10]. Figure 1a is an industrial instance, and Figure 1b shows a randomly-generated instance. Edges between variables within the same community are assigned a distinct colour, one per community. White edges represent inter-community edges, and red edges resulting from conflict clauses. As is evident, the industrial instance has lot more distinct communities that can be neatly partitioned, while the randomly generated instances typically have lots inter-community edges.

SATGraf can present the evolution of a formula by interacting with modified versions of SAT solvers. Currently only MiniSAT is supported, however C source code is included in the project to ease integration with other solvers. MiniSAT interacts with SATGraf's evolution mechanism by notifying it when a variable changes value – either by decision or propagation — and when new conflict clauses are added. SATGraf then updates the graph by either hiding, showing or colouring edges and nodes, or by redrawing the graph (if the user requests it). This allows users to observe the overall evolution of the structure of the formula, but also to see how each community is affected during solution. SATGraf is open source and available at [11,12] with an easy install version available at [13]. The project was developed in Java and has a modified version of MiniSAT included.

3   Results

SATGraf has been tested on several industrial, hard combinatorial, and randomly-generated formulas from the 2013 SAT competition [10]. The time taken to display the community structure of a single instance grows with the size of the input formula. This is to be expected due to the nature of the community detection and placement algorithms — which is the most time-consuming component. The resulting graphs, using the OL community detection and FR layout algorithms, can be seen in Figure 1. The community structure of the industrial instance has much better modularity than the one for the randomly-generated instance. This can be verified both visually and through the modularity measure: the industrial instance has a modularity of 0.77, while the randomly-generated one has a modularity of 0.16. Their solve times using MiniSAT are also different; The industrial instance takes 0.076 and the randomly-generated instance times out after 5000 seconds. SATGraf has been found to be efficient when viewing a number of different SAT instances, the largest observed containing approximately 450,000 variables and 1.4 million clauses. However, this utilised the "grid" layout algorithm. Unfortunately neither the number of variables, nor the number of clauses provide an accurate representation of the running time of SATGraf, as it is the number of edges that drives most of the workload. As such, a single clause containing 500 variables, will be more intensive than a 40,000 3-CNF formula.

SATGraf's evolution feature is partly shown in two pictures in Figure 2. The SAT instance here is obtained from a feature model [14] called *Fiasco* that can be downloaded from the SATGraf website [15]. A GIF of the entire evolution of *Fiasco* can also be found here. We chose this SAT formula since it is a good representation of an industrial application of SAT solvers. Furthermore, this instance is small enough so that we can actually show, in a timely manner, how the SAT solver dynamically morphs its graph (the instance and the generated learnt clauses). Finally, the solvers [16,17] solved this formula without generating too many conflicts, and thus it was easier to make sense of the evolution of the graph of this instance.

Observing the evolution showed an interesting trend. Namely, the removal of entire communities during the solving process. This evolution can be seen when going from the graph of the original SAT formula in Figure 2a, to the graph after the solver generates the first conflict shown in Figure 2b. It is easy to see that some of the communities have completely disappeared by the absence of their associated colour, i.e., the corresponding clauses have been satisfied.

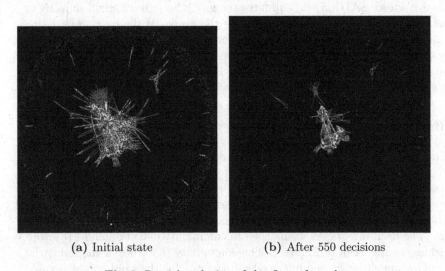

(a) Initial state   (b) After 550 decisions

Fig. 2. Partial evolution of the *fiasco* formula

3.1   Observation #1: VSIDS Chooses High-Centrality Bridge Variables

Whilst visualising industrial instances using SATGraf, we found that VSIDS was consistently choosing decision variables that have a high number of inter-community edges, which we call high-centrality *bridge variables*. For example in the SAT competition formula *aes_16_10_keyfind_3* 98% of the first 5000 decision variables had inter-community edges. This leads us to suspect that VSIDS is discovering the community structure implicitly whilst solving. We have found similar results for other industrial and hard combinatorial instances from the SAT competition. It is possible that the decision variables consistently being inter-community variables is either a random artefact of the VSIDS heuristic, or simply that a large number of the variables in the formula had inter-community edges. However, we do not believe this to be the case. In fact after 5000 decisions, 66% of the decision variables had more inter-community edges than intra-community edges. This observation presents a conjecture that can be validated independently through modifications to the VSIDS algorithm and empirical measurements, which is the subject of current research in our group.

3.2   Observation #2: VSIDS Moves Infrequently Between Communities

In addition to our previous observations, we observed that a high percentage of decision variables occurred within the same communities. When considering the *aes_16_10_keyfind_3* SAT formula, 80% of the decision variables were chosen from the same community as the previous decision variable. This would support the hypothesis that formulas which have a good community structure are sometimes solved one community at a time.

3.3   Observation #3: Backtracking May Incur Unnecessary Overhead

Whilst visualising the SAT formula *toybox* on SATGraf, we found that despite the high level of separability of the formula (mostly distinct, unconnected communities), backtracking caused variables that were unconnected to the conflict variables (either directly or transitively) to be reset. In most SAT formulas of interest, the communities will not be as clearly separated as in the *toybox* example. However, we present the conjecture that in some situations the backtracking of CDCL solvers results in more work being done than is necessary. We suggest that a potential solution to this would be a selective backtracking algorithm, that determines which variables are affected by a backtrack. While this would require additional time during solution to determine affected variables, in instances with higher solve times, it could prove effective.

Table 1. Comparison of SAT Visualization tools

Tool	Interactive	Evolution	Community	3D	Implication
DPVis[18]	✓	✓	✗	✗	✓
GraphInsight[19]	✓	✗	✗	✓	✗
iSat[20]	✗	✓	✗	✗	✗
GraphViz[21]	✗	✗	✗	✗	✗
SATGraf	✓	✓	✓	✗	✓

4   Related Work

SATGraf is the only tool that we know of that has both visualization capabilities to view the "user-defined structure" of SAT instances and evolution feature that shows how this structure is morphed during solution. While other tools [18–21] have visualization or evolution capabilities, they do not allow for user defined structure, nor do they show how the solver morphs this structure. Instead, the choice of graph structure of SAT instances is hard-coded in these tools. Additionally, we support community structure, while the tools we compare against do not. Table 1 highlights the differences between visualization tools that we found. Those differences range across a handful of categories such as interactive (ability to hide/show nodes, edges or other structural information); evolution (ability to see the evolution of the SAT formula); structure (ability to display the community (or any other) structure); 3D (three dimensional capability) and implication (can generate the implication graph). **DPVis** [18], is the closest to SATGraf in terms of features. It is a graphing tool designed to expose how a CDCL solver morphs a SAT instance as it is being solved. It offers a number of features such as multiple layout algorithms, the ability to set specific values on literals displayed in the graph, and performing unit propagation. However, unlike SATGraf it does not allow the user to specify the formula structure (e.g., community structure), nor does it allow the user to specify a non-included real-world solver as the evolution engine. **DPVis** uses a built-in implementation of the DPLL algorithm, along with a hard coded interface to MiniSAT to display the evolution, whereas SATGraf uses a user provided real world solver. While currently only two solvers support this technique (MiniPure and MiniSAT), it is possible for the user to implement this on any solver, using the provided API. Each tool presented in Table 1 has different strengths and weaknesses. However, the only tool that can accomplish visualizing additional structure of a SAT formula, both in its original state and while being solved by a SAT solver, is SATGraf.

5   Conclusion

SATGraf presents a way to visualise a SAT instance's community structure. Furthermore, SATGraf has the ability to dynamically graph the community structure of a CDCL SAT solver's progress while solving a SAT formula. These features were shown to be unique to SATGraf when compared to various similar tools. These new capabilities yielded hypotheses regarding the correlation between the community structure of input instances and performance of CDCL SAT solvers. We found that the better the modularity is, the less time the SAT solver needs, and the CDCL SAT solver often seems to solve SAT formulas one community at a time.

References

1. Clarke, E., Talupur, M., Veith, H., Wang, D.: SAT based predicate abstraction for hardware verification. In: Giunchiglia, E., Tacchella, A. (eds.) SAT 2003. LNCS, vol. 2919, pp. 78–92. Springer, Heidelberg (2004)
2. Biere, A., Cimatti, A., Clarke, E.M., Fujita, M., Zhu, Y.: Symbolic model checking using sat procedures instead of bdds. In: Proceedings of the 36th Annual ACM/IEEE Design Automation Conference, pp. 317–320. ACM (1999)
3. Newsham, Z., Ganesh, V., Fischmeister, S., Audemard, G., Simon, L.: Impact of community structure on SAT solver performance. In: Sinz, C., Egly, U. (eds.) SAT 2014. LNCS, vol. 8561, pp. 252–268. Springer, Heidelberg (2014)
4. Ansótegui, C., Giráldez-Cru, J., Levy, J.: The community structure of SAT formulas. In: Cimatti, A., Sebastiani, R. (eds.) SAT 2012. LNCS, vol. 7317, pp. 410–423. Springer, Heidelberg (2012)
5. Clauset, A., Newman, M.E.J., Moore, C.: Finding community structure in very large networks. Physical Review E **70**(6), 066111 (2004)
6. Zhang, W., Pan, G., Wu, Z., Li, S.: Online community detection for large complex networks. In: Proceedings of the Twenty-Third International Joint Conference on Artificial Intelligence, pp. 1903–1909. AAAI Press (2013)
7. Newman, M.E.J., Girvan, M.: Finding and evaluating community structure in networks (2003). http://arxiv.org/pdf/cond-mat/0308217.pdf (last viewed December 2013)
8. Kamada, T., Kawai, S.: A general framework for visualizing abstract objects and relations. ACM Trans. Graph. **10**(1), 1–39 (1991)
9. Fruchterman, T.M.J., Reingold, E.M.: Graph drawing by force-directed placement. Software: Practice and Experience **21**(11), 1129–1164 (1991)
10. SAT competition 2013 (2013). http://satcompetition.org/2013/ (last viewed January 2014)
11. Newsham, Z., Lindsay, W., Liang, J., Ganesh, V., Fischmeister, S., Czarnecki, K.: Satgraf sat formula visualization tool. http://bitbucket.org/znewsham/satgraf
12. Newsham, Z., Lindsay, W., Liang, J., Ganesh, V., Fischmeister, S., Czarnecki, K.: Satgraf structure source. http://bitbucket.org/znewsham/satlib
13. Newsham, Z., Lindsay, W., Liang, J., Ganesh, V., Fischmeister, S., Czarnecki, K.: Satgraf visualisation executable. https://bitbucket.org/znewsham/satgraf/downloads/satgraf.zip
14. Kang, K.C., Cohen, S.G., Hess, J.A., Novak, W.E., Peterson, A.S.: Feature-oriented domain analysis (FODA) feasibility study. Technical report, DTIC Document (1990)
15. Newsham, Z., Lindsay, W., Liang, J., Ganesh, V., Fischmeister, S., Czarnecki, K.: Satgraf: Results (2014). http://satbench.uwaterloo.ca/satgraf/index (last viewed January 2015)
16. Taiwan, T., Wang, H.: Minipure (2013). http://edacc4.informatik.uni-ulm.de/SC13/solver-description-download/134 (last viewed January 2014)
17. Een, N., Sörensson, N.: Minisat: a SAT solver with conflict-clause minimization. In: SAT 2005 (2005)
18. Sinz, C., Dieringer, E.-M.: DPvis – a tool to visualize the structure of SAT instances. In: Bacchus, F., Walsh, T. (eds.) SAT 2005. LNCS, vol. 3569, pp. 257–268. Springer, Heidelberg (2005)

19. Nicolini, C., Dallachiesa, M.: Graphinsight: An interactive visualization system for graph data exploration. http://www.graphinsight.com
20. Orbe, E., Areces, C., Infante-López, G.: iSat: structure visualization for SAT problems. In: Bjørner, N., Voronkov, A. (eds.) LPAR-18 2012. LNCS, vol. 7180, pp. 335–342. Springer, Heidelberg (2012)
21. Bilgin, A., Ellson, J., Gansner, E., Smyrna, O., Hu, Y., North, S.: Graphviz - graph visualization software. http://www.graphviz.org/

Hints Revealed

Jonathan Kalechstain[1,2]([✉]), Vadim Ryvchin[1], and Nachum Dershowitz[2]

[1] Design Technology Solutions Group, Intel Corporation, Haifa, Israel
{kalechstain,vadimryv}@gmail.com
[2] School of Computer Science, Tel Aviv University, Tel Aviv, Israel
nachumd@tau.ac.il

Abstract. We propose a notion of *hints*, clauses that are not necessarily consistent with the input formula. The goal of adding hints is to speed up the SAT solving process. For this purpose, we provide an efficient general mechanism for hint addition and removal. When a hint is determined to be inconsistent, a hint-based partial resolution-graph of an unsatisfiable core is used to reduce the search space. The suggested mechanism is used to boost performance by adding generated hints to the input formula. We describe two specific hint-suggestion methods, one of which increases performance by 30% on satisfiable SAT '13 competition instances and solves 9 instances not solved by the baseline solver.

1 Introduction

Modern backtrack search-based SAT solvers are indispensable in a broad variety of applications [3]. In a classical SAT interface, the solver is given one formula in conjunctive normal form (CNF) and determines whether it is satisfiable or not. Performance of SAT solvers has improved dramatically over the past years [15]. The main advancements came as result of developing new heuristics for existing conflict-driven clause-learning (CDCL) solver techniques, like deletion strategies, decision heuristics, and restart strategies (plus preprocessing and in-processing).

In this work, we propose and investigate a novel method for cutting the search space explored by the SAT solver so as to help it reach a solution faster. The idea is to add *hints*, clauses that are not necessarily "correct", in the sense that they are not necessarily implied by the original input formula.

We call our hint-addition platform HSAT (Hint Sat), and present two variants that have been implemented in HAIFAMUC [13]. HAIFAMUC is an adaptation of MINISAT 2.2 [4], which we will henceforth refer to it as BASE.

The addition of hints H to the original formula F creates an extended formula F'. Hints can, of course, affect the satisfiability of the formula. As long as H is implied by F, the extended formula F' will be equi-satisfiable with the original F (either both are satisfiable or neither is). This means that if F is satisfiable but F' is not, then there must be a contradiction between the added hints and the original formula.

In HSAT, we try to solve only the extended formula F'. In case it is satisfiable, we are done, and the solver declares that the original formula was likewise satisfiable. Otherwise, the extended formula is unsatisfiable, in which case we

© Springer International Publishing Switzerland 2015
M. Heule and S. Weaver (Eds.): SAT 2015, LNCS 9340, pp. 71–87, 2015.
DOI: 10.1007/978-3-319-24318-4_7

need to understand whether the hints are the cause of unsatisfiability, that is, whether any hint is a necessary part of the proof of the empty clause. This is accomplished by an examination of the resolution graph that is built during the run of the solver on F'. In [14], the authors presented an efficient way (their "optimization **A**") of saving a partial resolution with respect to a given subset of input clauses. We use this ability to restrict tracking so that only the effects of hints are recorded in the partial graph. Marking clauses to track their origin is an old idea used in Chaff [8] and later reintroduced in [17], and is well adapted to cases when tracking of clauses is required. When the extended formula is unsatisfiable, we check the cone of the empty clause. If it includes a hint, then the status of the original formula remains unknown and additional operations are required (like deletion of the hints). Otherwise, the original formula is unsatisfiable, and we are done. Handling of inconsistent clauses was done in several other applications, like parallel solving [6,7]; our solution differs, having the ability to track the full effect of the partial resolution tree.

In case the result is unknown and the UNSAT core contains only one hint, an additional optimization can be made by using the UNSAT core of the partial resolution graph. Suppose the UNSAT core contains only hint h, then h must contradict F, and $\neg h$ is implied by F. As $\neg h$ is, in this circumstance, a set (conjunction) of unit clauses, each literal in h can be negated and added as a fact to F, which will increase the number of facts and reduce the search space to be explored. This optimization can be generalized to include all graph dominators in the partial resolution graph. (See Theorem 1 below.)

We introduce two heuristics for hint generation. The first, "Avoiding Failing Branches" (AFB), is a purely deterministic hint-addition method. The main idea behind it is the same idea that drives restarts in modern SAT solvers, namely, the possibility that the solver is spending too much time on "bad branches", branches that do not contain the satisfying assignment to the problem. Our motivation is to prevent the solver from entering branches that have already been explored. In our algorithm, we describe an *explored branch* that is a subset of decision variables. We pick the most conflict-active decisions and add a hint that explicitly precludes choosing that set again. In this approach, we keep a score for each literal. The score is boosted every time a clause containing it participates in a conflict. The literals with the highest scores are added to a hint in their negated form. The hint is then added right after a restart, and the same set of active decision variables will never be chosen unless the hint is removed. This approach leads to significantly improved solver times for satisfiable instances.

A second heuristic, "Randomize Hints" (RH), draws a given number of random assignments, and tries to create a set of hints that will contradict the instance. When the solver concludes unsatisfiability, all dominators of the partial resolution graph are extracted, and all literals in all dominators are added as facts in their negated form.

We continue in the next section with the formalization and various preliminaries. Section 3 presents the HSAT algorithm, and, in Sect. 5, we demonstrate its correctness. The two heuristic hint-generation methods of Sect. 4 are empirically evaluated in Sect. 6. We conclude and discuss future work in Sect. 7.

This paper contains several contributions. An efficient generic mechanism is introduced to add hints, the goal of which is to speed up the solver. It is based on the ability to remove clauses and all the facts derived from them. In HSAT, we use the partial resolution graph of BASE to remove the hints and their effect in case of an unsatisfiable conclusion. In [9–11] and later in [14], it was shown that the alternative, using selector variables for clause removal [5, 12], is inferior to the use of the resolution graph. We extend the path-strengthening technique published in [10]. Instead of using only immediate children of the removed clauses, we use all dominators in the partial resolution graph provided in BASE. We introduce two algorithms for hint generation, one of them (AFB) increasing performance for satisfiable instances by 19–30%.

2 Preliminaries

We presume some basic knowledge of the Boolean Satisfiability Problem and CDCL SAT solvers [3]. Let φ be a CNF formula $c_1 \wedge c_2 \cdots \wedge c_m$. We write $c_i \in \varphi$ if $\varphi = c_1 \wedge \cdots \wedge c_i \wedge \cdots \wedge c_m$. Each clause $c = \ell_1 \vee \ell_2 \vee \cdots \vee \ell_k$ is a disjunction of literals, and each literal ℓ_i is either a variable v or its negation $\neg v$. We write $\ell_j \in c_i$ if $c_i = \ell_1 \vee \cdots \vee \ell_j \vee \cdots \vee \ell_k$. In what follows, V denotes the set of variables occurring in φ, and $n = |V|$.

For two clauses $c_i = v \vee c$ and $c_j = \neg v \vee c'$, both involving the same variable $v \in V$, their binary *resolvent* is

$$Resol(c_i, c_j) \triangleq c \vee c'.$$

A conflict occurs when several solver decisions and subsequent implications result with a clause being unsatisfiable. In CDCL SAT solvers, a clause preventing the last conflicting set of decisions is created and added; it is referred to as a *conflict clause*. In [8], it was shown that the best clause is the one created by finding the cut in the implication graph that includes the Unique-Implication-Point (UIP) closest to the conflict. That cut corresponds to a number of binary resolutions performed on clauses that are inside the cut or intersect it. For example, Fig. 1 illustrates the cut and the clauses c_4, c_5, c_6 that participated in the resolutions that derived the conflict.

If φ is a formula and H is a set of hint clauses, then by $\varphi \wedge H$ we mean their conjunction: $\varphi \wedge \bigwedge_{h \in H} h$, which we will call a *hint-extended formula*.

In HSAT, we use a *resolution graph* to determine why $\varphi \wedge H$ is unsatisfiable, when it is, by extracting the *UNSAT core*.

Definition 1 (Hyper-Resolution). *Let $c_1, c_2, \ldots, c_i$ be all the clauses (from the implication graph) that participated in the binary resolutions that created the first UIP conflict clause U. The Hyper-Resolution function*

$$Hyper(c_1, c_2, \ldots, c_i) \triangleq U$$

yields that resulting conflict clause U.

$$c_1 = \left(\sim v_1 \vee v_2 \right)$$

$$c_2 = \left(\sim v_1 \vee v_3 \vee v_9 \right)$$

$$c_3 = \left(\sim v_2 \vee \sim v_3 \vee v_4 \right)$$

$$c_4 = \left(\sim v_4 \vee v_5 \vee v_{10} \right)$$

$$c_5 = \left(\sim v_4 \vee v_6 \vee v_{11} \right)$$

$$c_6 = \left(\sim v_5 \vee \sim v_6 \right)$$

$$c_7 = \left(v_1 \vee v_7 \vee \sim v_{12} \right)$$

$$c_8 = \left(v_1 \vee v_8 \right)$$

$$c_9 = \left(\sim v_7 \vee \sim v_8 \vee \sim v_{13} \right)$$

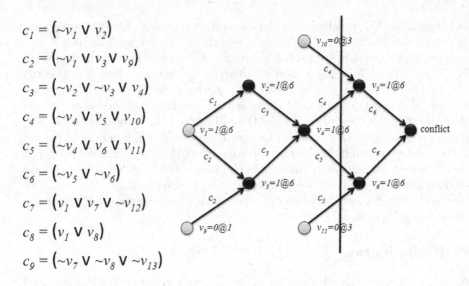

Fig. 1. Conflict analysis graph. The grey nodes represent decision variables while the black nodes represent propagated values. The vertical line is the first UIP cut.

Writing $v_i = a@b$ means that v_i was assigned to a at decision level b.

As mentioned in Sect. 1, we use a partial resolution graph to generate hints. This graph is used to determine whether there exists a directed path from H to an empty clause.

Definition 2 (Resolution Graph). *The Resolution Graph $G = (V, E)$ is defined recursively as follows:*

$$V := \varphi \cup H \cup \{Hyper(c_1, \ldots, c_m) \mid c_1, \ldots, c_m \in V\}$$
$$E := \{(c_i, Hyper(c_1, \ldots, c_i, \ldots, c_m)) \mid c_1, \ldots c_m \in V\}.$$

In words, the vertices are the initial clauses and hints closed under hyper-resolution and the edges point from participating clauses to their hyper-resolvent.

Determining whether a path exists from H to the empty clause is possible by saving only the part relevant to hints. The partial resolution graph will consist only of hints or conflict clauses that were derived by some hint. To do so, we start just with the hints and define the relevant hint-based *Partial Resolution Graph* as follows:

Definition 3 (Partial Resolution Graph). *The Partial Resolution Graph $G_P = (V_P, E_P)$ is defined recursively as follows:*

$$V_P := H \cup \{Hyper(c_1, \ldots, c_i, \ldots, c_m) \mid c_i \in V_P, c_1, \ldots, c_{i-1}, c_{i+1}, \ldots, c_m \in V\}$$
$$E_P := \{(c_i, Hyper(c_1, \ldots, c_i, \ldots, c_m)) \mid c_i \in V_P, c_1, \ldots, c_{i-1}, c_{i+1}, \ldots, c_m \in V\}.$$

In words, the vertices are the hints closed under hyper-resolution and the edges point from participating clauses to their hyper-resolvent. Figure 2 contains an

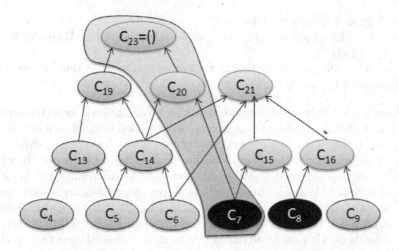

Fig. 2. Resolution graph. Black nodes are the set H. Blue nodes are the hyper resolvents of V_P. The grey nodes are the nodes in $V \setminus V_P$.

example illustrating Definitions 2,3. The nodes c_4, c_5, c_6, c_9 form the set φ, while c_7, c_8 are the hints in H. The entire graph represents G while the black and blue nodes and the edges between them are the restricted graph G_P.

Having defined the partial resolution graph, we are interested in isolating the proof of unsatisfiability. To do so, we define the UNSAT core (UC) of a resolution graph.

Definition 4 (UNSAT Core). *The UNSAT core is a subset UC of $\varphi \cup H$ that is backward reachable from the empty clause in G.*

We are interested in finding that part of UC that is relevant to the hints:

Definition 5 (Relevant UNSAT Core). *The Relevant UnsatCore is the intersection $RC = H \cap UC$.*

We refer to the set of all dominator points (a vertex that lies along every path) between RC and the empty clause in an UNSAT proof as $Dominator_{RC}$.

The negation $\neg h$ of a clause $h = \ell_1 \vee \ell_2 \cdots \vee \ell_k$ is the set (conjunction) of its negated literals $\neg \ell_1 \wedge \neg \ell_2 \cdots \wedge \neg \ell_k$, viewed as unit facts.

3   Hint Addition

We proffer a general platform for adding clauses without worrying that they might be inconsistent with the input formula. These "hint" clauses can be created using prior knowledge about the formula's origins or from information garnered during SAT solving, as explained in Sect. 4. Our solution enjoys several benefits:

1. No additional literals are added.
2. We delay the effect of hints by using techniques from HAIFAMUC as described in [14].
3. "Bad" hints, hints participating in the empty clause derivation, are used for search space reduction.

We use a resolution-graph-based solution to avoid the need for extra literals and to enable further optimizations in case the extended formula is unsatisfiable on account of the hints. In addition, we want to prevent any aggressive intervention of hints in the SAT solver's solution process, by using hints only when necessary, which is achieved by delaying their use. We discriminate in favor of the use of ordinary clauses because conflicts derived from hints are not necessarily consistent with the formula. The same motivation underlies modern Minimal Unsatisfiable Set (MUS) and Group Minimal Unsatisfiable Set (GMUS) solvers, which prefer to use clauses already known to be in the minimal core, to keep that as small as possible. Because of the similarity between hints and core clauses in MUS and GMUS solvers, we base our solver on HAIFAMUC and use the optimization techniques described in [14]. These techniques allow us to prioritize ordinary clauses over hints and therefore reduce the run-time effect of hints.

The optimizations relevant to hints are the following:

1. Maintain only the partial resolution proof of clauses derived from the added hints. This prevents the keeping of the whole resolution proof in the memory and significantly reduces the memory footprint of the solver.
2. Selective clause minimization. Clause minimization [2,16] is a technique for shrinking conflict clauses. If the learned clause is not derived from the hints, then during shrinking we prevent the use of hints in the minimization. The result is that no additional dependencies on hints are added even at the expense of longer learned clauses.
3. Postponed propagation over hints. This optimization is performed by changing the order of BCP (Binary Constraint Propagation). BCP first runs over ordinary (non-hint) clauses, and only if no conflict is found does it run over hints. The motivation is to prefer conflicts caused by ordinary clauses.
4. Selective learning of hints and selective backtracking. Both optimizations change the learning scheme by reducing the number of clauses effected by hints in case an ordinary clause can be learned.

We denote these optimization techniques collectively as HMUCOPT.

One of the benefits of using a resolution-graph method is the availability of clause relation information, which can be used in case the extended formula is unsatisfiable on account of hints. In [10], a *path strengthening* technique was presented in relation to the MUS problem solution. It uses the partial resolution graph and is used to check whether a clause c is part of the MUS. Checking whether $c \in$ MUS can be done by checking if the formula is unsatisfiable without using c. If it is, then c cannot be part of the minimal core. To speed up the SAT solver run, the negation of the clause is added to the SAT Solver as assumptions. Path strengthening extends this set of assumptions by analyzing the resolution

---

Algorithm 1 HSAT– Solves an extended formula, negates dominators and cleans hints' effects.

Input: *instance* – Boolean formula in CNF form
          H – Initial set of Hints (in our case $\emptyset$)
Output: SAT or UNSAT (ignore TIMEOUT)

1: **while** TRUE **do**
2:     $model := Solve(instance \wedge H)$ ▷ New $h \in H$ can be added in $Solve()$
3:     **if** $model \neq$ NULL **then**
4:         **return** SAT   ▷ We have the model
5:     **else**
6:         $RC := GetRC()$
7:         **if** $RC.Size() = 0$ **then**
8:             **return** UNSAT
9:         **else**
10:             $Dominator_{RC} := GetDominators(RC)$
11:             **for** each $D \in Dominator_{RC}$ **do**
12:                 $instance := instance \wedge Negate(D)$
13:             **for** each $c_i \in$ RC **do**
14:                 $RemoveClauses(c_i)$

---

graph. If c has only one derived clause in the cone of the empty clause, then the literals of this clause are added as assumptions as well. This operation is performed recursively until a clause with more than one child is reached. In HSAT, we extend this by using all dominators between the hint clause and the empty clause in the partial resolution graph.

Algorithm 1 introduces the general workflow of HSAT. Operation $Solve()$ is a modification of a generic SAT solver with several additions. First, it allows the addition of new hints and produces a partial resolution in case those hints are added. In addition, $Solve()$ contains an implementation of HMUCOPT. Operation $Solve()$ can return satisfiable or unsatisfiable. In the satisfiable case, we are done, as the solver found a satisfying assignment to the formula. In case the result is unsatisfiable, we check the RC (UNSAT core of hints) created by the hints. The extraction of RC is performed using $GetRC()$. If RC is empty, then the solver found a proof of the empty clause without relying on hints, so the original formula is unsatisfiable. Otherwise, we find all dominators of the RC using $GetDominators()$. (See Alg. 2 and the next paragraph.) For each dominator, we add its negation via $Negate()$ to the input formula and create a new $instance$, which goes back to $Solve()$. Before the next call to $Solve()$, we clean the effect of hints in G_P by means of $RemoveClauses()$. The correctness of Alg. 1 is justified by the observations of Sect. 5. As mentioned already, for $Solve()$ we use a modification of BASE, so all the optimizations HMUCOPT are used, as was introduced in [14]. This way, we ensure an increased chance of finding the solution without hints if such a solution is easy to find.

The operation $GetDominators()$ gets all nodes $v \in V_P$ such that all paths from H to the empty clause go through v. At first, we save all nodes

---

Algorithm 2 *GetDominators()* – Gets all dominators in G_P. This set will be negated in Alg. 1

Input: G_P – The Partial Resolution Graph
Input: *workList* – list of vertices. Initially set to RC
Output: $Dominator_{RC}$ – The dominators with respect to G_P

1: **while** *workList.Size()* > 0 **do**
2:     $v := GetAllParentsMarked(workList)$
3:     **if** *workList.Size()* $= 1$ **then**
4:         $Dominator_{RC}.Push(v)$ ▷ A dominator
5:     $Mark(v)$ ▷ v now marked
6:     **for** each $u \in Children(v)$ **do**
7:         **if** $\neg IsMarked(u)$ **then**
8:             $workList.Push(u)$
9:     $workList.Remove(v)$
10: **return** $Dominator_{RC}$

---

from RC in a list called *workList*. The algorithm iterates until *workList* is empty. We get from the list some $v \in V_P$ that has all parent marked using $GetAllParentsMarked(workList)$. Note that since RC has no parents, all members of RC have all parents marked. If the size of *workList* is 1, then v is a dominator and we push it into $Dominator_{RC}$. We mark v using $Mark(v)$ and push all its unmarked children into *workList*. Note that the empty clause is a child too.

4   Hint Creation Algorithms

Heuristics for hint generation can vary from completely random selection to a purely deterministic selection algorithm.

4.1   Avoiding Failing Branches

In this section, we present a deterministic heuristic for hint creation based on the restart strategy and conflicts. We call this heuristic *Avoiding Failing Branches* (AFB). The idea is to track the most conflict-active decisions in the explored branch and add a hint that explicitly prevents choosing that set again. If a restart took place, it is reasonable to assume heuristically that the last explored branch is less likely to contain the satisfying assignment.

In AFB, we keep an array of variable activity to determine the most conflict-active decisions. When the solver encounters a conflict, we update the scores of all variables responsible for the conflict. We will explain what "responsible" means shortly.

The decision to add hints is taken upon backtracking. If the backtracking is actually a restart, then the most active literals are chosen to participate in a hint, which is added right after the restart. Because the literals are added in

Algorithm 3 Update the score of a variable after a conflict. The score is updated for all decision variables in the first UIP, and for all variables in the reason clause for non-decision variables.

```
1:  Analyze() {
2:  ···
3:  U := ComputeFirstUip()
4:  ···
5:  for each ℓ ∈ U do
6:      v := Var(ℓ)                                      ▷ v is the variable of ℓ
7:      if DecisionVariable(v) then
8:          variableScores[v] := variableScores[v] + 1
9:      else
10:         c_v := Reason(v)
11:         for each ℓ' ∈ c_v do
12:             v' := Var(ℓ')
13:             variableScores[v'] := variableScores[v'] + 1
14: ...
15: }
```

their negated form, all explored branches containing the set of literals in the hint will not be re-explored.

In Alg. 3, which is implemented within the function *Analyze()* of MINISAT 2.2 [4], we update the score of the variables participating in a conflict. For this purpose, we keep an array of variables (*variableScores*), which is updated for all literals that are in the first UIP (U) computed in *ComputeFirstUip()*. We then iterate all variables $v \in U$. If v is a decision variable (*DecisionVariable(v)*), we increment its score by one. Otherwise, we take the reason for v being assigned ($c_v := Reason(v)$) and increment the score for all variables in c_v.

In Fig. 1, the first UIP node is $U = v_{10} \lor \neg v_4 \lor v_{11}$. Alg. 3 will first compute U and iterate through all its literals. The scores of decision variables v_{10}, v_{11} are increased in line 8; v_4 is not a decision variable, so its reason, c_3, is computed in line 10. The score of variables v_2, v_3, v_4 is increased in line 13.

The hints are added in the function *CancelUntil()* of MINISAT 2.2 [4]. If a restart is decided upon, we use the information acquired by Alg. 3 to choose the most active literals to participate in the hint. A literal ℓ is chosen to participate if *variableScores[Var(ℓ)]* is greater than some threshold θ. The integer *conflict* is the number of conflicts since *Solve()* was called. Three magic numbers, $\alpha \in [0..1]$, $x \in \mathbb{N}$, $y \in \mathbb{N}$, also appear in Alg. 4. They are used in the following fashion:

1. A literal ℓ is added to the hint if *variableScores[Var(ℓ)]* $> \alpha \times conflicts = \theta$.
2. We observed that, as time passes, it's advisable to increase θ.
3. Parameter x was added as a minimal threshold to prevent adding hints too "quickly". The idea is to prevent hints from being used when easy instances are solved.
4. Parameter y is used to ensure that new hints are not too small. Small hints can be too influential in the search procedure.

Algorithm 4 AFB hint addition – Adds a hint built of all negated literals with a score exceeding θ.

1: $CancelUntil(backtrackLevel)\{$
2: **if** $backtrackLevel > 0$ **then**   ▷ This is not a restart
3:     Performing backtracking until $backtrackLevel$...
4:         Upon freeing variable v:
5:             $variableScores[v] := 0$
6:         ...
7: **else**   ▷ This is a restart
8:     **if** $conflicts > x$ **then**
9:         **for** each **decision** variable v with decision ℓ **do**
10:             **if** $variableScores[v] > \alpha \times conflicts$ **then**
11:                 $hint.Push(\neg l)$
12:     **for** each variable v with decision ℓ **do**
13:         $variableScores[v] := 0$
14:     Perform backtracking until $backtrackLevel = 0$...
15:     **if** $hint.Size() > y$ **then**
16:         $AddClause(hint)$
17:     $hint.Clear()$
18: ...
19: $\}$

We maintain a vector of literals, *hint*, to store the clause that might form the future hint. Function $AddClause()$ adds the hint to the input instance.

4.2   Randomized Hints

We introduce next a completely random selection algorithm for hint creation, based on random assignments and satisfiability checking. We call this heuristic *Randomize Hints* (RH). In this algorithm, we use random assignments to see if we can learn literals that are likely untrue, that is, if chosen, a conflict is reached. We add these literals to form a new hint, that will hopefully lead the solver to an unsatisfiable conclusion. This hint is then negated, and the explored search space is reduced.

The randomized hint is created before HSAT is called. First, k random assignments are drawn, each with uniform distribution over $\{0,1\}^n$. These assignments are then checked on every clause. If some clause is unsatisfied, we bump the grade of all literals in the clause. We keep a vector of grades, *literalsGrades()*, and track the maximal graded literals that will be chosen to participate in the hint. We encourage the solver to pick the literals of the hint as decisions by increasing the activity of the variables involved in MINISAT's $VarBumpActivity(v)$.

The following functions and variables are used in Alg. 5 for random hints:

1. $DrawRandomAssignments(num)$ creates num random assignments.
2. $ClauseSatisfied(c, \sigma)$ returns true iff $\sigma(c) = $ TRUE.
3. $PopMax()$ returns and removes the literal with the highest score.

Algorithm 5 Create randomized hints – draws random assignments and boosts score for all literals in a clause unsatisfied by an assignment. The literals with the highest scores are chosen to participate in hints.

Input: *sizeOfHint* – Size of the hint
Input: *assignments* – Number of assignments to
      draw

1: *DrawRandomAssignments*(*assignments*)
2: **for** each Assignment σ **do**
3:     **for** each Clause c **do**
4:         **if** $\neg ClauseSatisfied(c, \sigma)$ **then**
5:             **for** each literal $\ell \in c$ **do**
6:                 *literalsGrades*[l] := *literalsGrades*[l] + 1
7:                 *VarBumpActivity*(*Var*(ℓ))
8: **for** $i \in [0..sizeOfHint - 1]$ **do**
9:     *hint*[i] := *literalsGrades*.*PopMax*()
10: *AddClause*(*hint*)
11: *hint*.*Clear*()

---

5   Theoretical Basis

For completeness, a few observations are in place, which should serve to convince readers that correctness is being maintained.

Proposition 1. *For any formula* φ, *a set of hints* H *and assignment* $\sigma : V \to \{0,1\}$ *of truth values to the variables of* $\varphi \wedge H$,

$$\sigma(\varphi \wedge H) \Rightarrow \sigma(\varphi).$$

Proposition 2. *For any formula* φ *and set of hints* H,

$$\varphi \wedge H \in UNSAT \Rightarrow \varphi \wedge \neg H \equiv \varphi.$$

By $\neg H$, we mean $\bigvee_{h \in H} \neg h$.

Proof. If $\varphi \wedge H \in UNSAT$, then $\neg(\varphi \wedge H)$, which is equivalent to $\varphi \Rightarrow \neg H$. ∎

From Proposition 2, we establish the following:

Proposition 3. *Given* $\varphi \wedge H \in UNSAT$ *and* $|H| = 1$ *where* $h = \ell_1 \vee \ell_2 \cdots \vee \ell_k$

$$\varphi \wedge \neg \ell_1 \wedge \neg \ell_2 \wedge \cdots \wedge \neg \ell_k \equiv \varphi.$$

This observation is critical for HSAT. In this case, k new facts are learned, which helps reduce the fraction of the search space that gets explored.

As mentioned earlier, this idea can be generalized to include all dominators.

Theorem 1. *If* $\varphi \wedge H$ *is unsatisfiable, then* $\varphi \equiv \varphi \wedge \neg D$ *for every* $D \in Dominator_{RC}$.

Proof. Since $D \in Dominator_{RC}$, it is sufficient to prove $\varphi \wedge D \in UNSAT$. By Proposition 2, $\varphi \equiv \varphi \wedge \neg D$. ∎

6 Experimental Results

6.1 AFB Results: SAT 2013

We compare now the performance of HSAT, with and without heuristic AFB. We find that hints have a positive effect for satisfiable instances but cause a moderate degradation for unsatisfiable ones. The positive results for satisfiable instances are in line with our presumption that, if a restart takes place, it is heuristically likelier that the satisfying assignment to the problem lies on another branch.

We ran over 150 *satisfiable* instances from SAT 2013, but the results reported below refer only to the 113 that were fully solved by at least one solver within half an hour. All of the instances are publicly available at [1]. We implemented all the algorithms in BASE [14], which is built on top of MINISAT 2.2 [4]. The code is public and available at [13]. For the experiments, we used machines running Intel® Xeon® processors with 3Ghz CPU frequency and 32GB of memory.

Table 1. AFB performance results for SAT 2013: satisfiable (left) and unsatisfiable (right) instances. Run-time is in minutes.

SAT	BASE	AFB
Run-time	990	**697**
Unsolved (by one)	9	**4**

UNSAT	BASE	AFB
Run-time	**727**	779
Unsolved (by one)	**2**	3

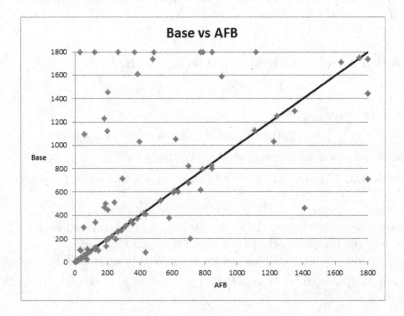

Fig. 3. Comparing AFB to BASE.

Table 1 displays a 30% improvement in overall runtime for satisfiable instances. Furthermore, there are 9 instances solved by AFB that are not solved by the base solver, compared to 4 instances solved by BASE but not by AFB.

In addition, 130 *unsatisfiable* instances from SAT 2013 were tested; the reported results refer only to the 60 that were fully solved by at least one solver within the 30-minute time limit. Table 1 shows a 7% degradation in overall runtime for unsatisfiable instances.

Figure 3 presents BASE vs. AFB. The diagonal $y = x$ emphasizes the superiority of AFB. Figure 4 presents the time differential between BASE and AFB. On average, AFB solves one of these problem instances $2\frac{1}{2}$ minutes faster than the baseline. The graphs refer to satisfiable instances only.

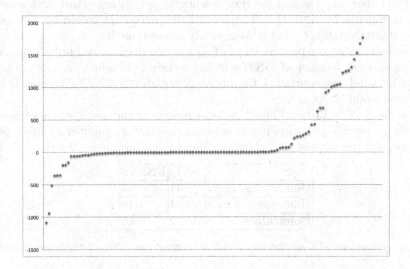

Fig. 4. The time difference (in seconds) between BASE and AFB

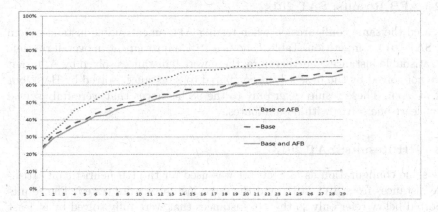

Fig. 5. Comparing percentage of instances solved by BASE and AFB.

Table 2. AFB performance results for SAT 2014: satisfiable instances. Run-time is in minutes.

SAT	BASE	AFB
Run-time	833	**681**
Unsolved (by one)	9	**2**

Figure 5 shows three curves, plotted at one minute intervals. The lower curve (A) is the percentage of instances solved by BASE and AFB both; the middle (B) is the percentage of instances solved by BASE the upper (C) is the percentage solved by either one. The gap $B - A$ represents the percentage of instances solved by BASE but not by AFB; $C - B$ represents the percentage solved by AFB but not by BASE. Notably, $C - B$ is consistently larger than $B - A$.

We observe that the positive effect of AFB is due to successful branch cutting by hints and not because of HSAT's ability to negate dominators. Most of the hints added did not contradict the instance, so HSAT's UNSAT core abilities were not helpful in AFB.

For the SAT 2013 benchmark, we also measured the average size of hints (number of literals participating in a hint), the average number of hints per instance, and the number of dominators found in all instances:

	SAT	UNSAT
Hint average size	34	43
Hints per instance	0.84	1.16
Dominators	2	15

Hints were used in 34% of the satisfiable instances and 39% of the unsatisfiable cases.

6.2   AFB Results: SAT 2014

We used the same configuration when testing AFB on satisfiable instances from the SAT 2014 competition. Table 2 shows a 19% improvement in overall runtime for satisfiable instances. Furthermore, there were 9 instances solved by AFB that were not solved by the base solver, compared to 2 instances solved by BASE but not by AFB. These results refer only to the 98 instances that were fully solved by at least one solver within 30 minutes.

6.3   RH Results: SAT 2013

The same configuration as in Sect. 6.1 was used for the RH heuristic on satisfiable instances from SAT 2013, and the same instances were tested. The results reported below refer only to the 116 instances that were fully solved by at least one solver within 30 minutes. Table 3 shows a 8% improvement in overall runtime for satisfiable instances.

Table 3. RH performance results: satisfiable (left) and unsatisfiable (right) instances. Run-time is in minutes.

SAT	BASE	RH
Run-time	1080	**988**
Unsolved (by one)	12	12

UNSAT	BASE	RH
Run-time	**757**	888
Unsolved (by one)	**3**	9

We were admittedly surprised to see that satisfiable instances were solved faster because of "good" hints, hints that do not contradict the input. We were surprised because we tried to build hints that would contradict the input and have the negation of dominators drive the solution.

In addition, 130 *unsatisfiable* instances from SAT 2013 were tested; the results below refer only to the 60 that were fully solved by at least one solver within the 30-minute time limit. Table 3 shows a 15% degradation in overall runtime for unsatisfiable instances.

Combining the two heuristics, AFB and RH, as though they would run in parallel for half an hour on the SAT 2013 benchmark, we obtain 16 SAT instances that are solved for which BASE times out, versus 2 that only BASE solves, and 5 UNSAT instances that BASE fails on, versus 3 only by BASE.

7   Discussion

We have introduced a new paradigm and platform, called HSAT, with which one can speed up SAT solving by means of added clauses. It enables the addition of "hint" clauses that are not necessarily derivable from the original formula but which can nevertheless help the solver reach a solution faster. HSAT avoids the addition of new literals, using instead a partial resolution graph to keep track of the effect of hints. We have seen that the AFB hint heuristic, which causes the prover to avoid retaking the most conflict-active decisions, outperforms the (hintless) baseline system and introduces a significant improvement in the solver core. On a benchmark of 280 instances, 150 of which are satisfiable: AFB achieved a 30% runtime improvement over the baseline and solved 9 instances not solved by the baseline prover.

Though these results are very encouraging, we have reason to believe that future work can lead to further improvements. For example, we tried to increment conflict decision variable scores by an amount that is inversely proportional to its depth in the proof tree, so those closer to the root (which have greater impact) get greater weight. This approach did not work for the thresholds we looked at, but might work for others. Another example is that our hint heuristics do not work well for unsatisfiable instances, the main reason being that there are usually no dominator clauses, in which case unsatisfiability does not drive the subsequent search very well. In this case, the incremental running of Alg. 1 just adds overhead. An interesting avenue for research would be to design hints that create multiple dominators or that lead the solver to a contradiction faster.

There are an endless number of ways to create hints, and many places in the process to add them; so far we have only explored a few options. It is likely that there remain even more interesting ways to create good hints for satisfiable instances and, hopefully, for unsatisfiable ones, too.

Acknowledgments. The suggestion to exploit hints originated with Ofer Strichman. The authors would like to thank Alex Nadel and Ofer Strichman for their advice and reading of a draft of this paper. This work is part of the first author's M.Sc. thesis at Tel Aviv University.

References

1. SAT competition (2013). http://satcompetition.org/2013/downloads.shtml
2. Beame, P., Kautz, H.A., Sabharwal, A.: Towards understanding and harnessing the potential of clause learning. CoRR abs/1107.0044 (2011). http://arxiv.org/abs/1107.0044
3. Biere, A., Heule, M.J.H., van Maaren, H., Walsh, T. (eds.): Handbook of Satisfiability, Frontiers in Artificial Intelligence and Applications, vol. 185. IOS Press, February 2009
4. Eén, N., Sörensson, N.: An extensible SAT-solver. In: Giunchiglia, E., Tacchella, A. (eds.) SAT 2003. LNCS, vol. 2919, pp. 502–518. Springer, Heidelberg (2004)
5. Eén, N., Sörensson, N.: Temporal induction by incremental SAT solving. Electr. Notes Theor. Comput. Sci. **89**(4), 543–560 (2003). http://dx.doi.org/10.1016/S1571-0661(05)82542–3
6. Hyvärinen, A.E.J., Junttila, T., Niemelä, I.: Grid-based SAT solving with iterative partitioning and clause learning. In: Lee, J. (ed.) CP 2011. LNCS, vol. 6876, pp. 385–399. Springer, Heidelberg (2011). http://dx.doi.org/10.1007/978-3-642-23786-7_30
7. Lanti, D., Manthey, N.: Sharing information in parallel search with search space partitioning. In: Nicosia, G., Pardalos, P. (eds.) LION 7. LNCS, vol. 7997, pp. 52–58. Springer, Heidelberg (2013). http://dx.doi.org/10.1007/978-3-642-44973-4_6
8. Moskewicz, M.W., Madigan, C.F., Zhao, Y., Zhang, L., Malik, S.: Chaff: engineering an efficient SAT solver. In: Proceedings of the 38th Design Automation Conference (DAC), Las Vegas, NV, pp. 530–535. ACM, June 2001. http://doi.acm.org/10.1145/378239.379017
9. Nadel, A.: Boosting minimal unsatisfiable core extraction. In: Bloem, R., Sharygina, N. (eds.) Proceedings of 10th International Conference on Formal Methods in Computer-Aided Design (FMCAD), Lugano, Switzerland, pp. 221–229. IEEE, October 2010. http://ieeexplore.ieee.org/xpls/abs_all.jsp?arnumber=5770953
10. Nadel, A., Ryvchin, V., Strichman, O.: Efficient MUS extraction with resolution. In: Formal Methods in Computer-Aided Design (FMCAD), Portland, OR, pp. 197–200. IEEE, October 2013. http://ieeexplore.ieee.org/xpl/freeabs_all.jsp?arnumber=6679410
11. Nadel, A., Ryvchin, V., Strichman, O.: Accelerated deletion-based extraction of minimal unsatisfiable cores. JSAT **9**, 27–51 (2014). https://satassociation.org/jsat/index.php/jsat/article/view/116
12. Oh, Y., Mneimneh, M.N., Andraus, Z.S., Sakallah, K.A., Markov, I.L.: AMUSE: a minimally-unsatisfiable subformula extractor. In: Malik, S., Fix, L., Kahng, A.B. (eds.) Proceedings of the 41st Design Automation Conference (DAC), San Diego, CA, pp. 518–523. ACM, June 2004. http://doi.acm.org/10.1145/996566.996710

13. Ryvchin,   V.:   HaifaMUC.   https://www.dropbox.com/s/uhxeps7atrac82d/ Haifa-MUC.7z
14. Ryvchin, V., Strichman, O.: Faster extraction of high-level minimal unsatisfiable cores. In: Sakallah, K.A., Simon, L. (eds.) SAT 2011. LNCS, vol. 6695, pp. 174–187. Springer, Heidelberg (2011)
15. Sakallah, K.A., Simon, L. (eds.): SAT 2011. LNCS, vol. 6695. Springer, Heidelberg (2011)
16. Sörensson, N., Biere, A.: Minimizing learned clauses. In: Kullmann, O. (ed.) SAT 2009. LNCS, vol. 5584, pp. 237–243. Springer, Heidelberg (2009). http://dx.doi.org/10.1007/978-3-642-02777-2_23
17. Strichman, O.: Accelerating bounded model checking of safety properties. Formal Methods in System Design **24**(1), 5–24 (2004). http://dx.doi.org/10.1023/B:FORM.0000004785.67232.f8

Mining Backbone Literals in Incremental SAT
A New Kind of Incremental Data

Alexander Ivrii[1], Vadim Ryvchin[2], and Ofer Strichman[3]([⊠])

[1] IBM Research Lab, Haifa, Israel
alexi@il.ibm.com
[2] Design Technology Solutions, Intel Co., Haifa, Israel
rvadim@tx.technion.ac.il
[3] Information Systems Engineering, IE, Technion, Haifa, Israel
ofers@ie.technion.ac.il

Abstract. In incremental SAT solving, information gained from previous similar instances has so far been limited to learned clauses that are still relevant, and heuristic information such as activity weights and scores. In most settings in which incremental satisfiability is applied, many of the instances along the sequence of formulas being solved are unsatisfiable. We show that in such cases, with a P-time analysis of the proof, we can compute a set of literals that are logically implied by the next instance. By adding those literals as assumptions, we accelerate the search.

1 Introduction

Incremental SAT solving is used in numerous applications, including Bounded Model Checking [5], SMT solving [10], unsat core extraction [7], high-level (group) UNSAT core extraction [13], and model-checking via IC3 [6]. In all these applications a sequence of closely related SAT formulas is being solved, and by saying that they are solved incrementally we mean that the SAT solver retains information between subsequent calls, in order to expedite the overall process. Originally there was *clause sharing* [16–18], which means that learnt clauses at step i, which are still relevant for step $i + 1$, are retained. As of MiniSat [8] most competitive solvers support *assumptions*, which enables them to support a far more general incrementality mechanism, which not only retains relevant learned clauses, but also heuristic information such as literal scores and variable activity. Assuming that consecutive instances are similar, this information saves time. Furthermore, only the added clauses and the assumptions have to be communicated to the solver at each iteration, thus saving time on parsing and reloading the formula to memory.

In this article we show that whenever an instance ψ in the sequence of solved formulas is proven to be unsatisfiable, we can extract from the proof constants that can be added to the next instance ψ', hence offering a new type of data transfer in incremental solving. Our context is a CDCL solver such as Minisat [8], which is augmented with in-memory proof logging, or is at least capable of

© Springer International Publishing Switzerland 2015
M. Heule and S. Weaver (Eds.): SAT 2015, LNCS 9340, pp. 88–103, 2015.
DOI: 10.1007/978-3-319-24318-4_8

producing a proof on demand. There is an overhead associated with maintaining the proof in memory, and generic SAT solvers typically refrain from doing it. In some applications such proof logging is necessary, however, like solvers used in interpolation-based model-checking [2,11], the minimal-core extraction tool HAIFAMUC [14,15], and the incremental solver described in [14]. For this work we chose to focus on HAIFAMUC, as it is open source. HAIFAMUC maintains in memory a *partial* proof, and more specifically the part of the proof that is rooted at clauses that will potentially be removed in future instances, which is all that we need for our technique.

In graph terms, a proof is a directed graph in which nodes represent clauses and edges represent the antecedent relation between them, i.e., the parents of a clause are its antecedents. When the empty clause is reachable from some roots of the graph we call the proof a refutation. The edges maintained by solvers, including HAIFAMUC, correspond to *hyper resolution* inference. Such an inference has multiple clauses $c_1 \ldots c_n$ as antecedents, and one clause c as a consequent, if and only if *there exists* a *binary* resolution proof of c from $c_1 \ldots c_n$. For simplicity we will refer to the nodes in the resolution proof simply as the clauses that they represent. For a root clause c, let $cone(c)$ denote the maximal reachable subgraph that is rooted at c. For a set of clauses C, we overload $cone$ by defining $cone(C) = \bigcup_{c \in C} cone(c)$.

Our technique exploits the following observation, which was originally made by Nadel in his thesis [12]:

Observation 1. *Let π be a refutation. Then every vertex cut in π represents an inconsistent set of clauses.*

Intuitively this observation is correct because every set of clauses that forms a cut (i.e., separates the roots from the empty clause), implies the empty clause, and hence must be unsatisfiable. A consequent of this observation is

Corollary 1. *Let π be a refutation, and let C be a set of root clauses in its core. Let vc be a vertex cut in $cone(C)$, i.e., vc separates C from the empty clause. Then*

$$\pi \setminus cone(C) \Rightarrow \bigvee_{cl \in vc} \neg cl . \tag{1}$$

To see why (1) is correct, observe first that $\pi \setminus cone(C)$ and $cone(C)$ represent a partition of π. Falsely assume, then, that there exists an assignment α such that $\alpha \models \pi \setminus cone(C) \wedge \bigwedge_{cl \in vc} cl$. Take any subset of clauses from $\pi \setminus cone(C)$ that together with vc form a cut in π. Since α satisfies both that subset and vc, then α satisfies a full cut in π, which contradicts Observation 1.

The cut vc is of course not unique in $cone(C)$. Let $Cuts(C)$, then, be the set of *all* cuts in $cone(C)$. We can now generalize (1) to

$$\pi \setminus cone(C) \Rightarrow \bigwedge_{vc \in Cuts(C)} \bigvee_{cl \in vc} \neg cl . \tag{2}$$

How is (2) relevant to faster incremental SAT solving? Consider the case that we have an unsatisfiable instance ψ accompanied with a refutation π, and in the next iteration we would like to remove a set of root clauses C and add another set of clauses C'. This means that we now need to check

$$\psi' \equiv C' \wedge (\pi \setminus cone(C)) . \tag{3}$$

According to (2) we can check instead the equivalent formula

$$\psi' \wedge \bigwedge_{vc \in Cuts(C)} \bigvee_{cl \in vc} \neg cl . \tag{4}$$

For simplicity of the discussion we will ignore from hereon the added clauses C', since they do not affect the correctness argument.

There is no a priory reason to believe that checking (4) is easier than checking ψ' itself, however. First, the set $Cuts$ can be exponential in size and generally cannot be computed efficiently; second, even if we have this set or part thereof, adding a disjunction of terms which is, while being logically implied by the formula, not emanating from the search itself, is not likely to accelerate the search. But suppose that instead of considering all cuts, we only consider *singleton* cuts, i.e., those that include a single clause. In such a case (4) amounts to adding the negation of those clauses, or in other words adding constants to the formula, which *are* likely to accelerate the search.

In the case of deletion-based MUC extraction, it is rather easy to find singleton cuts, because it is always the case that $|C| = 1$: each iteration corresponds to removing a single candidate clause and checking the satisfiability of the remaining formula. This property is used in a technique called *path strengthening* (PS) [14], which helps finding minimal unsatisfiable cores faster. It is illustrated in Fig. 1.

To see how PS finds those cuts, let c be the removed clause, i.e. $C = c$. PS focuses on the subgraph of $cone(c)$ that leads to the empty clause:

Definition 1 (Rhombus). *Let c be a root clause in the core of a refutation π. Then the rhombus of c, denoted $\Diamond_\pi(c)$, is the clauses in $cone(c)$ that are on some path from c to the empty clause $\bot$.*

We overload $\Diamond_\pi$ to a set of clauses in the natural way: $\Diamond_\pi(C) = \bigcup_{c \in C} \Diamond_\pi(c)$.

PS finds in linear time a maximal 'chain' of clauses $c_0, c_1, \ldots, c_n$ in $\Diamond_\pi(c_0)$ such that $c_0 \equiv c$, for $0 \leq i < n$, c_i is a parent of c_{i+1}, and, finally, c_n is the only clause in this chain that has multiple children in $\Diamond_\pi(c_0)$. Note that this means that each of them is a *dominator* in $\Diamond_\pi(c_0)$ with respect to the empty clause. It then checks $\pi \setminus cone(c)$ under the assumptions $\{\neg l \mid \exists i \in [1..n]. \ l \in c_i\}$. These extra assumptions, empirically, accelerate the search [14]. PS can be seen as an extension of *redundancy removal*, which was used in the MUC-finder MUSER2 [3], that only uses $\neg c_0$ as assumptions.

Whereas PS is based on a sequence of clauses $c_0, \ldots, c_n$ as explained above, in this article we show that there are many more literals that can be easily found and assumed in the next instance. In particular, we observe that

Observation 2. *The negation of every literal that appears in every path from C to the empty clause is implied by ψ'.*

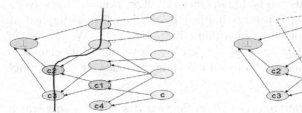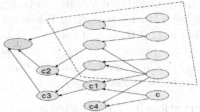

Fig. 1. Demonstrating path strengthening (PS) [14]. Left: given a refutation π, every vertex cut corresponds to a set of clauses that must be contradictory, since together they imply the empty clause. Right: consequently, when removing a clause and its cone ($cone(c) = \{c, c1..c4, \bot\}$), the rest of the formula, which here is marked with a dashed polygon, cannot be satisfied with a model that also satisfies a vertex cut in $\Diamond_\pi(c) = \{c, c1..c3, \bot\}$. Each of c and $c1$ constitute a (singleton) cut in $\Diamond_\pi(c)$. PS exploits this property, and adds their negation as assumptions when solving $\psi' \equiv \pi \setminus cone(c)$. Identifying such a chain of clauses takes linear time.

To see why, consider such a literal l and the set of clauses $S = (l \vee A_1), (l \vee A_2), \ldots$ in $\Diamond_\pi(C)$ that it appears in, where $A_1, A_2, \ldots$ are disjunctions of literals. By definition, $S \in Cuts(C)$. Then according to (2), we have

$$\psi' \Rightarrow \bigvee_{cl \in S} \neg cl \tag{5}$$

or equivalently

$$\psi' \Rightarrow (\neg l \wedge \neg A_1) \vee (\neg l \wedge \neg A_2) \vee \ldots , \tag{6}$$

from which we can conclude

$$\psi' \Rightarrow \neg l \wedge (\neg A_1 \vee \neg A_2 \vee \ldots) \tag{7}$$

and then

$$\psi' \Rightarrow \neg l . \tag{8}$$

Hence, $\neg l$ is implied by ψ', which means that we can add it as an assumption when solving ψ'. We will show in Sect. 2 a P-time algorithm for detecting such literals, and analyze its complexity. We call this extended technique *Mining Backbone Literals*, or MBL for short. The term *backbone literals* of a formula appeared in the past in multiple contexts such as phase transition, maxSAT and optimization problems (see a recent survey in [9]). It is defined as literals that are satisfied in all models of the formula. Generally deciding whether a literal has this property is NP-complete [9]), which underlines the benefit we gain from analyzing the proof of ψ. Note that in contrast to all the prior works mentioned in the above survey, we are not interested in finding the complete set of such literals.

Finally, even more literals can be found efficiently if we consider the solver's *state*. Let Con be the set of *constants* implied by ψ', i.e., literals implied at decision level 0 (those, of course, are backbone literals as well). We say that a path in the resolution graph is *satisfied* by Con if at least one of its clauses is satisfied by Con. More literals can be found based on the following observation, which extends Observation 2:

Observation 3. *The negation of every literal that appears in every path which is not satisfied by Con from C to the empty clause is implied by ψ'.*

A special case is when all paths from C to the empty clause are satisfied by Con. Then $\psi' \Rightarrow false$ and we can immediately declare ψ' to be unsatisfiable. We will prove the correctness of this technique, which we call MBL-inline, or MBL_i for short, and discuss variations thereof in Sec. 2.2.

Note that PS as described earlier finds only a subset of what MBL (and MBL_i) finds, because any literal in the clauses $c_0, \dots, c_n$ that PS finds has the property mentioned in Observation 2: it appears in every path from $C \equiv c_0$ to the empty clause. Furthermore, in contrast to PS which is only relevant when $|C| = 1$, MBL addresses the general case, where C is arbitrary, and hence is relevant for general incremental SAT solving.

A Preview of Our Empirical Results. As a case study we experimented with this technique in the context of finding minimal unsatisfiable cores (MUC) based on clause-deletion. Our new technique MBL finds on average five times more assumption literals comparing to PS, and does so in a negligible amount of time.

Having PS as our base-line, MBL reduces the run-time of MUC extraction by 6%-7% on average only, depending on the benchmark set. With additional optimizations targeted on MUC extraction that increase the benefit of finding these extra literals, we reach an improvement of 10%. We will describe those optimizations in Sec. 3. This is, admittedly, a modest improvement. Our analysis of the data reveals two possible reasons for this.

First, our data shows that extra assumptions have diminishing value, since at some point many of them are already implied by previous assumptions and constants (constants are literals implied at decision level 0), and are therefore redundant. The ratio of implied assumptions depends on the order in which they are given to the solver, but since this order is rather arbitrary in our case, we measured the redundancy with

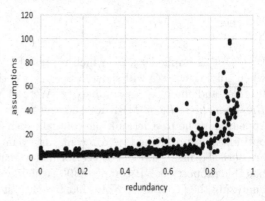

Fig. 2. Assumptions have a diminishing value.

the default ordering. In our benchmark sets redundancy varies a lot between instances. In the SAT'11 and SAT02$_\beta$ benchmark sets (each with hundreds of instances) we measured $\approx 40\%$ and $\approx 22\%$ redundancy on average, respectively. We witnessed a positive correlation between the number of assumptions and the redundancy ratio. The scatter graph in Fig. 2 shows this data for the SAT02$_\beta$ benchmark set. One can observe that when the number of assumptions per SAT call is high, the redundancy also goes up, and approaches 1.

Second, there are particular characteristics of deletion-based MUC extraction, which make it vulnerable, in terms of run-time, to assumptions. We will discuss those in detail in Sect. 3. These characteristics are not present in other known popular domains of incremental satisfiability, such as bounded model checking, so it is reasonable to assume that a larger benefit will be observed in such domains.

In the next section we present an algorithm for computing MBL and a variation thereof called MBL$_i$. In Sect. 3 we will describe in some detail our case study of MUC extraction: how it is applied and what did we learn from our experiments in this domain.

2   Mining Backbone Literals

As before let C denote the set of root clauses that are removed when progressing from ψ to ψ'. We assume that ψ is unsatisfiable, and that C is in the core of the refutation π. In our implementation each node n in $\Diamond_\pi(C)$ is a structure with four fields:

- n.clause
- n.LitSet — a set of literals that are present in all paths from C to n. Initialized to $\emptyset$;
- n.NumOfParents — the number of parents n has in $\Diamond_\pi(C)$. Initialized to 0.
- n.visited — whether n was visited before in the traversal. Initialized to false.

Alg. 1 presents MBL-GET, which computes the MBL literals — those that are present on each path from the root nodes C to the empty clause. In the end of the algorithm, the following relation holds for each node n in $\Diamond_\pi(C)$:

$$\text{n.LitSet} = \left(\bigcap_{\text{p} \in parents(\text{n})} \text{p.LitSet} \right) \cup \text{n.clause} . \tag{9}$$

In words, the literals that appear on every path from C to n, are those that appear on every path to each of its parents, and the literals in the clause of n itself.

MBL-GET begins by calling COUNTPARENTS, which simply updates n.NumOfParents with the number of parents n has inside the $\Diamond_\pi(C)$. The function CHILDREN(P), which is called in line 6, returns the set of children p has in $\Diamond_\pi(C)$. MBL-GET maintains a queue of nodes, initialized in line 14 to the clauses in C. A node n enters this queue only after its n.LitSet was fully computed (in the case of a root node, its literal set is the clause itself, as can be seen in line 11). From (9) it is clear that we can compute this set only after such sets were computed for all of n's parents. This is why we need n.NumOfParents: we use it to guarantee that a node enters the queue only after all its parents were processed — see lines 24 and 26. We refer to the currently processed node as the parent, and denote it by p, as can be seen in line 16. We iterate through p's children, intersect their current literal sets with p.LitSet in line 23, and if p is the last parent of a child node n, then we add in line 25 n.clause (see right element of (9)) and enqueue n. The last node to be processed — see line 17 — is the empty clause. Its literal-set is the result of this procedure. The negation of each of these literals can be added to ψ', the next formula to be solved, without changing its satisfiability.

Algorithm 1. COUNTPARENTS updates n.NumOfParents for each node n with the number of parents in $\Diamond_\pi(C)$. MBL-GET computes the set of literals that are on every path from C to the empty clause. Those are the literals that are present in each clause along some vertex cut of $\Diamond_\pi(C)$.

```
 1: function COUNTPARENTS(nodeSet C)
 2:     NodeQueue Q;                                        ▷ A queue of nodes
 3:     Q.enqueue(C);
 4:     while Q is not empty do
 5:         p = Q.dequeue();
 6:         for each node n in children(p) do
 7:             n.NumOfParents++;
 8:             if n.NumOfParents = 1 then Q.enqueue(n); ▷ Ensures n enters Q once


 9: function MBL-GET(nodeSet C)
10:     CountParents(C);
11:     for each c in C do
12:         c.LitSet = c.Clause;
13:     NodeQueue Q;                                        ▷ A queue of nodes
14:     Q.enqueue(C);
15:     while Q is not empty do
16:         p = Q.dequeue();
17:         if p.clause.isEmpty() then return p.LitSet;
18:         for each node n in children(p) do
19:             n.NumOfParents--;
20:             if !n.visited then
21:                 n.LitSet = p.LitSet;
22:                 n.visited = true;
23:             else n.LitSet.intersect(p.LitSet);
24:             if n.NumOfParents = 0 then                 ▷ All parents handled
25:                 n.LitSet.union(n.clause);
26:                 Q.enqueue(n);
```

Complexity. Let N, E be the number of nodes and edges in $\diamondsuit_\pi(C)$, respectively, and let K be the size of the largest clause in $\diamondsuit_\pi(C)$. The maximal size of the literal set is the number of different literals in $\diamondsuit_\pi(C)$, which we will denote by L. Each of the clauses have to be sorted for the union and intersection operations (lines 23 and 25), which takes $O(N(K \log K))$. Intersection takes not more than $O(L)$, and there are $O(E)$ intersections. The overall complexity is thus $E \cdot L + N \cdot K \log K$.

Example 1. Fig. 3 shows a possible $\diamondsuit_\pi(C)$ for C having a single clause $c = (1\ 2\ \text{-}3)$ on the left, and the corresponding n.LitSet computed for each node n by MBL-GET on the right. Note that $\diamondsuit_\pi(C)$ is part of a hyper-resolution proof, hence the internal nodes may have additional incoming edges from outside of $\diamondsuit_\pi(C)$, which are not shown in the figure. The algorithm returns n.LitSet for n being the empty clause in line 17, namely the literal set (1 2 -3 5) in this case. These are the literals that appear in all paths from C to the empty clause. In case of $|C| = 1$, as it is in this example, the output always includes the literals of the root clause itself. Here the prefix of clauses under c that has a single parent in the cone of c is just c itself. Hence with PS we would return in this case (1 2 -3) only, while missing the literal 5.

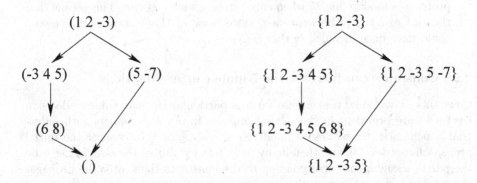

Fig. 3. Left: a partial (hyper) resolution graph corresponding to $\diamondsuit_\pi(1\ 2\ \text{-}3)$. Right: At each node n, showing n.LitSet at the end of MBL-GET.

2.1   Optimizations and Implementation Details

Some additional details about how we implemented MBL-GET in practice follow:

- **Optimization I.** In case $|C| = 1$, we collect the literals in the initial chain of clauses (the same chain that was used in PS and was described in Sect. 1) — these are known to be part of the end result. Then, rather than propagating them down the resolution graph, we just save them and add them to the set of literals that is eventually returned by MBL-GET. In the example resolution graph of Fig. 3, this optimization amounts to *not* propagating (1 2 -3) down the graph, rather adding these literals only at the very end, hence saving some of the cost of intersection in line 23. Since the overall algorithm is linear in the size of the graph, the importance of this optimization is admittedly marginal.

- **Optimization II.** Cutoff values: we collected statistics that show that when the width of $\diamondsuit_\pi(C)$ is large, only few MBL literals, if at all, are detected; furthermore, the wider $\diamondsuit_\pi(C)$ is, the longer it takes to compute the literals[1]. We therefore terminate early the computation if either the number of children of C in $\diamondsuit_\pi(C)$ crosses a certain threshold Th_1, or the width of $\diamondsuit_\pi(C)$ crosses another threshold Th_2 (i.e., the size of the queue in line 4). Our implementation currently sets these values by default to $Th_1 = 400$ and $Th_2 = 500$, based on (limited) experiments. If one of the thresholds is crossed we revert to PS (we can do it in our case because, recall, for MUC extraction it is always the case that $|C| = 1$. In the general case the fallback solution can be to not report assumptions to the next instance).

- **Handling Deleted Clauses.** The description of MBL-GET so far ignored the fact that not all clauses are available along the resolution graph due to clause deletion. Our solver maintains the graph itself (the arcs), but the clauses themselves are possibly deleted from memory. Our solution to this problem is to skip line 25 when encountering such a clause. This means that the end result may be weakened, since some of the literals in that clause could have been included in that set.

2.2   Constants can Increase the Number of Assumptions

Constants — variables that are forced to a particular Boolean value at decision level 0 — are prevalent in the solving process. In our experiments with industrial benchmarks we witnessed hundreds of constants already after the initial propagation at level 0, and then many more learned during the search. Let σ be the partial assignment corresponding to the constants right after the propagation at decision level 0 (note that we are not restricting ourselves to the initial propagation; the solver may backtrack to level 0 many times during the search). By definition of constants, any assignment that satisfies the formula, if such an assignment exists, is an extension of σ. This fact can be used to increase the set of MBL literals as hinted in the introduction (see Observation 3).

In the traversal of $\diamondsuit_\pi(C)$ by MBL-GET, we ignore the set of clauses $\{c' \mid \sigma \models c'\}$, i.e., the set of clauses that are satisfied by the constants. This effectively 'cuts' paths to the empty clause, and hence potentially increases the set of MBL literals, because we intersect the literal sets in line 23 of less clauses. To apply this optimization, right after line 18 we check if `n.clause` is satisfied by σ, and if yes then we `continue`, i.e., jump to the next iteration of the `for` loop.

---

[1] It was shown in [4] that wide proofs are necessarily long, although there the reference was to resolution and not hyper resolution. If we consider the rank of each clause in the hyper-resolution proof to be bounded, then this result applies here as well.

Theorem 1. *Let l denote a literal in $\mathtt{n.LitSet}$, where $\mathtt{n}$ is the empty clause, in the revised MBL-GET described above. Then $\psi' \Rightarrow \neg l$.*

Proof. Observe that the set of clauses S in $\Diamond_\pi(C)$ that contain l, and the set S' of nodes that were ignored by the revised procedure as described above, form a cut in $\Diamond_\pi(C)$. Falsely assume the existence of an assignment α such that $\alpha \models \psi' \wedge l$. Then $\alpha \models \bigwedge_{cl \in S \cup S'} cl$, because it satisfies l and it coincides with σ on the constants. Take any subset of nodes in ψ' that together with $S \cup S'$ form a cut in π. Since that subset is also satisfied by α, then we have a cut in π that is satisfied by α, which contradicts Observation 1. $\qquad\square$

As an illustration, consider again Example 1, and suppose that -7 is a constant implied by the formula. Then the clause (5 -7) is satisfied and hence ignored, and then the literal set reaching the empty clause is larger: (1 2 -3 4 5 6 8).

Note that the MBL literals may imply other constants, hence this idea can be applied iteratively, until convergence. Empirically it happens quite often that with this process all paths to the empty clause are cut by constants, a case in which we immediately return UNSAT, since every assignment that satisfies ψ' also satisfies a cut in $\Diamond_\pi(C)$.

We implemented several variations of this basic idea, and let the user control them via flags. One flag controls whether it is indeed applied iteratively until convergence, or only once. Another flag determines whether to apply this in the beginning of the search after propagation at level 0, or, at the other extreme, every time the solver backtracks to level 0 with a new learned constant. In the experiments presented in Sec. 3.2, we refer to this technique with the first flag turned off and the second turned on, as *MBL-inline*, or MBL_i for short.

3   A Case Study: Using MBL in the Context of MUC Extraction

We implemented the ideas described in the previous section in HAIFAMUC [14, 15], a minimal unsat core extractor, which is based on resolution and hence fits our needs. In the following subsection we will describe the core algorithm of HAIFAMUC, which will help us later, in sect. 3.2, to explain the experimental results.

3.1   How HAIFAMUC Extracts Minimal Unsatisfiable Cores

HAIFAMUC is a deletion-based minimal unsatisfiable core extractor that maintains (parts of) the proof in memory. Algorithm 2 describes in pseudo-code a simplified version of its main loop (see [14] for the full version), still based on PS. For MBL simply replace the called function in line 19 with MBL-GET. The implementation of MBL_i is more complicated as it is intertwined in the search engine of the solver, so we will not present it here in pseudocode. The code is rather self-explanatory so let us only emphasize the role of the Boolean variable

LastProofOK. This variable is set to false if the last unsatisfiability proof relied on assumptions. In such a case we cannot perform PS, because we do not have a proof of the empty clause (we only have a proof of the negation of the assumptions). An actual proof of the empty clause is necessary because without it we cannot compute a path with the properties explained in Sect. 1. Examining the logs one can frequently see an unsat proof based on assumptions, and then a long sequence of SAT results, none of which can enjoy the benefit of PS. This leads us to:

Observation 4. *Reaching an unsat answer based on assumptions, while generally being faster than without them, has two drawbacks:*

1. *PS (and similarly MBL) cannot be used until another iteration results in a full proof, and*
2. *Only the candidate clause c is removed rather than everything outside the cone (compare line 7 to lines 10–12).*

This observation helps understanding the results that will be presented next.

Algorithm 2. The basic main algorithm of HAIFAMUC with PS. In the first iteration the condition in line 5 is true and in line 6 it is false.

Input: Unsat formula ψ.
Output: A MUC of ψ.

1: $assumptions = \emptyset$;
2: Mark all ψ's clauses as 'unknown';
3: **while** true **do**
4:     $\langle res, \pi \rangle := SAT(\psi, assumptions)$;
5:     **if** res = UNSAT **then**
6:         **if** $assumptions$ used in proof **then**
7:             Mark c as 'not in MUC';
8:             $LastProofOK$ = false;
9:         **else**
10:             **for** each root outside of $core(\pi)$ **do**
11:                 Mark root as 'not in MUC';
12:                 Remove $cone(root)$ from ψ;
13:             $LastProofOK$ = true;
14:     **else**   ▷ SAT
15:         Mark c as 'MUC';   ▷ c is now in core
16:         Add $TmpRemoved$ back to ψ;
17:     **if** no clause is marked as 'unknown' **then return** clauses marked as 'MUC';
18:     Select c from roots marked as 'unknown';
19:     **if** $LastProofOK$ **then** $assumptions$ = PATHSTRENGTHENING(c);
20:     **else** $assumptions = \neg c$;
21:     $TmpRemoved = cone(c)$;
22:     $\psi = \psi \setminus TmpRemoved$;

3.2 Experiments

Results with the 2011 MUC Competition Benchmarks. We took the 200 benchmarks of the 2011 MUC-extraction track[2] (the mus/ subdirectory in the archive), the only competition ever held in this track, and removed four of them that could not be solved by any of the techniques in 15 minutes. According to the remaining 196 benchmarks — see Fig. 4 (left) — there is 6% improvement in the run-time of MBL comparing to PS, our base line, whereas MBL_i has a 0.5% negative effect. When considering the run-time per benchmark family — see Fig. 4 (right) — it is evident that the degradation in results of MBL_i happens only in one family (the 'abstraction-refinement-intel' set)[3], whereas in the other families it either improves or has marginal effect on the results. Returning to the table in the left of the figure: 'literals' is the number of added assumption literals, 'iterations' is the number of iterations until a MUC is found, and 'lit/iter' is the ratio between them; 'unsat by assump.' is the number of iterations in which the solver returned unsat based on the assumptions; 'overhead' is the *total* run time spent on finding the literals (i.e., for all iterations); 'longest call' is the *longest* SAT call (in sec.) in every CNF instance (not including the initial run) of Alg. 2, and finally '% implied assump.' is the percentage of literals that their value was already implied by previous literals and constants in the formula by the time we tried to apply them.

Measure	PS	MBL	MBL_i
success	196	196	196
time	21.5	**20.3**	21.7
literals	2770.8	11377.2	9886.1
iterations	1180.6	1188.5	1203.7
lit / iter	2.3	9.6	8.2
unsat by assump.	226.5	232.4	236.2
decisions (M)	30.7	30.6	47.3
overhead	0.0	0.3	0.3
longest call	0.5	0.4	0.4
% implied assump.	0.4	0.5	0.5

Bench. Family	PS	MBL	MBL_i
fdmus-v100	16.7	16.8	16.7
abs-ref-intel	26.3	**22.3**	36.7
atpg	0.0	0.0	0.0
bmc-aerielogic	4.6	4.4	4.4
bmc-default	6.3	6.3	6.3
design-debugging	1.1	1.1	**1.0**
equivalence-checking	59.9	55.3	**43.3**
fpga-routing	58.2	**50.3**	50.8
hardware-verification	48.4	47.7	**44.9**
product-configuration	0.0	0.0	0.0
software-verification	63.1	**57.6**	64.9
Total	21.5	20.3	21.7

Fig. 4. Results with the 2011 MUC track benchmarks. Left: various measures. Right: Run-time (sec.) by family;

The number of assumption literals that we add per SAT invocation is 4.1X larger with MBL comparing to with PS. Interestingly it is only 3.5X for MBL_i, despite the fact that MBL_i searches more aggressively for such literals. The reason for this phenomenon is related to the first item of Observation 4: more

---

[2] Officially that track was called MUS, for minimal unsatisfiable set.

[3] In fact it happens because of one benchmark in this family – an extreme outlier.

assumptions increase the probability that the proof relies on them, which in turn shuts off the search for such assumptions in future iterations, until there is an unsat proof without assumptions. Indeed observe that 'unsat by assump.' is higher for MBL_i. Also observe that the number of iterations is higher with MBL_i, which relates to the second part of Observation 4.

Analysis of the run-time of MBL: according to the 'overhead' row, the total amount of time spent on computing the MBL literals is less than half a second per CNF instance. Dividing it by the number of iterations shows that it takes on average around $3 * 10^{-4}$ seconds per SAT call. Recall that we bound the search for MBL literals to prevent spending too much time in case the proof is large (see optimization II in Sect. 2), hence the theoretical exponential upper-bound on the size of the proof is irrelevant: the overhead is always small. Indeed examining the data further reveals that it takes less than 10^{-3} sec. in 95% of the cases and less than $2 * 10^{-2}$ seconds in all cases. For comparison, we analysed the run time of the solver itself. The average time of a SAT call (not including the initial call) in the case of MBL is 0.012 sec., which is about 40X longer than the time it takes to compute the MBL literals. In these instances the SAT run-time is particularly short[4]. Hence the relative overhead is expected to be smaller on harder instances.

Analysis of implied assumptions. We already mentioned in the introduction that more assumptions typically implies more redundancy, i.e., a larger portion of them are implied by other assumptions. To quantify the connection between these two figures, we computed their *Spearman correlation* ρ. This is a frequently-used measurement for checking *monotonicity* between two arrays, i.e., for two arrays $A1, A2$ of equal size, $\rho(A1, A2) = 1$ if and only if for all i, j, when $A1[i] > A1[j]$ then $A2[i] > A2[j]$; $\rho(A1, A2) = -1$ if the opposite relation holds, and $\rho(A1, A2) = 0$ if there is no relation between the two arrays. With the above benchmark set we computed $\rho = 0.57$, which shows a strong connection between these two figures.

Results with the SAT 2002_β Competition Benchmarks. The SAT 2002_β competition benchmark set contains over 500 CNF application instances. After removing those that are SAT and benchmarks that cannot be solved with any of our parameters within 15 minutes, we were left with 216 instances. The table in Fig. 5 presents these results. The 5 columns on the right repeat the results for those 52 benchmarks in the set that at least in one of the parameters sets, they impose a SAT call that takes over a second. The table shows a gain of 5% and 6% to MBL and MBL_i in the first set, respectively, and a gain of 3% and 7% to MBL and MBL_i in the second set, respectively. The increased benefit matches our intuition that extra assumptions can help more in hard SAT calls.

---

[4] This is expected for this benchmark set, because only easy CNF instances were selected for inclusion in this set to begin with.

Measure	all			longest > 1		
	PS	MBL	MBL_i	PS	MBL	MBL_i
success	215	215	215	52	52	52
time	62.3	59.3	**58.9**	264.5	258.1	**245.9**
literals	3321.9	5778.1	5925.3	9484.5	32960.1	31628.3
iterations	1032.4	1037.1	1044.4	3817.0	3909.9	3867.7
lit/iter	3.2	5.6	5.7	2.5	8.4	8.2
unsat by assump.	284	285.9	287	1296.1	1386.0	1331.42
decisions (M)	10.0	10.3	10.6	68.7	69.8	67.7
overhead	0.0	0.1	0.2	0.0	0.8	1.2
longest call	2.0	1.9	1.8	9.8	9.3	8.9
% implied lit.	0.1	0.1	0.1	0.0	0.2	0.2

Fig. 5. Results on the $sat02_\beta$ benchmarks, rounded.

3.3   Mitigating the Side-Effect of Assumptions on MUC Search

As described in Observation 4, in MUC extraction extra assumptions can have indirect negative impact on the run-time. We now describe several heuristics that we experimented with, to mitigate this undesired side-effect.

- *Delayed activation.* Activate assumptions (in any one of the three methods) and the search for additional assumptions (with MBL_i), only after running for a while without assumptions. If the solver is able to find a solution within this delay, we gain by having a proper proof.
- *Continue execution after detection of UNSAT by assumptions.* When reaching UNSAT based on the assumptions, ignore the fact that we know the result is unsat, and continue the search for a bounded amount of time (defined by a 'budget' R) with the hope that the SAT solver will detect UNSAT without using the assumptions. The value of the assumptions with such a strategy is limited to the SAT case, and as a fallback solution when it takes too much time to find a proof without them.

In our implementation both the delay and the budget R are defined in terms of the number of restarts. We experimented with two variation of the second optimization above: either reset R to 0 for every SAT call, or not. The latter makes sense in the context of MUC, because the earlier iterations, when they end with UNSAT, are able to remove large parts of the proof via an analysis of the core, hence we want to spend time for finding a proof without assumptions. Indeed in all the

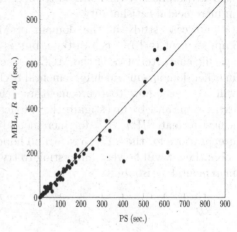

empirically winning configurations the latter option was selected. The table in Fig. 6 shows results of the best configurations that we found for this benchmark set. The best configuration shows 10% improvement over the base-line, and one more solved instance. The scatter graph above on the right shows that most of the benefit is in the hard instances.

The default of HAIFAMUC is now set to the above best configuration, and its source code is available in [1]. Spreadsheets with detailed results can be found at the same location.

Measure	PS, R=120	MBL, R=160	MBL$_i$, $R = 80$, Delay=5	MBL$_i$, R=40
success	215	**216**	215	215
time	56.9	**56.5**	57.0	**56.1**
literals	3057.3	11820.9	3162.6	7940.6
iterations	969.9	998.7	979.4	1003.7
lit/iter	3.2	11.8	3.2	7.9
unsat by assump.	211.1	212.6	223.5	245.0
decisions (M)	10.1	10.5	10.0	10.2
overhead	0.00	0.38	0.03	0.21
longest call	1.83	2.15	1.73	1.67
% implied assump.	0.14	0.27	0.15	0.24

Fig. 6. Results on the SAT02$_\beta$ benchmarks, rounded, with various configurations of the heuristics to mitigate the negative side-effect of extra assumptions. A timeout is counted as 900 sec.

4　Conclusions and Future Work

We presented a technique called Mining Backbone Literals (MBL) and variations thereof, that potentially accelerates incremental satisfiability. It is based on analysing resolution refutations for the purpose of finding literals that are logically implied by the consecutive SAT call. This is a new type of data transfer in incremental satisfiability.

Our case study in the domain of MUC extraction showed only modest improvement ($\approx$10%) in run-time, but as we explained this is mostly related to specific characteristics of the MUC-extraction algorithm, which are not present in other domains in which incremental SAT is used. We hope that future research will a) reveal how to overcome these characteristics in deletion-based MUC extraction, and b) investigate the impact of MBL in other domains. Recently a new format ('IPASIR') for incremental solving was suggested as part of the preparations for the SAT'15 race, and hopefully standard benchmarks will follow soon; then it will be very interesting to try our methods in the domain of general incremental satisfiability.

References

1. HAIFAMUC. http://ie.technion.ac.il/ofers/haifasolvers/
2. Bar-Ilan, O., Fuhrmann, O., Hoory, S., Shacham, O., Strichman, O.: Reducing the size of resolution proofs in linear time. STTT **13**(3), 263–272 (2011)
3. Belov, A., Marques-Silva, J.: MUSer2: An efficient MUS extractor. J. on Satisfiability, Boolean Modeling and Computation (JSAT) **8**(1/2), 123–128 (2012)
4. Ben-Sasson, E., Wigderson, A.: Short proofs are narrow - resolution made simple. J. ACM **48**(2), 149–169 (2001)
5. Biere, A., Cimatti, A., Clarke, E., Zhu, Y.: Symbolic model checking without BDDs. In: Cleaveland, W.R. (ed.) TACAS 1999. LNCS, vol. 1579, pp. 193–207. Springer, Heidelberg (1999)
6. Bradley, A.R.: SAT-based model checking without unrolling. In: Jhala, R., Schmidt, D. (eds.) VMCAI 2011. LNCS, vol. 6538, pp. 70–87. Springer, Heidelberg (2011)
7. Dershowitz, N., Hanna, Z., Nadel, A.: A scalable algorithm for minimal unsatisfiable core extraction. In: Biere, A., Gomes, C.P. (eds.) SAT 2006. LNCS, vol. 4121, pp. 36–41. Springer, Heidelberg (2006)
8. Eén, N., Sörensson, N.: An extensible SAT-solver. In: Giunchiglia, E., Tacchella, A. (eds.) SAT 2003. LNCS, vol. 2919, pp. 502–518. Springer, Heidelberg (2004)
9. Janota, M., Lynce, I., Marques-Silva, J.: Algorithms for computing backbones of propositional formulae. AI Commun. **28**(2), 161–177 (2015)
10. Kroening, D., Strichman, O.: Decision procedures - an algorithmic point of view. Theoretical Computer Science. Springer-Verlag, February 2008 (to be published)
11. McMillan, K.L.: Interpolation and SAT-based model checking. In: Hunt Jr, W.A., Somenzi, F. (eds.) CAV 2003. LNCS, vol. 2725, pp. 1–13. Springer, Heidelberg (2003)
12. Nadel, A.: Understanding and Improving a Modern SAT Solver. PhD thesis, Tel Aviv University, Tel Aviv, Israel, August 2009
13. Nadel, A.: Boosting minimal unsatisfiable core extraction. In: Bloem, R., Sharygina, N. (eds.) FMCAD (2010)
14. Nadel, A., Ryvchin, V., Strichman, O.: Efficient MUS extraction with resolution. In: FMCAD, pp. 197–200. IEEE (2013)
15. Ryvchin, V., Strichman, O.: Faster extraction of high-level minimal unsatisfiable cores. In: Sakallah, K.A., Simon, L. (eds.) SAT 2011. LNCS, vol. 6695, pp. 174–187. Springer, Heidelberg (2011)
16. Shtrichman, O.: Sharing information between instances of a propositional satisfiability (SAT) problem, December 2000. US provisional patent (60/257,384). Later became patent US2002/0123867 A1
17. Shtrichman, O.: Pruning techniques for the SAT-based bounded model checking problem. In: Margaria, T., Melham, T.F. (eds.) CHARME 2001. LNCS, vol. 2144, pp. 58–70. Springer, Heidelberg (2001)
18. Whittemore, J., Kim, J., Sakallah, K.: Satire: a new incremental satisfiability engine. In: IEEE/ACM Design Automation Conference (DAC) (2001)

Constructing SAT Filters with a Quantum Annealer

Adam Douglass, Andrew D. King[✉], and Jack Raymond

D-Wave Systems Inc., 3033 Beta Avenue,
Burnaby, BC V5G 4M9, Canada
{adouglass,aking,jraymond}@dwavesys.com

Abstract. SAT filters are a novel and compact data structure that can be used to quickly query a word for membership in a fixed set. They have the potential to store more information in a fixed storage limit than a Bloom filter. Constructing a SAT filter requires sampling diverse solutions to randomly constructed constraint satisfaction instances, but there is flexibility in the choice of constraint satisfaction problem. Presented here is a case study of SAT filter construction with a focus on constraint satisfaction problems based on MAX-CUT clauses (Not-all-equal 3-SAT, 2-in-4-SAT, etc.) and frustrated cycles in the Ising model. Solutions are sampled using a D-Wave quantum annealer, and results are measured against classical approaches. The SAT variants studied are of interest in the context of SAT filters, independent of the solvers used.

Keywords: SAT filter · Quantum annealing · Ising model · Maximum cut · Sampling · Constraint satisfaction problem

1 Introduction

Weaver et al. [37] recently presented *SAT filters* as an efficient data structure by which a set can be filtered with no false negatives and a false positive rate approaching zero as the situation requires. The construction of a SAT filter with high space efficiency and low false positive rate requires finding many solutions to a random n-variable instance of a constraint satisfaction problem (CSP), which was originally chosen to be k-SAT for $k \geq 3$ [37]. The effectiveness of the filter with respect to storage requirements, or *efficiency*, requires a low probability that a randomly generated clause would be satisfied by every solution in the filter. Consequently the solutions must satisfy various independence tests, e.g. they should not differ by only a few bitflips.

This application is a natural fit for a D-Wave Two (DW2) quantum annealer, which is capable of quickly sampling many low-energy states of a spin system in the Ising model [18,35] using open-system quantum annealing [1,3,9]. As the basis of the filter one can choose a CSP that is readily expressed in the Ising model, for example Not-all-equal 3-SAT [17,20]. In the context of filters and elsewhere, CSP solution sampling is a problem of both theoretical and practical interest [6,16].

© D-Wave Systems Inc. 2015
M. Heule and S. Weaver (Eds.): SAT 2015, LNCS 9340, pp. 104–120, 2015.
DOI: 10.1007/978-3-319-24318-4_9

This paper reports on a case study using a D-Wave quantum annealer to solve SAT instances and construct SAT filters, providing comparisons with classical solvers and comparing filters generated using several SAT variants, the choice of which has great effect on the characteristics of the resulting filter. Although processor size restricts the study to relatively small instances, the study gives a first look at several "exotic" SAT filters and an early application of sampling optima via quantum annealing.

Section 2 discusses the preliminaries of SAT filters and *blocked* SAT filters, which involve constructing a SAT filter for each bucket of a dense hash table. Section 3 discusses the expression and solution of CSPs in the Ising model. Section 4 discusses the CSPs used to construct filters. Sections 5 and 6 review the results of two cases in this study; the first investigates the construction of a filter with many blocks of fixed size, while the second investigates how filters and solver performance evolve for various CSPs as the number of variables increases. Finally Section 7 offers conclusions.

2   Filters, SAT Filters and Blocked SAT Filters

Given a domain W and a subset $X \subset W$, the *set membership problem* is to determine if an element x of W is in X. A *filter* is a data structure that allows fast set membership queries, possibly with false positives but not false negatives. That is, if $x \in X$ then the filter will always return $F(x) = 1$ (*maybe*). If $x \notin X$ then the filter will (deterministically) return either $F(x) = 1$ (*maybe*) or $F(x) = 0$ (*no*). Here it is assumed that $|X| \ll |W|$, and therefore the *false positive rate* p of the filter can be defined as the probability that a randomly selected $x \in W$ has $F(x) = 1$.

The aim is to minimize the false positive rate of F while minimizing storage requirements (in bits, denoted by $|F|$), construction time, and query time. Given a storage limit of $|F|$ bits, the information-theoretic limit on the false positive rate is given by the equation

$$\frac{-\log_2(p)}{|F|/|X|} \leq 1. \tag{1}$$

This result appears as Theorem 4.1 in [37], summarizing results in [36]. Here the numerator represents the *bits of cut-down* and the denominator represents the average storage bits per keyword in X. The *efficiency* of the filter F is therefore defined as

$$\mathcal{E}(F) = \frac{-\log_2(p)}{|F|/|X|}. \tag{2}$$

The standard tool for this situation is a *Bloom filter* [2]. A Bloom filter F is constructed for X with false positive rate p using parameters r and n derived from $|X|$ and p. The filter consists of a bitstring $B = (b_1, \ldots, b_n)$ and r hash functions $h_1, \ldots, h_r$, each of which hashes an element of W (ideally uniformly) to an element of $\{1, \ldots, n\}$. To construct the filter, initialize all bits of B to zero.

Then for each $x \in X$ and hash function h_i set $b_{h_i(x)} = 1$. To query $F(x)$, simply check if $\prod_{i=1}^{r} b_{h_i(x)} = 1$. If so, then $F(x) = 1$. If not, then $F(x) = 0$. Optimal choices of n and r will, for sufficiently large X, give a filter F with efficiency $\mathcal{E}(F) \approx \ln(2) \approx 0.69$.

2.1   SAT Filters

It is possible to exceed the efficiency of a Bloom filter using *SAT filters*, introduced recently by Weaver et al. [37]. A SAT filter is, like a Bloom filter, a space-efficient data structure used for set membership testing with one-way error[1]. Unlike Bloom filters, SAT filters are *offline*: new elements cannot be added to the filter after the filter is built. And unlike any online filters [24], SAT filters can achieve efficiency arbitrarily close to 1. SAT filters can be constructed using a variety of constraints, and the constraint used typically imposes an upper bound on efficiency (see Section 4.4). This section describes a 3-SAT filter for illustrative purposes.

During construction of a Bloom filter with $r = 1$, each keyword $x \in X$ is hashed to a bit in B. During construction of a 3-SAT filter with $r = 1$, each keyword $x_i \in X$ is hashed to a uniformly chosen 3-SAT clause $C_i = h(x_i)$ over n variables, where n is given an appropriate value with respect to $|X|$. Here a clause is constructed by choosing three distinct variables at random, and negating each one independently with probability $\frac{1}{2}$. The conjunction of these random clauses gives a random 3-SAT formula:

$$\mathcal{F} = \bigwedge_{i}^{|X|} C_i. \tag{3}$$

Random 3-SAT instances have a satisfiability phase transition near 4.26 clauses per variable [8]. This means that if $|X| \leq (4.26 - \epsilon)n$ for $\epsilon > 0$, and $|X|$ is sufficiently large, then $\mathcal{F}$ is satisfiable with high probability [27]. Storage of the 3-SAT filter F consists of storing the hash function h and a collection $S = (s_1, \ldots, s_k)$ of solutions to $\mathcal{F}$. $F(x)$ is queried by checking whether the clause $C = h(x)$ is satisfied by each truth assignment in S.

If $r = 1$, then $|F| = n$ (here hash functions are not included in the analysis of storage cost, so we store $r = 1$ bits per variable) and the false positive rate of a 3-SAT filter is 7/8. Since $|F|/|X| \gtrsim 1/4.26$, this gives the efficiency

$$\mathcal{E}(F) = \frac{-\log_2(7/8)}{n/|X|} \lesssim 0.81. \tag{4}$$

The false positive rate given by a single solution is generally too high. Weaver et al. provided two ways of dealing with this issue [37]. The first is to construct a *multi-instance* filter, which stores one solution to each of multiple formulae. The second is to construct a *single-instance* filter, which stores multiple solutions to

---

[1] A variety of alternatives to Bloom filters have been proposed, e.g. [11,28–30].

a single formula. The former suffers from increased query time, since multiple hashes must be evaluated. The latter suffers from decreased efficiency, since multiple solutions will generally not be perfectly independent. The severity of this dependence depends heavily on the SAT variant used.

2.2   Blocked SAT Filters

In certain situations it is impractical to find solutions to a SAT formula on n variables and $|X|$ clauses – this case study is limited to CSPs on at most 110 variables (typically fewer) due to the limited size of the D-Wave processor. This situation calls for a *blocked* filter [30], which uses a *blocking hash function* $\hat{h} : W \rightarrow \{1, \ldots, b\}$ that divides X into b blocks

$$X_i = \{x \in X \mid \hat{h}(x) = i\}. \tag{5}$$

A filter F_i is then constructed for each block X_i, and to query $x \in W$ for membership in X, x is queried for membership in $X_{\hat{h}(x)}$ using $F_{\hat{h}(x)}$.

This blocked filter, like a normal filter, has one-way error. Under the assumption that $\hat{h}$ maps members of W to $\{1, \ldots, b\}$ equiprobably, the false positive rate p is the mean of the b false positive rates $\{p_i\}_{i=1}^{b}$ of the individual filters. More rigorously, letting $W_i = \{x \in W \mid \hat{h}(x) = i\}$,

$$p = \sum_{i=1}^{m} \frac{|W_i|}{|W|} p_i. \tag{6}$$

Even in the case where $r = 1$, blocked filters can adversely affect efficiency because of the variance in bin sizes under $\hat{h}$ [21,30]. This issue is considered in Section 5.

3   Solving CSPs in the Ising Model with a D-Wave Processor

This paper contains results from a D-Wave Two (DW2) quantum annealing processor using a *Washington W3* chip operating over 1097 of 1152 configured qubits in a $\mathcal{C}_{12}$ *Chimera* layout [4]. All runs use the minimum anneal time of 20µs.

D-Wave quantum annealing processors are designed to minimize the energy of an Ising spin configuration. Input to the processor consists of an *Ising Hamiltonian* (h, J), where $h \in \mathbb{R}^N$ is a vector of *local fields* and $J \in \mathbb{R}^{N \times N}$ is a matrix of *couplings*, which may be assumed to be symmetric. The *energy* of a spin configuration $s \in \{-1, 1\}^N$ is defined as

$$E(s) = E(h, J, s) = s^T J s + s^T h. \tag{7}$$

The output of an *anneal* (i.e. a hardware run) of the processor is a low-energy state s, which consists of an Ising spin (either -1 or 1) for each *qubit*.

In a D-Wave processor, not all pairs of qubits are coupled, and therefore the set of nonzero entries of J must respect the physical constraints of the processor. One can view (h, J) as a set of vertex and edge weights, respectively, of the *qubit connectivity graph*, whose vertices correspond to qubits and whose edges correspond to couplers. The qubit connectivity graph for the processor used in this report is shown in Fig. 1.

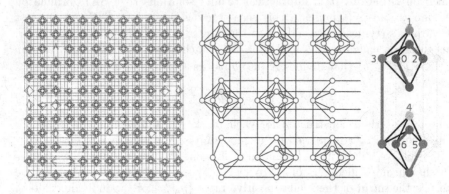

Fig. 1. The 1097-qubit hardware graph (left) of the processor used, with detail (center) of the bottom-left corner. This graph is a subgraph of the 1152-qubit 12×12 Chimera grid. The irregular pattern of connectivity was designed so that high-connectivity problems can be effectively emulated. (right, color online) An embedded problem of two interacting clauses is shown $(2in4(x_0, x_1, x_2, x_3) \wedge 2in4(x_3, x_4, x_5, x_6))$. Each variable is represented by one or more qubits; thicker lines indicate two variables acting as a single logical variable (achieved by energy penalties causing qubit behavior to coincide). Given this contraction the variables interact in two cliques through the remaining edges.

A state s minimizing $E(h, J, s)$ is called a *ground state* of (h, J); all other states are called *excited states*. A lowest-energy excited state is called a *first excited state*. Given a ground state s_0 and a first excited state s_1, $g := E(s_1) - E(s_0)$ is the *minimum final gap* of the Hamiltonian.

3.1   Constraint Satisfaction Problems in the Ising Model

Let f be a Boolean function with an n-dimensional binary range. For simplicity, assume $f : \{-1, 1\}^n \to \{0, 1\}$. Now suppose an n-dimensional Ising Hamiltonian (h, J) satisfies $E(s) = x$ if $f(s) = 0$, and $E(s) \geq x + g$ if $f(s) = 1$. Then (h, J) *encodes* f with gap g.

Now take a collection of Boolean functions

$$\{ f_i : \{-1, 1\}^n \to \{0, 1\} \}_{i=1}^{k}$$

and a collection of Hamiltonians $\{(h_i, J_i)\}_{i=1}^k$ such that (h_i, J_i) encodes f_i with gap g_i. Then the Hamiltonian $(\sum_{i=1}^k h_i, \sum_{i=1}^k J_i)$ encodes $\max_i f_i$ with gap $\min_i g_i$.

This fact makes the Ising model suitable for application to constraint satisfaction problems [15,19,25]. In particular, it is straightforward to formulate certain SAT variants as Ising problems in such a way that they can be solved by DW2: Let $f(s) = 0$ represent satisfaction; max therefore represents conjunction in this context.

3.2   Graph Minors in the Ising Model

Random instances of the SAT variants considered here are constructed as a weighted subgraph G of either a complete graph K_n or a complete bipartite graph $K_{\frac{n}{2}, \frac{n}{2}}$. In general, these instances cannot be solved directly using the D-Wave processor's native connectivity. Rather, G must be embedded in the hardware graph G_H as a *graph minor*, which amounts to transforming G_H into G by the operations of *edge contraction, vertex deletion*, and *edge deletion* [10]. Edge and vertex deletion can be realized trivially in the Ising model by setting couplers to $J = 0$. Edge contraction can be realized by setting couplers to $J = -\kappa$, where κ is large, and the coupling therefore compels a set of qubits to act as a single *logical qubit*. The choice of the parameter κ is nontrivial. In this paper, runs on minor-embedded instances are optimized over five possible choices of κ. See [7,20,35] for further discussion. Embedding G into G_H is achieved here using a specialized heuristic algorithm [5]. In Fig. 1, edge contraction of a 2-cell Chimera graph allows representation of a 2in4SAT Ising problem described a pair of 4-cliques, anti-ferromagnetically coupled, sharing one variable.

4   SAT Variants in the Ising Model

Although k-SAT was previously used to construct SAT filters [37], it is not amenable to representation in the Ising model. To represent a SAT relations k binary variables with only pairwise relations available is not possible for $k \geq 3$ without the introduction of additional *ancillary* variables to mediate the interactions. To represent an instance on n variables and m clauses requires at least m ancillary variables, so an instance near the phase transition has many more ancillary Ising variables than original variables. Not-all-equal 3-SAT (*NAE3SAT*) is an NP-complete SAT variant that does not present the same obstacle [14,17,20]. This section first discusses issues surrounding NAE3SAT, then moves on to further extensions. Ultimately we are interested in sampling satisfying assignments, which for the problem classes presented is NP-hard [16].

4.1   NAE3SAT

An NAE3SAT clause consists of three literals, and is satisfied precisely if at least one literal is true and one literal is false. Thus a clause ensures that three literals are not all equal. This requirement is easily expressed in the Ising model. Consider the Ising problem

$$\min_{s_1,s_2,s_3 \in \{-1,1\}} (s_1 s_2 + s_2 s_3 + s_1 s_3).$$

This Hamiltonian has six ground states with energy -1, corresponding to states in which not all variables are equal. The other two states have energy 3, meaning that this Hamiltonian encodes a NAE3SAT clause with gap 4. The qubit connectivity graph of a NAE3SAT clause is a triangle with a coupling of $+1$ on each edge.

Monotone NAE3SAT. Unlike 3-SAT, NAE3SAT remains NP-complete when all literals are non-negated – in this case the problem is equivalent to hypergraph 2-coloring [23]. This variant is called *monotone NAE3SAT* or *MNAE3SAT*. Asymptotically, NAE3SAT and MNAE3SAT have very similar characteristics. For small systems, however, marginal gains in efficiency can be realized from the fact that not all solutions satisfy the same number of possible clauses. These small-system considerations are relevant due to limitations imposed by the D-Wave processor. All MNAE3SAT instances studied here have 45 variables. Instances on up to around 65 variables can be embedded consistently.

4.2   MAX-CUT SAT Variants

The Ising formulation of an NAE3SAT clause can be thought of as an unweighted MAX-CUT problem on a triangle, i.e. a K_3. Replacing K_3 with larger cliques gives a sequence of SAT variants with different properties: NAE3SAT, 2in4SAT, 2or3in5SAT, 3in6SAT, etc. These variants can more generally be called kMCSAT for $k = 3, 4, 5, 6, \ldots$. A randomly constructed kMCSAT clause contains a random k-set of variables, each one of which is negated independently with probability $\frac{1}{2}$; the coupling on the edge between two variables is -1 if precisely one is negated, and is $+1$ otherwise.

When k is even, kMCSAT clauses are locally inflexible, since no single bit flip connects any two solutions. When these clauses are agglomerated in a random manner, a *global* rearrangement of the state space is required on the core of the graph to move between solutions. In other words, solutions are isolated. If the core of the graph is the graph itself (i.e. if every variable is in at least two clauses) then a typical kMCSAT instance will be *locked*, meaning that solutions are not only isolated, but separated pairwise by $O(\log n)$ bit flips [38]. By contrast, when k is odd there can be local rearrangements that connect solutions, and large clusters of closely related (in Hamming space) solutions are to be expected. Large sets of this type are undesirable in the construction of SAT filters, because two closely related solutions will only store slightly more information than one solution alone.

4.3   Bipartite 4FLSAT

To this point, each SAT variant presented operates implicitly over the complete graph K_n. That is, any set of k variables can be chosen to be in a clause. Further, there is yet another perspective on NAE3SAT: a clause can be thought of as a *frustrated loop* (cycle, in graph theoretic terms) in the Ising model. Each NAE3SAT clause is represented in the Ising model by a cycle of couplings with values ± 1, containing an odd number of antiferromagnetic (positive, repulsive) couplings. So a random NAE3SAT clause is generated as a random frustrated 3-cycle from the complete graph.

In a new variant called *bipartite 4FLSAT (B4FLSAT)*, each clause is a frustrated 4-cycle in the complete bipartite graph $K_{\frac{n}{2},\frac{n}{2}}$. These instances are closely related to the frustrated loop instances that were the subject of recent benchmarking work [15,19] where the instances are constructed in the native Chimera topology using frustrated loops of varying length. A B4FLSAT clause is satisfied if and only if the Ising state corresponding to the truth assignment minimizes the energy of the clause in the Ising model.

$$\min_{s_1,s_2,s_3,s_4 \in \{-1,1\}} \left(-s_1 s_2 - s_2 s_3 - s_3 s_4 + s_1 s_4 \right).$$

Explicitly, the truth assignments satisfying a B4FLSAT clause with no negated literals are TTTT, TTTF, TTFF, TFFF, FFFF, FFFT, FFTT, and FTTT. A 4FLSAT clause has a false positive rate of $1/2$. B4FLSAT instances are unlocked, like kMCSAT for odd k.

4.4   Threshold Analysis

Despite the small size of the problems investigated here, it is useful to understand the asymptotic structure of the solution space that governs the quality of attainable filters and scaling of sampling methods. A common feature of the SAT variants described here is the presence of a *satisfiability* transition and a *dynamical* transition. The satisfiability threshold is the ratio α_s of constraints to variables separating, asymptotically almost surely, regimes with and without solutions. The satisfiability threshold thus implies a bound $\mathcal{E}_s$ on the maximum (asymptotically achievable) efficiency, since any filter built from an unsatisfiable instance will have false positive rate 1. The dynamical transition is, by contrast, the ratio α_d at which the solution space shatters into disconnected components. This shattering is related to hardness for sampling and optimization, although the relationship is complicated and an active area of research [22].

A simple rigorous upper bound on the satisfiability threshold (α_{1MM}) is obtained by a first moment method [27]; equivalently, α_{1MM} gives the information-theoretic upper bound on α_s in the context of filters. Transitions are also approximable by statistical physics techniques, namely the *1RSB energetic cavity method* and the *reconstruction on trees method* [26,27]. The thresholds indicate that building efficient SAT filters requires sampling in the shattered phase, where local search methods will often struggle. Table 1 gives estimates on the transitions and maximum efficiency of NAE3SAT, 3in6SAT, and B4FLSAT.

Table 1. Thresholds and efficiency of the three CSPs studied in Section 6

Problem	Dynamical α_d	Satisfiability α_s	Annealed α_{1MM}	Efficiency $\mathcal{E}_s$
NAE3SAT	1.50	2.11	2.409	0.88
3in6SAT	0.48	0.57	0.596	0.96
B4FLSAT	0.50	0.78	1	0.78

4.5   Software Solvers

The next sections compare performance of the D-Wave processor against three classical software approaches. The first two, WalkSAT [33] and Dimetheus [12, 13], are SAT solvers, and are applied directly to the SAT problems in question (after naive clause-wise conversion to CNF). The third, Selby's solver [32], is a specialized implementation of the Hamze-de Freitas-Selby algorithm, written to solve Ising problems on D-Wave's native Chimera architecture. The version used, like in previous work [19], is modified to make it act more analogously to the D-Wave processor.

Both SAT solvers were run naively with the aim of generating 1000 solutions to each SAT instance. WalkSAT was run with default command line parameters and 1000 random restarts. Dimetheus was run 1000 times with command line parameters -guide 0 and -cdclSelectDirRule 0. For both Selby and DW2, samples were mapped from the embedded Chimera space to the SAT space via majority vote, then quenched to a local minimum (see [20] for further explanation).

5   Case Study 1: A Blocked MNAE3SAT Filter

This section describes the construction of a 3500-block MNAE3SAT for 262,144 randomly generated 16-byte numbers (UUIDs). The MNAE3SAT instance for each block was constructed over 45 variables, giving a mean clause-to-variable ratio of $\bar{\alpha} \approx 1.664$. This is very close to the minimum requirement for matching the efficiency of a Bloom filter (ignoring small-system MNAE3SAT considerations discussed in Section 4), i.e. $\log(2)/\log_2(3/4)$. Of the 3500 MNAE3SAT instances, 77 were unsatisfiable.

For each MNAE3SAT instance, 24,000 (not necessary optimal) samples were drawn from DW2, 10,000 from Selby, and 1000 each from WalkSAT and Dimetheus. WalkSAT and Dimetheus successfully returned 1000 solutions for each satisfiable instance, while the number of solutions returned by DW2 and Selby varied substantially (see Fig. 2). Variation in the numbers reflect different operational modes of the algorithms, and the probability of each solver's output of being a valid solution (low for DW2 and Selby). In order to avoid conditioning filter construction on the number of solutions returned, each instance drew from a set of solutions whose size was limited by the minimum number of solutions returned by any solver. For most blocks this was Dimetheus and WalkSAT.

Fig. 3 shows the distribution of α and the number of solutions for these blocks (computed using sharpSAT [34]), highlighting a challenge in constructing

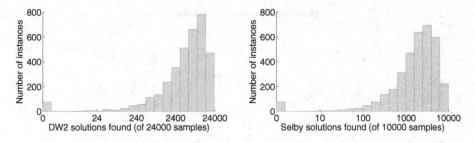

Fig. 2. Solutions returned by DW2 (left) and Selby (right)

efficient blocked filters. In order to ensure that almost all blocks are satisfiable, the number of blocks must be high enough that almost the entire binomial distribution of α is below the satisfiability threshold α_s. The effect seen here declines in importance as the size of blocks increases, and can be mitigated in several ways that trade efficiency for query time or similar [21]. This paper does not explore the question further.

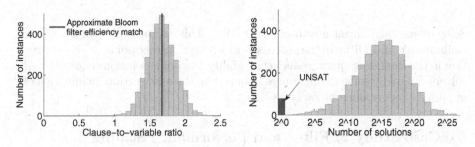

Fig. 3. Clause-to-variable ratios (left) and solution counts (right) for 3500 MNAE3SAT instances. A line indicates $-\log(2)/\log_2(3/4)$, the ratio at which a single-solution NAE3SAT filter matches the optimal efficiency of a Bloom filter

Fig. 4 shows the decline in efficiency for the overall filter as more and more solutions are stored per block, up to 20. Two methods of selecting r SAT solutions for a filter block are used. In the *online* approach, r random solutions are selected from the multiset of solutions returned by the solver in question. This reflects the situation where the solver returns solutions that are independent of the previously returned solutions, and the user wants r solutions as fast as possible. In the *greedy offline* approach, each solution is iteratively selected subject to minimizing the false positive rate of the filter at that point. This reflects the more realistic situation in which the user spends a certain amount of time generating SAT solutions, then constructs a filter greedily with a subset of the solutions at hand.

It is interesting that in this context Dimetheus starts out strong and declines in efficacy, while the opposite is true for WalkSAT. For online filter construction,

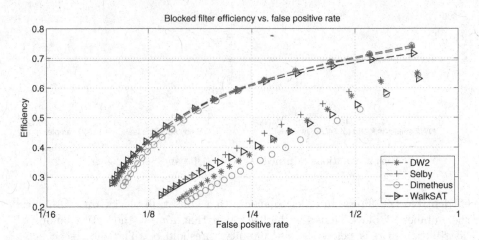

Fig. 4. (Color online) Blocked filter efficiency versus false positive rate for MNAE3SAT filter at mean clause-to-variable ratio of 1.66. Upper data with lines represents offline (greedy) filter construction. Lower data represents online filter construction. The efficiency of an ideal Bloom filter, 0.69, is indicated

Selby offers a significant advantage over DW2. This can be explained by current calibration nonidealities in the processor, which tilt the processor in a certain direction in the Hamming space, reducing its ability to sample solutions equitably. All solvers can easily solve the instances in question; the next section includes a look at performance scaling for various CSPs.

6   Case Study 2: Filter and Performance Scaling

The second part of the case study investigates how the construction of filters evolves as SAT instances grow. Considered here are random NAE3SAT, 3in6SAT, and B4FLSAT instances of increasing size. For each CSP the testbed contains 10 instances of each size shown, using only even sizes for B4FLSAT. Each instance is generated near the midpoint between the dynamical and satisfiability thresholds; NAE3SAT, 3in6SAT, and B4FLSAT use target ratios of 1.8, 0.55, and 0.70, respectively. For each size of each SAT variant, the ten instances were used as blocks in the construction of a blocked SAT filter.

For each SAT instance in this section, only 20,000 samples were drawn using DW2, representing a total anneal time of 0.4s. As before, 10,000 samples were drawn using Selby, and 1000 were drawn using each of WalkSAT and Dimetheus. False positive rates for the filters were estimated using a Monte Carlo approach.

Fig. 5 shows the evolution of filter efficiency versus false positive rate as the number of solutions used for each block increases from 1 to 20. The dependence on the properties of the CSP used are clear. 3in6SAT filters should show potential for very efficient filters that suffer from a low clause-to-variable ratio and the requirement of solving a hard CSP, but large filters reflect the low number of

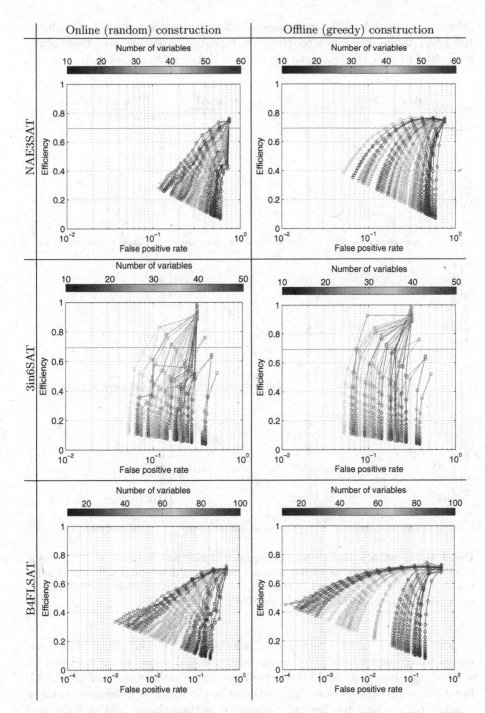

Fig. 5. (Color online) Evolution of filter efficiency/false positive rate tradeoff for filters constructed by DW2.

solutions returned by DW2 and more so Selby on these instances. Although the B4FLSAT filters maintain a relatively consistent false positive rate, these filters are bound to have efficiency at most 0.78, as discussed in Section 4.4.

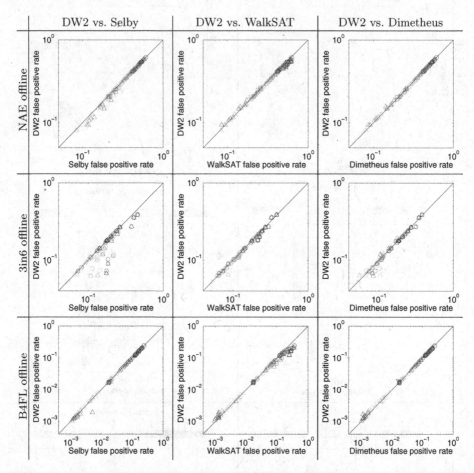

Fig. 6. (Color online) Instance-wise comparison of false positive rates for DW2 versus classical software solvers for offline filter construction. Filters using 3, 6, and 12 solutions are denoted by ○, □, and △ respectively. Colors/shades denote problem size as in Fig. 5. Points below the diagonal favor DW2.

Fig. 6 gives a direct comparison of false positive rate for offline filters generated using output from DW2 versus each classical solver, using 3, 6, and 12 solutions per block. WalkSAT's relatively poor performance on small B4FLSAT instances seems to be a bona fide weakness of the solver as a sampler on these small systems, which may have troublesome structural characteristics. False positive rates for DW2 would likely improve given refined calibration, which is currently underway. Overall, the four solvers give results of comparable quality.

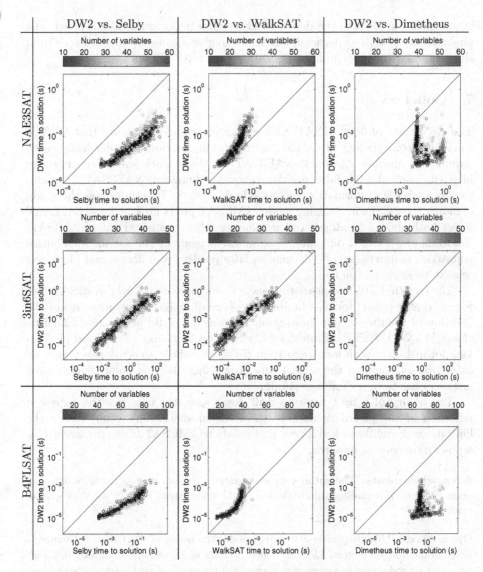

Fig. 7. (Color online) Instance-wise comparison of solution time for DW2 versus classical software solvers. × markers indicate median for a given size.

6.1 Scaling of Solution Time

Fig. 7 shows how the time required by each solver to draw a solution scales with the number of variables. Dimetheus, being a multi-mode solver that sometimes resorts to exhaustive search, is likely entering this slow mode prematurely for small instances of NAE3SAT and B4FLSAT. Solution time of DW2 versus WalkSAT is particularly interesting for 3in6SAT, as explained in the conclusion. Although both solvers find the B4FLSAT instances easy, it would be interesting to know if there is an explanation behind the sharp inflection point in the lower-middle panel. The analysis here is quite superficial, but serves to give a counterpoint to Fig. 5 from the perspective of problem difficulty rather than filter efficiency.

7   Conclusions

The case study of 3500 MNAE3SAT instances provides evidence that DW2 is capable of constructing filters competitively, but solution of larger instances is required to improve efficiency. NAE3SAT, at these block sizes, can only give high efficiency at high false positive rates. Furthermore, NAE3SAT filter efficiency reaches an information-theoretic limit at 0.88. This motivates the study of more exotic CSPs in the Ising model. kMCSAT filters for $k \geq 4$ and B4FLSAT filters both show advantages and weaknesses when compared with NAE3SAT. Tradeoffs are between difficulty, maximum efficiency at the satisfiability phase transition, maintenance of efficiency as false positive rate drops, and block size relative to keyword count.

The fact that DW2 appears to show a scaling advantage over WalkSAT (see Fig. 7, center panel) for this limited set of small 3in6SAT instances raises the question of whether or not multi-qubit cotunneling could play a role for large-clause MAX-CUT CSPs. Embedded 3in6SAT problems may be a good place to look for instances with long relaxation times, i.e. where an advantage might be gained by lengthening the anneal time from 20µs. To this point, finding such instances has been a challenge [15,31,35].

The ability of a SAT filter to drill down to a low false positive rate is a reflection of the global richness of the solution space. This study shows the limitations of small-blocked filters, and points to kMCSAT filters for larger k as a possible avenue of research.

Acknowledgments. The authors are very grateful to Sean Weaver for helpful discussions and advice regarding this work and to the anonymous referees for their careful readings.

References

1. Albash, T., Vinci, W., Mishra, A., Warburton, P.A., Lidar, D.A.: Consistency tests of classical and quantum models for a quantum annealer. Physical Review A **91**(4), 042314 (2015)
2. Bloom, B.H.: Space/time trade-offs in hash coding with allowable errors. Communications of the ACM **13**(7), 422–426 (1970)

3. Boixo, S., Smelyanskiy, V.N., Shabani, A., Isakov, S.V., Dykman, M., Denchev, V.S., Amin, M., Smirnov, A., Mohseni, M., Neven, H.: Computational role of collective tunneling in a quantum annealer. arXiv preprint arXiv:1411.4036 (2014)

4. Bunyk, P., Hoskinson, E., Johnson, M., Tolkacheva, E., Altomare, F., Berkley, A., Harris, R., Hilton, J., Lanting, T., Przybysz, A., et al.: Architectural considerations in the design of a superconducting quantum annealing processor. IEEE Transactions on Applied Superconductivity (2014)

5. Cai, J., Macready, W., Roy, A.: A practical heuristic for finding graph minors. arXiv preprint arXiv:1406.2741 (2014)

6. Chakraborty, S., Meel, K.S., Vardi, M.Y.: A scalable and nearly uniform generator of SAT witnesses. In: Sharygina, N., Veith, H. (eds.) CAV 2013. LNCS, vol. 8044, pp. 608–623. Springer, Heidelberg (2013)

7. Choi, V.: Minor-embedding in adiabatic quantum computation: I. The parameter setting problem. Quantum Information Processing 7(5), 193–209 (2008)

8. Crawford, J.M., Auton, L.D.: Experimental results on the crossover point in random 3-SAT. Artificial Intelligence 81(1), 31–57 (1996)

9. Dickson, N., et al.: Thermally assisted quantum annealing of a 16-qubit problem. Nature Communications 4, May 1903, January 2013. http://www.ncbi.nlm.nih.gov/pubmed/23695697

10. Diestel, R.: Graph Theory. Graduate Texts in Mathematics, vol. 173, 4th edn. Springer (2012)

11. Fan, B., Andersen, D.G., Kaminsky, M., Mitzenmacher, M.D.: Cuckoo filter: practically better than Bloom. In: Proceedings of the 10th ACM International on Conference on Emerging Networking Experiments and Technologies, pp. 75–88. ACM (2014)

12. Gableske, O.: Dimetheus. In: SAT Competition 2014: Solver and Benchmark Descriptions, pp. 29–30 (2014)

13. Gableske, O.: An Ising model inspired extension of the product-based MP framework for SAT. In: Sinz, C., Egly, U. (eds.) SAT 2014. LNCS, vol. 8561, pp. 367–383. Springer, Heidelberg (2014)

14. Garey, M.R., Johnson, D.S.: Computers and Intractability. W.H. Freeman (1979)

15. Hen, I., Albash, T., Job, J., Rønnow, T.F., Troyer, M., Lidar, D.: Probing for quantum speedup in spin glass problems with planted solutions (2015). arXiv preprint arXiv:1502.01663v2

16. Jerrum, M., Valiant, L., Vazirani, V.: Random generation of combinatorial structures from a uniform distribution. Theoretical Computer Science 43, 169–188 (1986)

17. Jiménez, A., Kiwi, M.: Computational hardness of enumerating groundstates of the antiferromagnetic Ising model in triangulations. Discrete Applied Mathematics (2014)

18. Johnson, M., Amin, M., Gildert, S., Lanting, T., Hamze, F., Dickson, N., Harris, R., Berkley, A., Johansson, J., Bunyk, P., et al.: Quantum annealing with manufactured spins. Nature 473(7346), 194–198 (2011)

19. King, A.D.: Performance of a quantum annealer on range-limited constraint satisfaction problems (2015). arXiv preprint arXiv:1502.02098v1

20. King, A.D., McGeoch, C.C.: Algorithm engineering for a quantum annealing platform. arXiv preprint arXiv:1410.2628 (2014)

21. Krimer, E., Erez, M.: The power of $1+\alpha$ for memory-efficient Bloom filters. Internet Mathematics 7(1), 28–44 (2011)

22. Krzakala, F., Zdeborová, L.: Phase transitions and computational difficulty in random constraint satisfaction problems. In: Journal of Physics: Conference Series, vol. 95, p. 012012. IOP Publishing (2008)
23. Lovász, L.: Coverings and colorings of hypergraphs. In: Proc. 4th Southeastern Conference on Combinatorics, Graph Theory, and Computing, pp. 3–12. Utilitas Mathematica Publishing, Winnipeg (1973)
24. Lovett, S., Porat, E.: A lower bound for dynamic approximate membership data structures. In: 2013 IEEE 54th Annual Symposium on Foundations of Computer Science, pp. 797–804. IEEE (2010)
25. Lucas, A.: Ising formulations of many NP problems. Frontiers in Physics **2**(5) (2014)
26. Mézard, M., Montanari, A.: Reconstruction on trees and spin glass transition. Journal of Statistical Physics **124**(6), 1317–1350 (2006)
27. Mézard, M., Montanari, A.: Information, Physics, and Computation. Oxford University Press (2009)
28. Pagh, A., Pagh, R., Rao, S.S.: An optimal Bloom filter replacement. In: Proceedings of the Sixteenth Annual ACM-SIAM Symposium on Discrete Algorithms, pp. 823–829. Society for Industrial and Applied Mathematics (2005)
29. Porat, E.: An optimal Bloom filter replacement based on matrix solving. In: Frid, A., Morozov, A., Rybalchenko, A., Wagner, K.W. (eds.) CSR 2009. LNCS, vol. 5675, pp. 263–273. Springer, Heidelberg (2009)
30. Putze, F., Sanders, P., Singler, J.: Cache-, hash- and space-efficient Bloom filters. In: Demetrescu, C. (ed.) WEA 2007. LNCS, vol. 4525, pp. 108–121. Springer, Heidelberg (2007)
31. Rønnow, T., Wang, Z., Job, J., Boixo, S., Isakov, S., Wecker, D., Martinis, J., Lidar, D., Troyer, M.: Defining and detecting quantum speedup. Science **345**(6195), 420–424 (2014)
32. Selby, A.: Efficient subgraph-based sampling of Ising-type models with frustration. arXiv preprint arXiv:1409.3934v1 (2014)
33. Selman, B., Kautz, H., Cohen, B., et al.: Local search strategies for satisfiability testing. Cliques, Coloring, and Satisfiability: Second DIMACS Implementation Challenge **26**, 521–532 (1993)
34. Thurley, M.: sharpSAT – counting models with advanced component caching and implicit BCP. In: Biere, A., Gomes, C.P. (eds.) SAT 2006. LNCS, vol. 4121, pp. 424–429. Springer, Heidelberg (2006)
35. Venturelli, D., Mandrà, S., Knysh, S., O'Gorman, B., Biswas, R., Smelyanskiy, V.: Quantum optimization of fully-connected spin glasses. arXiv preprint arXiv:1406.7553 (2014)
36. Walker, A.: Filters. Undergraduate thesis, Haverford College, Haverford, PA (2007)
37. Weaver, S.A., Ray, K.J., Marek, V.W., Mayer, A.J., Walker, A.K.: Satisfiability-based set membership filters. Journal on Satisfiability, Boolean Modeling and Computation **8**, 129–148 (2014)
38. Zdeborová, L., Mézard, M.: Locked constraint satisfaction problems. Physical Review Letters **101**(7), 078702 (2008)

#∃SAT: Projected Model Counting

Rehan Abdul Aziz[✉], Geoffrey Chu, Christian Muise, and Peter Stuckey

National ICT Australia, Victoria Laboratory,
Department of Computing and Information Systems,
The University of Melbourne, Melbourne, Australia
raziz@student.unimelb.edu.au

Abstract. Model counting is the task of computing the number of assignments to variables V that satisfy a given propositional theory F. The model counting problem is denoted as #SAT. Model counting is an essential tool in probabilistic reasoning. In this paper, we introduce the problem of model counting projected on a subset of original variables that we call *priority* variables $P \subseteq V$. The task is to compute the number of assignments to P such that there exists an extension to *non-priority* variables $V \setminus P$ that satisfies F. We denote this as #∃SAT. Projected model counting arises when some parts of the model are irrelevant to the counts, in particular when we require additional variables to model the problem we are counting in SAT. We discuss three different approaches to #∃SAT (two of which are novel), and compare their performance on different benchmark problems.

1 Introduction

Model counting is the task of computing the number of models of a given propositional theory, represented as a set of clauses (SAT). Often, instead of the original model count, we are interested in model count projected on a set of variables P.

Given a problem on variables P, we may need to introduce additional variables to encode the constraints on the variables P into Boolean clauses in the propositional theory F. Counting the models of F does not give the correct count if the new variables are not *functionally defined* by the original variables P. Thankfully, most methods of encoding constraints introduce new variables that are functionally defined by original variables, but there are cases where the most efficient encoding of constraints does not enjoy this property. Hence we should consider projected model counting for these kinds of problems.

Alternatively, in the counting problem itself, we may only be interested in some of the variables involved in the problem. Unless the interesting variables functionally define the uninteresting variables, we need projected model counting. An example is in *evaluating* robustness of a given solution. The goal is to count the changes that can be made to a subset of variables in the solution

NICTA is funded by the Australian Government as represented by the Department of Broadband, Communications and the Digital Economy and the Australian Research Council through the ICT Centre of Excellence program.

M. Heule and S. Weaver (Eds.): SAT 2015, LNCS 9340, pp. 121–137, 2015.
DOI: 10.1007/978-3-319-24318-4_10

such that it still remains a solution (possibly after allowing some repairs, e.g. in supermodels of a propositional theory [10]). The variables representing change are priority variables. In our benchmarks, we consider an example from the planning domain, where we are interested in robustness of a given partially ordered plan to the initial conditions, i.e., we want to count the number of initial states, such that the given partially ordered plan still reaches the given goal state(s).

Projected model counting is a challenging problem that has received little attention. It is in #P^{NP}. If all the variables are priority variables, then it becomes a #SAT problem (#P), and if all variables are non-priority variables, then it reduces to SAT (NP). There has been little development of specialized algorithms for projected model counting in the literature. Some dedicated attempts at solving the problem are presented in [13] and [8]. In the latter, the primary motivation is solution enumeration, and not counting. Closely related problems are projection or *forgetting* in formulas that are in deterministic decomposable negation normal form (d-DNNF [4]) [5], and *Boolean quantifier elimination* [3,11,19].

In this paper, we present three different approaches for projected model counting.

- The first technique is straight-forward and its basic idea is to modify DPLL-based model counters to search first on the priority variables, followed by finding only a single solution for the remaining problem. This technique is not novel and has been proposed in [13]. It has also been suggested in [15] in a slightly different context. Unlike [13] which uses external calls to MINISAT to check satisfiability of non-priority components, we handle all computations within the solver.
- The second approach is a significant extension of the algorithm presented in [8]. The basic idea is that every time a solution S is found, we generalize it by greedily finding a subset of literals S' that are sufficient to satisfy all clauses of the problem. By adding $\neg S'$ as a clause, we save an exponential amount of search that would visit all extensions of S'. This extension conveniently blends in the original algorithm of [8], which has the property that the number of blocking clauses are polynomial in the number of priority variables at any time during the search.
- Our third technique is a novel idea which reuses model counting algorithms: computing the d-DNNF of the original problem, forgetting the non-priority variables in the d-DNNF, converting the resulting DNNF to CNF, and counting the models of this CNF.

We compare these three techniques on different benchmarks to illustrate their strengths and weaknesses.

2   Preliminaries

We consider a finite set $\mathcal{V}$ of propositional variables. A literal l is a variable $v \in \mathcal{V}$ or its negation $\neg v$. The negation of a literal $\neg l$ is $\neg v$ if $l = v$ or v if $l = \neg v$. Let $var(l)$ represent the variable of the literal, i.e., $var(v) = var(\neg v) = v$. A clause is a set of literals that represents their disjunction, we shall write in parentheses $(l_1, \ldots, l_n)$. For any formula (e.g. a clause) C, let $vars(C)$ be the set of variables appearing in C. A formula F in conjunctive normal form (CNF) is a conjunction of clauses, and we represent it simply as a set of clauses. An assignment θ is a set of literals, such that if $l \in \theta$, then $\neg l \notin \theta$. We shall write them using set notation. Given an assignment θ then $\neg \theta$ is the clause $\bigvee_{l \in \theta} \neg l$. Given an assignment θ over $\mathcal{V}$ and set of variables P then $\theta_P = \{l \mid l \in \theta, var(l) \in P\}$

Given an assignment θ, the *residual* of a CNF F w.r.t. θ is written $F|_\theta$ and is obtained by removing each clause C in F such that there exists a literal $l \in C \cap \theta$, and simplifying the remaining clauses by removing all literals from them whose negation is in θ. We say that an assignment θ is a solution *cube*, or simply a cube, of F iff $F|_\theta$ is empty. The size of a cube θ, $size(\theta)$ is equal to $2^{|\mathcal{V}| - |\theta|}$. A solution in the classical sense is a cube of size 1. The model count of F, written, $ct(F)$ is the number of solutions of F.

We consider a set of priority variables $\mathcal{P} \subseteq \mathcal{V}$. Let the non-priority variables be $\mathcal{N}$, i.e., $\mathcal{N} = \mathcal{V} \setminus \mathcal{P}$. Given a cube θ' of formula F, then $\theta \equiv \theta'_P$ is a *projected cube* of F. The size of the projected cube is equal to $2^{|\mathcal{P}| - |\theta|}$. The projected model count of F, $ct(F, \mathcal{P})$ is equal to the number of projected cubes of size 1. The projected model count can also be defined as the number of assignments θ s.t. $vars(\theta) = \mathcal{P}$ and there exists an assignment θ' s.t. $vars(\theta') = \mathcal{N}$ and $\theta \cup \theta'$ is a solution of F.

A Boolean formula is in *negation normal form* (NNF) iff the only subformulas that have negation applied to them are propositional variables. An NNF formula is *decomposable* (DNNF) iff for all conjunctive formulae $c_1 \wedge \cdots \wedge c_n$ in the formula, the sets of variables of conjuncts are pairwise disjoint, $vars(c_i) \cap var(c_j) = \emptyset, 1 \leq i \neq j \leq n$. Finally, a DNNF is *deterministic* (d-DNNF) if for all disjunctive formulae $d_1 \vee \cdots \vee d_n$ in the formula, the disjuncts are pairwise logically inconsistent, $d_i \wedge d_j$ is unsatisfiable, $1 \leq i \neq j \leq n$. A d-DNNF is typically represented as a tree or DAG with inner nodes and leaves being OR/AND operators and literals respectively. Model counting on d-DNNF can be performed in polynomial time (in d-DNNF size) by first computing the satisfaction probability and then multiplying the satisfaction probability with total number of assignments. Satisfaction probability can be computed by evaluating the arithmetic expression that we get by replacing each literal with 0.5, $\vee$ with $+$ and $\wedge$ with $\times$ in the d-DNNF.

3 Model Counting

In this section we review two algorithms for model counting that are necessary for understanding the remainder of this paper. For a more complete treatment of model counting algorithms, see [12].

3.1 Solution Enumeration Using SAT Solvers

In traditional DPLL-algorithm [6], once a decision literal is retracted, it is guaranteed that all search space extending the current assignment has been exhausted. Due to this, we can be certain that the search procedure is complete and does not miss any solution. This is not true, however, for modern SAT solvers [14] that use random restarts and First-UIP backjumping. In the latter, the search backtracks to the last point in search where the learned clause is asserting, and that might mean backjumping over valid solution space. It is not trivial to infer from the current state of the solver which solutions have already been seen and therefore, to prevent the search from finding an already visited solution θ, SAT solvers add the blocking clause $\neg\theta$ in the problem formulation as soon as θ is found.

3.2 DPLL-style Model Counting

One of the most successful approaches for model counting extends the DPLL algorithm (see [2,16,18]). Such model counters borrow many useful features from SAT solvers such as nogood learning, watched literals and backjumping etc to prune parts of search that have no solution. However, they have three additional important optimizations that make them more efficient at model counting as compared to solution enumeration using a SAT solver. A key property of all these optimizations is that their implementation relies on actively maintaining the residual formula during the search. This requires visiting all clauses in the worst case at every node in the search tree.

Say we are solving F and the current assignment is θ. The first optimization in model counting is cube detection; as soon as the residual is empty, we can stop the search and increment our model count by $size(\theta)$. This avoids continuing the search to visit all extensions of the cube since all of them are solutions of F. The second optimization is *caching* [1] which reuses model counts of previously encountered sub-problems instead of solving them again as follows. Say we have computed the model count below θ and it is equal to c, we store c against $F|_\theta$. If, later in the search, our assignment is θ' and $F|_\theta = F|_{\theta'}$, then we can simply increment our count by c by looking up the residual. The third optimization is *dynamic decomposition* and it relies on the following property of Boolean formulas: given a formula G, if (clauses of) G can be split into $G_1, \ldots, G_n$ such that $vars(G_i) \cap vars(G_j) = \emptyset, 1 \leq i \neq j \leq n$, and $\bigcup_{i \in 1..n} vars(G_i) = vars(G)$, then $ct(G) = ct(G_1) \times \ldots \times ct(G_n)$. Model counters use this property and split the residual into disjoint components and count the models of each component and multiply them to get the count of the residual. Furthermore, when used with caching, the count of each component is stored against it so that if a component appears again in the search, then we can retrieve its count instead of computing it again.

4   Projected Model Counting

In this section, we present three techniques for projected model counting.

4.1   Restricting Search to Priority Variables

This algorithm works by slightly modifying the DPLL-based model counters as follows. First, when solving any component, we only allow search decisions on non-priority variables if the component does not have any priority variables. Second, if we find a cube for a component, then the size of that cube is equal to 2 to the power of number of priority variables in the component. Finally, as soon as we find a cube for a component, we recursively mark all its *parent* components (components from earlier decision levels whose decomposition yielded the current component), that do not have any priority variables as solved. As a result, the count of 1 from the last component is propagated to all parent components whose clauses are exclusively on non-priority variables. Essentially, we store the fact that such components are *satisfiable*.

Example 1. Consider the formula F with priority variables $\mathcal{P} = \{p, q, r\}$ and non-priority variables $\mathcal{N} = \{x, y, z\}$.

$$(\neg q, x, \neg p), (\neg r, \neg y, z), (r, \neg z, \neg p), (z, y, \neg p, r), (r, z, \neg y, \neg p), (p, q)$$

Here is the trace of a possible execution using the algorithm in this subsection. We represent a component as a pair of (unfixed) variables and residual clauses.
1a. Decision p. The problem splits into $C_1 = (\{q, x\}, \{(\neg q, x)\})$ and $C_2 = (\{r, y, z\}, \{(\neg r, \neg y, z), (r, \neg z), (z, y, r), (r, z, \neg y)\})$.
2a. We solve C_1 first. Decision $\neg q$. We get $C_3 = (\{x\}, \emptyset)$ and $ct(C_3, \mathcal{P}) = 1$ (trivial), we backtrack to C_1. 2b. Decision q, propagates x, and it is a solution. We backtrack and set $ct(C_1, \mathcal{P}) = ct(C_3, \mathcal{P}) + 1 = 2$.
2c. Now, we solve C_2. Decision r gives $C_4 = (\{y, z\}, \{(\neg y, z)\})$.
3a. Decision z, we get $C_5 = (\{y\}, \emptyset)$ and $ct(C_5, \mathcal{P}) = 1$. We backtrack to level C_2 setting $ct(C_4, \mathcal{P}) = 1$ since the last decision was a non-priority variable.
2d. Decision $\neg r$ fails (propagates z, y, $\neg y$). We set $ct(C_2, \mathcal{P}) = ct(C_4, \mathcal{P}) = 1$ and backtrack to root F to try the other branch.
1b. Decision $\neg p$, propagates q and gives $C_6 = (\{x\}, \emptyset)$ and $C_7 = (\{r, y, z\}, \{(\neg r, \neg y, z)\})$. We note that $ct(C_6, \mathcal{P}) = 1$ (trivial) and move on to solve C_7.
2e. Decision $\neg r$ gives $C_8 = (\{y\}, \emptyset)$ and $C_9 = (\{z\}, \emptyset)$ with counts 1 each. We go back to C_7 to try the other branch.
2f. Decision r gives $C_{10} = (\{y, z\}, \{(\neg y, z)\})$ which is the same as C_4 which has the count of 1. Therefore, $ct(C_7, \mathcal{P}) = ct(C_8, \mathcal{P}) \times ct(C_9, \mathcal{P}) + ct(C_4, \mathcal{P}) = 2$. All components are solved, and there are no more choices to be tried, we go back to root to get the final model count.

A visualization of the search is shown in Figure 1. The overall count is $ct(F, \mathcal{P}) = ct(C_1, \mathcal{P}) \times ct(C_2, \mathcal{P}) + ct(C_6, \mathcal{P}) \times ct(C_7, \mathcal{P}) = 4$.   □

4.2   Blocking Seen Solutions

This approach extends the projected model counting algorithm given in [8], which has been implemented in the ASP solver CLASP [7,9]. The algorithm is

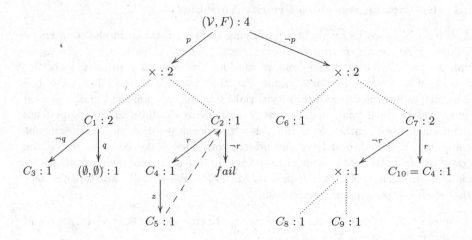

Fig. 1. A visualization of the search tree for model counting with priority variables. Nodes are marked with residual clauses and counts. Dotted edges indicate dynamic decomposition, dashed edged indicate backjumps over non-priority decisions.

originally for model enumeration, not model counting, and therefore, it suffers in instances where there are small number of cubes, but the number of extensions of these cubes to solutions is large. We present a modification of the algorithm that does not have this shortcoming. But first, let us briefly summarize the motivation behind the algorithm and its technical details.

The motivation presented in [8] is absence of any specialized algorithm in SAT (as well as ASP) for model enumeration on a projected set of variables, and the apparent flaws in the following two straight-forward approaches for model enumeration. The first is essentially the approach in 4.1 without dynamic decomposition, caching, and cube detection, i.e., to search on variables in $\mathcal{P}$ first and check for a satisfying extension over $\mathcal{N}$. This interference with the search can be exponentially more expensive in the worst case, although this approach is not compared against other methods in the experiment. The second approach is to keep track of solutions that have been found and for each explored solution θ, add the *blocking clause* $\neg\theta_{\mathcal{P}}$ (this is also presented in [13], although the algorithm restarts and calls MINISAT by adding the clause each time a solution is found). In the worst case, the number of solutions can be exponential in $|\mathcal{P}|$, and this approach, as experiments confirm, can quickly blow up in space. Note that, as opposed to the learned clauses which are redundant w.r.t. the original CNF and can be removed any time during the search, the blocking clauses need to be stored permanently, and cannot be removed naively.

The algorithm of [8] runs in polynomial space and works as follows. At any given time during its execution, the search is divided into *controlled* and *free* search. The free part of the search runs as an ordinary modern DPLL-based SAT solver would run with backjumping, conflict-analysis etc. In the controlled part of the search, the decision literals are strictly on variables in $\mathcal{P}$ and how they are chosen is described shortly. Following the original convention, let bl represent the last level of controlled search space. Initially, it is equal to 0. Every time a solution θ (with projection θ_P) is found, the search jumps back to bl, selects a literal x from θ_P that is unfixed (at bl), and forces it to be the next decision. It increments bl by 1, adds the blocking clause $\neg\theta_P$ and most importantly, *couples* the blocking clause with the decision x in the sense that when we backtrack from x and try (force) $\neg x$, $\neg\theta_P$ can be removed from memory as it is satisfied by $\neg x$. Additionally, backtracking in the controlled region is provably designed to disallow skipping over any solution. Therefore, when we try $\neg x$, all solutions under x will have been explored. Furthermore, with $\neg x$, all subsequent blocking clauses that were added will have been satisfied since all of them include $\neg x$. This steady removal of clauses ensures that the number of blocking clauses at any given time is in $O(|\mathcal{P}|)$.

We now describe how we extend the above algorithm by adding solution minimization to it. We keep a global solution count, initially set to 0. Once a solution θ is found, we generalize (minimize) the solution as shown in the procedure shrink Figure 2. We start constructing the new solution cube S by adding all current decisions from $1 \ldots bl$. Then, for each clause in the problem (C in pseudo-code) and current blocking clauses (B), we intersect it with the current assignment. If the intersection contains a literal whose variable is in $\mathcal{N}$ or S, we skip the clause, otherwise, we add one priority literal from the intersection in S (we choose one with the highest frequency in the original CNF). After visiting all clauses, we use $\neg S$ as a blocking clause instead of the one generated by the algorithm above ($\neg\theta_P$). Finally, we add $2^{|\mathcal{P}|-|S|}$ to the global count. The rest of the algorithm remains the same. Note that the decision literals from the controlled part of the search are necessary to add in the cube, since the algorithm in [8] assumes that once a controlled decision is retracted, all the blocking clauses that were added below it are satisfied. This could be violated by our solution minimization if we do not add controlled decisions to S.

Example 2. Consider the CNF in Example 1. Initially, the controlled search part is empty, $B = \emptyset$ and $bl = 0$ as per the original algorithm. Say CLASP finds the solution: $\{p, \neg q, x, z, r, \neg y\}$. shrink produces the generalized solution: $S = \{r, p\}$ by parsing the clauses $(r, z, \neg p)$ and (p, q) respectively (all other clauses can be satisfied by non-priority literals). We increment the model count by 2 ($2^{3-|S|}$), store the blocking clause $\neg S = (\neg r, \neg p)$ and increment bl by 1. Say, we pick r, due to the added blocking clause, it propagates $\neg p$, which propagates q. Say that CLASP now finds the solution $\{r, \neg p, q, \neg y, z, x\}$. In shrink, we start by including r in S since that is a forced decision, and then while parsing the clauses, we get $S = \{r, \neg p, q\}$. Note that if we didn't have to include the blocking clause $(\neg r, \neg p)$, then we could get away with $S = \{r, q\}$ which would be wrong since that shares the solution $\{r, q, p\}$ with the previous cube. We increment the count to 3 and cannot force any other decision, so we try the decision $\neg r$ in the controlled part. At the same time, upon backtracking, we remove all blocking clauses from B, so it is now empty. Say CLASP finds the solution $\{\neg r, \neg p, q, x, \neg y, z\}$, shrink gives

```
shrink(θ)
    S := {}                                    % universal solution cube
    for (i ∈ 1 ... bl)
        S.add(dec(i))                          % add decision to S
    for (c ∈ C ∪ B)
        f := false
        for (l ∈ C)
            if (l ∈ θ) and (l ∈ N or l ∈ S)
                f := true
                break
        if (f = false)                         % if nothing makes the clause true already
            let p ∈ θ ∩ C : var(p) ∈ P         % pick p with highest freq.
            S.add(p)                           % add literal to cube
    ct := ct + 2^(|P|−|S|)
    B.add(¬S)
```

Fig. 2. Pseudo-code for shrinking a solution θ of original clauses C and blocking clauses B to a solution cube S, adding its count and a blocking clause to prevent its reoccurrence.

$S = \{\neg r, \neg p, q\}$. We increment the count to 4, and when we add $\neg S$ as a blocking clause, there are no more solutions under $\neg r$. Therefore, our final count is 4. The visualization for this example is given in Figure 3.   □

4.3 Counting Models of Projected d-DNNF

As mentioned in Section 2, it is possible to do model counting on d-DNNF in polynomial time (in the size of the d-DNNF), however, once we perform projection on $\mathcal{P}$ (or *forgetting on* $\mathcal{N}$ [5]) by replacing all literals whose variables are in $\mathcal{N}$ with *true*, the resulting logical formula is not deterministic anymore and model counting is no longer tractable (see [5]).

In this approach, we first compute the d-DNNF of F, then project away the literals from the d-DNNF whose variables are in $\mathcal{N}$, convert this projected DNNF back to CNF, and then count the models of this CNF. The pseudo-code is given in Figure 4. The conversion from d-DNNF to CNF is formalized in the procedure d2c, which takes as its input a d-DNNF (as a list of nodes *Nodes*) and returns a CNF C. It is assumed that *Nodes* is topologically sorted, i.e., the children of all nodes appear before their parents. d2c maps nodes to literals in the output CNF with the dictionary *litAtNode*. It also maps introduced (Tseitin) variables to expressions that they represent in a map *litWithHash*. v represents the index of the next Tseitin variable to be created. d2c initializes its variables with the method init(). Next, it visits each node n, and checks its type. If it is a literal and if it is a non-priority variable, then it is replaced with *true* (projected away), otherwise, the node is simply mapped to the literal. If n is an AND or an OR node, then we get corresponding literals of its children from the method simplify. We compute the hash to see if we can reuse some previous introduced variable instead of introducing a new one. If not, then we create a new variable through

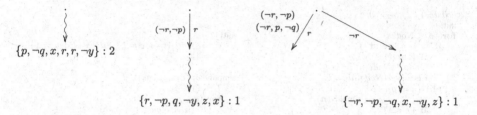

Fig. 3. A visualization of counting models via blocking solutions. The curly arcs indicate free search, ending in a solution, with an associated count. The controlled search is indicated by full arcs, and blocking clauses associated with controlled search decisions are shown on arcs.

the method Tseitin which also posts the corresponding equivalence clauses in C. Finally, we post a clause that says that the literal for the root (which is the last node) should be true. The method simplify essentially maps all the children nodes to their literals. Furthermore, if one of the literals is *true* and the input is an OR-node, it returns a list containing a true literal. For an AND node, it filters all the true literals from the children.

The next theorem shows that the method described in this section for projected model counting is correct.

Theorem 1. $ct(C) = ct(F, \mathcal{P})$

Proof (sketch). The entire algorithm transforms the theory from F to C by producing 2 auxiliary states: the d-DNNF of F (let us call it D) and the projection of this d-DNNF (let us call this $D_{\mathcal{P}}$). By definition, F and D are logically equivalent. On the other end, notice that the models of $D_{\mathcal{P}}$ and C are in one-to-one correspondence. Although the two are not logically equivalent due to the addition of Tseitin variables, it can be shown that these variables do not introduce any extra model nor eliminate any existing model since they are simply functional definitions of variables in $\mathcal{P}$ by construction (as a side note, the only reason for introducing these variables is to efficiently encode $D_{\mathcal{P}}$ as CNF, otherwise, C and $D_{\mathcal{P}}$ would be logically equivalent). Furthermore, we can show that the simplifications (replacing *true* $\vee E$ with *true* and *true* $\wedge E$ with E) in the procedure simplify, and reusing Tseitin variables (through hashing) also do not affect the bijection. This just leaves us with the task of establishing bijection between the models of D and $D_{\mathcal{P}}$, which, fortunately, has already been done in [4]. Theorem 9 in the paper says that replacing non-priority literals with true literals in a d-DNNF is a proper projection operation, and Lemma 3 establishes logical equivalence between D and $D_{\mathcal{P}}$ modulo variables in $\mathcal{P}$.

Example 3. Consider the formula F with priority variables p, q and non-priority variables x, y, z:

$$(\neg x, p), (q, \neg x, y), (\neg p, \neg y, \neg z, q), (x, q), (\neg q, p)$$

```
d2c(Nodes)
    init()
    for (n ∈ Nodes)                                  init()
        if (n is a literal l)                            C = (), litAtNode = {}, litWithHash = {}
            if (var(l) ∈ 𝒩)                              v := |𝒱|
                litAtNode[n] := true
            else litAtNode[n] := l                   simplify(op(c₁, ..., cₖ))
        elif (n = op(c₁, ..., cₖ))                       L = ()
            (l₁, ..., lⱼ) := simplify(n)                 for (c ∈ c₁, ..., cₖ)
            if (j = 1)                                       if (litAtNode[c] = true)
                litAtNode[n] := l₁                               if (op = OR) return (true)
            else                                             else
                h := hash(op, (l₁, ..., lⱼ))                     L.add(litAtNode[c])
                if (litWithHash.hasKey(h))               return L
                    litAtNode[n] := litWithHash[h]
                else                                 Tseitin(op, (l₁, ..., lⱼ))
                    v' := Tseitin(op, (l₁, ..., lⱼ))     Add clauses v ⇔ op(l₁, ..., lⱼ) in C
                    litAtNode[n] := v'                   v := v + 1
                    litWithHash[h] := v'                 return v - 1
    C.add({litAtNode[Nodes.last()]})
    return C
```

Fig. 4. Pseudo-code for projected model counting via counting models of CNF encoding of projected DNNF

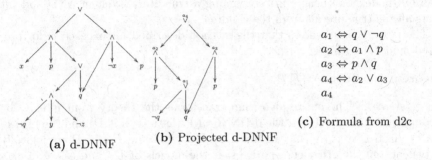

(a) d-DNNF   (b) Projected d-DNNF   (c) Formula from d2c

$a_1 \Leftrightarrow q \vee \neg q$
$a_2 \Leftrightarrow a_1 \wedge p$
$a_3 \Leftrightarrow p \wedge q$
$a_4 \Leftrightarrow a_2 \vee a_3$
a_4

Fig. 5. Example of application of d2c

The projected model count is 2 $((p, q)$ and $(p, \neg q))$.

Figure 5 shows the initial d-DNNF (5a), the DNNF obtained by replacing all non-priority literals by *true* and simplifying (5b) and the d2c translation of the projected DNNF (5c). Notice that if we perform model counting naively on the projected DNNF, we get a count of 3 since we double count the model (p, q). The satisfaction probability is:

$$\left(\frac{1}{2} + \frac{1}{2}\right) \times \frac{1}{2} + \left(\frac{1}{2} \times \frac{1}{2}\right) = \frac{3}{4}$$

From satisfaction probability, we get the wrong model count $2^2 \times \frac{3}{4} = 3$. However, if we count the models of the translated formula in (5c), we get the correct count of 2.   □

5   Experiments

We compare the following solvers on various benchmarks: CLASP in its projection mode (CLASP), our extension of clasp with cube minimization (#CLASP), model counting with searching on priority variables first (DSHARP_P), and counting models of projected DNNF (D2C). In each row of the following tables, $|\mathcal{P}|$ is the number of priority variables. T and D represent the execution time and number of decisions taken by the solver. R is a parameter to gauge the quality of cubes computed by #CLASP, the higher it is, the better. It is equal to $\log_2(\frac{\#sols}{\#cubes})$. A value of 0 indicates that all solution cubes computed have size 1, while the maximum value is equal to the number of priority variables, which is the unique case when there is only one cube and every assignments to priority variables is a solution. R essentially quantifies the advantage over enumeration, the less constrained a problem is, and the more general the cubes are, the higher the advantage. S is the size (in bytes) of the CNF computed by D2C that is subsequently given to the solver SHARPSAT for model counting. The timeout for all experiments is 10 minutes. All times are shown in seconds. The experiments were run on NICTA's HPC cluster. [1]

5.1   Uniform Random 3-SAT and Boolean Circuits

Table 1 shows the results from uniform random 3-SAT and random Boolean circuits. In this table, for each problem instance, we show how the solvers perform as we increase the number of priority variables. A "..." after a row means that every solver either ran out of time or memory for all subsequent number of priority variables until the next one shown. For each instance, a row is added that provides the following information about it: name, number of solutions as reported by DSHARP, number of variables and clauses and time and decisions taken by DSHARP. Note that this time should be added to the time of D2C in order to get the actual time of D2C approach.

Let us look at the results form uniform random 3-SAT. All instances have 100 variables, and the number of clauses is varied. We try clause-to-variable ratios of 1, 1.5, 2, 3 and 4. Note that for model counting, the difficulty peaks at the ratio of approximately 1.5 [12]. For the first 3 instances, #CLASP is the clear winner while CLASP also does well, DSHARP_P lags behind both, and D2C does not even work since the original instance cannot be solved by DSHARP. For #CLASP, as we increase the number of clauses, the cube quality decreases due to the problem becoming more constrained and cube minimization becoming less effective. For 300 clauses, we see a significant factor coming into play for DSHARP_P. The original instance is solved by DSHARP. As we increase the number of priority variables until nearly the middle, the performance of DSHARP_P degrades, but after 50 priority variables, it starts getting better. This is because the degradation due to searching on priority variables first becomes less significant and the

---

[1] All benchmarks and solvers are available at: http://people.eng.unimelb.edu.au/pstuckey/countexists

search starts working more naturally in its VSADS mode [17]. D2C also solves two rows in this instance but is still largely crippled as compared to other solvers. Finally, with 400 clauses, we are well past the peak difficulty and the number of models is small enough to be enumerated efficiently by CLASP. All solvers finish all rows of this instance in less than .15 seconds. We tried the same ratios for 200 and 300 variables. For 200 variables, we saw the same trend, although the problem overall becomes harder and the number of solved rows decreases. For 300 variables, the problem becomes significantly harder to be considered a suitable benchmark.

The Boolean circuits are generated with n variables as follows: we keep a set initialized with the n original variables, then as long as the set is not a singleton, we randomly pick an operator o (AND, OR, NOT), remove random operands V from the set, create a new variable v and post the constraints $v \leftrightarrow o(V)$ and put v back in the set. The process is repeated c times. In the table, we show the results where n is 30, and c is 1,5,10. Note that a higher value of c means that the problem is more constrained. Overall, for all instances, DSHARP_P is the superior approach, followed by CLASP; and D2C is better than #CLASP in $c = 1$ but the converse is true for higher values of c. All solvers find $c = 5$ to be the most difficult instance. We saw similar trends for different values of n that have appropriate hardness with same values of c.

5.2   Planning

Table 2 summarizes performance of different projected model counting algorithms on checking robustness of partially ordered plans to initial conditions. We take five planning benchmarks: depots, driver, rovers, logistics, and storage. For each benchmark, we have two variants, one with the goal state fixed and one where the goal is relaxed to be any viable goal (shown with a capital A in the table representing *any* goal). For the two variants, the priority variables are defined such that by doing projected model counting, we count the following. For the first problem, we count the number of initial states the given plan can achieve the given goal from. For the second problem, we count the number of initial states plus all goal configurations that the given plan works for. Each row in the table represents the summary of 10 instances of same size. The first 3 columns show the instance parameters. For each solver, ✓shows how many instance the solver was able to finish within time and memory limits. All other solver parameters are averages over finished instances. Another difference from the previous table is that we have added the execution time of DSHARP in D2C and DSHARP time is shown in parenthesis. There was no case in which only DSHARP finished and the remaining steps of D2C did not finish.

Overall, DSHARP_P solves the most instances (42), followed by #CLASP (41), D2C (34), and finally CLASP which solves only 4 instances from the storage benchmark, and otherwise suffers due to the inability to detect cubes. DSHARP_P and D2C only fail on all instances in 2 benchmarks while #CLASP fails in 4, so they are more robust in that sense. For D2C, the running time is largely taken by producing the d-DNNF and the second round of model counting is relatively cheaper. The cube quality of #CLASP is quite significant for all instances that it solves.

Table 1. Results from random uniform 3-SAT and Boolean circuits

$|\mathcal{P}|$	#	CLASP T	D	#CLASP T	D	R	DSHARP_P T	D	D2C T	D	S
		UF #=— $|V|$=100 $|C|$=100 T=— D=—									
5	32	0	2271	0	291	3.00	.04	1150	—	—	—
10	1024	.01	71309	0	533	7.19	.99	35927	—	—	—
15	32768	.40	2023146	0	1888	10.30	7.92	370034	—	—	—
25	2.7e+07	345.57	1.4e+09	0	10584	17.36	—	—	—	—	—
35	1.8e+10	—	—	.02	62016	24.32	—	—	—	—	—
50	1.9e+14	—	—	107.75	7.1e+07	27.04					
...											
		UF #=— $|V|$=100 $|C|$=150 T=— D=—									
5	32	0	1937	0	247	3.00	.03	1286	—	—	—
10	1024	.02	54077	.01	2933	4.19	.68	27142	—	—	—
15	32768	.42	1767073	0	2101	9.96	31.63	1057551	—	—	—
25	2.1e+07	270.91	8.7e+08	.34	393130	11.25	—	—	—	—	—
35	2.8e+09	—	—	24.98	1.5e+07	12.84	—	—	—	—	—
...											
		UF #=— $|V|$=100 $|C|$=200 T=— D=—									
5	32	0	1354	0	259	2.42	.04	1304	—	—	—
10	1024	.01	47771	0	1370	5.25	1.12	37596	—	—	—
15	30712	.43	1408296	0	4659	8.29	37.07	874826	—	—	—
25	1.8e+07	218.20	6.4e+08	1.69	1801261	8.54	—	—	—	—	—
...											
		UF #=2.603e+11 $|V|$=100 $|C|$=300 T=31.44 D=571163									
5	32	0	986	0	646	0.75	.05	671	40.92	31	865K
10	970	.02	25441	.02	11450	1.20	2.07	14146	102.63	969	1.6M
15	12990	.22	290973	.11	61663	2.30	12.74	144211	—	—	6.7M
25	226117	3.84	3432170	1.66	464908	3.21	57.57	808670	—	—	15M
35	5126190	49.02	6.6e+07	15.65	3367386	4.67	161.67	2834211	—	—	38M
50	—	—	—	—	—	—	—	—	—	—	61M
65	1.6e+09	—	—	—	—	—	70.80	1552565	—	—	89M
75	2.0e+10	—	—	—	—	—	70.74	1330586	—	—	104M
85	2.9e+10	—	—	—	—	—	50.18	780597	—	—	113M
100	2.6e+11	—	—	—	—	—	28.62	571163	—	—	134M
		UF #=45868 $|V|$=100 $|C|$=400 T=.05 D=244									
5	7	0	1078	0	907	0.49	.08	219	.01	6	549
10	25	0	1308	0	1103	0.94	.14	322	.01	17	1.3K
15	32	0	1582	0	1242	1.09	.12	376	.01	21	3.1K
25	105	.01	2242	0	1290	2.32	.09	373	.01	34	3.4K
35	246	.01	3068	0	1338	2.52	.06	363	.01	107	8.6K
50	952	.01	6737	.01	2241	3.24	.05	361	.02	202	14K
65	3417	.01	16388	.01	2889	4.41	.05	262	.09	321	21K
75	7964	.04	26979	.02	2845	5.32	.05	250	.04	426	22K
85	13274	.04	36445	.02	3993	5.18	.05	237	.06	563	26K
100	45868	.11	46623	.03	4639	6.74	.04	244	.07	688	31K

Table 1. (continued)

$|\mathcal{P}|$	#	CLASP		#CLASP			DSHARP_P		D2C		
		T	D	T	D	R	T	D	T	D	S
				$n=30$ $c=1$ #=9.657e+08 $|V|$=99 $|C|$=167 T=0 D=111							
5	16	0	409	0	292	0.54	0	113	.01	4	1.1K
9	160	0	3418	0	1184	1.54	0	143	0	8	1.2K
14	552	0	9305	0	5030	1.00	0	82	.01	22	3.8K
24	248960	1.16	2718019	1.49	833682	2.07	0	130	.01	169	6.1K
34	1621760	6.13	1.2e+07	6.45	1999088	3.13	.01	111	.05	656	9.7K
49	3.9e+07	104.26	1.9e+08	353.25	1.4e+07	4.78	.01	162	.10	1393	14K
64	1.5e+08	394.21	4.6e+08	—	—	—	0	129	.18	2143	18K
74	4.4e+08	—	—	—	—	—	.01	108	.21	2982	20K
84	7.2e+08	—	—	—	—	—	0	99	.20	2624	24K
99	9.7e+08	—	—	—	—	—	.01	111	.20	2517	27K
				$n=30$ $c=5$ #=9.426e+07 $|V|$=389 $|C|$=867 T=288.45 D=1036363							
19	12192	.16	155331	.75	146058	0.00	16.48	120619	—	—	5.8M
38	208716	2.57	1882991	15.23	1985136	0.00	95.37	834705	—	—	47M
58	1.2e+07	93.69	3.7e+07	—	—	—	—	—	—	—	100M
97	3.3e+07	248.76	6.9e+07	—	—	—	427.85	1509308	—	—	171M
136	6.1e+07	428.89	9.1e+07	—	—	—	—	—	—	—	252M
...											
291	9.3e+07	—	—	—	—	—	300.78	985065	—	—	574M
330	9.4e+07	—	—	—	—	—	299.02	1074927	—	—	672M
389	9.4e+07	—	—	—	—	—	308.84	1036363	—	—	783M
				$n=30$ $c=10$ #=5066 $|V|$=766 $|C|$=1771 T=.32 D=1400							
38	282	.01	1196	.03	1797	0.00	.31	1412	.08	256	36K
76	1618	.02	3479	.12	5434	0.00	.40	1600	.81	2046	137K
114	2581	.03	4984	.21	7953	0.00	.52	1787	1.69	3702	173K
191	4948	.05	5558	.52	12243	0.00	.54	1904	4.98	7519	330K
268	5066	.07	5458	.63	12235	0.00	.38	1784	6.69	10508	478K
383	5066	.09	5513	.85	12356	0.00	.69	1975	9.27	12528	698K
497	5066	.08	5253	1.47	12471	0.00	.39	1680	12.46	11818	1.1M
574	5066	.11	5211	1.68	12358	0.00	.52	1616	14.85	11911	1.1M
651	5066	.09	5500	1.21	12546	0.00	.36	1500	13.42	11807	1.4M
766	5066	.09	5072	2.06	12389	0.00	.33	1400	21.61	11644	1.6M

Table 2. Results from robustness of partially ordered plans to initial conditions

Instance				CLASP			#CLASP				DSHARP_P			D2C			
Name	$|\mathcal{V}|$	$|C|$	$|\mathcal{P}|$	✓	T	D	✓	T	D	R	✓	T	D	✓	T	D	S
depotsA	9402	211901	224	0	—	—	0	—	—	—	1	24.82	92206	1	8.34 (6.98)	1813	154K
depots	9211	211796	111.8	0	—	—	2	4.16	4.43e+6	31.54	1	24.74	91909	1	7.71 (6.67)	1642	149K
driverA	2068	12798	135	0	—	—	0	—	—	—	5	36.01	27104.8	3	164.73 (161.42)	68.33	150.5K
driver	1999	12700	68	0	—	—	10	0.31	1.23e+5	51.7	5	15.8	1.45e+4	3	118.67 (116.00)	29.3	109K
logisticsA	18972	324568	447	0	—	—	0	—	—	—	0	—	—	0	—	—	—
logistics	18702	324352	224	0	—	—	6	33.52	1.81e6	165.09	0	—	—	0	—	—	—
roversA	3988	27634	209	0	—	—	0	—	—	—	5	69.92	51965	3	1.11 (1.06)	53.33	5.37K
rovers	3851	27535	104	0	—	—	10	0.30	36769.7	88.16	5	76.26	52245.4	3	1.04 (1.01)	12.67	3.4K
storageA	915	3465	93	1	454.2	2.5e9	3	43.81	3.89e7	18.01	10	49.04	47112.50	9	103.35 (78.60)	1964.2	440.21K
storage	851	3420	47	3	15.05	7.87e7	10	0.05	30686	30.47	10	15.48	12444	9	57.1 (53.46)	625.67	254.58K

6   Related Work and Conclusion

The area of Boolean Quantifier Elimination (BQE) seems closely related to projected model counting. Although the goal in BQE is to produce a non-priority variables free representation (usually CNF), some algorithms can also be adapted for projected model counting. Of particular interest are techniques in [3] and [11]. The algorithm of [3] finds cubes in decreasing (increasing) order of cube size (number of literals in cube). While this approach does not require cube minimization, it does not run in polynomial space, and if a problem only has large cubes, then significant time might be wasted searching for smaller ones. The second interesting approach, with promising results, is given in [11], which uses a DPLL-style search and also decomposes the program at each step like the approach described in 4.1. A good direction for future work is to investigate how well these techniques lend themselves to projected model counting and whether there is any room for integration with the ideas presented in this paper.

In this paper we compare four algorithms for projected model counting. We see that each algorithm can be superior in appropriate circumstances:

- When the number of solutions is small then CLASP [8] is usually the best.
- When the number of solution cubes is much smaller than solutions, and there is not much scope for component caching, then #CLASP is the best.
- When component caching and dynamic decomposition are useful then DSHARP_P is the best.
- Although D2C is competitive, it rarely outperforms both #CLASP and DSHARP_P. Having said that, D2C approach has another important aspect besides projected model counting. It is a method to perform projection on a d-DNNF without losing determinism. This can be done by computing the d-DNNF of the CNF produced by the d2c procedure (instead of model counting), and then simply forgetting the Tseitin variables (replacing with *true*). It can be shown that this operation preserves determinism. Furthermore, our experiments show that the last model counting step takes comparable time to computing the first d-DNNF in most cases (and in many cases, takes significantly less time), which means that the approach is an efficient way of performing projection on a d-DNNF.
  While we use d-DNNF for D2C approach, it is possible to use other, less succinct, languages like Ordered Binary Decision Diagrams (OBDDs). We leave the comparison with other possible knowledge compilation-based approaches for projected model counting as future work.

As the problem of projected model counting is not heavily explored, there is significant scope for improving algorithms for it. A simple improvement would be to portfolio approach to solving the problem, combining all four of the algorithms, to get something close to the best of each of them.

References

1. Bacchus, F., Dalmao, S., Pitassi, T.: DPLL with Caching: A new algorithm for #SAT and Bayesian Inference. Electronic Colloquium on Computational Complexity (ECCC) **10**(003) (2003)
2. Bacchus, F., Dalmao, S., Pitassi, T.: Solving #SAT and Bayesian Inference with Backtracking Search. Journal of Artificial Intelligence Research (JAIR) **34**, 391–442 (2009)
3. Brauer, J., King, A., Kriener, J.: Existential quantification as incremental SAT. In: Gopalakrishnan, G., Qadeer, S. (eds.) CAV 2011. LNCS, vol. 6806, pp. 191–207. Springer, Heidelberg (2011)
4. Darwiche, A.: Decomposable negation normal form. J. ACM **48**(4), 608–647 (2001)
5. Darwiche, A., Marquis, P.: A knowledge compilation map. Journal of Artificial Intelligence Research (JAIR) **17**, 229–264 (2002)
6. Davis, M., Logemann, G., Loveland, D.: A machine program for theorem proving. Communications of the ACM **5**(7), 394–397 (1962)
7. Gebser, M., Kaufmann, B., Neumann, A., Schaub, T.: Conflict-driven answer set solving. In: Proceedings of the 20th International Joint Conference on Artificial Intelligence, p. 386. MIT Press, January 2007
8. Gebser, M., Kaufmann, B., Schaub, T.: Solution enumeration for projected boolean search problems. In: van Hoeve, W.-J., Hooker, J.N. (eds.) CPAIOR 2009. LNCS, vol. 5547, pp. 71–86. Springer, Heidelberg (2009)
9. Gebser, M., Kaufmann, B., Schaub, T.: Conflict-driven answer set solving: From theory to practice. Artificial Intelligence **187**, 52–89 (2012)
10. Ginsberg, M.L., Parkes, A.J., Roy, A.: Supermodels and robustness. In: Proceedings of the Fifteenth National Conference on Artificial Intelligence and Tenth Innovative Applications of Artificial Intelligence Conference, AAAI 1998, IAAI 1998, July 26–30, 1988, Madison, Wisconsin, USA, pp. 334–339 (1998)
11. Goldberg, E., Manolios, P.: Quantifier elimination by dependency sequents. In: Formal Methods in Computer-Aided Design (FMCAD), pp. 34–43, October 2012
12. Gomes, C.P., Sabharwal, A., Selman, B.: Model counting (2008)
13. Klebanov, V., Manthey, N., Muise, C.: SAT-based analysis and quantification of information flow in programs. In: Joshi, K., Siegle, M., Stoelinga, M., D'Argenio, P.R. (eds.) QEST 2013. LNCS, vol. 8054, pp. 177–192. Springer, Heidelberg (2013)
14. Moskewicz, M.W., Madigan, C.F., Zhao, Y., Zhang, L., Malik, S.: Chaff: engineering an efficient SAT solver. In: Proceedings of the 38th Design Automation Conference, pp. 530–535. ACM (2001)
15. Palacios, H., Bonet, B., Darwiche, A., Geffner, H.: Pruning conformant plans by counting models on compiled d-dnnf representations. In: Proceedings of the Fifteenth International Conference on Automated Planning and Scheduling (ICAPS 2005), June 5–10, 2005, Monterey, California, USA, pp. 141–150 (2005)
16. Sang, T., Bacchus, F., Beame, P., Kautz, H.A., Pitassi, T.: Combining component caching and clause learning for effective model counting. In: Proceedings of the 7th International Conference on Theory and Applications of Satisfiability Testing (SAT 2004) (2004)
17. Sang, T., Beame, P., Kautz, H.: Heuristics for fast exact model counting. In: Bacchus, F., Walsh, T. (eds.) SAT 2005. LNCS, vol. 3569, pp. 226–240. Springer, Heidelberg (2005)

18. Thurley, M.: sharpSAT – counting models with advanced component caching and implicit BCP. In: Biere, A., Gomes, C.P. (eds.) SAT 2006. LNCS, vol. 4121, pp. 424–429. Springer, Heidelberg (2006)
19. Zengler, C., Küchlin, W.: Boolean quantifier elimination for automotive configuration – a case study. In: Pecheur, C., Dierkes, M. (eds.) FMICS 2013. LNCS, vol. 8187, pp. 48–62. Springer, Heidelberg (2013)

Computing Maximal Autarkies with Few and Simple Oracle Queries

Oliver Kullmann[1] and Joao Marques-Silva[2(✉)]

[1] Computer Science Department, Swansea University, Swansea, UK
[2] INESC-ID, IST, University of Lisbon, Lisbon, Portugal
jpms@tecnico.ulisboa.pt

Abstract. We consider the algorithmic task of computing a *maximal autarky* for a clause-set F, i.e., a partial assignment which satisfies every clause of F it touches, and where this property is destroyed by adding any non-empty set of further assignments. We employ SAT solvers as oracles here, and we are especially concerned with minimising the number of oracle calls. Using the standard SAT oracle, $\log_2(n(F))$ oracle calls suffice, where $n(F)$ is the number of variables, but the drawback is that (translated) cardinality constraints are employed, which makes this approach less efficient in practice. Using an extended SAT oracle, motivated by the capabilities of modern SAT solvers, we show how to compute maximal autarkies with $2\sqrt{n(F)}$ simpler oracle calls, by a novel algorithm, which combines the previous two main approaches, based on the autarky-resolution duality and on SAT translations.

1 Introduction

A well-known application area of SAT solvers is the analysis of over-constrained systems, i.e. systems of constraints that are inconsistent. A number of computational problems can be related with the analysis of over-constrained systems. These include minimal explanations of inconsistency, and minimal relaxations to achieve consistency. Pervasive to these computational problems is the problem of computing a "maximal autarky" of a propositional formula, since clauses satisfied by an autarky cannot be included in minimal explanations of inconsistency or minimal relaxations to achieve consistency. In the experimental study [26] it was realised that using as few SAT calls as possible, via cardinality-constraints, performs much worse than using a linear number of calls. To use only a sublinear number of calls, without using cardinality constraints, is the goal of this paper.

Given a satisfiable clause-set F and a partial assignment φ, in general $\varphi * F$, the result of the application (instantiation) of φ to F, might be unsatisfiable. φ is an *autarky* for (arbitrary) F iff every clause C of F touched by φ (i.e., $\mathrm{var}(C) \cap \mathrm{var}(\varphi) \neq \emptyset$) is satisfied by φ (i.e., $\exists x \in C : \varphi(x) = 1$). Now if F is satisfiable,

This work is partially supported by SFI PI grant BEACON (09/IN.1/I2618), FCT grant POLARIS (PTDC/EIA-CCO/123051/2010) and national funds through FCT with reference UID/CEC/50021/2013.

M. Heule and S. Weaver (Eds.): SAT 2015, LNCS 9340, pp. 138–155, 2015.
DOI: 10.1007/978-3-319-24318-4_11

then also $\varphi * F$ is satisfiable, since due to the autarky property holds $\varphi * F = \{C \in F : \mathrm{var}(C) \cap \mathrm{var}(\varphi) = \emptyset\} \subseteq F$. Thus "autarky reduction" $F \rightsquigarrow \varphi * F$ can take place (satisfiability-equivalently). An early use of autarkies is [4], for the solution of 2-SAT. The notion "autarky" was introduced in [30] for faster k-SAT decision, which can be seen as an extension of [4]. For an overview of such uses of autarkies for SAT solving see [8]. Besides such incomplete usage (using only autarkies "at hand"), the complete search for "all" autarkies (or the "strongest" one) is of interest. Either with (clever) exponential-time algorithms, or for special classes of clause-sets, where polynomial-time is possible, or considering only restricted forms of autarkies to enable polynomial-time handling; see [11] for an overview. In [18, 19] autarky theory is generalised to non-boolean clause-sets.

Finitely many autarkies can be composed to yield another autarky, which satisfies precisely the clauses satisfied by (at least) one of them; this was first observed in [31]. So complete autarky reduction for a clause-set F, elimination of clauses satisfied by some autarky as long as possible, yields a unique sub-clause-set, called the *lean kernel* $\mathrm{N_a}(F) \subseteq F$, as introduced in [14] and further studied in [16]; we note that $F \in \mathcal{SAT} \Leftrightarrow \mathrm{N_a}(F) = \top$, where $\top$ is the empty clause-set. Clause-sets without non-trivial autarkies are called *lean*, and are characterised by $\mathrm{N_a}(F) = F$; the set of all lean clause-sets is called $\mathcal{LEAN}$, and was shown to be coNP-complete in [16]. A *maximal autarky* for F is one which can not be extended; note that a maximal autarky φ always exist, where $\varphi = \langle \rangle$, the empty partial assignment, iff F is lean. An autarky φ is maximal iff $\mathrm{var}(\varphi) = \mathrm{var}(F) \setminus \mathrm{var}(\mathrm{N_a}(F))$. Thus $\mathrm{var}(F) \setminus \mathrm{var}(\mathrm{N_a}(F))$ is called the *largest autarky var-set*. For a maximal autarky φ the result of the autarky reduction is $\mathrm{N_a}(F)$, while any autarky which yields $\mathrm{N_a}(F)$ is called *quasi-maximal*.

Algorithmic problems associated with autarkies. The basic algorithmic problems related to general "autarky systems", which allow to specialise the notion of autarky, for example in order to enable polynomial-time computations, are discussed in [11, Section 11.11.6]. Regarding *decision problems*, for this paper only one problem is relevant here, namely AUTARKY EXISTENCE, deciding whether a clause-set F has a non-trivial autarky; the negation is LEAN, deciding whether $F \in \mathcal{LEAN}$. An early oracle-result is [16, Lemma 8.6], which shows, given an oracle for LEAN, how to compute LEAN KERNEL with at most $n(F)$ oracle calls (for all "normal autarky systems", using the terminology from [11, Section 11.11]). We are concerned in this paper with the *functional problems*, where the four relevant problems are as follows, also stating the effort for checking a solution:

NON-TRIVIAL AUTARKY: Find some non-trivial autarky (if it exists; otherwise return the empty autarky). Checking an autarky is in P.

QUASI-MAXIMAL AUTARKY or MAXIMAL AUTARKY: Find a (quasi-)maximal autarky; by a trivial computation, from a quasi-maximal autarky we can compute a maximal one. Checking that φ is a quasi-maximal autarky for F means checking that φ is an autarky (easy), and that $\varphi * F$ is lean, and so checking is in coNP. A quasi-maximal autarky can be computed by repeated calls to NON-TRIVIAL AUTARKY (until no non-trivial autarky exists anymore).

NON-TRIVIAL VAR-AUTARKY: Find the var(iable)-set of some non-trivial autarky (if it exists; otherwise return the empty set). Checking that V is the variable-set of an autarky means checking that $F[V]$, the restriction of F to V, is satisfiable, thus checking is in NP.

(QUASI-)MAXIMAL VAR-AUTARKY or LEAN KERNEL: Compute the largest autarky var-set (or a quasi-maximal one), or compute the lean kernel; all three tasks are equivalent by trivial computations. Checking that V is the largest autarky var-set means checking that $F[V]$ is satisfiable and that $\{C \in F : \operatorname{var}(F) \cap V = \emptyset\}$ is lean, so checking is in D^P ([32]). The solution to MAXIMAL VAR-AUTARKY or to LEAN KERNEL is unique and always exists. The var-set of a quasi-maximal autarky can be computed by repeated calls to NON-TRIVIAL VAR-AUTARKY.

Just having the var-set of the autarky φ enables us to perform the autarky reduction $F \rightsquigarrow \varphi * F$, namely $\varphi * F = \{C \in F : \operatorname{var}(C) \cap \operatorname{var}(\varphi) = \emptyset\}$, but from the var-set $\operatorname{var}(\varphi)$ in general we can not derive the autarky φ itself, which is needed to provide a certificate for the autarky-property. For example, F is satisfiable iff $\operatorname{var}(F)$ is the largest autarky var-set, and in general without further hard work it is not possible to obtain the satisfying assignment from (just) the knowledge that F is satisfiable. An interesting case is discussed in [21, Subsection 4.3] and (in greater depth) in [22, Section 10], where we can compute a certain autarky reduction in polynomial-time, but it is not known how to find the autarky (efficiently). So NON-TRIVIAL VAR-AUTARKY is weaker than NON-TRIVIAL AUTARKY, and MAXIMAL VAR-AUTARKY is weaker than MAXIMAL AUTARKY. We tackle in this paper the hardest problem, MAXIMAL AUTARKY.

To obtain a complexity calibration, we can consider the computational model where polynomial-time computation and (only) one oracle call is used. Then MAXIMAL VAR-AUTARKY is equivalent to PARALLEL SAT, which has as input a list $F_1, \ldots, F_m$ of clause-sets, and as output m bits deciding satisfiability of the inputs: On the one hand, given these $F_1, \ldots, F_m$, make them variable-disjoint and input their union to the MAXIMAL VAR-AUTARKY oracle — F_i is satisfiable iff $\operatorname{var}(F_i)$ is contained in the largest autarky var-set. On the other hand it is an easy exercise to see, that for example via the translation $F \rightsquigarrow t(F)$ used in this paper, introduced as Γ_2 in [26], we can compute the largest autarky var-set by inputting $t(F) \cup \{\{v_1\}\}, \ldots, t(F) \cup \{\{v_n\}\}$ to PARALLEL-SAT, where $\operatorname{var}(F) = \{v_1, \ldots, v_n\}$. Similarly it is easy to see that MAXIMAL AUTARKY is equivalent to PARALLEL FSAT (here now also the satisfying assignments are computed).

General approaches for the lean kernel. See [11, Section 11.10] for an overview. A fundamental method for computing a (quasi-)maximal autarky, strengthened in this paper, uses the *autarky-resolution duality* ([14, Theorem 3.16]): the variables in the largest autarky var-set are precisely the variables not usable in any resolution refutation. The basic algorithm, reviewed as algorithm $\mathcal{A}_0$ in Definition 5 in this paper (with a refined analysis), was first given in [15] and somewhat generalised in [11, Theorem 11.10.1]; see [20] for a discussion and some experimental results. A central concept is, what in this paper we call an *extended SAT oracle* $\mathcal{O}_{01}$, which for a satisfiable input outputs a satisfying assignment, while $\mathcal{O}_{01}$ on an unsatisfiable input outputs the variables used by some resolution refutation. In order to also accommodate polynomial-time results, the oracle $\mathcal{O}_{01}$ may get its inputs from a class $\mathcal{C}$ of clause-sets, which is stable (closed) under removal of variables. However, for the new algorithm of this paper (Algorithm $\mathcal{A}_{01}$ presented in Theorem 1), we do not consider classes $\mathcal{C}$ as for $\mathcal{A}_0$, since the input is first transformed, and then also some clauses are added, which would complicate the requirements on $\mathcal{C}$. The other main method to compute autarkies uses reduction to SAT problems, denoted by $F \rightsquigarrow t(F)$ in this paper, where the solutions of $t(F)$ correspond to the autarkies of F. This was started by [25], and further extended first in [11, Subsection 11.10.4], and then in [26], which contains a thorough discussion of the various reductions. The basic algorithm here is $\mathcal{A}_1$ (Definition 8), which iteratively extracts autarkies via the translation until reaching the lean kernel. When combined with cardinality constraints and binary search, indeed $\log_2 n$ oracle calls are sufficient; see Algorithm $\mathcal{A}_{bs}$ (Definition 9). But these cardinality constraints make the tasks much harder for the SAT oracle. The new algorithm $\mathcal{A}_{01}$ of this paper (Definition 10) indeed combines the two basic approaches $\mathcal{A}_0, \mathcal{A}_1$, by applying the autarky-resolution duality to the translation and using a more clever choice of "steering clauses" to search for autarkies. To better understand this combination of approaches, all four algorithms $\mathcal{A}_0$, $\mathcal{A}_1$, $\mathcal{A}_{bs}$ and $\mathcal{A}_{01}$, are formulated in a unified way, striving for elegance *and* precision. One feature is, that the input is updated in-place, which not only improves efficiency, but also simplifies the analysis considerably.

Related literature. When for $\mathcal{C}$ (as above) the extended SAT oracle $\mathcal{O}_{01}$ runs in polynomial time, then by [11, Theorem 11.10.1] the algorithm $\mathcal{A}_0$ computes a quasi-maximal autarky in polynomial time. The basic applications to 2-CNF, HORN, and the case that every variable occurs at most twice, are reviewed in [11, Section 11.10.9]. The other known polytime results regarding computation of the lean kernel use the *deficiency*, as introduced in [6], and further studied in [14]). Here the above algorithm $\mathcal{A}_0$ can not be employed, since crossing out variables can increase this measure (see [18, Section 10] for a discussion). [13, Theorem 4.2] shows that the lean kernel is computable in polynomial time for bounded (maximal) deficiency. In [5] the weaker result, that SAT is decidable in polynomial time for bounded maximal deficiency, has been shown, and strengthened later in [36] to fixed-parameter tractability, which is unknown for the computation of the lean kernel. [18, Theorem 10.3] shows that also a maximal autarky can be computed in polynomial time for bounded maximal deficiency, and this for generalised non-boolean clause-sets, connecting to constraint satisfaction.

The connection to the field of hypergraph 2-colouring, the problem of deciding whether one can colour the vertices of a hypergraph with two colours, such that monochromatic hyperedges are avoided, has been established in [17]; see [11, Section 11.12.2] and [22, Subsection 1.6] for overviews. Exploiting the solution of a long-outstanding open problem by [29,33], the lean kernel is computable in polynomial time by [17] for classes of clause-sets, which by [22, Subsection 1.6], via the translation of SAT problems into hypergraph 2-colourability problems, strongly generalises the polytime results (discussed above) for maximal deficiency of clause-sets (partially proven, partially conjectured).

Autarkies have a hidden older history in the field of *Qualitative Matrix Analysis (QMA)*, which yields potential applications of autarky algorithms in economics and elsewhere. QMA was initiated by [35], based on the insight that in economics often the magnitude of a quantity is irrelevant, but only the *sign* matters. So *qualitative solvability* of systems of equations and/or inequalities is considered, a special property of such systems, namely that changes of the coefficients, which leave their signs invariant, do not change the signs of the solutions. For a textbook, concentrating on the combinatorial theory, see [2], while a recent overview is [7]. The very close connections to autarky theory have been realised in [16, Section 5] (motivated by [3]), and further expanded in [17]; see [11, Subsection 11.12.1] for an overview. While preparing this paper we came across [9], which introduces "weak satisfiability", which is *precisely* the existence of a non-trivial autarky. It is shown ([9, Theorem 5]), that weak satisfiability is NP-complete; this is the earliest known proof of $\mathcal{LEAN}$ being coNP-complete. Apparently these connections to SAT have not been pursued further. The central notions in the early history of QMA were "S-matrix" and "L-matrix", which by [16] are essentially the variable-clause matrices of certain sub-classes of $\mathcal{LEAN}$. Unaware of these connections, [10, Theorem 1.2] showed directly that recognition of L-matrices is coNP-complete. Lean clause-sets correspond to "L^+-matrices" introduced in [23], and the decomposition of a clause-set into the lean kernel and the largest autark sub-clause-set now becomes a triangular matrix decomposition into an L^+-matrix and the remainder ([23, Lemma 3.3]).

Applications. See [20] for a general discussion of various redundancy criteria in clause-sets. Identification of maximal autarkies finds application in the analysis of over-constrained systems, for example autark clauses cannot be included in MUSes (minimally unsatisfiable sub-clause-sets) and so, by minimal hitting set duality, cannot be included in MCSes (minimal corrections sets, whose removal leads to a satisfiable clause-set). As discussed above, via the computation of a maximal autarky we can compute basic matrix decompositions of QMA; apparently due to the lack of efficient implementations, at least the related subfield of QMA (which is concerned with NP-hard problems) had yet little practical applications, and the efficient algorithms for computing maximal autarkies via SAT (and extensions) might be a game changer here.

Overview. In Section 2 we provide all background. Section 3 discusses oracles $(\mathcal{O}, \mathcal{O}_1, \mathcal{O}_0, \mathcal{O}_{01})$, and reviews the first basic algorithm $\mathcal{A}_0$ (Definition 5), analysed in Lemma 2. Section 4 introduces the basic translation $F \rightsquigarrow t(F)$, where $t(F)$ expresses autarky-search for F, and proves various properties. The second basic algorithm $\mathcal{A}_1$ is reviewed in Definition 8 and analysed in Lemma 4. Algorithm $\mathcal{A}_{bs}$ is given in Definition 9, using cardinality constraints (translated into CNF). The use of "steering clauses", collected into a set P of positive clauses, is discussed in Subsection 4.2, with the main technical result Corollary 2, which shows that variables involved in a resolution refutation of $t(F) \cup P$ can not be part of the largest autarky var-set of F. The novel algorithm $\mathcal{A}_{01}$ finally is introduced in Section 5, first using an unspecified P (Definition 10), and then instantiating this scheme in Theorem 1 to obtain at most $2\sqrt{n(F)}$ many calls to $\mathcal{O}_{01}$. We conclude in Section 6 by presenting conjectures and open problems.

2   Preliminaries

We use $\mathbb{N} = \{n \in \mathbb{Z} : n \geq 1\}$ and $\mathbb{N}_0 = \mathbb{N} \cup \{0\}$. The powerset of a set X is denoted by $\mathbb{P}(X)$, while $\mathbb{P}_f(X) := \{X' \in \mathbb{P}(X) : X' \text{ finite }\}$. Maps are sets of ordered pairs, and so for maps f, g the relation $f \subseteq g$ says, that $f(x) = g(x)$ holds for each x in the domain of f, which is contained in the domain of g.

We have the set $\mathcal{VA}$ of variables, with $\mathbb{N} \subseteq \mathcal{VA}$, and the set $\mathcal{LIT}$ of literals, with $\mathcal{VA} \subset \mathcal{LIT}$. The complementation operation is written $x \in \mathcal{LIT} \mapsto \overline{x} \in \mathcal{LIT}$, and fulfils $\overline{\overline{x}} = x$. On $\mathbb{N}$ the complementation is arithmetical negation, and thus $\mathbb{Z} \setminus \{0\} \subseteq \mathcal{LIT}$. Every literal is either a variable or a complemented variable; forgetting the possible complementation is done by the projection var : $\mathcal{LIT} \to \mathcal{VA}$. For $L \subseteq \mathcal{LIT}$ we use $\overline{L} := \{\overline{x} : x \in L\}$ and $\mathrm{lit}(L) := L \cup \overline{L}$. A clause is a finite set $C \subset \mathcal{LIT}$ of literals with $C \cap \overline{C} = \emptyset$, while a clause-set is a finite set of clauses; the set of all clause-sets is denoted by $\mathcal{CLS}$. The empty clause is denoted by $\bot := \emptyset$, the empty clause-set by $\top := \emptyset \in \mathcal{CLS}$. Furthermore $p\text{-}\mathcal{CLS} := \{F \in \mathcal{CLS} : \forall C \in F : |C| \leq p\}$ for $p \in \mathbb{N}_0$.

For a clause C we define $\mathrm{var}(C) := \{\mathrm{var}(x) : x \in C\}$, while for a clause-set F we define $\mathrm{var}(F) := \bigcup_{C \in F} \mathrm{var}(C)$. We use the following measures: $n(F) := |\mathrm{var}(F)| \in \mathbb{N}_0$ is the number of variables, $c(F) := |F| \in \mathbb{N}_0$ is the number of clauses, $\ell(F) := \sum_{C \in F} |C| \in \mathbb{N}_0$ is the number of literal occurrences.

A partial assignment is a map $\varphi : V \to \{0,1\}$ for some finite $V \subset \mathcal{VA}$, where we write $\mathrm{var}(\varphi) := V$, while the set of all partial assignments is denoted by $\mathcal{PASS}$. A special partial assignment is the empty partial assignment $\langle\rangle := \emptyset \in \mathcal{PASS}$. Furthermore we use $\mathrm{lit}(\varphi) := \mathrm{lit}(\mathrm{var}(\varphi))$, and extend φ to $\mathrm{lit}(\varphi)$ via $\varphi(\overline{v}) = 1 - \varphi(v)$ for $v \in \mathrm{var}(\varphi)$. For $\varepsilon \in \{0,1\}$ we define $\varphi^{-1}(\varepsilon) := \{x \in \mathrm{lit}(\varphi) : \varphi(x) = \varepsilon\}$.

The application $\varphi * F \in \mathcal{CLS}$ of $\varphi \in \mathcal{PASS}$ to $F \in \mathcal{CLS}$ is defined as $\varphi * F := \{C \setminus \varphi^{-1}(0) : C \in F \wedge C \cap \varphi^{-1}(1) = \emptyset\}$. Then $\mathcal{SAT} := \{F \in \mathcal{CLS} \mid \exists \varphi \in \mathcal{PASS} : \varphi * F = \top\}$, and $\mathcal{USAT} := \mathcal{CLS} \setminus \mathcal{SAT}$.

The restriction of $F \in \mathcal{CLS}$ to $V \subseteq \mathcal{VA}$ is defined as $F[V] := \{C \cap \mathrm{lit}(V) : C \in F\} \setminus \{\bot\} \in \mathcal{CLS}$, i.e., removal of clauses $C \in F$ with $\mathrm{var}(C) \cap V = \emptyset$, and restriction of the remaining clauses to variables in V.

Finally we use $\mathcal{CLS}(V) := \{F \in \mathcal{CLS} : \mathrm{var}(F) \subseteq V\}$, $\mathcal{PASS}(V) := \{\varphi \in \mathcal{PASS} : \mathrm{var}(\varphi) \subseteq V\}$ and $\mathcal{TASS}(V) := \{\varphi \in \mathcal{PASS} : \mathrm{var}(\varphi) = V\}$ ("total assignments") for $V \subseteq \mathcal{VA}$.

Now to autarkies; this paper is essentially self-contained, but if more information is desired, see the handbook chapter [11]. A partial assignment $\varphi \in \mathcal{PASS}$ is an *autarky for* $F \in \mathcal{CLS}$ iff for all $C \in F$ with $\mathrm{var}(\varphi) \cap \mathrm{var}(C) \neq \emptyset$ holds $\varphi * \{C\} = \top$ iff $\forall C \in F : \varphi * \{C\} \in \{\top, \{C\}\}$; the set of all autarkies for F is denoted by $\mathrm{Auk}(F) \subseteq \mathcal{PASS}$. The empty partial assignment $\langle\rangle$ is an autarky for every $F \in \mathcal{CLS}$, and in general we call an autarky φ for F *trivial* if $\mathrm{var}(\varphi) \cap \mathrm{var}(F) = \emptyset$. For $\top$ as well as $\{\bot\}$ every partial assignment is a trivial autarky. Note that every satisfying assignment for F is also an autarky for F, and it is a trivial autarky iff $F = \top$. Another simple but useful property is that φ is an autarky for $\bigcup_{i \in I} F_i$ for a finite family $(F_i)_{i \in I}$ of clause-sets iff φ is an autarky for all F_i, $i \in I$. We also note that φ is an autarky for F iff φ is an autarky for $F \cup \{\bot\}$ iff φ is an autarky for $F \setminus \{\bot\}$ (for autarkies the empty clause is invisible). In general it is best to allow that autarkies assign non-occurring variables, but it is also needed to have a notation which disallows this; following [11, Definition11.9.1]:

Definition 1. *For $F \in \mathcal{CLS}$ let $\mathbf{Auk^r(F)} := \mathrm{Auk}(F) \cap \mathcal{PASS}(\mathrm{var}(F))$ ('r" like "restricted" or "relevant"), while by $\mathbf{var(Auk^r(F))} := \bigcup_{\varphi \in \mathrm{Auk^r}(F)} \mathrm{var}(\varphi)$ we denote the **largest autarky-var-set**.*

$\mathcal{LEAN} \subset \mathcal{USAT} \cup \{\top\}$ is the set of $F \in \mathcal{CLS}$ such that $\mathrm{Auk^r}(F) = \{\langle\rangle\}$, while the *lean kernel* of $F \in \mathcal{CLS}$, denoted by $\mathrm{N_a}(F) \subseteq F$, is the largest element of $\mathcal{LEAN}$ contained in F (it is easy to see that $\mathcal{LEAN}$ is closed under finite union). We have $\mathrm{var}(\mathrm{Auk^r}(F)) \cup \mathrm{var}(\mathrm{N_a}(F)) = \mathrm{var}(F)$ and $\mathrm{var}(\mathrm{Auk^r}(F)) \cap \mathrm{var}(\mathrm{N_a}(F)) = \emptyset$. See [11, Subsection 11.8.3] for various characterisations of the lean kernel.

Definition 2. *For $F \in \mathcal{CLS}$ let $\mathbf{n_A(F)} := |\mathrm{var}(\mathrm{Auk^r}(F))| \in \mathbb{N}_0$ be the number of variables in the largest autarky-var-set and $\mathbf{n_L(F)} := |\mathrm{var}(\mathrm{N_a}(F))| \in \mathbb{N}_0$ be the number of variables in the lean kernel.*

So $n(F) = n_A(F) + n_L(F)$. On the finite set $\mathrm{Auk^r}(F)$ we have a natural partial order given by inclusion. There is always the smallest element $\langle\rangle \in \mathrm{Auk^r}(F)$, while the maximal elements of $\mathrm{Auk^r}(F)$ are called *maximal autarkies* for F. For maximal autarkies φ, ψ holds $\mathrm{var}(\varphi) = \mathrm{var}(\psi) = \mathrm{var}(\mathrm{Auk^r}(F))$; here we use that the composition of autarkies is again an autarky, i.e., for autarkies φ, ψ for F there is an autarky θ for F with $\varphi * (\psi * F) = \psi * (\varphi * F) = \theta * F$.

Definition 3. *Let $\mathbf{Auk{\uparrow}(F)} \subseteq \mathrm{Auk^r}(F)$ be the set of maximal autarkies.*

A *quasi-maximal autarky* for F is an $\varphi \in \mathrm{Auk^r}(F)$ with $\varphi * F = \mathrm{N_a}(F)$. By supplying arbitrary values for the missing variables we obtain efficiently a maximal autarky from a quasi-maximal autarky.

3   Oracles

The main computational task considered in this paper is the computation of some element of $\mathrm{Auk{\uparrow}}(F)$ for inputs $F \in \mathcal{CLS}$. Our emphasis is on the number of calls to an "oracle", which solves NP-hard problems, while otherwise the computations are in polynomial time. The **NP (-SAT) oracle** $\mathcal{O} : \mathcal{CLS} \to \{0,1\}$ just maps $F \in \mathcal{CLS}$ to 1 in case of $F \in \mathcal{SAT}$, and to 0 otherwise. As we will see in Example 4, for deciding leanness, one call suffices. For a **(standard) SAT oracle** $\mathcal{O}_1 : \mathcal{CLS} \to \{0\} \cup (\{1\} \times \mathcal{PASS})$, the SAT solver also returns a satisfying assignment, and then also a non-trivial autarky can be returned in case of non-leanness. As introduced in [15], we consider here a strengthened oracle $\mathcal{O}_{01}$, to return something also for unsatisfiable inputs. Recall that a *tree resolution refutation* for $F \in CLS$ is a binary tree, where the nodes are labelled with clauses, such that the leaves are labelled by (some) clauses of F (the "axioms"), while the root is labelled with $\perp$, and such that for each inner node, with children labelled by clauses C, D, we have $C \cap \overline{D} = \{x\}$ for some $x \in LIT$, while the label of that inner node is $(C \setminus \{x\}) \cup (D \setminus \{\overline{x}\})$.

Definition 4. *An **extended SAT oracle** is a map $\mathcal{O}_{01} : CLS \to \{0,1\} \times (\mathbb{P}_f(VA) \cup PASS)$, which for input $F \in USAT$ returns $(0, \mathrm{var}(F'))$ for some $F' \subseteq F$, such that there is a tree refutation using as axioms precisely F', and for $F \in SAT$ returns $(1, \varphi)$ for some $\varphi \in PASS(\mathrm{var}(F))$ and $\varphi * F = \top$. If we don't need the satisfying assignment, then we use $\mathcal{O}_0 : CLS \to \{1\} \cup (\{0\} \times \mathbb{P}_f(VA))$.*

In the following we will indicate the type of oracle by using one of $\mathcal{O}_0, \mathcal{O}_1, \mathcal{O}_{01}$. See [11, Subsection 11.10.3] for a short discussion how to efficiently integrate the computations for $\mathcal{O}_0, \mathcal{O}_{01}$ into a SAT solver, both look-ahead ([8]) and CDCL solvers ([28]). It is important to notice here that we do not need a full resolution refutation, but only the variables involved in it. The above use of *tree* resolution is only a convenient way of stating the condition that all axioms are actually used in the refutation. Furthermore, there is no need for any sort of minimisation of the refutation, as we see by the following lemma.

Lemma 1. *If for $F \in CLS$ holds $\mathcal{O}_0(F) = (0, V)$, then $V \cap \mathrm{var}(\mathrm{Auk}^r(F)) = \emptyset$.*

Proof: As shown in [14, Lemma 3.13], for any autarky $\varphi \in \mathrm{Auk}(F)$ and any clause C touched by φ there is no tree resolution refutation of F using C. $\square$

So the more clauses are involved in the resolution refutation (i.e., the larger V), the more variables we can exclude from the largest autarky-var-set, and thus minimising resolution refutation in general will be counter-productive. One known approach to compute a maximal autarky of $F \in CLS$, as reviewed in [11, Subsection 11.10.3] (especially Theorem 11.10.1 there), is based on the full *autarky-resolution duality* ([14, Theorem 3.16]): the variables involved in some autarky of F are altogether, i.e., $\mathrm{var}(\mathrm{Auk}^r(F)) = \mathrm{var}(F) \setminus \mathrm{var}(\mathrm{N_a}(F))$, precisely the variables not usable by some tree resolution refutation of F. So the algorithm, called $\mathcal{A}_0(F)$ here, iteratively removes variables not usable in an autarky and clauses consisting solely of such variables, via Lemma 1, until a satisfying assignment φ is found (which must happen eventually), and φ is then a quasi-maximal (due to autarky-resolution duality):

Definition 5. *For input $F \in CLS$, the algorithm $\mathcal{A}_0(F)$, using oracle $\mathcal{O}_{01}$ and computing a partial assignment φ, performs the following computation:*

1. *While $\mathrm{var}(F) \neq \emptyset$ do:*
   (a) *Compute $\mathcal{O}_{01}(F)$, obtaining $(0, V)$ resp. $(1, \varphi)$.*
   (b) *In case of $(0, V)$, let $F := F[\mathrm{var}(F) \setminus V]$.*
   (c) *In case of $(1, \varphi)$, let $F := \top$.*
2. *Return φ.*

Lemma 2 ([14]). *For $F \in \mathcal{CLS}$ the algorithm $\mathcal{A}_0(F)$ computes a quasi-maximal autarky for F, using at most $\min(n_L(F) + 1, n(F))$ calls of oracle $\mathcal{O}_{01}$.*

The best case for algorithm $\mathcal{A}_0(F)$ in terms of the number of oracle calls is given for $F \in \mathcal{SAT}$, where just one call suffices. For the worst-case $F \in \mathcal{LEAN}$ on the other hand $\mathcal{A}_0(F)$ might use $n(F)$ oracle calls:

Example 1. Let $F := \big\{ \{1\}, \{-1\}, \{2\}, \{-2\}, \ldots, \{n\}, \{-n\} \big\}$ for $n \in \mathbb{N}_0$. We have $F \in \mathcal{LEAN}$, and each loop iteration will remove exactly one pair $\{i\}, \{-i\}$, until all clauses are removed.

4   The Basic Translation

We now review the translation $t : \mathcal{CLS}(\mathcal{VA}_0) \to \mathcal{CLS}$ from [26], called Γ_2 there, which represents the search for an autarky φ for $F \in \mathcal{CLS}(\mathcal{VA}_0)$ as a SAT problem $t(F)$; here $\mathcal{VA}_0$ is the set of *primary variables*, while the variables in $\mathcal{VA} \setminus \mathcal{VA}_0$ are used as *auxiliary variables*. The translation $t(F)$ uses two types of variables, the primary variables $v \in \mathrm{var}(F)$ themselves, where $v \mapsto 1$ *now* means $v \in \mathrm{var}(\varphi)$, and for every $v \in \mathrm{var}(F)$ two auxiliary variables $t(v), t(\overline{v})$, where $t(x) \mapsto 1$ for $x \in \mathrm{lit}(F)$ means $\varphi(x) = 1$. In other words, the three possible states of a variable $v \in \mathrm{var}(F)$ w.r.t. the partial assignment φ, namely "unassigned" ($v \notin \mathrm{var}(\varphi)$), "set true" ($\varphi(v) = 1$), "set false" ($\varphi(v) = 0$), are represented by three of the four states of assigned variables $t(v), t(\overline{v})$, namely "unassigned" is $t(v), t(\overline{v}) \mapsto 0$, "set true" is $t(v) \mapsto 1, t(\overline{v}) \mapsto 0$, and "set false" is $t(v) \mapsto 0, t(\overline{v}) \mapsto 1$. The variable v *in the translation* $t(F)$ just acts as an indicator variable, showing whether v is involved in the autarky or not. We have then three types of clauses in $t(F)$: the *autarky clauses* for $C \in F$ and $x \in C$, stating that if x gets false by the autarky, then some other literal of C must get true, plus the *AMO (at-most-one) clauses* for $t(v), t(\overline{v})$ and the *connection* between v and $t(v), t(\overline{v})$. It is useful for argumentation to have the more general form $t_V(F)$, where only φ with $\mathrm{var}(\varphi) \subseteq V$ are considered:

Definition 6. *We assume a set $\mathbb{N} \subseteq \mathcal{VA}_0 \subset \mathcal{VA}$ of "primary variables" together with an injection $t : \mathrm{lit}(\mathcal{VA}_0) \to \mathcal{VA}$, yielding the "auxiliary variables", such that $\mathcal{VA}_0 \cap t(\mathrm{lit}(\mathcal{VA}_0)) = \emptyset$ and $\mathcal{VA}_0 \cup t(\mathrm{lit}(\mathcal{VA}_0)) = \mathcal{VA}$. For $V \subseteq \mathcal{VA}_0$ let $V' := V \cup t(\mathrm{lit}(V))$. In general we define an equivalence relation on $\mathcal{VA}$, where every equivalence class contains (precisely) three elements, namely $v, t(v), t(\overline{v})$ for $v \in \mathcal{VA}_0$. A set $V \subseteq \mathcal{VA}$ is **saturated**, if for $v \in V$ and every equivalent v' holds $v' \in V$. The **saturation** $V \subseteq \boldsymbol{V'} \subseteq \mathcal{VA}$ of $V \subseteq \mathcal{VA}$ is the saturation under this equivalence relation, i.e., addition of all equivalent variables.*

  Now the translation $t_V : \mathcal{CLS}(\mathcal{VA}_0) \to \mathcal{CLS}(V')$ for $V \in \mathbb{P}_f(\mathcal{VA}_0)$ has the following clauses for $t_V(F)$:

  *I for $C \in F$ and $x \in C$ with $\mathrm{var}(x) \in V$ the **autarky clause** $\{\overline{t(\overline{x})}\} \cup \{t(y) : y \in C \setminus \{x\}, \mathrm{var}(y) \in V\}$ (i.e., $t(\overline{x}) \to \bigvee_{y \in C \setminus \{x\}, \mathrm{var}(y) \in V} t(y)$);*

  *II for each $v \in V$ the **AMO-clause** $\{\overline{t(v)}, \overline{t(\overline{v})}\}$;*

III *for each* $v \in V$ *the clauses of* $v \leftrightarrow (t(v) \vee t(\overline{v}))$, *i.e., the three clauses* $\{\overline{v}, t(v), t(\overline{v})\}, \{\overline{t(v)}, v\}, \{\overline{t(\overline{v})}, v\}$ *(the **indicator clauses**).*

Especially $\boldsymbol{t(F)} := t_{\mathrm{var}(F)}(F)$ *for* $F \in \mathcal{CLS}(\mathcal{VA}_0)$.

For $F \in \mathcal{CLS}(\mathcal{VA}_0)$ and $V \in \mathbb{P}_{\mathrm{f}}(\mathcal{VA}_0)$ holds $\mathrm{var}(t_V(F)) = V' = V \cup t(\mathrm{lit}(V))$, $V \cap t(\mathrm{lit}(V)) = \emptyset$, and $n(t(F)) = 3n(F)$, $c(t(F)) = \ell(F) + 4n(F)$. Due to the four AMO- and indicator-clauses, every satisfying assignment for $t_V(F)$ must be total, that is, for $\varphi \in \mathcal{PASS}$ with $\varphi * t_V(F) = \top$ holds $\mathrm{var}(t_V(F)) \subseteq \mathrm{var}(\varphi)$.

Example 2. For $F = \big\{ \{1\}, \{-1\}, \ldots, \{n\}, \{-n\} \big\}$ as in Example 1, we have $2n$ autarky clauses, which are $\{\overline{t(i)}\}$ for $i \in \{-n, \ldots, n\} \setminus \{0\}$.

Partial assignments φ on the primary variables are translated to assignments on the primary+auxiliary variables via $t_{0,V}(\varphi)$ (assigning unassigned variables to 0 in the translation) and $t(\varphi)$ (leaving them unassigned), while the backwards direction goes via via $t^{-1}(\varphi)$:

Definition 7. *For* $V \in \mathbb{P}_{\mathrm{f}}(\mathcal{VA}_0)$ *we define a translation* $\boldsymbol{t_{0,V}} : \mathcal{PASS}(V) \rightarrow \mathcal{TASS}(V')$ *for* $\varphi \in \mathcal{PASS}(V)$ *by* $t_{0,V}(\varphi)(v) = 1 \Leftrightarrow v \in \mathrm{var}(\varphi)$ *for* $v \in V$, *while* $t_{0,V}(\varphi)(t(x)) = 1 \Leftrightarrow \mathrm{var}(x) \in \mathrm{var}(\varphi) \wedge \varphi(x) = 1$ *for* $x \in \mathrm{lit}(V)$.

The translation $\boldsymbol{t} : \mathcal{PASS}(\mathcal{VA}_0) \rightarrow \mathcal{PASS}$ *for* $\varphi \in \mathcal{PASS}(\mathcal{VA}_0)$ *is the partial assignment, where* $\mathrm{var}(t(\varphi))$ *is the saturation of* $\mathrm{var}(\varphi)$, *while* $t(\varphi)(v) = 1$ *for* $v \in \mathrm{var}(\varphi)$, *and* $t(\varphi)(t(x)) = 1 \Leftrightarrow \varphi(x) = 1$ *for* $x \in \mathrm{lit}(\varphi)$.

In the other direction, any partial assignment $\varphi \in \mathcal{PASS}$ *with* $\mathrm{var}(\varphi)$ *saturated yields a partial assignment* $\boldsymbol{t^{-1}(\varphi)} \in \mathcal{PASS}(\mathcal{VA}_0)$ *with* $\mathrm{var}(t^{-1}(\varphi)) := \varphi^{-1}(1) \cap \mathcal{VA}_0$ *and* $t^{-1}(\varphi)(v) = \varphi(t(v))$ *for* $v \in \mathrm{var}(t^{-1}(\varphi))$.

As already stated, $t_{0,V}(\varphi)$ makes explicit which variables are unassigned by φ, namely assigning them with 0, and thus it needs to know V, while $t(\varphi)$ just leaves them unassigned. We have $t^{-1}(t_{0,V}(\varphi)) = t^{-1}(t(\varphi)) = \varphi$.

Example 3. $t_V(F) \in \mathcal{SAT}$ for $F \in \mathcal{CLS}(\mathcal{VA}_0)$ and $V \in \mathbb{P}_{\mathrm{f}}(\mathcal{VA}_0)$, since for $t_{0,V}(\langle\rangle) = \langle v \rightarrow 0 : v \in V\rangle \cup \langle t(x) \rightarrow 0 : x \in \mathrm{lit}(V)\rangle$ we have $t_{0,V}(\langle\rangle) * t_V(F) = \top$.

$t(F)$ does its job, i.e., its solutions represent all the autarkies of F:

Lemma 3 ([26]). *Consider* $F \in \mathcal{CLS}(\mathcal{VA}_0)$ *and* $V \in \mathbb{P}_{\mathrm{f}}(\mathcal{VA}_0)$.

1. *If* $\mathcal{O}_1(t_V(F)) = (1, \varphi)$, *then* $t^{-1}(\varphi) \in \mathrm{Auk}^{\mathrm{r}}(F) \cap \mathcal{PASS}(V)$.
2. $t_{0,V}(\varphi) * t_V(F) = \top$ *for* $\varphi \in \mathrm{Auk}^{\mathrm{r}}(F) \cap \mathcal{PASS}(V)$.

Before discussing the usage of $t(F)$, we remark that the variables $\mathrm{var}(F) \subseteq \mathrm{var}(t(F))$ are used purely for a more convenient discussion, while for a practical application they would be dropped, and the translation called Γ_3 in [26] would be used (except possibly for Algorithm $\mathcal{A}_{\mathrm{bs}}$ defined later, which uses cardinality constraints): the variables of $t(F)$ then would be just $t(\mathrm{lit}(F))$, and the clauses would be the autarky- and AMO-clauses (only). In our applications $v \in \mathrm{var}(F)$ occurs in the translations only positively, and would be replaced by the two positive literals $t(v), t(\overline{v})$ (together).

4.1 Basic Usages

Example 4. A simple algorithm for finding a non-trivial autarky for $\text{var}(F) \neq \emptyset$ evaluates $\mathcal{O}_1(t(F) \cup \{\text{var}(F)\})$. By Lemma 3 we get, that if the solver returns 0, then $F \in \mathcal{LEAN}$, while if $(1, \varphi)$ is returned, then $t^{-1}(\varphi)$ is a non-trivial autarky for F (the non-triviality is guaranteed by the additional clause $\text{var}(F)$).

Algorithm $\mathcal{A}_1(F)$, computing a maximal autarky, iterates the algorithm from Example 4; the details are as follows, where we formulate the algorithm in such a way that it has the same basic structure as $\mathcal{A}_0$ (recall Definition 5) and our novel algorithm $\mathcal{A}_{01}$ (to be given in Definition 10):

Definition 8. *For input $F \in \mathcal{CLS}(\mathcal{VA}_0)$ the algorithm $\mathbf{\mathcal{A}_1(F)}$, using oracle $\mathcal{O}_1$ and computing a partial assignment φ, performs the following computation:*

1. $\varphi := \langle \rangle$, $P := \{\text{var}(F)\}$, $F := t(F)$.
2. *While* $\text{var}(P) \neq \emptyset$ *do:*
   (a) *Compute* $\mathcal{O}_1(F \cup P)$, *obtaining 0 resp. $(1, \psi)$.*
   (b) *In case of 0, let $P := \top$ and $F := \top$.*
   (c) *In case of $(1, \psi)$, let $\psi' := t^{-1}(\psi)$, and update $P := P[\text{var}(P) \setminus \text{var}(\psi')]$, $F := t(\psi') * F$, and $\varphi := \varphi \cup \psi'$.*
   In words: obtain the autarky ψ' from ψ, remove the variables of ψ' from P and F, and add ψ' to the result-autarky φ.
3. *Return φ.*

Lemma 4. *For $F \in \mathcal{CLS}(\mathcal{VA}_0)$ the algorithm $\mathcal{A}_1(F)$ computes $\varphi \in \text{Auk}\!\uparrow(F)$, using at most $\min(n_A(F) + 1, n(F))$ calls of oracle $\mathcal{O}_1$.*

Proof: The algorithm always terminates, and moreover for the number $m \geq 0$ of executions of the while-body we have $m \leq \min(n_A(F)+1, n(F))$, since in each round P gets reduced by some variables from an autarky (due to the choice of P). Let F_{-1} be the input, let $F_0 := t(F_{-1})$, and let F_i for $i = 1, \ldots, m$ be the current F after execution of i-th iteration; similarly, let P_0 be the original value of P, and let P_i be the current P after the i-th iteration, and let $\varphi_0 := \langle \rangle$, and let φ_i be the value of φ after the i-th iteration. Finally, let V_i for $i = 1, \ldots, m$ be $\text{var}(P_i)$ in case of 0 resp. the value of $\text{var}(\psi')$ after round i, and let $W_0 := \text{var}(F_{-1})$, and let $W_i := W_{i-1} \setminus V_i$ for $i = 1, \ldots, m$. Inductively we show that $F_i = t_{W_i}(\varphi_i * F_{-1})$ for $i \in \{0, \ldots, m\}$, where φ_i is an autarky for F_{-1} by Lemma 3, Part 1, and $P_i = P_0[W_i]$ for $i \in \{1, \ldots, m\}$, where $W_m = \emptyset$. Variables only vanish as part of some autarky for F_{-1}, and thus $\varphi_i \in \text{Auk}\!\uparrow(F_{-1}[W_0 \setminus W_i])$ for $i \in \{0, \ldots, m\}$. $\square$

The best case for algorithm $\mathcal{A}_1(F)$ in terms of the number of oracle calls is given for $F \in \mathcal{LEAN}$, where just one call suffices. For the worst-case $F \in \mathcal{SAT}$ however, $\mathcal{A}_1(F)$ might use $n(F)$ oracle calls:

Example 5. Let $F := \{\{1\}, \ldots, \{n\}\} \in \mathcal{SAT}$ for $n \in \mathbb{N}_0$. In the worst case (depending on the answers of $\mathcal{O}_1$), in each call only one unit-clause $\{i\}$ is removed.

The algorithm realising the currently best number of calls to $\mathcal{O}_1$ uses SAT-encodings of cardinality constraints (see [34]); different from the literature, we follow our general scheme and iteratively apply the autarkies found:

Definition 9. *For input $F \in \mathcal{CLS}(\mathcal{VA}_0)$ the algorithm $\mathcal{A}_{\mathbf{bs}}(F)$, using oracle $\mathcal{O}_1$ and computing a partial assignment φ, performs the following computation:*

1. $\varphi := \langle\rangle$, $n := n(F)$, $V := \mathrm{var}(F)$, $F := t(F)$ *(n is an upper bound on the size of a maximal autarky, V is the set of variables potentially used by it).*
2. *While $n \neq 0$ do:*
   (a) $m := \lceil \frac{n}{2} \rceil$; *let G be a CNF-representation of the cardinality constraint* "$\sum_{v \in V} v \geq m$"; *compute $\mathcal{O}_1(F \cup G)$, obtaining 0 resp. $(1, \psi)$.*
   (b) *In case of 0, let $n := m - 1$.*
   (c) *In case of $(1, \psi)$, let $\psi' := t^{-1}(\psi)$, and update $n := n - n(\psi')$, $V := V \setminus \mathrm{var}(\psi')$, $F := t(\psi') * F$, and $\varphi := \varphi \cup \psi'$.*
3. *Return φ.*

As it should be obvious by now:

Lemma 5. *For $F \in \mathcal{CLS}(\mathcal{VA}_0)$ the algorithm $\mathcal{A}_{\mathbf{bs}}(F)$ computes $\varphi \in \mathrm{Auk}{\uparrow}(F)$, using at most $\lceil \log_2(n(F)) \rceil$ calls of oracle $\mathcal{O}_1$ (for $n(F) > 0$).*

That the upper bound of Lemma 5 is attained, can be seen again with Example 5 (in the worst case). We remark that if we allow calls to Partial MaxSAT (see [24] for an overview), then just one call is enough (as used in [25]), and that without cardinality constraints, namely using $t(F)$ as the hard clauses and $\{v\}$ for $v \in \mathrm{var}(F)$ as the soft clauses. Indeed, as shown in [26, Proposition1], this translation has a unique "minimal correction set" (MCS), i.e., a unique minimal subset of the soft clauses, whose removal yields a satisfiable clause-set, and so any MCS-solver can be used (just one call).

4.2 Adding Positive "Steering" Clauses

Generalising the use of P in Algorithm $\mathcal{A}_1$, we consider some positive clause-set P over $\mathrm{var}(F)$ (i.e., $P \subseteq \mathbb{P}(\mathrm{var}(F))$), and use $t(F) \cup P \in \mathcal{CLS}$ to gain larger autarkies. Note that the elements of P require variables to be in the autarky, and so in general P should contain several shorter clauses, while for $\mathcal{A}_1$ we just used one full clause (containing all variables). If the oracle then yields unsatisfiability, this is no longer the end of the search (due to the lean kernel been reached), since the clauses of P involved in the refutation might not involve all remaining variables. The extended oracle is now needed to tell us which clauses of P were used. To do so, we first note that autarkies for F yield autarkies for $t(F) \cup P$ (where for a simpler algorithm we allow P to contain variables not in $t(F)$):

Lemma 6. *Consider $F \in \mathcal{CLS}(\mathcal{VA}_0)$ and $P \in \mathbb{P}_{\mathrm{f}}(\mathbb{P}_{\mathrm{f}}(\mathcal{VA}_0))$. For $\varphi \in \mathrm{Auk}^{\mathrm{r}}(F)$ we have $t(\varphi) \in \mathrm{Auk}^{\mathrm{r}}(t(F) \cup P)$.*

Proof: $t(\varphi)$ is an autarky for P, since $t(\varphi)$ does not set variables from $\mathrm{var}(F)$ to 0. By Lemma 3, Part 2, we get that $t_0(\varphi)$ is a satisfying assignment for $t(\varphi)$; now $t(\varphi)$ just unsets all triples $v, t(v), t(\overline{v})$ with $v \notin \mathrm{var}(\varphi)$, where $t_0(\varphi)$ sets these three variables to 0. Thus obviously $t(\varphi)$ is also an autarky for the AMO-clauses and the indicator clauses. Assume an autarky clause D for $C \in F$ and $x \in C$, touched by $t(\varphi)$ but not satisfied. Thus there is $y \in C$ with $\mathrm{var}(x) \notin \mathrm{var}(\varphi)$ and $\varphi(y) = 0$; since φ is an autarky, there is $y' \in C$ with $\varphi(y') = 1$, whence $t(\varphi)(t(y')) = 1$ with $t(y') \notin C$, contradicting the assumption. $\qquad\square$

Thus the saturation of the largest autarky-var-set of F is contained in the largest autarky-var-set for $t(F) \cup P$:

Corollary 1. *Consider* $F \in \mathcal{CLS}(\mathcal{VA}_0)$ *and* $P \in \mathbb{P}_{\mathrm{f}}(\mathbb{P}_{\mathrm{f}}(\mathcal{VA}_0))$. *Then the set* $\mathrm{var}(\mathrm{Auk}^{\mathrm{r}}(t(F) \cup P))$ *is saturated and contains* $\mathrm{var}(\mathrm{Auk}^{\mathrm{r}}(F))$.

Proof: It remains to show that $\mathrm{var}(\mathrm{Auk}^{\mathrm{r}}(t(F) \cup P))$ is saturated, and this follows by just considering the AMO-clauses and the indicator clauses: If v is assigned, then also $t(v), t(\overline{v})$ need to be assigned for an autarky, while if one of $t(v), t(\overline{v})$ is assigned, then also v needs to be assigned. $\qquad\square$

Using Lemma 1, we obtain the main insight, that if the oracle yields $(0, V)$ for $t(F) \cup P$, then none of the elements of V are in the largest autarky-var-set:

Corollary 2. *If for* $F \in \mathcal{CLS}(\mathcal{VA}_0)$ *and* $P \in \mathbb{P}_{\mathrm{f}}(\mathbb{P}_{\mathrm{f}}(\mathcal{VA}_0))$ *the oracle yields* $\mathcal{O}_0(t(F) \cup P) = (0, V)$, *then* $V' \cap \mathrm{var}(\mathrm{Auk}^{\mathrm{r}}(F)) = \emptyset$ *(recall Definition 6 for* V'*).*

5   The New Algorithm

We now present the novel algorithm scheme $\mathcal{S}_{01}(F, P)$, combining algorithms $\mathcal{A}_0$ (Definition 5) and $\mathcal{A}_1$ (Definition 8), which takes as input $F \in \mathcal{CLS}$ and additionally $P \subseteq \mathbb{P}(\mathrm{var}(F))$, and computes some autarky $\varphi \in \mathrm{Auk}^{\mathrm{r}}(F)$; for our current best generic instantiation we specify P in Theorem 1, obtaining algorithm $\mathcal{A}_{01}(F)$.

Definition 10. *For inputs* $F \in \mathcal{CLS}(\mathcal{VA}_0)$ *and* $P \subseteq \mathbb{P}(\mathrm{var}(F))$, *the algorithm* $\mathcal{S}_{01}(\boldsymbol{F}, \boldsymbol{P})$, *using oracle* $\mathcal{O}_{01}$ *and computing a partial assignment* φ, *performs the following computation (using the saturation* V' *as in Definition 6):*

1. *$\varphi := \langle \rangle$, $F := t(F)$.*
2. *While $\mathrm{var}(P) \neq \emptyset$ do:*
   - *(a) Compute $\mathcal{O}_{01}(F \cup P)$, obtaining $(0, V)$ resp. $(1, \psi)$.*
   - *(b) In case of $(0, V)$, let $V := V'$, $P := P[\mathrm{var}(P) \setminus V]$, $F := F[\mathrm{var}(F) \setminus V]$.*
   - *(c) In case of $(1, \psi)$, let $\psi' := t^{-1}(\psi)$, and update $P := P[\mathrm{var}(P) \setminus \mathrm{var}(\psi')]$, $F := t(\psi') * F$, and $\varphi := \varphi \cup \psi'$.*
3. *Return φ.*

While $\bot \in P$ is of no real use, it doesn't cause a problem for the algorithm, and will be removed from P in the first round by the restriction (whether the implicit resolution refutation of $t(F) \cup P$ chooses $\bot$ as the refutation or not).

Lemma 7. *For $F \in \mathcal{CLS}(\mathcal{VA}_0)$ and $P \subseteq \mathbb{P}(\mathrm{var}(F))$ the algorithm $\mathcal{S}_{01}(F, P)$ computes an autarky $\varphi \in \mathrm{Auk}^{\mathrm{r}}(F)$. If $\mathrm{var}(P) = \mathrm{var}(F)$, then $\varphi \in \mathrm{Auk}{\uparrow}(F)$.*

Proof: The proof extends the proof of Lemma 4, by extending the handling of the case $\mathcal{O}_{01}(F \cup P) = (0, V)$. The algorithm always terminates, since in each round P gets reduced. Let $m \geq 0$ be the number of executions of the while-body. Let F_{-1} be the input, let $F_0 := t(F_{-1})$, and let F_i for $i = 1, \ldots, m$ be the current F after execution of i-th iteration; similarly, let P_0 be the input-value of P, and let P_i be the current P after the i-th iteration, and let $\varphi_0 := \langle \rangle$, and let φ_i be the value of φ after the i-th iteration. Finally, let V_i for $i = 1, \ldots, m$ be the value of V resp. $\mathrm{var}(\psi')$ after round i, and let $W_0 := \mathrm{var}(F_{-1})$, and let $W_i := W_{i-1} \setminus V_i$ for $i = 1, \ldots, m$. Inductively we show that $F_i = t_{W_i}(\varphi_i * F_{-1})$ for $i \in \{0, \ldots, m\}$, where φ_i is an autarky for F_{-1} by Lemma 3, Part 1, and $P_i = P_0[W_i]$ for $i \in \{1, \ldots, m\}$. Since variables vanish from P only by restriction, we have $V_1 \cup \ldots V_m \supseteq \mathrm{var}(P)$, and thus $W_m \subseteq W_0 \setminus \mathrm{var}(P)$. Variables only vanish, if either they are realised as not being element of $\mathrm{var}(\mathrm{Auk}^{\mathrm{r}}(F_{-1}))$ (Corollary 2), or as part of some autarky for F_{-1}. So $\varphi_i \in \mathrm{Auk}{\uparrow}(F_{-1}[W_0 \setminus W_i])$ for $i \in \{0, \ldots, m\}$, and if $\mathrm{var}(P) = \mathrm{var}(F_{-1})$, then φ_m is a maximal autarky for F_{-1}. □

If instead of an unrestricted (maximal) autarky $\varphi \in \mathrm{Auk}^{\mathrm{r}}(F)$ we want to compute a (maximal) autarky $\varphi \in \mathrm{Auk}^{\mathrm{r}}(F)$ with $\mathrm{var}(\varphi) \subseteq V$ for some given $V \subseteq \mathcal{VA}$, then we may just replace the input F by $F[V]$ (or we choose P with $\bigcup P = V$, and restrict the result).

Example 6. The simplest cases for computing maximal autarkies use (I) $P = \{\mathrm{var}(F)\}$ or (II) $P = \{\{v\} : v \in \mathrm{var}(F)\}$. In Case I, we essentially obtain $\mathcal{A}_1$ (Definition 8), and $\mathcal{S}_{01}(F, P)$ produces autarkies until the lean kernel is reached, so we only have SAT-answers with one final UNSAT-answer. In Case II, the scheme becomes very similar to $\mathcal{A}_0$ (Definition 5), and we remove elements of P until we obtain the variables of $\mathrm{var}(\mathrm{Auk}^{\mathrm{r}}(F))$, and so we only have UNSAT-answers with one final SAT answer. If $F \in \mathcal{LEAN}$, then in Case I only one call of the oracle is needed (as in Example 4), while in Case II, for F as in Example 1 we need $n(F)$ oracle calls. On the other hand, if $F \in \mathcal{SAT}$, then in Case I, for F as in Example 5 we need $n(F)$ oracle calls, while in Case II only one call of the oracle is needed.

A more intelligent use of $\mathcal{S}_{01}$ employs a better P, to mix the SAT- and UNSAT-answers of the oracle.

Lemma 8. *For $F \in \mathcal{CLS}(\mathcal{VA}_0)$ and $P \subseteq \mathbb{P}(\mathrm{var}(F))$ with $P \in p\text{-}\mathcal{CLS}$ $(p \in \mathbb{N}_0)$, algorithm $\mathcal{S}_{01}(F, P)$ uses at most $\min(p, n_{\mathrm{A}}(F)) + \min(c(P), n_{\mathrm{L}}(F))$ oracle calls.*

Proof: Every oracle call removes at least one clause from P (in the unsat-case), since $t_V(F) \in \mathcal{SAT}$, or one variable from all clauses of P (in the sat-case). □

So we need to minimise the sum of the number of clauses in P and the maximal clause-length, which is achieved by using disjoint clauses of size $\sqrt{n(F)}$; by Lemmas 7, 8 we obtain:

Theorem 1. *Consider $F \in \mathcal{CLS}(\mathcal{VA}_0)$. Choose $P' \subseteq \mathbb{P}(\mathrm{var}(F))$ such that P' is a partitioning of $\mathrm{var}(F)$ (the elements are pairwise disjoint and non-empty, the union is $\mathrm{var}(F)$) with $\forall V \in P' : |V| \leq \lceil \sqrt{n(F)} \rceil$ and $c(P') \leq \lceil \sqrt{n(F)} \rceil$.*

Such a partitioning P' can be computed in linear time. Algorithm $\mathcal{A}_{01}(F) := \mathcal{S}_{01}(F, P')$ computes a maximal autarky for F, using at most $\min(s, n_A(F)) + \min(s, n_L(F)) \leq 2s$ calls of $\mathcal{O}_{01}$, where $s := \lceil \sqrt{n(F)} \rceil \in \mathbb{N}_0$.

Up to the factor, the upper bound of Theorem 1 is attained:

Example 7. For F as in Example 2 as well as F as in Example 5 we need now $\lceil \sqrt{n(F)} \rceil$ oracle calls (in the worst-case).

6   Conclusion and Outlook

We reviewed the algorithms $\mathcal{A}_0, \mathcal{A}_1, \mathcal{A}_{bs}$ for computing maximal autarkies, using a unified scheme, and presented the new algorithm $\mathcal{A}_{01}$. We are employing four different types of oracles: $\mathcal{O}$ is the basic oracle, just indicating satisfiability resp. unsatisfiability, $\mathcal{O}_0$ in the unsatisfiable case yields the set of variables used by some resolution refutation, $\mathcal{O}_1$ in the satisfiable case yields a satisfying assignment, while $\mathcal{O}_{01}$ combines these capabilities. We investigated in some depth the translation $F \rightsquigarrow t(F)$, which encodes the autarky search for F. The complexities of the four algorithms are summarised as follows (with slight inaccuracies), stating the number and type of oracle calls and the call-instances:

- $\mathcal{A}_0(F)$: $n_L(F)$ calls of $\mathcal{O}_{01}$, subinstances of F.
- $\mathcal{A}_1(F)$: $n_A(F)$ calls of $\mathcal{O}_1$, subinstances of $t(F)$ plus one large positive clause.
- $\mathcal{A}_{01}(F)$: $\sqrt{n(F)}$ calls of $\mathcal{O}_{01}$, subinstances of $t(F)$ plus positive clauses.
- $\mathcal{A}_{bs}(F)$: $\log_2(n(F))$ calls of $\mathcal{O}_1$, subinstances of $t(F)$ plus one varying cardinality constraint in CNF-representation.

Question 1. As we can see from Examples 6, 7, the choice P' from Theorem 1, instantiating the scheme $\mathcal{S}_{01}$ and yielding $\mathcal{A}_{01}$, can be improved at least in special cases. Are more intelligent choices of P possible, heuristically, for special classes, or even in general? The optimal choice (hard to compute) is $P := \{\mathrm{var}(N_a(F))\} \cup \{\{v\} : v \in \mathrm{var}(\mathrm{Auk}^r(F))\}$, which needs two oracle calls.

Question 2. We conjecture the number $\Omega(\sqrt{n(F)})$ of oracle calls from Theorem 1 to be optimal in general, but the question here is, how to formalise the restrictions to the input of oracle $\mathcal{O}_{01}$ (so that for example the SAT translations of cardinality constraints are excluded). With these restrictions in place, we also conjecture that when only using oracle $\mathcal{O}_1$ (as algorithm $\mathcal{A}_1$ does (Definition 8)), that then in general $\Omega(n(F))$ many calls are needed.

Question 3. How do $\mathcal{A}_0, \mathcal{A}_1, \mathcal{A}_{01}, \mathcal{A}_{bs}$ compare to each other? Are they pairwise incomparable? Is their oracle usage optimal under suitable constraints?

Question 4. In this paper we concentrated on the hardest functional task: What about the complexity of the computation of the lean kernel, when using oracles $\mathcal{O}, \mathcal{O}_0, \mathcal{O}_1, \mathcal{O}_{01}$? Do we need less calls than for computing maximal autarkies?

Only one precise conjecture on lower bounds for the computation of maximal autarkies seems possible currently:

Conjecture 1. The computation of a maximal autarky for input $F \in \mathcal{CLS}$, when using a SAT oracle $\mathcal{O}_1$, in general needs $\Omega(\log_2(n(F)))$ many calls; possibly one can even show that for every (deterministic) algorithm there exists an instance needing at least $\log_2(n(F))$ many calls.

Finally we remark that for the considerations of this paper more fine-grained complexity notions for function classes and their oracle usage are needed. Function classes just using NP-oracles (only returning yes/no) have been studied starting with [12], while a systematic study of "function oracles" has been started in [27], using "witness oracles"; we note that $\mathcal{O}_0, \mathcal{O}_{01}$ are not such witness oracles (we can not easily check the returned var-sets).

References

1. Biere, A., Heule, M.J.H., van Maaren, H., Walsh, T. (ed.): Handbook of Satisfiability, volume 185 of Frontiers in Artificial Intelligence and Applications. IOS Press, February 2009
2. Brualdi, R.A., Shader, B.L.: Matrices of sign-solvable linear systems, volume 116 of Cambridge Tracts in Mathematics. Cambridge University Press (1995). ISBN 0-521-48296-8. doi:10.1017/CBO9780511574733
3. Davydov, G., Davydova, I.: Tautologies and positive solvability of linear homogeneous systems. Annals of Pure and Applied Logic **57**(1), 27–43 (1992). doi:10.1016/0168-0072(92)90060-D
4. Even, S., Itai, A., Shamir, A.: On the complexity of timetable and multicommodity flow problems. SIAM Journal Computing **5**(4), 691–703 (1976). doi:10.1137/0205048
5. Fleischner, H., Kullmann, O., Szeider, S.: Polynomial-time recognition of minimal unsatisfiable formulas with fixed clause-variable difference. Theoretical Computer Science **289**(1), 503–516 (2002). doi:10.1016/S0304-3975(01)00337-1
6. Franco, J., Van Gelder, A.: A perspective on certain polynomial-time solvable classes of satisfiability. Discrete Applied Mathematics **125**(2–3), 177–214 (2003). doi:10.1016/S0166-218X(01)00358-4
7. Hall, F.J., Li, Z.: Sign pattern matrices. In: Hogben, L. (ed.), Handbook of Linear Algebra, Discrete Mathematics and Its Applications, pp. 33:1-33:21. Chapman & Hall/CRC, 2007. ISBN 1-58488-510-6. doi:10.1201/9781420010572.ch33
8. Heule, M.J.H., van Maaren, H.: Look-ahead based SAT solvers. In: Biere et al. [1], chapter 5, pp. 155–184. doi:10.3233/978-1-58603-929-5-155
9. Klee, V., Ladner, R.: Qualitative matrices: Strong sign-solvability and weak satisfiability. In: Greenberg, H.J., Maybee, J.S. (eds.), Computer-Assisted Analysis and Model Simplification, pp. 293–320 (1981). Proceedings of the First Symposium on Computer-Assisted Analysis and Model Simplification, University of Colorado, Boulder, Colorado, March 28, 1980. doi:10.1016/B978-0-12-299680-1.50022-7
10. Klee, V., Ladner, R., Manber, R.: Signsolvability revisited. Linear Algebra and its Applications **59**, 131–157 (1984). doi:10.1016/0024-3795(84)90164-2
11. Büning, H.K., Kullmann, O.: Minimal unsatisfiability and autarkies. In : Biere et al. [1], chapter 11, pp. 339–401. doi:10.3233/978-1-58603-929-5-339

12. Krentel, M.W.: The complexity of optimization problems. Journal of Computer and System Sciences **36**(3), 490–509 (1988). doi:10.1016/0022-0000(88)90039-6
13. Kullmann, O.: An application of matroid theory to the SAT problem. In: Proceedings of the 15th Annual IEEE Conference on Computational Complexity, pp. 116–124, July 2000. doi:10.1109/CCC.2000.856741
14. Kullmann, O.: Investigations on autark assignments. Discrete Applied Mathematics **107**, 99–137 (2000). doi:10.1016/S0166-218X(00)00262-6
15. Kullmann, O.: On the use of autarkies for satisfiability decision. In: Kautz, H., Selman, B. (ed.), LICS 2001 Workshop on Theory and Applications of Satisfiability Testing (SAT 2001), volume 9 of Electronic Notes in Discrete Mathematics (ENDM), pp. 231–253. Elsevier Science, June 2001. doi:10.1016/S1571-0653(04)00325-7
16. Kullmann, O.: Lean clause-sets: Generalizations of minimally unsatisfiable clause-sets. Discrete Applied Mathematics **130**, 209–249 (2003). doi:10.1016/S0166-218X(02)00406-7
17. Kullmann, O.: Polynomial time SAT decision for complementation-invariant clause-sets, and sign-non-singular matrices. In: Marques-Silva, J., Sakallah, K.A. (eds.) SAT 2007. LNCS, vol. 4501, pp. 314–327. Springer, Heidelberg (2007)
18. Kullmann, O.: Constraint satisfaction problems in clausal form I: Autarkies and deficiency. Fundamenta Informaticae **109**(1), 27–81 (2011). doi:10.3233/FI-2011-428
19. Kullmann, O.: Constraint satisfaction problems in clausal form II: Minimal unsatisfiability and conflict structure. Fundamenta Informaticae **109**(1), 83–119 (2011). doi:10.3233/FI-2011-429
20. Kullmann, O., Lynce, I., Marques-Silva, J.: Categorisation of clauses in conjunctive normal forms: minimally unsatisfiable sub-clause-sets and the lean kernel. In: Biere, A., Gomes, C.P. (eds.) SAT 2006. LNCS, vol. 4121, pp. 22–35. Springer, Heidelberg (2006)
21. Kullmann, O., Zhao, X.: On variables with few occurrences in conjunctive normal forms. In: Sakallah, K.A., Simon, L. (eds.) SAT 2011. LNCS, vol. 6695, pp. 33–46. Springer, Heidelberg (2011)
22. Kullmann, O., Zhao, X.: Bounds for variables with few occurrences in conjunctive normal forms. Technical Report arXiv:1408.0629v3 [math.CO], arXiv, November 2014. http://arxiv.org/abs/1408.0629
23. Lee, G.-Y., Shader, B.L.: Sign-consistency and solvability of constrained linear systems. The Electronic Journal of Linear Algebra, 4:1–18, August 1998. http://emis.matem.unam.mx/journals/ELA/ela-articles/4.html
24. Li, C.M., Manyà, F.: MaxSAT, hard and soft constraints. In: Biere et al. [1], chapter 19, pp. 613–631. doi:10.3233/978-1-58603-929-5-613
25. Liffiton, M.H., Sakallah, K.A.: Searching for autarkies to trim unsatisfiable clause sets. In: Kleine Büning, H., Zhao, X. (eds.) SAT 2008. LNCS, vol. 4996, pp. 182–195. Springer, Heidelberg (2008)
26. Marques-Silva, J., Ignatiev, A., Morgado, A., Manquinho, V., Lynce, I.: Efficient autarkies. In: Schaub, T., Friedrich, G., O'Sullivan, B. (ed.), 21st European Conference on Artificial Intelligence (ECAI 2014), volume 263 of Frontiers in Artificial Intelligence and Applications, pp. 603–608. IOS Press (2014). doi:10.3233/978-1-61499-419-0-603
27. Marques-Silva, J., Janota, M.: On the query complexity of selecting few minimal sets. Technical Report TR14-031, Electronic Colloquium on Computational Complexity (ECCC), March 2014. http://eccc.hpi-web.de/report/2014/031/

28. Marques-Silva, J.P., Lynce, I., Malik, S.: Conflict-driven clause learning SAT solvers. In: Biere et al. [1], chapter 4, pp. 131–153. doi:10.3233/978-1-58603-929-5-131

29. McCuaig, W.: Pólya's permanent problem. The Electronic Journal of Combinatorics, 11, 2004. #R79, 83 p. http://www.combinatorics.org/ojs/index.php/eljc/article/view/v11i1r79

30. Monien, B., Speckenmeyer, E.: Solving satisfiability in less than 2^n steps. Discrete Applied Mathematics **10**(3), 287–295 (1985). doi:10.1016/0166-218X(85)90050-2

31. Okushi, F.: Parallel cooperative propositional theorem proving. Annals of Mathematics and Artificial Intelligence **26**(1–4), 59–85 (1999). doi:10.1023/A:1018946526109

32. Papadimitriou, C.H., Wolfe, D.: The complexity of facets resolved. Journal of Computer and System Sciences **37**(1), 2–13 (1988). doi:10.1016/0022-0000(88)90042-6

33. Robertson, N., Seymour, P.D., Thomas, R.: Permanents, Pfaffian orientations, and even directed circuits. Annals of Mathematics **150**(3), 929–975 (1999). doi:10.2307/121059

34. Roussel, O., Manquinho, V.: Pseudo-boolean and cardinality constraints. In: Biere et al. [1], chapter 22, pp. 695–733. doi:10.3233/978-1-58603-929-5-695

35. Samuelson, P.A.: Foundations of Economic Analysis. Harvard University Press (1947)

36. Szeider, S.: Minimal unsatisfiable formulas with bounded clause-variable difference are fixed-parameter tractable. Journal of Computer and System Sciences **69**(4), 656–674 (2004). doi:10.1016/j.jcss.2004.04.009

HordeSat: A Massively Parallel Portfolio SAT Solver

Tomáš Balyo[✉], Peter Sanders, and Carsten Sinz

Karlsruhe Institute of Technology (KIT), Karlsruhe, Germany
{tomas.balyo,peter.sanders,carsten.sinz}@kit.edu

Abstract. A simple yet successful approach to parallel satisfiability
(SAT) solving is to run several different (a portfolio of) SAT solvers on
the input problem at the same time until one solver finds a solution. The
SAT solvers in the portfolio can be instances of a single solver with different configuration settings. Additionally the solvers can exchange information usually in the form of clauses. In this paper we investigate whether
this approach is applicable in the case of massively parallel SAT solving.
Our solver is intended to run on clusters with thousands of processors,
hence the name HordeSat. HordeSat is a fully distributed portfolio-based
SAT solver with a modular design that allows it to use any SAT solver
that implements a given interface. HordeSat has a decentralized design
and features hierarchical parallelism with interleaved communication and
search. We experimentally evaluated it using all the benchmark problems
from the application tracks of the 2011 and 2014 International SAT Competitions. The experiments demonstrate that HordeSat is scalable up to
hundreds or even thousands of processors achieving significant speedups
especially for hard instances.

1 Introduction

Boolean satisfiability (SAT) is one of the most important problems of theoretical
computer science with many practical applications in which SAT solvers are used
in the background as high performance reasoning engines. These applications
include automated planning and scheduling [21], formal verification [22], and
automated theorem proving [10]. In the last decades the performance of state-
of-the-art SAT solvers has increased dramatically thanks to the invention of
advanced heuristics [25], preprocessing and inprocessing techniques [19] and data
structures that allow efficient implementation of search space pruning [25].

The next natural step in the development of SAT solvers was parallelization.
A very common approach to designing a parallel SAT solver is to run several
instances of a sequential SAT solver with different settings (or several different
SAT solvers) on the same problem in parallel. If any of the solvers succeeds
in finding a solution all the solvers are terminated. The solvers also exchange
information mainly in the form of learned clauses. This approach is referred to

P. Sanders – This research was partially supported by DFG project SA 933/11-1.

M. Heule and S. Weaver (Eds.): SAT 2015, LNCS 9340, pp. 156–172, 2015.
DOI: 10.1007/978-3-319-24318-4_12

as portfolio-based parallel SAT solving and was first used in the SAT solver ManySat [14]. However, so far it was not clear whether this approach can scale to a large number of processors.

Another approach is to run several search procedures in parallel and ensure that they work on disjoint regions of the search space. This explicit search space partitioning has been used mainly in solvers designed to run on large parallel systems such as clusters or grids of computers [9].

In this paper we describe HordeSat – a scalable portfolio-based SAT solver and evaluate it experimentally. Using efficient yet thrifty clause exchange and advanced diversification methods, we are able to keep the search spaces largely disjoint without explicitly splitting search spaces. Another important feature of HordeSat is its modular design, which allows it to be independent of any concrete search engines. HordeSat uses Sat solvers as black boxes communicating with them via a minimalistic interface.

Experiments made using benchmarks from the application tracks of the 2011 and 2014 Sat Competitions [3] show that HordeSat can outperform state-of-the-art parallel SAT solvers on multiprocessor machines and is scalable on computer clusters with thousands of processors. Indeed, we even observe superlinear average speedup for difficult instances.

2 Preliminaries

A *Boolean variable* is a variable with two possible values *True* and *False*. By a *literal* of a Boolean variable x we mean either x or $\overline{x}$ (*positive* or *negative literal*). A *clause* is a disjunction (OR) of literals. A *conjunctive normal form* (CNF) *formula* is a conjunction (AND) of clauses. A clause can be also interpreted as a set of literals and a formula as a set of clauses. A truth assignment ϕ of a formula F assigns a truth value to its variables. The assignment ϕ satisfies a positive (negative) literal if it assigns the value True (False) to its variable and ϕ satisfies a clause if it satisfies any of its literals. Finally, ϕ satisfies a CNF formula if it satisfies all of its clauses. A formula F is said to be satisfiable if there is a truth assignment ϕ that satisfies F. Such an assignment is called a *satisfying assignment*. The satisfiability problem (SAT) is to find a satisfying assignment of a given CNF formula or determine that it is unsatisfiable.

Conflict Driven Clause Learning. Most current complete state-of-the-art SAT solvers are based on the conflict-driven clause learning (CDCL) algorithm [23]. In this paper we will use CDCL solvers only as black boxes and therefore we provide only a very coarse-grained description. For a detailed discussion of CDCL refer to [5]. In Figure 1 we give a pseudo-code of CDCL. The algorithm performs a depth-first search of the space of partial truth assignments (assignDecisionLiteral, backtrack – unassigns variables) interleaved with search space pruning in the form of unit propagation (doUnitPropagation) and learning new clauses when the search reaches a conflict state (analyzeConflict, addLearnedClause). If a conflict cannot be resolved by backtracking then the

```
                CDCL (CNF formula F)
    CDCL0        while not all variables assigned do
    CDCL1            assignDecisionLiteral
    CDCL2            doUnitPropagation
    CDCL3            if conflict detected then
    CDCL4                analyzeConflict
    CDCL5                addLearnedClause
    CDCL6                backtrack or return UNSAT
    CDCL7        return SAT
```

Fig. 1. Pseudo-code of the conflict-driven clause learning (CDCL) algorithm

formula is unsatisfiable. If all the variables are assigned and no conflict is detected then the formula is satisfiable.

3   Related Work

In this section we give a brief description of previous parallel SAT solving approaches. A much more detailed listing and description of existing parallel solvers can be found in recently published overview papers such as [15,24].

Parallel CDCL – Pure Portfolios. The simplest approach is to run CDCL several times on the same problem in parallel with different parameter settings and exchanging learned clauses. If there is no explicit search space partitioning then this approach is referred to as the pure portfolio algorithm. The first parallel portfolio SAT solver was ManySat [14]. The winner of the latest (2014) Sat Competition's parallel track – Plingeling [4] is also of this kind.

The motivation behind the portfolio approach is that the performance of CDCL is heavily influenced by a high number of different settings and parameters of the search such as the heuristic used to select a decision literal. Numerous heuristics can be used in this step [25] but none of them dominates all the other heuristics on each problem instance. Decision heuristics are only one of the many settings that strongly influence the performance of CDCL solvers. All of these settings can be considered when the diversification of the portfolio is performed. For an example see ManySat [14]. Automatic configuration of SAT solvers in order to ensure that the solvers in a portfolio are diverse is also studied [30].

Exchanging learned clauses grants an additional boost of performance. It is an important mechanism to reduce duplicate work, i.e., parallel searches working on the same part of the search space. A clause learned from a conflict by one CDCL instance distributed to all the other CDCL instances will prevent them from doing the same work again in the future.

The problem related to clause sharing is to decide how many and which clauses should be exchanged. Exchanging all the learned clauses is infeasible especially in the case of large-scale parallelism. A simple solution is to distribute all the clauses that satisfy some conditions. The conditions are usually related to the length of the clauses and/or their glue value [1]. An interesting technique called "lazy clause exchange" was introduced in a recent paper [2]. We leave the adaptation of this technique to future work however, since it would make the design of our solver less modular. Most of the existing pure portfolio SAT solvers are designed to run on single multi-processor computers. An exception is CL-SDSAT [17] which is designed for solving very difficult instances on loosely connected grid middleware. It is not clear and hard to quantify whether this approach can yield significant speedups since the involved sequential computation times would be huge.

Parallel CDCL – Partitioning The Search Space Explicitly. The classical approach to parallelizing SAT solving is to split the search space between the search engines such that no overlap is possible. This is usually done by starting each solver with a different fixed partial assignment. If a solver discovers that its partial assignment cannot be extended into a solution it receives a new assignment. Numerous techniques have presented how to manage the search space splitting based on ideas such guiding paths [9], work stealing [20], and generating sufficiently many tasks [11]. Similarly to the portfolio approach the solvers exchange clauses.

Most of the previous SAT solvers designed for computer clusters or grids use explicit search space partitioning. Examples of such solvers are GridSAT [9], PM-SAT [11], GradSat [8], C-sat [26], ZetaSat [6] and SatCiety [28]. Experimentally Comparing HordeSat with those solvers is problematic, since these solvers are not easily available online or they are implemented for special environments using non-standard middleware. Nevertheless we can get some conclusions based on looking at the experimental sections of the related publications.

Older grid solvers such as GradSat [8], PM-SAT [11] SatCiety [28], ZetaSat [6] and C-sat [26] are evaluated on only small clusters (up to 64 processors) using small sets of older benchmarks, which are easily solved by current state-of-the-art sequential solvers and therefore it is impossible to tell how well do they scale for a large number of processors and current benchmarks. The solver GridSAT [9] is run on a large heterogeneous grid of computers containing hundreds of nodes for several days and is reported to solve several (at that time) unsolved problems. Nevertheless, most of those problems can now be solved by sequential solvers in a few minutes. Speedup results are not reported. A recent grid-based solving method called Part-Tree-Learn [16] is compared to Plingeling and is reported to solve less instances than Plingeling. This is despite the fact that in their comparison the number of processors available to Plingeling was slightly less [16].

To design a successful explicit partitioning parallel solver, complex load balancing issues must be solved. Additionally, explicit partitioning clearly brings runtime and space overhead. If the main motivation of explicit partitioning is to ensure that the search-spaces explored by the solvers have no overlap, then we believe that the extra work does not pay off and frequent clause sharing is enough to approximate the desired behavior [1]. Moreover, in [18] the authors argue that plain partitioning approaches can increase the expected runtime compared to pure portfolio systems. They prove that under reasonable assumptions there is always a distribution that results in an increased expected runtime unless the process of constructing partitions is ideal.

4   Design Decisions

In this section we provide an overview of the high level design decisions made when designing our portfolio-based SAT solver HordeSat.

Modular Design. Rather than committing to any particular SAT solver we design an interface that is universal and can be efficiently implemented by current state-of-the-art SAT solvers. This results in a more general implementation and the possibility to easily add new SAT solvers to our portfolio.

Decentralization. All the nodes in our parallel system are equivalent. There is no leader or central node that manages the search or the communication. Decentralized design allows more scalability and also simplifies the algorithm.

Overlapping Search and Communication. The search and the clause exchange procedures run in different (hardware) threads in parallel. The system is implemented in a way that the search procedure never waits for any shared resources at the expense of losing some of the shared clauses.

Hierarchical Parallelization. HordeSat is designed to run on clusters of computers (nodes) with multiple processor cores, i.e., we have two levels of parallelization. The first level uses the shared memory model to communicate between solvers running on the same node and the second level relies on message passing between the nodes of a cluster.

The details and implementation of these points are discussed below.

5   Black Box for Portfolios

Our goal is to develop a general parallel portfolio solver based on existing state-of-the-art sequential CDCL solvers without committing to any particular solver. To achieve this we define a C++ interface that is used to access the solvers in the portfolio. Therefore new SAT solvers can be easily added just by implementing this interface. By *core solver* we will mean a SAT solver implementing the interface.

In this section we describe the essential methods of the interface. All the methods are required to be implemented in a thread safe way, i.e., safe execution

---

[1] According to our experiments only 2-6% of the clauses are learned simultaneously by different solvers in a pure portfolio, which is an indication that the overlap of search-spaces is relatively small.

by multiple threads at the same time must be guaranteed. First we start with the basic methods which allow us to solve formulas and interrupt the solver.

void addClause(vector<int> clause): This method is used to load the initial formula that is to be solved. The clauses are represented as lists of literals which are represented as integers in the usual way. All the clauses must be considered by the solver at the next call of `solve`.

SatResult solve(): This method starts the search for the solution of the formula specified by the `addClause` calls. The return value is one of the following `SatResult = {SAT, UNSAT, UNKNOWN}`. The result `UNKNOWN` is returned when the solver is interrupted by calling `setSolverInterrupt()`.

void setSolverInterrupt(): Posts a request to the core solver instance to interrupt the search as soon as possible. If the method `solve` has been called, it will return `UNKNOWN`. Subsequent calls of `solve` on this instance must return `UNKNOWN` until the method `unsetSolverInterrupt` is called.

void unsetSolverInterrupt(): Removes the request to interrupt the search.

Using these four methods, a simple portfolio can be built. When using several instances of the same deterministic SAT solver, some diversification can be achieved by adding the clauses in a different order to each solver.

More options for diversification are made possible via the following two methods. A good way of diversification is to set default *phase* values for the variables of the formula, i.e., truth values to be tried first. These are then used by the core solver when selecting decision literals. In general many solver settings can be changed to achieve diversification. Since these may be different for each core solver we define a general method for diversification which the core solver can implement in its own specific way.

void setPhase(int var, bool phase): This method is used to set a default phase of a variable. The solver is allowed to ignore these suggestions.

void diversify(int rank, int size): This method tells the core solver to diversify its settings. The specifics of diversification are left to the solver. The provided parameters can be used by the solver to determine how many solvers are working on this problem (`size`) and which one of those is this solver (`rank`). A trivial implementation of this method could be to set the pseudo-random number generator seed of the core solver to `rank`.

The final three methods of the interface deal with clause sharing. The solvers can produce and accept clauses. Not all the learned clauses are shared. It is expected that each core solver initially offers only a limited number of clauses which it considers most worthy of sharing. The solver should increase the number of exported clauses when the method `increaseClauseProduction` is called. This can be implemented by relaxing the constraints on the learned clauses selected for exporting.

void addLearnedClause(vector<int> clause): This method is used to add learned clauses received from other solvers of the portfolio. The core solver can decide when and whether the clauses added using this method are actually considered during the search.

void setLearnedClauseCallback(LCCallback* callback): This method is used to set a callback class that will process the clauses shared by this solver. To export a clause, the core solver will call the void write(vector<int> clause) method of the LCCallback class. Each clause exported by this method must be a logical consequence of the clauses added using addClause or addLearnedClause.

void increaseClauseProduction(): Inform the solver that more learned clauses should be shared. This could mean for example that learned clauses of bigger size or higher glue value [1] will be shared.

The interface is designed to closely match current CDCL SAT solvers, but any kind of SAT solver can be used. For example a local search SAT solver could implement the interface by ignoring the calls to the clause sharing methods.

For our experiments we implemented the interface by writing binding code for MiniSat [29] and Lingeling [4]. In the latter case no modifications to the solver were required and the binding code only uses the incremental interface of Lingeling. As for MiniSat, the code has been slightly modified to support the three clause sharing methods.

6   The Portfolio Algorithm

In this section we describe the main algorithm used in HordeSat. As already mentioned in section 4 we use two levels of parallelization. HordeSat can be viewed as a multithreaded program that communicates using messages with other instances of the same program. The communication is implemented using the Message Passing Interface (MPI) [12]. Each MPI process runs the same multithreaded program and takes care about the following tasks:

- Start the core solvers using solve. Use one fresh thread for each core solver.
- Read the formula and add its clauses to each core solver using addClause.
- Ensure diversification of the core solvers with respect to the other processes.
- Ensure that if one of the core solvers solves the problem all the other core solvers and processes are notified and stopped. This is done by using setSolverInterrupt for each core solver and sending a message to all the participating processes.
- Collect the exported clauses from the core solvers, filter duplicates and send them to the other processes. Accept the exported clauses of the other processes, filter them and distribute them to the core solvers.

The tasks of reading the input formula, diversification, and solver starting are performed once after the start of the process. The communication of ending and clause exchange is performed periodically in rounds until a solution is found. The main thread sleeps between these rounds for a given amount of time specified as a parameter of the solver (usually around 1 second). The threads running the core solvers are working uninterrupted during the whole time of the search.

6.1 Diversification

Since we can only access the core solvers via the interface defined above, our only tools for diversification are setting phases using the `setPhase` method and calling the solver specific `diversify` method.

The `setPhase` method allows us to partition the search space in a semi-explicit fashion. An explicit search space splitting into disjoint subspaces is usually done by imposing phase restrictions instead of just recommending them. The explicit approach is used in parallel solvers utilizing guiding paths [9] and dynamic work stealing [20].

We have implemented and tested the following diversification procedures based on literal phase recommendations.

- *Random.* Each variable gets a phase recommendation for each core solver randomly. Note that this is different from selecting a random phase each time a decision is made for a variable in the CDCL procedure.
- *Sparse.* Each variable gets a random phase recommendation on exactly one of the host solvers in the entire portfolio. For the other solvers no phase recommendation is made for the given variable.
- *Sparse Random.* For each core solver each variable gets a random phase recommendation with a probability of $(\#solvers)^{-1}$, where $\#solvers$ is the total number of core solvers in the portfolio.

Each of these can be used in conjunction with the `diversify` method whose behavior is defined by the core solvers. As already mentioned we use Lingeling and MiniSat as core solvers. In case of MiniSat, we implemented the `diversify` method by only setting the random seed. For Lingeling we copied the diversification algorithm from Plingeling [4], which is the multi-threaded version of Lingeling based on the portfolio approach and the winner of the parallel application track of the 2014 SAT Competition [3]. In this algorithm 16 different parameters of Lingeling are used for diversification.

6.2 Clause Sharing

The clause sharing in our portfolio happens periodically in rounds. Each round a fixed sized (1500 integers in the implementation) message containing the literals of the shared clauses is exchanged by all the MPI processes in an all-to-all fashion. This is implemented by using the `MPI_Allgather` [12] collective communication routine defined by the MPI standard.

Each process prepares the message by collecting the learned clauses from its core solvers. The clauses are filtered to remove duplicates. The fixed sized message buffer is filled up with the clauses, shorter clauses are preferred. Clauses that did not fit are discarded. If the buffer is not filled up to its full capacity then one of the core solvers of the process is requested to increase its clause production by calling the `increaseClauseProduction` method.

The detection of duplicate clauses is implemented by using Bloom filters [7]. A Bloom filter is a space-efficient probabilistic set data structure that allows false-positive matches, which in our case means that some clauses might be considered to be duplicates even if they are not. The usage of Bloom filters requires a set of hash functions that map clauses to integers. We use the following hash function which ensures that permuting the literals of a clause does not change its hash value.

$$H_i(C) = \bigoplus_{\ell \in C} \ell \cdot primes[abs(\ell \cdot i) \bmod |primes|]$$

where $i > 0$ is a parameter we are free to choose, C is a clause, $\oplus$ denotes bitwise exclusive-or, and *primes* is an array of large prime numbers. Literals are interpreted as integers in the usual way, i.e., x_j as j and $\overline{x}_j$ as $-j$.

Each MPI process maintains one Bloom filter g_x for each of its core solvers x and an additional global one g. When a core solver x exports a learned clause C, the following steps are taken.

- Clause C is added to g_x.
- If $C \notin g$, C is added to g as well as into a data structure e for export.
- If several core solvers concurrently try to access e, only one will succeed and the new clauses of the other core solvers are ignored. This way, we avoid contention at the shared resource e and rather ignore some clauses.

After the global exchange of learned clauses, the incoming clauses need to be filtered for duplicates and distributed to the core solvers. The first task is done by using the global Bloom filter g. For the second task we utilize the thread local filters g_x to ensure that each of them receives only new clauses.

All the Bloom filters are periodically reset, which allows the repeated sharing of clauses after some time. Our initial experiments showed that this approach is more beneficial than maintaining a strict "no duplicate clauses allowed"-policy.

Overall, there are three reasons why a clause offered by a core solver can get discarded. One is that it was duplicate or wrongly considered to be duplicate due to the probabilistic nature of Bloom filters. Second is that another core solver was adding its clause to the data structure for global export at the same time. The last reason is that it did not fit into the fixed size message sent to the other MPI processes. Although important learned clauses might get lost, we believe that this relaxed approach is still beneficial since it allows a simpler and more efficient implementation of clause sharing.

7   Experimental Evaluation

To examine our portfolio-based parallel SAT solver HordeSat we did experiments with two kinds of benchmarks. We used the benchmark formulas from the application tracks of the 2011 and 2014 SAT Competitions [3] (545 instances) [2]

---

[2] Originally we only used the 2014 instances. A reviewer suggested to try the 2011 instances also, conjecturing that they would be harder to parallelize. Surprisingly, the opposite turned out to be true.

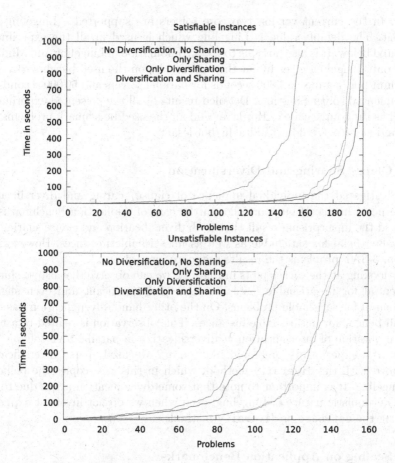

Fig. 2. The influence of diversification and clause sharing on the performance of Horde-Sat using Lingeling (16 processes with 1 thread each) on random 3-SAT problems.

and randomly generated 3-SAT formulas (200 sat and 200 unsat instances). The random formulas have 250–440 variables and 4.25 times as many clauses, which corresponds to the phase transition of 3-SAT problems [27].

The experiments were run on a cluster allowing us to reserve up to 128 nodes. Each node has two octa-core Intel Xeon E5-2670 processors (Sandy Bridge) with 2.6 GHz and 64 GB of main memory. Therefore each node has 16 cores and the total number of available cores is 2048. The nodes communicate using an InfiniBand 4X QDR Interconnect and use the SUSE Linux Enterprise Server 11 (x86_64) (patch level 3) operating system. HordeSat was compiled using g++ (SUSE Linux) 4.3.4 [gcc-4_3-branch revision 152973] with the "-O3" flag.

If not stated otherwise, we use the following parameters: The time of sleeping between clause sharing rounds is 1 second. The default diversification algorithm is the combination of "sparse random" and the native diversification of the core solver. In the current version two core solvers are supported – Lingeling and MiniSat. The default value is Lingeling which is used in all the experiments presented below. It is also possible to use a combination of Lingeling and MiniSat. Using only Lingeling gives by far the best results on the used benchmarks. The time limit per instance is 1 000 seconds for parallel solvers and 50 000 seconds for the sequential solver Lingeling. Detailed results of all the presented experiments as well as the source code of HordeSat and all the used benchmark problems can be found at http://baldur.iti.kit.edu/hordesat.

7.1 Clause Sharing and Diversification

We investigated the individual influence of clause sharing and diversification on the performance of our portfolio. In the case of application benchmarks we obtained the unsurprising result that both diversification and clause sharing are highly beneficial for satisfiable as well as unsatisfiable instances. However, for random 3-SAT problems the results are more interesting.

By looking at the cactus plots in Figure 2 we can observe that clause sharing is essential for unsatisfiable instances while not significant and even slightly detrimental for satisfiable problems. On the other hand, diversification has only a small benefit for unsatisfiable instances. This observation is related to a more general question of intensification vs diversification in parallel SAT solving [13].

For the experiments presented in Figure 2 we used sparse diversification combined with the `diversify` method, which in this case copies the behavior of Plingeling. It is important to note that some diversification arises due to the non-deterministic nature of Lingeling, even when we do not invoke it explicitly by using the `setPhase` or `diversify` methods.

7.2 Scaling on Application Benchmarks

In parallel processing, one usually wants good scalability in the sense that the speedup over the best sequential algorithm goes up near linearly with the number of processors. Measuring scalability in a reliable and meaningful way is difficult for SAT solving since running times are highly nondeterministic. Hence, we need careful experiments on a large benchmark set chosen in an unbiased way. We therefore use the application benchmarks of the 2011 and 2014 Sat Competitions. Our sequential reference is Lingeling which won the most recent (2014) competition. We ran experiments using 1,2,4,. . . ,512 processes with four threads each, each cluster nodes runs 4 processes. The results are summarized in Figure 3 using cactus plots. We can observe that increased parallelism is always beneficial for the 2011 benchmarks. In the case of all the benchmarks the benefits beyond 32 nodes are not apparent.

From a cactus plot it is not easy to see whether the additional performance is a reasonable return on the invested hardware resources. Therefore Table 1 summarizes that information in several ways in order to quantify the overall scalability of HordeSat on the union of the 2011 and 2013 benchmarks. We compute speedups for all the instances solved by the parallel solver. For instances

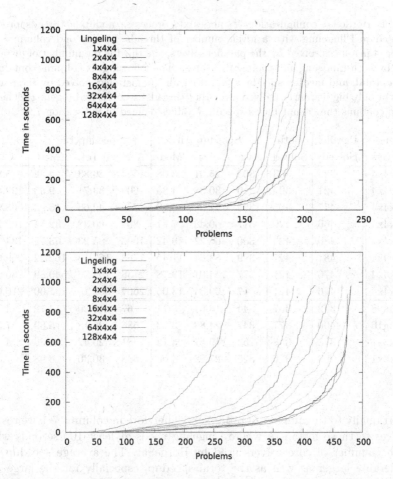

Fig. 3. The impact of doubling the number of processors on the runtime and the number solved problems for the 2011 and the union of 2011 and 2013 application instances. The labels represent (#nodes)x(#processes/node)x(#threads/process).

not solved by Lingeling within its time limit $T = 50\,000s$ we generously assume that it would solve them if given $T + \epsilon$ seconds and use the runtime of T for speedup calculation. Column 4 gives the average of these values. We observe considerable superlinear speedups *on average* for all the configurations tried. However, this average is not a very robust measure since it is highly dependent on a few very large speedups that might be just luck. In Column 5 we show the total speedup, which is the sum of sequential runtimes divided by the sum of parallel runtimes and Column 6 contains the median speedup.

Nevertheless, these figures treat HordeSat unfairly since most instances are actually too easy for investing a lot of hardware. Indeed, in parallel computing, it is usual to analyze the performance on many processors using *weak scaling* where one increases the amount of work involved in the considered instances

Table 1. HordeSat configurations (#nodes)x(#processes/node)x(#threads/process) compared to Plingeling with a given number of threads. The second column is the number of instances solved by the parallel solvers, the third is the number of instances solved by both Lingeling and the parallel solver. The following six columns contain the average, total, and median speedups for either all the instances solved by the parallel solvers or only big instances (solved after 10(#threads) seconds by Lingeling). The last column contains the "count based speedup" values defined in Subsection 7.2.

Core Solvers	Parallel Solved	Both Solved	Speedup All			Speedup Big			CBS
			Avg.	Tot.	Med.	Avg.	Tot.	Med.	
1x4x4	385	363	303	25.01	3.08	524	26.83	4.92	5.86
2x4x4	421	392	310	30.38	4.35	609	33.71	9.55	22.44
4x4x4	447	405	323	41.30	5.78	766	49.68	16.92	68.90
8x4x4	466	420	317	50.48	7.81	801	60.38	32.55	102.27
16x4x4	480	425	330	65.27	9.42	1006	85.23	63.75	134.37
32x4x4	481	427	399	83.68	11.45	1763	167.13	162.22	209.07
64x4x4	476	421	377	104.01	13.78	2138	295.76	540.89	230.37
128x4x4	476	421	407	109.34	13.05	2607	352.16	867.00	216.69
pling8	372	357	44	18.61	3.11	67	19.20	4.12	4.77
pling16	400	377	347	24.83	3.53	586	26.18	5.89	7.34
1x8x1	373	358	53	19.57	3.13	81	20.42	4.36	4.79
1x16x1	400	376	325	27.78	4.06	548	30.30	6.98	7.34

proportionally to the number of processors. Therefore in columns 7–9 we restrict ourselves to those instances where Lingeling needs at least $10p$ seconds where p is the number of core solvers used by HordeSat. The average speedup gets considerably larger as well as the total speedup, especially for the large configurations. The median speedup also increases but remains slightly sublinear. Figure 4 shows the distribution of speedups for these instances.

Another way to measure speedup robustly is to compare the times needed to solve a given number of instances. Let T_1 (T_p) denote the per instance time limits of the sequential (parallel) solver (50 000s (1 000s) in our case). Let n_1 (n_p) denote the number of instances solved by the sequential (parallel) solver within time T_1 (T_p). If $n_1 \geq n_p$ ($n_1 < n_p$) let T_1' (T_p') denote the smallest time limit for the sequential (parallel) solver such that it solves n_p (n_1) instances within the time limit T_1' (T_p'). We define the *count based speedup* (CBS) as

$$\text{CBS} = \begin{cases} T_1/T_p' & \text{if } n_1 < n_p \\ T_1'/T_p & \text{otherwise} . \end{cases}$$

The CBS scales almost linearly up to 512 cores and stagnates afterward. We are not sure whether this indicates a scalability limit of HordeSat or rather reflects a lack of sufficiently difficult instances – in our collection, there are only 65 eligible instances.

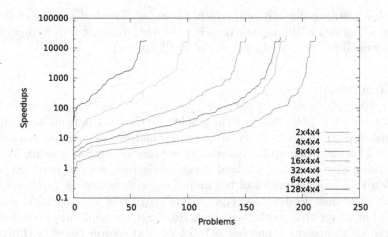

Fig. 4. Distribution of speedups on the "big instances" – the data corresponding to Columns 7–9 of Table 1.

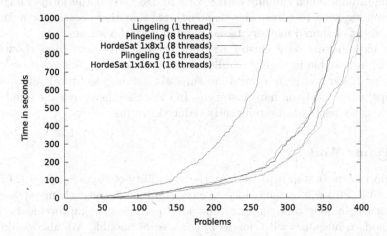

Fig. 5. Comparison of HordeSat and Plingeling with Lingeling on the 2011 and 2014 Sat Competition benchmarks.

7.3 Comparison with Plingeling

The most similar parallel SAT solver to our portfolio is the state-of-the-art solver Plingeling [4]. Plingeling is the winner of the parallel track of the 2014 SAT Competition. Both solvers are portfolio-based, they are using Lingeling and even some diversification code is shared. The main differences are in the clause sharing algorithms and that Plingeling does not run on clusters only single computers. For this reason we can compare the two solvers only on a single node. The results of this comparison on the benchmark problems of the 2011 and 2014 SAT Competition are displayed in Figure 5. Speedup values are given in Table 1.

Both solvers significantly outperform Lingeling. The performance of Horde-Sat and Plingeling is almost indistinguishable when running with 8 cores, while on 16 cores HordeSat gets slightly ahead of Plingeling.

8   Conclusion

HordeSat has the potential to reduce solution times of difficult yet solvable SAT instances from hours to minutes using hundreds of cores on commodity clusters. This may open up new interactive applications of SAT solving. We find it surprising that this was achieved using a relatively simple, portfolio based approach that is independent of the underlying core solver. In particular, this makes it likely that HordeSat can track future progress of sequential SAT solvers.

The Sat solver that works best with HordeSat for application benchmarks is Lingeling. Plingeling is another parallel portfolio solver based on Lingeling and it is also the winner of the most recent (2014) Sat Competition. Comparing the performance of HordeSat and Plingeling reveals that HordeSat is almost indistinguishable when running with 8 cores and slightly outperforms Plingeling when running with 16 cores. This demonstrates that there is still room for the improvement of shared memory based parallel portfolio solvers.

Our experiments on a cluster with up to 2048 processor cores show that HordeSat is scalable in highly parallel environments. We observed superlinear and nearly linear scaling in several measures such as average, total, and median speedups, particularly on hard instances. In each case increasing the number of available cores resulted in significantly reduced runtimes.

8.1   Future Work

An important next step is to work on the scalability of HordeSat for 1024 cores and beyond. This will certainly involve more adaptive clause exchange strategies. Even for single node configurations, low level performance improvements when using modern machines with dozens of cores seem possible. We also would like to investigate what benefits can be gained by having a tighter integration of core solvers by extending the interface. Including other kinds of (not necessarily CDCL – based) core solvers might also bring improvements.

When considering massively parallel SAT solving we probably have to move to even more difficult instances to make that meaningful. When this also means *larger* instances, memory consumption may be an issue when running many instances of a SAT solver on a many-core machine. Here it might be interesting to explore opportunities for sharing data structures for multiple SAT solvers or to decompose problems into smaller subproblems by recognizing their structure.

Acknowledgments. We would like to thank Armin Biere for fruitful discussion about the usage of the Lingeling API in a parallel setting.

References

1. Audemard, G., Simon, L.: Predicting learnt clauses quality in modern sat solvers. IJCAI **9**, 399–404 (2009)
2. Audemard, G., Simon, L.: Lazy clause exchange policy for parallel SAT solvers. In: Sinz, C., Egly, U. (eds.) SAT 2014. LNCS, vol. 8561, pp. 197–205. Springer, Heidelberg (2014)
3. Belov, A., Diepold, D., Heule, M.J., Järvisalo, M.: Sat competition (2014)
4. Biere, A.: Lingeling, plingeling and treengeling entering the sat competition 2013. In: Balint, A., Belov, A., Heule, M.J.H., Järvisalo, M. (ed.), Proceedings of SAT Competition 2013, vol. B-2013-1 of Department of Computer Science Series of Publications B, pp. 51–52. University of Helsinki, 2013 (2013)
5. Biere, A., Heule, M., van Maaren, H., Walsh, T.: Conflict-driven clause learning sat solvers. In: Handbook of Satisfiability, Frontiers in Artificial Intelligence and Applications, pp. 131–153 (2009)
6. Blochinger, W., Westje, W., Kuchlin, W., Wedeniwski, S.: Zetasat-boolean satisfiability solving on desktop grids. In: IEEE International Symposium on Cluster Computing and the Grid, 2005. CCGrid 2005, vol. 2, pp. 1079–1086. IEEE (2005)
7. Bloom, B.H.: Space/time trade-offs in hash coding with allowable errors. Communications of the ACM **13**(7), 422–426 (1970)
8. Chrabakh, W., Wolski, R.: Gradsat: A parallel sat solver for the grid. In: Proceedings of IEEE SC03 (2003)
9. Chrabakh, W., Wolski, R.: Gridsat: A chaff-based distributed sat solver for the grid. In: Proceedings of the 2003 ACM/IEEE conference on Supercomputing, p. 37. ACM (2003)
10. Flanagan, C., Joshi, R., Ou, X., Saxe, J.B.: Theorem proving using lazy proof explication. In: Hunt Jr, W.A., Somenzi, F. (eds.) CAV 2003. LNCS, vol. 2725, pp. 355–367. Springer, Heidelberg (2003)
11. Gil, L., Flores, P., Silveira, L.M.: Pmsat: a parallel version of minisat. Journal on Satisfiability, Boolean Modeling and Computation **6**, 71–98 (2008)
12. Gropp, W., Lusk, E., Doss, N., Skjellum, A.: A high-performance, portable implementation of the mpi message passing interface standard. Parallel computing **22**(6), 789–828 (1996)
13. Guo, L., Hamadi, Y., Jabbour, S., Sais, L.: Diversification and intensification in parallel SAT solving. In: Cohen, D. (ed.) CP 2010. LNCS, vol. 6308, pp. 252–265. Springer, Heidelberg (2010)
14. Hamadi, Y., Jabbour, S., Sais, L.: Manysat: a parallel sat solver. Journal on Satisfiability, Boolean Modeling and Computation **6**, 245–262 (2008)
15. Hölldobler, S., Manthey, N., Nguyen, V., Stecklina, J., Steinke, P.: A short overview on modern parallel sat-solvers. In: Proceedings of the International Conference on Advanced Computer Science and Information Systems, pp. 201–206 (2011)
16. Hyvärinen, A.E.J., Junttila, T., Niemelä, I.: Grid-Based SAT solving with iterative partitioning and clause learning. In: Lee, J. (ed.) CP 2011. LNCS, vol. 6876, pp. 385–399. Springer, Heidelberg (2011)
17. Hyvärinen, A.E., Junttila, T., Niemela, I.: Incorporating clause learning in grid-based randomized sat solving. Journal on Satisfiability, Boolean Modeling and Computation **6**, 223–244 (2014)
18. Hyvärinen, A.E.J., Manthey, N.: Designing scalable parallel SAT solvers. In: Cimatti, A., Sebastiani, R. (eds.) SAT 2012. LNCS, vol. 7317, pp. 214–227. Springer, Heidelberg (2012)

19. Järvisalo, M., Heule, M.J.H., Biere, A.: Inprocessing rules. In: Gramlich, B., Miller, D., Sattler, U. (eds.) IJCAR 2012. LNCS, vol. 7364, pp. 355–370. Springer, Heidelberg (2012)

20. Jurkowiak, B., Li, C.M., Utard, G.: A parallelization scheme based on work stealing for a class of sat solvers. Journal of Automated Reasoning **34**(1), 73–101 (2005)

21. Kautz, H.A., Selman, B., et al.: Planning as satisfiability. ECAI **92**, 359–363 (1992)

22. Kuehlmann, A., Paruthi, V., Krohm, F., Ganai, M.K.: Robust boolean reasoning for equivalence checking and functional property verification. IEEE Transactions on Computer-Aided Design of Integrated Circuits and Systems **21**(12) (2002)

23. Marques-Silva, J.P., Sakallah, K.A.: Grasp: A search algorithm for propositional satisfiability. IEEE Transactions on Computers **48**(5), 506–521 (1999)

24. Martins, R., Manquinho, V., Lynce, I.: An overview of parallel sat solving. Constraints **17**(3), 304–347 (2012)

25. Moskewicz, M.W., Madigan, C.F., Zhao, Y., Zhang, L., Malik, S.: Chaff: Engineering an efficient sat solver. In: Proceedings of the 38th Annual Design Automation Conference, pp. 530–535. ACM (2001)

26. Ohmura, K., Ueda, K.: c-sat: a parallel SAT solver for clusters. In: Kullmann, O. (ed.) SAT 2009. LNCS, vol. 5584, pp. 524–537. Springer, Heidelberg (2009)

27. Parkes, A.J.: Clustering at the phase transition. In: Proc. of the 14th Nat. Conf. on AI, pp. 340–345. AAAI Press/The MIT Press (1997)

28. Schulz, S., Blochinger, W.: Parallel sat solving on peer-to-peer desktop grids. Journal of Grid Computing **8**(3), 443–471 (2010)

29. Sorensson, N., Een, N.: Minisat v1.13 a sat solver with conflict-clause minimization. SAT 2005 (2005)

30. Xu, L., Hoos, H., Leyton-Brown, K.: Hydra: Automatically configuring algorithms for portfolio-based selection. AAAI Conference on Artificial Intelligence (2010)

Preprocessing for DQBF

Ralf Wimmer[(✉)], Karina Gitina, Jennifer Nist, Christoph Scholl,
and Bernd Becker

Albert-Ludwigs-Universität Freiburg, Freiburg im Breisgau, Germany
{wimmer,gitina,nistj,scholl,becker}@informatik.uni-freiburg.de

Abstract. For SAT and QBF formulas many techniques are applied in order to reduce/modify the number of variables and clauses of the formula, before the formula is passed to the actual solving algorithm. It is well known that these preprocessing techniques often reduce the computation time of the solver by orders of magnitude. In this paper we generalize different preprocessing techniques for SAT and QBF problems to dependency quantified Boolean formulas (DQBF) and describe how they need to be adapted to work with a DQBF solver core. We demonstrate their effectiveness both for CNF- and non-CNF-based DQBF algorithms.

1 Introduction

Many problems, practically relevant and at the same time hard from a complexity theoretic point of view, can be reduced to solving quantifier-free (SAT) or quantified (QBF) Boolean formulas. Such applications range, among many others, from verification and test of hard- and software [1,2] to planning [3], product configuration [4], and cryptanalysis [5]. During the last three decades, the development of very efficient algorithms to solve such formulas has paved the way from academic interest to industrial application of solver techniques. SAT-formulas with hundred thousands of variables and millions of clauses can be solved nowadays, with QBF about two orders of magnitude behind.

In this paper, we consider the more general, still practically relevant formalism of *dependency quantified Boolean formulas* (DQBF). "Standard" quantified Boolean formulas (in prenex normal form) have the restriction that each existential variable depends on all universal variables in whose scope it is. This restriction is relaxed for DQBF, which allows arbitrary dependencies at the cost of a higher complexity for the decision problem – for SAT it is NP-complete [6], for QBF PSPACE-complete [7], and for DQBF it is NEXPTIME-complete [8]. However, some applications like the verification of incomplete circuits [9] or the synthesis of safe controllers [10] require the higher expressiveness of DQBF. Therefore, first solvers for DQBF have been presented recently: IDQ [11] reduces the solution of a DQBF to the solution of a series of SAT instantiations. HQS [12] applies quantifier elimination to solve the formula.

This work was partly supported by the German Research Council (DFG) as part of the Transregional Collaborative Research Center AVACS (SFB/TR 14).

M. Heule and S. Weaver (Eds.): SAT 2015, LNCS 9340, pp. 173–190, 2015.
DOI: 10.1007/978-3-319-24318-4_13

Part of the success of SAT and QBF solving is due to efficient preprocessing of the formula under consideration. The goal of preprocessing is to simplify the formula by reducing/modifying the number of variables, clauses and quantifier alternations, such that it can be solved more efficiently afterwards. However, there is typically a trade-off between the number of variables and the number of clauses; e. g., eliminating variables by resolution can increase the number of clauses significantly, which in turn increases memory consumption and the cost of subsequent operations on the formula. Removing redundant clauses is also not always beneficial: search-based SAT and QBF solvers add implied clauses to the formula to drive the search away from unsatisfiable parts of the search space [13,14], which often reduces computation times considerably.

For SAT and QBF, efficient and effective preprocessing tools are available like SatELite [15], Coprocessor [16] for SAT and squeezeBF [17], bloqqer [18] for QBF. Both available DQBF solvers, however, still lack a preprocessing phase before the actual solving process. Due to the success of preprocessing in SAT and QBF, one can expect that preprocessing is beneficial for DQBF, too – even more because the actual solving process is more costly than for QBF. This raises the question which techniques can be generalized from SAT and QBF to DQBF. Which adaptations need to be made to make them correct for the more general formalism? After suitable adaptations have been found, the correctness proofs have to be re-done for DQBF carefully because for QBF they often exploit the fact that dependencies in QBF follow a linear order. But also techniques like the detection of backbone literals [19,20], which work for DQBF in the same way as for SAT and QBF, have to be re-thought: in SAT only incomplete, but cheap syntactic tests for the special case of unit literals are useful – determining backbone literals completely is as expensive as solving the SAT problem itself. For DQBF the situation is different as the decision problem is much harder. Even solving QBF approximations [9,21] of the formula at hand as an incomplete decision procedure can be beneficial. Additionally the higher flexibility regarding the dependency sets in DQBF makes some techniques more powerful compared to QBF and enables new techniques.

Taken together, in this paper for the first time preprocessing techniques are made available for DQBF solving. We generalize successful preprocessing techniques for QBF to DQBF like blocked clause elimination (BCE) [18,22], equivalence reasoning [17], structure extraction [23], and variable elimination by resolution [24]. All correctness proofs are available in an extended version of this paper [25]. We present experimental results which show the effectiveness of these techniques for DQBF. We demonstrate that the applied techniques have to be chosen depending on the solving techniques applied in the solver core. For example, BCE prevents an effective undoing of Tseitin transformation [26], which is used to transform a formula into conjunctive normal form (CNF). Therefore, it is better to disable BCE if the underlying solver core does not rely on a formula in CNF, and to use BCE if undoing Tseitin transformation is not possible because the solver core requires a formula in CNF. The experiments show that preprocessing both reduces the computation times and significantly increases the number of solved instances of both solvers, iDQ and HQS.

Structure of this paper. The next section introduces the necessary foundations of DQBF. Section 3 reviews incomplete, but cheap decision procedures for DQBF, Section 4 describes the preprocessing techniques for DQBF that we apply in our tool to simplify the DQBF at hand. Section 5 gives an experimental evaluation of the described techniques, and Section 6 concludes the paper.

2   Preliminaries

In this section, we briefly review the necessary foundations regarding dependency quantified Boolean formulas.

Let φ, κ be quantifier-free Boolean formulas over the set V of variables and $v \in V$. We denote by $\varphi[\kappa/v]$ the Boolean formula which results from φ by replacing all occurrences of v (simultaneously) by κ. For a set $V' \subseteq V$ we denote by $\mathcal{A}(V')$ the set of Boolean assignments for V', i.e., $\mathcal{A}(V') = \{\nu \,|\, \nu : V' \to \{0,1\}\}$.

Definition 1 (DQBF). *Let $V = \{x_1, \ldots, x_n, y_1, \ldots, y_m\}$ be a set of Boolean variables. A* dependency quantified Boolean formula *(DQBF) ψ over V has the form $\psi := \forall x_1 \forall x_2 \ldots \forall x_n \exists y_1(D_{y_1}^{\psi}) \exists y_2(D_{y_2}^{\psi}) \ldots \exists y_m(D_{y_m}^{\psi}) : \varphi$ where $D_{y_i}^{\psi} \subseteq \{x_1, \ldots, x_n\}$ for $i = 1, \ldots, m$ is the* dependency set *of y_i, and φ is a Boolean formula over V, the* matrix *of ψ.*

We often write $\psi = Q : \varphi$ with the quantifier prefix Q and the matrix φ. Throughout the whole paper we assume, unless explicitly stated differently, that a DQBF $\psi = Q : \varphi$ as in Definition 1 with φ in CNF is given. We denote its set of universal variables by $V_\forall^{\psi} = \{x_1, \ldots, x_n\}$ and its set of existential variables by $V_\exists^{\psi} = \{y_1, \ldots, y_m\}$. If we do not need to distinguish between existential and universal variables, we write $v \in V$. $Q \setminus \{v\}$ denotes the prefix that results from removing a variable $v \in V$ from Q together with its quantifier. If v is existential, then its dependency set is removed as well; if v is universal, then all occurrences of v in the dependency sets of existential variables are removed. Similarly we use $Q \cup \{\exists y(D_y^{\psi})\}$ to add existential variables to the prefix. The order in which the variables appear in the prefix is irrelevant. We introduce the dependency function $\mathrm{dep}_{\psi} : V \to 2^V$ by $\mathrm{dep}_{\psi}(v) = D_v^{\psi}$ if $v \in V_\exists^{\psi}$, and $\mathrm{dep}_{\psi}(v) = \{v\}$ for $v \in V_\forall^{\psi}$.

Definition 2 (Semantics of DQBF). *Let ψ be a DQBF with matrix φ as above. ψ is* satisfied *(written $\models \psi$) iff there are functions $s_{y_i} : \mathcal{A}(D_{y_i}^{\psi}) \to \{0,1\}$ for $1 \le i \le m$ such that replacing each y_i by (a Boolean expression for) s_{y_i} turns φ into a tautology. Then s_{y_i} is called a* Skolem function *for y_i.*

Two DQBFs ψ_1 and ψ_2 are *equivalent* iff $\models \psi_1 \Leftrightarrow \models \psi_2$ holds.

Definition 3 (QBF). *A* quantified Boolean formula *(QBF)[1] is a DQBF ψ such that $D_y^{\psi} \subseteq D_{y'}^{\psi}$ or $D_{y'}^{\psi} \subseteq D_y^{\psi}$ holds for any pair $y, y' \in V_\exists^{\psi}$ of existential variables.*

---

[1] We only consider closed QBFs in prenex form here, i.e., QBFs in which all variables are bound by a quantifer and in which the quantifiers precede the matrix.

In the following we assume that the matrix φ is given in *conjunctive normal form* (CNF). A formula is in CNF if it is a conjunction of *clauses*; a clause is a disjunction of *literals*, and a literal is either a variable v or its negation $\neg v$. We identify a formula in CNF with its set of clauses and a clause with its set of literals, e. g., we write $\{\{x_1, \neg x_2\}, \{x_2, \neg x_3\}\}$ for the formula $(x_1 \vee \neg x_2) \wedge (x_2 \vee \neg x_3)$. A clause C subsumes a clause C' iff $C \subseteq C'$. For a literal ℓ, $\mathrm{var}(\ell)$ denotes the corresponding variable, i. e., $\mathrm{var}(v) = \mathrm{var}(\neg v) = v$ and $\mathrm{dep}_\psi(\ell) = \mathrm{dep}_\psi(\mathrm{var}(\ell))$. Moreover, we define the "sign" sgn of a literal as $\mathrm{sgn}(v) = 1$ and $\mathrm{sgn}(\neg v) = 0$.

Each DQBF can be transformed such that the matrix is in CNF. While transforming the matrix directly into CNF can cause an exponential blow-up in size, Tseitin transformation [26] can do this with only a linear increase in size at the cost of additional existential variables. The idea is to introduce auxiliary existential variables that store the truth value of sub-expressions. Since the values of these variables are uniquely determined by the sub-expression, they can simply depend on all universal variables.

We assume that none of the clauses of the CNF φ under consideration is tautological, i. e., there is no variable v such that $\{v, \neg v\} \subseteq C$ for all $C \in \varphi$. The preprocessing operations we present check the modified or added clauses whether they are tautologies and, if this is the case, remove or ignore them.

Definition 4 (Resolution). *Let φ be a formula in CNF, ℓ a literal, and $C, C' \in \varphi$ clauses such that $\ell \in C$ and $\neg \ell \in C'$. The* resolvent *of C and C' w. r. t. to the pivot literal ℓ is given by $C \otimes_\ell C' := (C \setminus \{\ell\}) \cup (C' \setminus \{\neg \ell\})$.*

Resolvents are implied by the formula, i. e., if R is a resolvent of two clauses in φ, then φ and $\varphi \cup \{R\}$ are equivalent [27].

Currently, three solvers for DQBF have been proposed: An extension of the DPLL algorithm, typically applied for solving SAT and QBF formulas, has been described in [28]. However, no implementation thereof is available. The second solver is IDQ [11], which relies on a formula in CNF and uses instantiation-based solving, i. e., it reduces deciding a DQBF to deciding a series of SAT problems. Finally, there is the solver HQS [12], which applies quantifier elimination on And-Inverter Graphs (AIGs) to solve the formula. An AIG is essentially a circuit which consists of AND and inverter gates only. Although HQS reads the same CNF-based input format as IDQ, its back-end can handle Boolean formulas of arbitrary structure. We use both IDQ and HQS for the evaluation of the preprocessing techniques presented in the following.

3   Incomplete, but Cheap Decision Procedures

Before we present our preprocessing techniques for DQBF, we review an incomplete, but cheap decision procedure (called "filter") for DQBF. Our approach is as follows: First we apply preprocessing for DQBF, which is helpful for both the filter technique and the actual solver core. Then we run the filter technique, and only if it finishes with an inconclusive result, we apply the solver core. Experiments showed that it is beneficial to use a filter before the solving process.

The filter is based on *QBF approximations*: By using an appropriate quantifier prefix and the same matrix, a DQBF ψ can be over-approximated by a QBF $\Psi^\uparrow$ such that the unsatisfiability of $\Psi^\uparrow$ implies the unsatisfiability of ψ [9]. This is the case if $D_y^{\Psi^\uparrow} \supseteq D_y^\psi$ for all $y \in V_\exists^\psi$. Similarly one can construct an under-approximation $\Psi^\downarrow$ such that the satisfiability of $\Psi^\downarrow$ implies the satisfiability of ψ. As the under-approximation was inconclusive for all instances in our experiments, we focus on over-approximations to show the unsatisfiability of DQBFs. For the formal definitions of the approximations we refer the reader to [9].

Finkbeiner and Tentrup [21] improve these over-approximations by constructing a series of more and more precise QBF formulas. To make this possible they modify both the sets of variables and the matrix of the DQBF: The idea is to use $k \geq 1$ copies of the matrix and its variables. It is required that the existential variables are assigned consistently over all copies and that all copies of the matrix are satisfied. Consistent means that if the universal variables in the dependency set of an existential variable are assigned the same values in two copies, then the existential variables have to carry the same value. Since the sizes of the QBF instances grow considerably with increasing values of k, in most cases only values $k \leq 3$ are beneficial. For more details we refer the reader to [21].

4   Preprocessing Techniques for DQBF

In this section we describe techniques which can be applied to preprocess a DQBF. The proofs of the main theorems and lemmas are given in the extended version [25] of this paper.

4.1   Backbones, Monotonic and Equivalent Variables

Here we describe techniques which reduce both the number of variables in the formula and the number of clauses.

Unit and pure variables are well-known concepts from SAT and QBF solving. They can be replaced by constant values without influencing the formula's truth value. Typically a variable is defined as unit if the matrix contains a clause consisting only of this variable. A variable is pure if it occurs in the whole matrix either only positive or only negative:

Definition 5 (Unit and pure literals). *A literal ℓ is a* unit literal *if $\{\ell\} \in \varphi$; ℓ is a* pure literal *if $\neg\ell$ does not appear in any clause of φ.*

These are syntactic criteria that can be checked efficiently. This is necessary because in particular the detection of unit literals is one of the main operations of search-based SAT and QBF solvers as a part of Boolean constraint propagation (BCP).

For DQBF preprocessing, it is possible to use more expensive checks to determine variables which may be replaced by constants. Therefore we give a more general semantic definition:

Definition 6 (Backbones and monotonic variables). *A variable $v \in V$ is a positive (negative) backbone if $\varphi[0/v]$ ($\varphi[1/v]$, resp.) is unsatisfiable. A literal ℓ is a backbone, if $\ell = v$ and v a positive backbone, or if $\ell = \neg v$ and v a negative backbone. A variable $v \in V$ is positive (negative) monotonic if $\varphi[0/v] \wedge \neg\varphi[1/v]$ ($\varphi[1/v] \wedge \neg\varphi[0/v]$, resp.) is unsatisfiable.*

The following theorem states how we can exploit backbones and monotonic variables to reduce the size of the formula:

Theorem 1. *Let $\psi = Q : \varphi$ be a DQBF and $v \in V$ a backbone or a monotonic variable. If v is a positive or negative backbone and universal, ψ is unsatisfiable. Otherwise ψ is equivalent to ψ' where*

- *$\psi' = Q \setminus \{v\} : \varphi[1/v]$ if v is existential and either a positive backbone or positive monotonic, or v is universal and negative monotonic;*
- *$\psi' = Q \setminus \{v\} : \varphi[0/v]$ if v is existential and either a negative backbone or negative monotonic, or v is universal and positive monotonic.*

This theorem has been proven formally in [29]. Checks whether a variable is a backbone or monotonic can be done using a SAT solver. As already mentioned, in the SAT and QBF context typically efficient (sound but not complete) syntactic criteria are applied to detect backbones and monotonic variables. It is easy to show that unit literals are backbones and pure literals are monotonic.

Another cheap criterion to identify backbones uses the binary implication graph of a formula (which later also used to identify equivalent literals):

Definition 7. *Let $\varphi^2 = \{C \in \varphi \,|\, |C| = 2\}$ be the set of binary clauses. The binary implication graph of ψ is the directed graph $\mathrm{BIP}(\psi) = (L, E)$ with the set $L = \{v, \neg v \,|\, v \in V\}$ of literals as its set of nodes and $E = \{(v, \neg w), (\neg v, w) \,|\, \{v, w\} \in \varphi^2\}$ the set of edges.*

Then the following lemma holds:

Lemma 1. *A literal ℓ is a backbone if there is a path in $\mathrm{BIP}(\psi)$ from $\neg\ell$ to ℓ.*

If there is a path from literal ℓ to literal ℓ', we can derive the clause $\{\neg\ell, \ell'\}$ by resolution. In case of the lemma, the path from $\neg\ell$ to ℓ implies that we can derive the clause $\{\neg\neg\ell, \ell\} = \{\ell\}$. Since this is a resolvent of clauses in φ, it may be added to φ. Then we can apply Definition 5 to obtain the result.

Unit and pure literals, according to Definition 5, and backbones according to Lemma 1, can be determined efficiently by traversing the matrix or, respectively, the binary implication graph. Since solving a DQBF is much harder than solving a SAT (or even QBF) problem and the gain by eliminating one variable is larger, it often pays off to additionally use semantic checks (cf. Definition 6) for backbones and monotonic variables, which are based on solving a sequence of SAT problems. For backbones in the QBF context this observation has been made in [30].

Definition 8 (Equivalent literals). *The literals ℓ and μ are equivalent w. r. t. a propositional formula φ iff φ is equivalent to $\varphi \wedge (\ell \equiv \mu)$.*

Theorem 2. *Let ℓ and μ be equivalent literals. We assume, w. l. o. g., that* $\mathrm{sgn}(\ell) = 1$. *If* $\mathrm{var}(\ell), \mathrm{var}(\mu) \in V_\forall^\psi$, *then ψ is unsatisfiable. Otherwise, we assume w. l. o. g. that* $\mathrm{var}(\ell) \in V_\exists^\psi$. *If* $\mathrm{var}(\mu) \in V_\forall^\psi$ *and* $\mathrm{var}(\mu) \notin D_{\mathrm{var}(\ell)}^\psi$, *then ψ is unsatisfiable. If* $\mathrm{var}(\mu) \in V_\forall^\psi$ *and* $\mathrm{var}(\mu) \in D_{\mathrm{var}(\ell)}^\psi$, *then ψ is equivalent to* $Q \setminus \{\mathrm{var}(\ell)\} : \varphi[\mu/\ell]$. *If* $\mathrm{var}(\ell), \mathrm{var}(\mu) \in V_\exists^\psi$, *then ψ is equivalent to*

$$\psi' := \left(Q \setminus \{\mathrm{var}(\mu), \mathrm{var}(\ell)\}\right) \cup \left\{\exists\, \mathrm{var}(\mu)(D_{\mathrm{var}(\mu)}^\psi \cap D_{\mathrm{var}(\ell)}^\psi)\right\} : \varphi[\mu/\ell].$$

A proof can be found in the extended version [25] of this paper.

To detect equivalent literals, we exploit the following lemma:

Lemma 2. *Two literals ℓ, μ are equivalent if there is a path in* $\mathrm{BIP}(\psi)$ *from ℓ to μ and vice versa.*

We decompose $\mathrm{BIP}(\psi)$ into strongly connected components (SCCs) using Tarjan's SCC algorithm [31]. SCCs have the property that there is a path between each pair of nodes in an SCC. Therefore literals within one SCC are equivalent. They are replaced by one representative by applying Theorem 2. This procedure was described e. g., in [16, 32–35] for SAT preprocessing. Further equivalent literals can be found using structure extraction (see Section 4.5). Of course, even SAT checks based on Definition 8 may be beneficial in the DQBF context.

4.2   Reduction of Dependency Sets

In a DQBF, a universal variable $x \in V_\forall^\psi$ may be contained in the dependency set D_y^ψ of an existential variable $y \in V_\exists^\psi$, but actually, due to the structure of the matrix, the Skolem function for y does not need to exploit the information about x's value to satisfy the formula. If such a situation is detected, x can be removed from D_y^ψ. This potentially reduces the number of copies of variables, if universal expansion according to Theorem 5 is used for solving a DQBF.

An example for a situation when dependency sets may be reduced is when a circuit is transformed into CNF by Tseitin transformation. The dependency set D_y^ψ of a Tseitin variable y can be an arbitrary superset of the universal variables in its cone-of-influence. The variables in D_y^ψ that are not in the cone-of-influence of y can be removed from D_y^ψ without affecting the truth value of the formula.

Definition 9. *An existential variable $y \in V_\exists^\psi$ is independent of a universal variable $x \in V_\forall^\psi$ if either $x \notin D_y^\psi$ or replacing D_y^ψ by $D_y^\psi \setminus \{x\}$ does not change the truth value of ψ.*

Deciding whether two variables are independent has the same complexity as deciding the DQBF itself [36]. Therefore one resorts to sufficient criteria to show independence. The most simple ones are based on the incidence graph of the matrix:

The *variable-clause incidence graph* $G_{V,\varphi} = (V \cup \varphi, E)$ of the formula is an undirected graph with $E = \{\{v, C\} \in V \times \varphi \mid v \in C \vee \neg v \in C\}$.

Theorem 3 (Standard dependency scheme). *An existential variable $y \in V_\exists^\psi$ is independent of a universal variable $x \in V_\forall^\psi$ if there is no path in $G_{V,\varphi}$ from x to y, visiting only variables in $\{z \in V_\exists^\psi \mid x \in D_z^\psi\}$ in between.*

For a proof for this theorem, which generalizes a theorem from [36], see [25].

In the QBF context more powerful dependency schemes have been developed which can possibly identify more variables as independent, see, e. g., [36–40]. A generalization of these techniques will have an immediate benefit for DQBF solving by increasing the potential to save variable copies during universal expansion.

4.3   Universal Reduction, Resolution, and Universal Expansion

Universal reduction, resolution, and *universal expansion* are well-known techniques used during the solution of QBFs. Universal reduction removes a universal variable from a clause if the clause does not contain any existential variable which depends upon it. This technique has already been generalized to DQBF in [11, 41].

Lemma 3 (Universal reduction, [11, 41]). *Let $Q : \varphi \wedge C$ be a DQBF and $\ell \in C$ a universal literal such that for all $k \in C$ with $k \neq \ell$ we have $\mathrm{var}(\ell) \notin \mathrm{dep}_\psi(k)$. Then $Q : \varphi \wedge C$ and $Q : \varphi \wedge (C \setminus \{\ell\})$ are equivalent.*

For QBF resolution and universal reduction together are able to derive the empty clause iff the formula is unsatisfiable. This does not hold for DQBF [41]. Resolution in QBF formulas allows to eliminate an existential variable by replacing the clauses containing this variable with their resolvents. While adding resolvents is sound for DQBF as well, eliminating existential variables by resolution [15] only works under certain conditions. Here we give a set of sufficient conditions which allow variable elimination by resolution for DQBF. In particular when the formula is created by Tseitin transformation [26], variable elimination by resolution is applicable to a large subset of the formula's existential variables.

Theorem 4 (Variable elimination by resolution). *Let $y \in V_\exists^\psi$ be an existential variable of ψ. We partition φ into the sets $\varphi^y = \{C \in \varphi \mid y \in C\}$, $\varphi^{\neg y} = \{C \in \varphi \mid \neg y \in C\}$, and $\varphi^\emptyset = \varphi \setminus (C^y \cup C^{\neg y})$.*
If one of the following conditions is satisfied:
- *for all $C \in \varphi^y$ and all $k \in C$ we have $\mathrm{dep}_\psi(k) \subseteq \mathrm{dep}_\psi(y)$,*
- *for all $C' \in \varphi^{\neg y}$ and all $k \in C'$ we have $\mathrm{dep}_\psi(k) \subseteq \mathrm{dep}_\psi(y)$, or*
- *y is the defined variable of a functional definition, i. e., there are clauses encoding the relationship $y \equiv f(V')$ for some function f and arguments $V' \subseteq V \setminus \{y\}$, $\mathrm{dep}_\psi(v) \subseteq \mathrm{dep}_\psi(y)$ for all $v \in V'$ (cf. Sec. 4.5),*

then ψ is equivalent to $\psi' := Q \setminus \{y\} : \varphi^\emptyset \wedge \bigwedge_{C \in \varphi^y} \bigwedge_{C' \in \varphi^{\neg y}} C \otimes_y C'$.

Proof sketch. Resolvents are implied by the matrix, i. e., adding resolvents to the matrix yields an equivalent formula. If ψ is satisfied, then removing the clauses in φ^y and $\varphi^{\neg y}$ cannot make the formula unsatisfied, i. e., ψ' is satisfied.

Assume that ψ' is satisfied by Skolem functions s_z for $z \in V_\exists^\psi \setminus \{y\}$. We define $s_y := \neg\varphi^y[0/y][s_z/z \text{ for } z \in V_\exists^\psi \setminus \{y\}]$ in the first case, $s_y := \neg\varphi^{\neg y}[1/y][s_z/z$ for $z \in V_\exists^\psi \setminus \{y\}]$ in the second case, and $s_y := f(V')[s_z/z \text{ for } z \in V_\exists^\psi \setminus \{y\}]$ in the third case. It is not hard to show that s_y is an admissible Skolem function for y and that $\varphi[s_v/v \text{ for } v \in V_\exists^\psi]$ is indeed a tautology. Details can be found in the extended version [25] of this paper.   □

Theorem 4 does not provide a decision algorithm for arbitrary DQBFs, since it is possible that the conditions do not hold for any existential variable. Moreover, eliminating all existential variables fulfilling the conditions of Theorem 4 is in general not feasible because the number of clauses can grow considerably during elimination. We first create a list of variables that may be eliminated. For each such variable y we estimate the cost c_y of elimination, i. e., $c_y := \frac{|\varphi^0| + |\varphi^y| \cdot |\varphi^{\neg y}|}{|\varphi|}$. We eliminate one variable y with minimum cost provided that c_y is less than a user-specified factor $\varepsilon > 1$. After resolving variables we check for subsumed clauses, i. e., clauses C such that there is a clause C' with $C' \subseteq C$. Then C can be deleted [27].

Universal expansion [9, 41–43] is the corresponding method for eliminating universal variables. It is the main operation which the solver HQS [12] uses to transform the DQBF at hand into an equivalent QBF. This QBF can be solved by an arbitrary QBF solver.

Theorem 5 (Universal expansion). *Let $x_i \in V_\forall^\psi$, and $E_{x_i}^\psi = \{y_i \in V_\exists^\psi \mid x_i \in \mathrm{dep}_\psi(y_j)\}$. Then ψ is equivalent to*

$$(Q \setminus \{x_i\}) \cup \{\exists y_j'(D_{y_j}^\psi \setminus \{x_i\}) \mid y_j \in E_{x_i}^\psi\} : \varphi[1/x_i] \wedge \varphi[0/x_i][y_j'/y_j \text{ for all } y_j \in E_{x_i}^\psi].$$

A formal proof of this theorem is given, e. g., in [9]. In order to avoid unnecessary variable copies, we check using the standard dependency scheme (cf. Theorem 3) which existential variables actually depend on the expanded universal variable.

4.4 Blocked Clause Elimination

The concept of blocked clauses was introduced by Järvisalo et al. for SAT in [22] and later generalized to QBF by Biere et al. in [18]. Blocked clauses can be removed from a formula without changing its truth value. Before checking for blockedness, clauses can be extended by so-called hidden and covered literals [18, 44, 45]. This does not change the truth value of the formula, but increases the chance that a clause is blocked.

In this section, we first generalize the notion of blocked clauses to DQBF such that blocked clauses satisfy the same properties as in SAT and QBF. Then we investigate how to generalize hidden and covered literals to DQBF.

For a *QBF* $Q : \varphi \wedge C$, a clause C containing an existential literal $\ell \in C$ can be omitted (resulting in an equivalent formula), if 'ℓ is blocking for C', which means that for all $C' \in \varphi$ with $\neg\ell \in C'$ there is a variable k such that $\{k, \neg k\} \subseteq C \otimes_\ell C'$ and k precedes ℓ in the quantifier prefix (which means in DQBF notions:

$\mathrm{dep}_\psi(k) \subseteq \mathrm{dep}_\psi(\ell))$. In the QBF context the intuitive background of blocked clause elimination is simple: Consider a solving approach to QBF which always removes the innermost existential quantifiers (which depend on all universal ones) by resolution[2] and the innermost universal quantifiers (upon which no existential variable depends) by universal reduction until all quantifiers have been removed [24]. If ℓ is blocking for C, all resolvents resulting from C contain $\{k, \neg k\}$, i.e., are tautological, and their addition makes no contribution. The condition 'k precedes ℓ in the quantifier prefix' ensures that $\mathrm{var}(k)$ has not been removed before ℓ in the process sketched above, i.e., the reason $\{k, \neg k\}$ for the resolvents being tautological has not been removed. This implies that we can alternatively remove C from $\varphi \wedge C$ in the very beginning without changing the result of the solving process.

Fortunately, we can show that the notion of blocked clauses has a natural generalization to DQBF. However, the proof idea of blocked clause elimination sketched above does not work anymore, since in DQBF there is no linear order for the quantifiers such that 'removing quantifiers starting with the innermost' does not have a counterpart in DQBF; the correctness proof has to be re-done for DQBF carefully taking into account that arbitrary dependencies may be defined in a DQBF. We first give the generalized definition of blocked clauses:

Definition 10 (Blocked clauses). *Let $Q : \varphi \wedge C$ be a DQBF and C a clause with $\ell \in C$. Literal ℓ is a* blocking literal *for C if ℓ is existential and for all $C' \in \varphi$ with $\neg \ell \in C'$ there is a variable k such that $\{k, \neg k\} \subseteq C \otimes_\ell C'$ and $\mathrm{dep}_\psi(k) \subseteq \mathrm{dep}_\psi(\ell)$. A clause is* blocked *if it contains a blocking literal.*

Now we can prove results that are analogous to QBF and SAT.

Theorem 6 (Blocked clause elimination, BCE). *Let $Q : \varphi \wedge C$ be a DQBF with a blocked clause C. Then $Q : \varphi \wedge C$ and $Q : \varphi$ are equivalent.*

Proof sketch. The theorem can be shown by induction on the number $|\mathrm{dep}_\psi(\ell)|$ of ℓ's dependencies. The base case $\mathrm{dep}_\psi(\ell) = \emptyset$ works analogously to the QBF case, see [18]. For the induction step, we choose an arbitrary universal variable $x \in \mathrm{dep}_\psi(\ell)$ and eliminate it by universal expansion (see Theorem 5). In the resulting formula, ℓ and its copy ℓ' depend on one variable less. One can show that both copies of C in this formula are either blocked or tautological. Therefore they can be removed by the induction assumption. Un-doing the expansion step yields the result. A more detailed proof can be found in [25].   □

The purpose of the following techniques is to extend clauses by redundant literals. This increases the chance that the clause is blocked and can be deleted. If the extended clause is not blocked, the additional literals are removed again.

Definition 11 (Hidden literals). *Let $Q : \varphi \wedge C$ be a DQBF. A literal $\ell \notin C$ is a* hidden literal *for C if there is a clause $\{\ell_1, \ldots, \ell_n, \neg \ell\} \in \varphi$ such that $\{\ell_1, \ldots, \ell_n\} \subseteq C$.*

---

[2] Adding all possible resolvents with pivot variable v and then removing all clauses containing v or $\neg v$ corresponds to existential quantification of v.

Theorem 7 (Hidden literal addition, HLA). *Let $Q : \varphi \wedge C$ be a DQBF and ℓ a hidden literal for C. Then $Q : \varphi \wedge C$ and $Q : \varphi \wedge (C \cup \{\ell\})$ are equivalent.*

The idea of hidden literal addition is based on *self-subsuming resolution* [15]. The resolvent $(C \cup \{\ell\}) \otimes_\ell \{\ell_1, \ldots, \ell_n, \neg\ell\}$ is equal to C and subsumes $C \cup \{\ell\}$. Thus after adding the resolvent C, $C \cup \{\ell\}$ can be removed, leading to an equivalent formula. Note that the argument for hidden literal addition is based on a consideration of the matrix only, thus in this case the argumentation is exactly the same as for SAT and QBF.

This is in contrast to the 'covered literal addition' described in the following. For covered literals we need a careful generalization of the QBF definition together with a non-trivial proof of the generalization to DQBF.

Definition 12 (Covered literals). *Let $\psi = Q : \varphi \wedge C$ be a DQBF and let ℓ be an existential literal with $\ell \in C$. The set of resolution candidates for C w.r.t. ℓ is the set $R_\psi(C, \ell) = \{C' \in \varphi \mid \neg\ell \in C' \wedge \forall v \in V : (\{v, \neg v\} \subseteq C \otimes_\ell C' \Rightarrow \mathrm{dep}_\psi(v) \not\subseteq \mathrm{dep}_\psi(\ell))\}$.*

A literal k is a covered literal for C w.r.t. ℓ if $\mathrm{dep}_\psi(k) \subseteq \mathrm{dep}_\psi(\ell)$ and $k \in \bigcap R_\psi(C, \ell) \setminus \{\neg\ell\}$.

Theorem 8 (Covered literal addition, CLA). *Let $Q : \varphi \wedge C$ be DQBF and k a covered literal for C. Then $Q : \varphi \wedge C$ and $Q : \varphi \wedge (C \cup \{k\})$ are equivalent.*

Proof sketch. Assume that k is a covered literal for C w.r.t. ℓ. We show the theorem by induction on the number $|\mathrm{dep}_\psi(\ell)|$ of dependencies of ℓ. The induction base where $\mathrm{dep}_\psi(\ell) = \emptyset$ is similar to the QBF case (cf. [18]). For the induction step, we apply universal expansion of an arbitrary variable in $\mathrm{dep}_\psi(\ell)$ (see Theorem 5) to obtain a formula in which ℓ and its copy ℓ' both depend on one variable less. It is rather technical to show that adding k (k') to the copies of C in this formula leads to an equivalent formula, since these copies are either tautological or k (k') is a covered literal. By undoing the expansion step we obtain the desired result. For a detailed proof we refer to [25]. □

A rough basic intuition for covered literal addition is as follows: "If a literal k is already contained in all non-tautological resolvents of a clause C with pivot literal ℓ, then k may be added to C resulting in an equivalent formula." In addition to this basic idea we need the condition $\mathrm{dep}_\psi(k) \subseteq \mathrm{dep}_\psi(\ell)$ and a bigger set of resolution candidates $R_\psi(C, \ell) = \{C' \in \varphi \mid \neg\ell \in C' \wedge \forall v \in V : (\{v, \neg v\} \subseteq C \otimes_\ell C' \Rightarrow \mathrm{dep}_\psi(v) \not\subseteq \mathrm{dep}_\psi(\ell))\}$ instead of $R_\psi(C, \ell) = \{C' \in \varphi \mid \neg\ell \in C' \wedge \nexists v \in V : \{v, \neg v\} \subseteq C \otimes_\ell C'\}$ in order to be able to lead the (rather involved) proof of Theorem 8, see [25].

In order to reduce the size of the formula, we determine for each clause C the set H of hidden and the set K of covered literals. Then we check if $C \cup H \cup K$ is blocked or tautological. If this is the case, C is removed; otherwise C remains unchanged. This is iterated until we reach a fixed point.

Note that if a hidden or covered literal is universal, its addition can be helpful not only because it can make a clause blocked. If a CNF-based solver core uses elimination of universal variables to decide the formula, all clauses which contain an existential variable that depends on the eliminated universal variable have to be doubled [9]. If the clause contains the universal variable to be eliminated, one of these copies is satisfied and can therefore be omitted (cf. [46]).

4.5   Structure Extraction

The DQBF's matrix in CNF is often created from a circuit or a Boolean expression by Tseitin transformation [26], where a new existential variable v_e is created for each sub-expression e (or gate output). Clauses encoding the relationship $v_e \equiv e$ are added and the sub-expression e is replaced by the variable v_e. If a solver (like HQS) does not rely on a matrix in CNF, this transformation step can be undone. This removes all artificially introduced variables. Structure extraction is used in the QBF solver AIGsolve [23].

For example, a k-input AND gate $y \equiv \mathrm{AND}(\ell_1, \ldots, \ell_k)$ has a Tseitin encoding consisting of $(k+1)$ clauses $\{\neg y, \ell_1\}, \ldots, \{\neg y, \ell_k\}, \{y, \neg \ell_1, \ldots, \neg \ell_k\}$. In a functional definition $y \equiv f(\ell_1, \ldots, \ell_k)$, y is called the *defined variable*, f is the *definition* of y, and the clauses corresponding to the relationship $y \equiv f(\ell_1, \ldots, \ell_k)$ are the *defining clauses*.

Theorem 9. *Let $\psi = Q : \varphi$ be a DQBF and $\varphi^f \subseteq \varphi$ the defining clauses for the relationship $y \equiv f(\ell_1, \ldots, \ell_k)$. Then ψ is equivalent to $Q \setminus \{y\} : (\varphi \setminus \varphi^f)[f(\ell_1, \ldots, \ell_k)/y]$ if $y \in V_\exists^\psi$ and for $i = 1, \ldots, k$ we have $\mathrm{dep}_\psi(\ell_i) \subseteq \mathrm{dep}_\psi(y)$.*

Our implementation checks for defining clauses for (multi-input) (N)AND gates and 2-input XOR gates, both with arbitrarily negated inputs. We do not extract definitions that lead to cyclic dependencies.

Gate detection can be used as the last step of the preprocessing routine. If a relationship $y \equiv f(\ell_1, \ldots, \ell_k)$ is detected which does not lead to cyclic dependencies, we remove y from the prefix and the defining clauses from the matrix. We additionally use a data structure which assigns to each defined variable its definition. To create an AIG representation that can be passed to a non-CNF-based solver core like HQS, we convert the remaining clause into an AIG and then substitute the defined variables by their definitions.

The same structure extraction procedure can also be used to identify equivalent variables and unnecessary variable dependencies. For both purposes, the relationships are only detected, but neither are the defining clauses removed nor is the data structure that stores the relationships updated. Therefore this can also be used if the solver back-end requires a matrix in CNF: If there is the relationship $y \equiv f(\ell_1, \ldots, \ell_k)$ and $\bigcup_{i=1}^k \mathrm{dep}_\psi(\ell_i) \subsetneq D_y^\psi$, then D_y^ψ can be replaced by $\bigcup_{i=1}^k \mathrm{dep}_\psi(\ell_i)$. If two defined variables y, y' with the same definition are detected, i.e., $y \equiv f(\ell_1, \ldots, \ell_k)$ and $y' \equiv f(\ell_1, \ldots, \ell_k)$, then y and y' are equivalent and Theorem 2 can be applied to remove one of them.

5   Experimental Results

We have implemented the described techniques in C++ as a preprocessor for our DQBF solver HQS. To support other back-end solvers, too, it is able to write the resulting formula into a file in DQDIMACS format, which can be read by the currently only competing solver IDQ [11].

As benchmark instances we use 4381 formulas, resulting from the verification of incomplete circuits [9,11,21] and controller synthesis [10]. The synthesis benchmarks are those shipped with the tool *Demiurge 1.1.0* [10]. We used the encoding described in [10] to create a DQBF formulation.

All experiments were run on one Intel Xeon E5-2650v2 core at 2.60 GHz with 64 GB of main memory, running Ubuntu Linux 12.04 in 64-bit mode as operating system. We aborted all experiments whose computation time exceeded 900 seconds or which required more than 8 GB of memory. For solving QBFs, we use DepQBF 4.0 [47,48] with the QBF preprocessor bloqqer [18] (version 35) if the matrix is in CNF, and AIGsolve [23] if the matrix is given as an AIG.

We used two parameter settings for preprocessing, in the following called V_1 and V_2. Both use the detection of backbones (by syntactic and semantic checks), monotonic variables (by syntactic checks), and equivalent variables (both using the binary implication graph and structure extraction). We reduce the dependency sets of the existential variables using the standard dependency scheme and structure extraction. For these operations, the functional definitions are only detected, but neither are the defined variables replaced by their definition nor are the defining clauses removed.

- V_1 additionally enables structure extraction, which replaces the defined variables by their definitions. V_1 does not yield a CNF representation, but rather an And-Inverter Graph (AIG) [49] for the formula. Since IDQ requires a CNF representation of the matrix, V_1 can only be combined with HQS.

- V_2 applies BCE after adding hidden and covered literals and variable elimination by resolution ($\varepsilon = 1.1$), but disables structure extraction. V_2 yields a matrix in CNF; therefore it can be combined with both IDQ and HQS.

Table 1 shows the number of *solved instances* (out of 4381) for different combinations of preprocessing, filtering (see Section 3), and the HQS or IDQ solver cores. Preprocessing alone can only solve a small fraction of all instances (80 for V_1 and 57 for V_2). The filter solves already 935 instances for $k = 1$ (slightly more with higher values of k). The combination of preprocessing V_1 with the filter allows to decide 2459 instances (2240 with V_2). In spite

Table 1. Effect of preprocessing

Solver	Filter	Preproc.	Solved
none	$k = 1$	none	935
none	$k = 1$	V_1	2459
none	$k = 2$	V_1	2733
none	$k = 1$	V_2	2240
HQS	none	none	1537
HQS	none	V_1	3629
HQS	$k = 1$	V_1	3752
HQS	$k = 1$	V_1 + BCE	2174
HQS	$k = 2$	V_1	3737
HQS	$k = 1$	V_2	3542
IDQ	none	none	1073
IDQ	none	V_2	1378
IDQ	$k = 1$	none	1359
IDQ	$k = 1$	V_2	2714

of using bloqqer as preprocessor for simplifying the QBF over-approximations for filtering, doing DQBF preprocessing before reduces the solving times for the QBFs. Without DQBF preprocessing, solving the QBF approximation runs into a timeout frequently.

For HQS as solver back-end, the trend is similar: without preprocessing and filtering, HQS is able to solve 1537 instances, with V_1 preprocessing this number increases to 3629 instances, and if filtering is used thereafter, 3752 instances can be solved. We can also see that BCE largely prevents structure extraction: if all described techniques are enabled, only 2174 instances can be solved successfully. Increasing the value of k to 2 does not seem beneficial at least if a time limit of 15 min is used. For larger time limit, $k = 2$ can slightly increase the number of solved instances. Finally, if we combine V_2 with filtering ($k = 1$) and HQS, we can also observe a positive effect on the number of solved instances (3542); however, it is not as strong as with V_1, which includes structure extraction instead of BCE.

IDQ without filtering and preprocessing solves 1073 instances. This number is increased to 1378 by preprocessing (V_2) and to 1359 instances by filtering ($k = 1$). The combination with filtering and preprocessing yields 2714 solved instances.

In summary, the combination of filtering and preprocessing significantly increases the number of solved instances by a factor of up to 2.44 (for HQS) and 2.52 (for IDQ). The best results are obtained if the preprocessing techniques are chosen according to the solver core.

Now we focus on the *size of the instances* before and after preprocessing. Preprocessing variant V_2 reduces the number of clauses by 64 % on average, the number of existential variables by 76 % on average, but leaving the number of universal variables essentially unchanged. As preprocessing variant V_1 does not yield a CNF representation, we cannot compare the number of clauses. Instead we compare the size of the AIG representation of the matrix before and after preprocessing. V_1 reduces the number of existential variables by 97 % on average (including all Tseitin variables), the number of AIG nodes by 84 %, leaving the number of universal variables almost unchanged, too.

If the CNF structure of the matrix needs to be preserved (as in V_2) not all Tseitin variables can be removed by identifying functional definitions and by elimination by resolution, since this leads to a significant increase in size of the CNF. This effect is lessened by BCE, in particular if HLA and CLA are enabled.

Finally, we take a closer look at the *solving times* of the instances. For the instances which were solved with or without preprocessing and filtering, Fig. 1 compares the computation times when using only the solver core and when using the solver core after preprocessing and filtering. The times include everything from reading the input files to termination. The upper two pictures show HQS with V_1 (Fig. 1(a)) and V_2 (Fig. 1(b)) and filtering using $k = 1$, compared to HQS without preprocessing and filtering. Fig. 1(c) shows IDQ with V_2, compared to IDQ without preprocessing. In Fig. 1(d) we present the accumulated running times over all instances (unsolved instances contributing the time limit of 900 seconds) and the average running time of the solved instances.

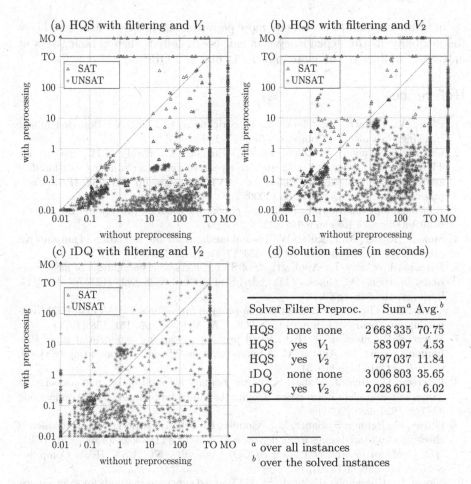

Fig. 1. Running times (in seconds) for HQS and IDQ with and without preprocessing.

In all three cases, preprocessing and filtering reduce the computation times for the vast majority of instances significantly, often by orders of magnitude. The very few exceptions in case of IDQ are instances that are very easy to solve such that the overhead for preprocessing exceeds the solving time. We can also observe that many instances, for which the solver core alone ran into a time out or memory out, can be solved successfully after preprocessing and filtering.

6   Conclusion

We have shown how preprocessing techniques for SAT and QBF can be generalized to DQBF. Experiments have demonstrated that they can reduce the running time of the actual solving process by orders of magnitude, both for CNF-based and non-CNF-based solver cores.

In future we want to investigate more powerful dependency schemes and how the flexibility in the dependency sets can be exploited when choosing sets of universal variables to eliminate in order to obtain a QBF.

References

1. Biere, A., Cimatti, A., Clarke, E.M., Strichman, O., Zhu, Y.: Bounded model checking. Advances in Computers **58**, 117–148 (2003)
2. Czutro, A., Polian, I., Lewis, M.D.T., Engelke, P., Reddy, S.M., Becker, B.: TIGUAN: thread-parallel integrated test pattern generator utilizing satisfiability analysis. In: International Conference on VLSI Design, pp. 227–232. IEEE Computer Society, New Delhi, India (2009)
3. Rintanen, J.: Constructing conditional plans by a theorem-prover. Journal of Artificial Intelligence Research **10**, 323–352 (1999)
4. Sinz, C., Kaiser, A., Küchlin, W.: Formal methods for the validation of automotive product configuration data. AI EDAM **17**(1), 75–97 (2003)
5. Mironov, I., Zhang, L.: Applications of SAT solvers to cryptanalysis of hash functions. In: Biere, A., Gomes, C.P. (eds.) SAT 2006. LNCS, vol. 4121, pp. 102–115. Springer, Heidelberg (2006)
6. Cook, S.A.: The complexity of theorem-proving procedures. In: Annual ACM Symposium on Theory of Computing (STOC), ACM Press, pp. 151–158 (1971)
7. Meyer, A.R., Stockmeyer, L.J.: Word problems requiring exponential time: Preliminary report. In: Annual ACM Symposium on Theory of Computing (STOC), pp. 1–9. ACM Press (1973)
8. Peterson, G., Reif, J., Azhar, S.: Lower bounds for multiplayer non-cooperative games of incomplete information. Computers and Mathematics with Applications **41**(7–8), 957–992 (2001)
9. Gitina, K., Reimer, S., Sauer, M., Wimmer, R., Scholl, C., Becker, B.: Equivalence checking of partial designs using dependency quantified Boolean formulae. In: IEEE Int'l Conf. on Computer Design (ICCD), Asheville, NC, USA, IEEE Computer Society, pp. 396–403 (2013)
10. Bloem, R., Könighofer, R., Seidl, M.: SAT-based synthesis methods for safety specs. In: McMillan, K.L., Rival, X. (eds.) VMCAI 2014. LNCS, vol. 8318, pp. 1–20. Springer, Heidelberg (2014)
11. Fröhlich, A., Kovásznai, G., Biere, A., Veith, H.: iDQ: Instantiation-based DQBF solving. In: Berre, D.L. (ed.) Int'l Workshop on Pragmatics of SAT (POS). EPiC Series, vol. 27, pp. 103–116. Vienna, Austria, EasyChair (2014)
12. Gitina, K., Wimmer, R., Reimer, S., Sauer, M., Scholl, C., Becker, B.: Solving DQBF through quantifier elimination. In: Int'l Conf. on Design, Automation and Test in Europe (DATE), Grenoble, France, IEEE (2015)
13. Jr., R.J.B., Schrag, R.: Using CSP look-back techniques to solve real-world SAT instances. In: Kuipers, B., Webber, B.L. (eds.): National Conference on Artificial Intelligence / Innovative Applications of Artificial Intelligence Conference (AAAI/IAAI), Providence, Rhode Island, USA, AAAI Press / The MIT Press, pp. 203–208 (1997)
14. Silva, J.P.M., Sakallah, K.A.: GRASP: A search algorithm for propositional satisfiability. IEEE Transactions on Computers **48**(5), 506–521 (1999)
15. Eén, N., Biere, A.: Effective preprocessing in SAT through variable and clause elimination. In: Bacchus, F., Walsh, T. (eds.) SAT 2005. LNCS, vol. 3569, pp. 61–75. Springer, Heidelberg (2005)

16. Manthey, N.: Coprocessor 2.0 – a flexible CNF simplifier. In: Cimatti, A., Sebastiani, R. (eds.) SAT 2012. LNCS, vol. 7317, pp. 436–441. Springer, Heidelberg (2012)

17. Giunchiglia, E., Marin, P., Narizzano, M.: sQueezeBF: an effective preprocessor for QBFs based on equivalence reasoning. In: Strichman, O., Szeider, S. (eds.) SAT 2010. LNCS, vol. 6175, pp. 85–98. Springer, Heidelberg (2010)

18. Biere, A., Lonsing, F., Seidl, M.: Blocked clause elimination for QBF. In: Bjørner, N., Sofronie-Stokkermans, V. (eds.) CADE 2011. LNCS, vol. 6803, pp. 101–115. Springer, Heidelberg (2011)

19. Kilby, P., Slaney, J.K., Thiébaux, S., Walsh, T.: Backbones and backdoors in satisfiability. In: Veloso, M.M., Kambhampati, S. (eds.): National Conference on Artificial Intelligence / Int'l Conf. on Innovative Applications of Artificial Intelligence (IAAI), Pittsburgh, Pennsylvania, USA, AAAI Press / The MIT Press, pp. 1368–1373 (2005)

20. Janota, M., Lynce, I., Marques-Silva, J.: Algorithms for computing backbones of propositional formulae. AI Communications 28(2), 161–177 (2015)

21. Finkbeiner, B., Tentrup, L.: Fast DQBF refutation. In: Sinz, C., Egly, U. (eds.) SAT 2014. LNCS, vol. 8561, pp. 243–251. Springer, Heidelberg (2014)

22. Järvisalo, M., Biere, A., Heule, M.: Blocked clause elimination. In: Esparza, J., Majumdar, R. (eds.) TACAS 2010. LNCS, vol. 6015, pp. 129–144. Springer, Heidelberg (2010)

23. Pigorsch, F., Scholl, C.: Exploiting structure in an AIG based QBF solver. In: Conf, I. (ed.) on Design, Automation and Test in Europe (DATE), pp. 1596–1601. IEEE, Nice, France (2009)

24. Biere, A.: Resolve and expand. In: H. Hoos, H., Mitchell, D.G. (eds.) SAT 2004. LNCS, vol. 3542, pp. 59–70. Springer, Heidelberg (2005)

25. Wimmer, R., Gitina, K., Nist, J., Scholl, C., Becker, B.: Preprocessing for DQBF (extended version). Reports of SFB/TR 14 AVACS number 110 (2015). http://www.avacs.org

26. Tseitin, G.S.: On the complexity of derivation in propositional calculus. Studies in Constructive Mathematics and Mathematical Logic Part 2, 115–125 (1970)

27. Biere, A., Heule, M., van Maaren, H., Walsh, T. (eds.): Handbook of Satisfiability. vol. 185 of Frontiers in Artificial Intelligence and Applications. IOS Press (2008)

28. Fröhlich, A., Kovásznai, G., Biere, A.: A DPLL algorithm for solving DQBF. In: Int'l Workshop on Pragmatics of SAT (POS), Trento, Italy (2012)

29. Gitina, K., Wimmer, R., Reimer, S., Sauer, M., Scholl, C., Becker, B.: Solving DQBF through quantifier elimination. Reports of SFB/TR 14 AVACS 107 (2015). http://www.avacs.org

30. Pigorsch, F., Scholl, C.: An AIG-based QBF-solver using SAT for preprocessing. In: Sapatnekar, S.S. (ed.) ACM/IEEE Design Automation Conference (DAC), pp. 170–175. ACM Press, Anaheim, CA, USA (2010)

31. Tarjan, R.E.: Depth-first search and linear graph algorithms. SIAM Journal on Computing 1(2), 146–160 (1972)

32. Brafman, R.I.: A simplifier for propositional formulas with many binary clauses. IEEE Transactions on Systems, Man, and Cybernetics, Part B 34(1), 52–59 (2004)

33. Gelder, A.V.: Toward leaner binary-clause reasoning in a satisfiability solver. Ann. Math. Artif. Intell. 43(1), 239–253 (2005)

34. Gershman, R., Strichman, O.: Cost-effective hyper-resolution for preprocessing CNF formulas. In: Bacchus, F., Walsh, T. (eds.) SAT 2005. LNCS, vol. 3569, pp. 423–429. Springer, Heidelberg (2005)

35. Heule, M.J.H., Järvisalo, M., Biere, A.: Efficient CNF simplification based on binary implication graphs. In: Sakallah, K.A., Simon, L. (eds.) SAT 2011. LNCS, vol. 6695, pp. 201–215. Springer, Heidelberg (2011)

36. Samer, M., Szeider, S.: Backdoor sets of quantified Boolean formulas. Journal of Automated Reasoning **42**(1), 77–97 (2009)

37. Samer, M.: Variable dependencies of quantified CSPs. In: Cervesato, I., Veith, H., Voronkov, A. (eds.) LPAR 2008. LNCS (LNAI), vol. 5330, pp. 512–527. Springer, Heidelberg (2008)

38. Lonsing, F., Biere, A.: Efficiently representing existential dependency sets for expansion-based QBF solvers. Electronic Notes in Theoretical Computer Science **251**, 83–95 (2009)

39. Van Gelder, A.: Variable independence and resolution paths for quantified boolean formulas. In: Lee, J. (ed.) CP 2011. LNCS, vol. 6876, pp. 789–803. Springer, Heidelberg (2011)

40. Slivovsky, F., Szeider, S.: Computing resolution-path dependencies in linear time. In: Cimatti, A., Sebastiani, R. (eds.) SAT 2012. LNCS, vol. 7317, pp. 58–71. Springer, Heidelberg (2012)

41. Balabanov, V., Chiang, H.K., Jiang, J.R.: Henkin quantifiers and Boolean formulae: A certification perspective of DQBF. Theoretical Computer Science **523**, 86–100 (2014)

42. Bubeck, U., Kleine Büning, H.: Dependency quantified horn formulas: models and complexity. In: Biere, A., Gomes, C.P. (eds.) SAT 2006. LNCS, vol. 4121, pp. 198–211. Springer, Heidelberg (2006)

43. Bubeck, U.: Model-based transformations for quantified Boolean formulas. Ph.D. thesis, University of Paderborn (2010)

44. Heule, M., Järvisalo, M., Biere, A.: Clause elimination procedures for CNF formulas. In: Fermüller, C.G., Voronkov, A. (eds.) LPAR-17. LNCS, vol. 6397, pp. 357–371. Springer, Heidelberg (2010)

45. Heule, M., Järvisalo, M., Biere, A.: Covered clause elimination. In: Voronkov, A., Sutcliffe, G., Baaz, M., Fermüller, C.G. (eds.): Int'l Conf. on Logic for Programming, Artificial Intelligence, and Reasoning (LPAR) (Short papers). vol. 13 of EPiC Series, Yogyakarta, Indonesia, EasyChair, pp. 41–46 (2010)

46. Heule, M.J.H., Seidl, M., Biere, A.: Blocked literals are universal. In: Havelund, K., Holzmann, G., Joshi, R. (eds.) NFM 2015. LNCS, vol. 9058, pp. 436–442. Springer, Heidelberg (2015)

47. Lonsing, F., Egly, U.: Incremental QBF solving by DepQBF. In: Hong, H., Yap, C. (eds.) ICMS 2014. LNCS, vol. 8592, pp. 307–314. Springer, Heidelberg (2014)

48. Lonsing, F., Biere, A.: DepQBF: A dependency-aware QBF solver. Journal on Satisfiability, Boolean Modelling and Computation **7**(2–3), 71–76 (2010)

49. Kuehlmann, A., Paruthi, V., Krohm, F., Ganai, M.K.: Robust Boolean reasoning for equivalence checking and functional property verification. IEEE Transactions on CAD of Integrated Circuits and Systems **21**(12), 1377–1394 (2002)

Incrementally Computing Minimal Unsatisfiable Cores of QBFs via a Clause Group Solver API

Florian Lonsing$^{(\boxtimes)}$ and Uwe Egly

Knowledge-Based Systems Group
Vienna University of Technology, Vienna, Austria
`florian.lonsing@tuwien.ac.at`
`http://www.kr.tuwien.ac.at/`

Abstract. We consider the incremental computation of minimal unsatisfiable cores (MUCs) of QBFs. To this end, we equipped our incremental QBF solver DepQBF with a novel API to allow for incremental solving based on clause groups. A clause group is a set of clauses which is incrementally added to or removed from a previously solved QBF. Our implementation of the novel API is related to incremental SAT solving based on selector variables and assumptions. However, the API entirely hides selector variables and assumptions from the user, which facilitates the integration of DepQBF in other tools. We present implementation details and, for the first time, report on experiments related to the computation of MUCs of QBFs using DepQBF's novel clause group API.

1 Introduction

Let $\psi = \hat{Q}.\phi$ be a QBF in *prenex CNF (PCNF)* where $\hat{Q} = Q_1 x_1 \ldots Q_n x_n$ with $Q_i \in \{\forall, \exists\}$ is the prefix containing quantified propositional variables x_i and ϕ is a quantifier-free CNF. Given a PCNF $\psi = \hat{Q}.\phi$, an *unsatisfiable core (UC)* of ψ is an unsatisfiable PCNF $\psi' = \hat{Q}'.\phi'$ such that $\hat{Q}' \subseteq \hat{Q}$ and $\phi' \subseteq \phi$. The prefix $\hat{Q}'$ is obtained from $\hat{Q}$ by deleting the quantified variables which do not occur in ϕ'. A *minimal unsatisfiable core (MUC)*[1] of ψ is an unsatisfiable core $\psi' = \hat{Q}'.\phi'$ of ψ where, for every $C \in \phi'$, the PCNF $\hat{Q}'.(\phi' \setminus \{C\})$ is satisfiable.

Incremental solving is crucial for the computation of MUCs in the context of propositional logic (SAT), e.g. [1,3,8,13,24,25,29]. Modifications of a CNF by adding and deleting clauses in incremental solving are typically implemented by *selector variables* and *assumptions* [2,9,10,17,20,21,23,26,30]. An added clause C is augmented by a fresh selector variable s so that actually $C \cup \{s\}$ is added. Via the solver API, the user assigns these variables as assumptions under which the CNF is solved to control whether a clause is effectively present in the CNF.

Different from the assumption-based approach, the SAT solver zChaff[2] [27] provides an API to modify the CNF by adding and removing *groups* (sets) of clauses. Clauses are associated with an integer ID of the group they belong to.

Supported by the Austrian Science Fund (FWF) under grant S11409-N23. We would like to thank Aina Niemetz and Mathias Preiner for helpful discussions.

[1] The terminology *minimal unsatisfiable subformula (MUS)* is equivalent to MUC.

[2] zChaff website (July 2015): https://www.princeton.edu/~chaff/zchaff.html

M. Heule and S. Weaver (Eds.): SAT 2015, LNCS 9340, pp. 191–198, 2015.
DOI: 10.1007/978-3-319-24318-4_14

In assumption-based incremental solving, clause groups may be emulated by augmenting all clauses in a group by the same selector variable. The user must specify the necessary assumptions via the API in all forthcoming solver invocations to enable and disable the right groups. In contrast to that, zChaff allows to delete groups by a single API function call. In terms of usability, we argue that incremental solving by a clause group API is less error-prone, more accessible to inexperienced users, and facilitates the integration of the solver in other tools.

We present a novel clause group API of our QBF solver DepQBF (version 4.0 or later)[3] in the style of zChaff. Different from zChaff, we implemented clause groups based on selector variables and assumptions to combine the conceptual simplicity of zChaff's API with state of the art assumption-based incremental solving. As a novel feature of our API, the handling of selector variables and assumptions is entirely carried out by the solver and is hidden from the user. Our approach is applicable to any SAT or QBF solver supporting assumptions. Based on the novel clause group API of DepQBF, we implemented a tool to compute MUCs of PCNFs, a problem which has not been considered so far. Results on benchmarks used in the QBF Gallery 2014 illustrate the applicability of the clause group API for MUC computation of PCNFs.

2   Implementing a Clause Group API

DepQBF is a solver for PCNFs based on the QBF-specific variant of the DPLL algorithm [6] with learning [12,18,32]. Since version 3.0 [20,21], DepQBF supports incremental QBF solving via an API to add and remove clauses in a stack-based way (cf. Fig. 3 in [21]). This API is suitable for solving incremental encodings where clauses added most recently tend to be removed again in subsequent solver calls, like reachability problems such as conformant planning [11] or bounded model checking [4,15]. The new clause group API of DepQBF, however, allows to add and delete clauses *arbitrarily*, which is necessary for the incremental computation of MUCs of PCNFs. We first present our novel approach to keeping selector variables invisible to the user, which is a unique feature of DepQBF. To this end, we distinguish between selector variables and variables in the encoding.

Let $S = \langle \psi_1, \ldots, \psi_n \rangle$ be a sequence of PCNFs. We consider variables over which the PCNFs ψ_i are defined as *user variables* because they are part of the problem encoding represented by S. When solving S incrementally, *selector variables* used to augment clauses in ψ_i are not part of the original encoding. Variables v are stored in an array VA indexed by an integer ID $id(v)$ of v such that $VA[id(v)] = v$. User and selector variables reside in separate sections of VA:

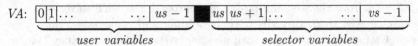

The total size of VA is vs. The user variable section has size us. The following invariants are maintained: $VA[id(v)] = v$ where $id(v) < us$ if v is a user variable

---

```
int main (int argc, char ** argv) {
  Solver *s = create();
  new_scope_at_nesting
    (s,QTYPE_FORALL,1);
  add(s,1);add(s,2);add(s,0);
  new_scope_at_nesting
    (s,QTYPE_EXISTS,2);
  add(s,3);add(s,4);add(s,0);

  ClauseGroupID id1 = new_cls_grp(s);
  open_cls_grp(s,id1);
  add(s,-1);add(s,-3);add(s,0);
  close_cls_grp(s,id1);

  ClauseGroupID id2 = new_cls_grp(s);
  open_cls_grp(s,id2);
  add(s,1);add(s,2);add(s,4);add(s,0);
  add(s,1);add(s,-4);add(s,0);
  close_cls_grp(s,id2);
  ...//continues on right column.

  ...//continued from left column.
  Result res = sat(s);
  assert(res == RESULT_UNSAT);
  ClauseGroupID *rgrps =
    get_relevant_cls_grps(s);
  assert(rgrps[0] == id2);
  reset(s);

  deactivate_cls_grp(s,rgrps[0]);
  res = sat(s);
  assert(res == RESULT_SAT);
  reset(s);

  activate_cls_grp(s,rgrps[0]);
  free(rgrps);

  delete_cls_grp(s,id1);
  res = sat(s);
  assert(res == RESULT_UNSAT);
  delete(s); }
```

Fig. 1. Clause group code example. Variables x_i are encoded as integers i. Given the PCNF $\psi := \forall x_1, x_2 \exists x_3, x_4. C_1 \wedge C_2 \wedge C_3$ with $C_1 = (\neg x_1 \vee \neg x_3), C_2 = (x_1 \vee x_2 \vee x_4), C_3 = (x_1 \vee \neg x_4)$, C_1 is put in group id1 and C_2, C_3 in group id2. An unsatisfiable core consisting only of group id2 is extracted from ψ. Deactivating group id2 results in the PCNF $\forall x_1 \exists x_3. C_1$. Activating id2 again and deleting id1 yields $\forall x_1, x_2 \exists x_4. C_2 \wedge C_3$.

and $us \leq id(v) < vs$ if v is a selector variable. If a new user variable v with $id(v) \geq us$ is added via the solver API, then VA is resized together with the user variable section. In this case the selector variables are assigned new, larger IDs and copied to a new position in VA. Then the literals of selector variables are renamed according to the newly assigned IDs in all (learned) clauses and cubes present in the current PCNF in a single pass. Resizing only the selector variable section of VA does not require assigning new IDs to selector variables. Similar to implementations of other SAT or QBF solvers, the user is responsible to avoid unnecessarily large user variable indices and thus avoid resizing VA.

The API of DepQBF prevents accessing selector variables in VA, which are hence invisible to the user. In contrast to traditional solver implementations, e.g. [10], where the user is responsible to maintain selector variables manually, the *internal* separation between user and selector variables allows to conveniently allocate and rename selector variables on the fly inside the solver and without any user interaction. This feature is particularly useful for solving dynamically generated sequences $S = \langle \psi_1, \ldots, \psi_n \rangle$ of PCNFs where the exact user variable IDs in each ψ_i are unknown at the beginning.

In the following, we present the novel clause group API of DepQBF along with the example shown in Fig. 1. A new clause group is created by calling new_cls_grp(), which returns a unique unsigned integer *cgid* as the ID of the group. Each time a new group *cgid* is created, *internally* a fresh selector variable *s* is allocated in the array *VA* and associated with the group *cgid*.

A group *cgid* must be opened by `open_cls_grp(cgid)` before clauses can be added to it. All clauses added via the API are associated with the currently opened group *cgid* by *internally* augmenting them with the selector variable *s* of group *cgid*. Groups must be closed by `close_cls_grp(cgid)`. When solving the current PCNF by `sat()`, *internally* the selector variables of all created groups are assigned *false* as assumptions to effectively activate the clauses in these groups.

Deleting a group by `delete_cls_grp(cgid)` invalidates its ID. When solving the current PCNF by `sat()`, *internally* the selector variables of all deleted groups are assigned *true* as assumptions to deactivate the clauses in all deleted groups and all learned clauses derived therefrom. Deleted clauses are physically removed from the data structures in a garbage collection phase if their number exceeds a certain threshold. Clauses which are added to the PCNF without opening a group by `open_cls_grp(cgid)` before are permanent and cannot be deleted.

In contrast to deletion, clause groups can also be *deactivated* by calling `deactivate_cls_grp(cgid)`. When solving the current PCNF by `sat()`, *internally* the selector variables of deactivated groups are assigned *true* similarly to deleted groups. However, clauses in deactivated groups are never removed from the data structures. Deactivated groups are activated again by `activate_cls_grp(cgid)`. Selector variables of activated groups are assigned *false* when solving the current PCNF.

DepQBF also allows for traditional incremental solving where the user handles selector variables manually [10]. Implementations of this approach like MiniSAT, for example, allow to physically delete clauses by first adding a unit clause containing a selector variable and then simplifying the formula based on unit clauses. This is in contrast to DepQBF where the formula is not simplified based on unit clauses to avoid the internal elimination of variables, which may be unexpected by the user.

If the current PCNF has been found unsatisfiable by `sat()`, then calling `get_relevant_cls_grps()` returns an array of the IDs of those groups which contain clauses used by the solver to determine unsatisfiability. The clauses in these groups amount to an unsatisfiable core of the PCNF. That core is obtained by *internally* collecting all selector variables relevant for unsatisfiability[4] and mapping them to the respective clause group IDs.

3   Computing Minimal Unsatisfiable Cores of QBFs

In contrast to theory [16], the computation of MUCs of PCNFs in *practice* has not been considered so far. Approaches to *nonminimal* UCs of PCNFs were presented in the context of checking Q-resolution refutations of PCNFs [31] and QMaxSAT [14]. For the first time we report on experiments related to the computation of MUCs of PCNFs. To this end, we implemented a tool to incrementally compute MUCs of PCNFs using the clause group API of DepQBF as follows.

---

[4] Similar to the function `analyzeFinal` in Minisat, for example.

Table 1. Statistics for unsatisfiable instances from the QBF Gallery 2014 where MUCs were successfully computed. Numbers of solved instances out of total ones are shown in parentheses. MUCs computed ($\#m$), total time to solve the initial unsatisfiable instances (ut) and to compute the MUCs (mt), total number of clauses in initial formulas ($|CNF|$) and in MUCs ($|MUC|$), total number of QBF solver calls ($\#c$), and the average ($\bar{r}$) and median ($\tilde{r}$) sizes of MUCs relative to the respective CNF sizes.

QBF Gallery Track	$\#m$	ut	mt	$|CNF|$	$|MUC|$	$\#c$	$\bar{r}$	$\tilde{r}$
applications (190 of 735):	182	6,304	7,941	4,744,494	73,206	81,631	6.1%	2.9%
QBFLIB (58 of 276):	46	1,009	2,264	323,497	34,777	36,888	14.1%	5.1%
preprocessing (38 of 243):	34	1,623	1,080	451,197	23,220	24,572	4.0%	2.2%

Given an unsatisfiable PCNF $\psi_0 = \hat{Q}.\phi$, first every single clause of ψ_0 is put in an individual clause group. Let $\psi := \psi_0$. The PCNF ψ is solved and a UC $\psi' = \hat{Q}'.\phi'$ is extracted by `get_relevant_cls_grps`. Then ψ is replaced by ψ' by deleting the clause groups which do not belong to ψ' from ψ. Given the updated $\psi = \hat{Q}.\phi$, every clause $C \in \phi$ is checked by solving the PCNF $\psi'' = \hat{Q}.(\phi \backslash \{C\})$. To this end, the group containing C is deactivated. If ψ'' is satisfiable then C is part of an MUC and hence C is activated again (C is a *transition clause* [25]). Otherwise, a UC ψ' of ψ'' is extracted, ψ is replaced by the UC ψ' like above, and again every clause in the updated ψ is checked. After every clause in the current ψ has been checked, the final ψ is an MUC of ψ_0. The number of solver calls in this well-known elimination-based algorithm is linear in the size of ψ_0 [13,24,25]. It applies iterative *clause set refinement* [3,8,28] by UCs. UCs are extracted by selector variables [1] in `get_relevant_cls_grps`, which is in contrast to extraction based on resolution proofs [28,29]. The algorithm is common to compute MUCs of CNFs but has not been applied to PCNFs so far.

Using our tool, we computed MUCs of instances from the *applications (AT)*, *QBFLIB (QT)*, and *preprocessing (PT)* tracks of the QBF Gallery 2014.[5] We preprocessed the instances from AT and QT using Bloqqer [5]. In total, we allowed 900s of wall clock time and seven GB of memory to solve an instance by DepQBF and to compute an MUC. Table 1 summarizes the results of our experiments[6] run on an AMD Opteron 6238 at 2.6 GHz under 64-bit Linux. MUCs were successfully computed for 95% of the solved unsatisfiable instances in AT (79% of QT and 89% of PT). On average, MUC computation took 43s in AT (49s in QT and 31s in PT). When increasing the total timeout to 3600s, then 186 MUCs were computed in AT (48 in QT and 36 in PT).

Iterative clause set refinement by UCs potentially reduces the number of solver calls. In the worst case, there is one solver call per each single clause in the initial PCNF ψ_0. However, on average there was one solver call per 58, 8, and 18 clauses in AT, QT, and PT, respectively.

The physical deletion of clauses not belonging to a MUC reduces the memory footprint and the run time. The plot below shows the sorted total run

---

[5] http://qbf.satisfiability.org/gallery/

[6] We refer to an appendix of this paper with additional experimental data [22].

times (y-axis) of the MUC workflow on instances in AT where MUCs were successfully computed (x-axis). If clauses are deleted by `delete_cls_grp` (UC-d) then 182 MUCs are computed but only 169 if clauses are permanently deactivated by `deactivate_cls_grp` instead (UC-nd). We attribute this effect to overhead caused by deactivated clauses still present in the data structures. Only 79 MUCs are computed without iterative clause set refinement by UCs using `get_relevant_cls_grps` and instead checking *every* clause in ψ_0 one by one (OBO). We made similar observations for QT and PT. On instances where an MUC was computed by both UC-d and UC-nd, in general UC-nd is slower (up to +316% on PT) and has a larger memory footprint (up to +70% on AT). The difference between UC-d and OBO is more pronounced, where in general OBO is slower (up to +4126% on PT) and has a larger memory footprint (up to +243% on AT).

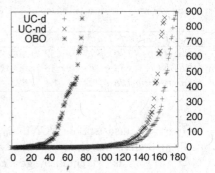

Our experiments show that physical deletion of clauses by `delete_cls_grp` (UC-d) and the extraction of UCs by `get_relevant_cls_grps` based on selector variables are crucial for the computation of MUCs of PCNFs. These features are provided directly by the novel clause group API of DepQBF.

4   Conclusion

We presented a novel API of our solver DepQBF for incremental QBF solving based on clause groups and its application to MUC computation. The clause group API is conceptually simple yet employs state of the art approaches to assumption-based incremental SAT solving. Improvements of assumption-based incremental solving [2,17,30] are also applicable to our implementation.

The API encapsulates the handling of selector variables and assumptions entirely inside the solver. This is a unique feature of DepQBF, which facilitates its integration in other tools. It is particularly useful for solving dynamically generated sequences of PCNFs where the exact variable IDs are unknown at the beginning. The clause group API is general and fits any search-based SAT and QBF solver capable of solving under assumptions.

A potential application of the clause group API is (M)UC extraction of PCNFs in core-guided QMaxSAT [14] and SMT, similar to SAT-based UC extraction in SMT [7]. Further, our API readily supports the extraction of high-level UCs [19,28,29] where, different from our experiments with MUC computation, multiple clauses are put in a clause group. We applied the novel clause group API of DepQBF to compute MUCs of PCNFs for the first time. Our results indicate the efficiency and applicability of our implementation. As future work, we want to integrate incremental preprocessing in DepQBF in a way where the implementation details are hidden by the API [30].

References

1. Asín, R., Nieuwenhuis, R., Oliveras, A., Rodríguez-Carbonell, E.: Efficient generation of unsatisfiability proofs and cores in SAT. In: Cervesato, I., Veith, H., Voronkov, A. (eds.) LPAR 2008. LNCS (LNAI), vol. 5330, pp. 16–30. Springer, Heidelberg (2008)
2. Audemard, G., Lagniez, J.-M., Simon, L.: Improving glucose for incremental SAT solving with assumptions: application to MUS extraction. In: Järvisalo, M., Van Gelder, A. (eds.) SAT 2013. LNCS, vol. 7962, pp. 309–317. Springer, Heidelberg (2013)
3. Belov, A., Lynce, I., Marques-Silva, J.: Towards efficient MUS extraction. AI Commun. **25**(2), 97–116 (2012)
4. Benedetti, M., Mangassarian, H.: QBF-based formal verification: experience and perspectives. JSAT **5**, 133–191 (2008)
5. Biere, A., Lonsing, F., Seidl, M.: Blocked clause elimination for QBF. In: Bjørner, N., Sofronie-Stokkermans, V. (eds.) CADE 2011. LNCS, vol. 6803, pp. 101–115. Springer, Heidelberg (2011)
6. Cadoli, M., Schaerf, M., Giovanardi, A., Giovanardi, M.: An algorithm to evaluate quantified boolean formulae and its experimental evaluation. J. Autom. Reasoning **28**(2), 101–142 (2002)
7. Cimatti, A., Griggio, A., Sebastiani, R.: A simple and flexible way of computing small unsatisfiable cores in SAT modulo theories. In: Marques-Silva, J., Sakallah, K.A. (eds.) SAT 2007. LNCS, vol. 4501, pp. 334–339. Springer, Heidelberg (2007)
8. Dershowitz, N., Hanna, Z., Nadel, A.: A scalable algorithm for minimal unsatisfiable core extraction. In: Biere, A., Gomes, C.P. (eds.) SAT 2006. LNCS, vol. 4121, pp. 36–41. Springer, Heidelberg (2006)
9. Eén, N., Sörensson, N.: An extensible SAT-solver. In: Giunchiglia, E., Tacchella, A. (eds.) SAT 2003. LNCS, vol. 2919, pp. 502–518. Springer, Heidelberg (2004)
10. Eén, N., Sörensson, N.: Temporal induction by incremental SAT solving. Electr. Notes Theor. Comput. Sci. **89**(4), 543–560 (2003)
11. Egly, U., Kronegger, M., Lonsing, F., Pfandler, A.: Conformant planning as a case study of incremental QBF solving. In: Aranda-Corral, G.A., Calmet, J., Martín-Mateos, F.J. (eds.) AISC 2014. LNCS, vol. 8884, pp. 120–131. Springer, Heidelberg (2014)
12. Giunchiglia, E., Narizzano, M., Tacchella, A.: Clause/Term resolution and learning in the evaluation of quantified boolean formulas. J. Artif. Intell. Res. (JAIR) **26**, 371–416 (2006)
13. Grégoire, É., Mazure, B., Piette, C.: On Approaches to Explaining Infeasibility of Sets of Boolean Clauses. In: ICTAI, pp. 74–83. IEEE Computer Society (2008)
14. Ignatiev, A., Janota, M., Marques-Silva, J.: Quantified maximum satisfiability. In: Järvisalo, M., Van Gelder, A. (eds.) SAT 2013. LNCS, vol. 7962, pp. 250–266. Springer, Heidelberg (2013)
15. Jussila, T., Biere, A.: Compressing BMC encodings with QBF. ENTCS **174**(3), 45–56 (2007)
16. Kleine Büning, H., Zhao, X.: Minimal false quantified boolean formulas. In: Biere, A., Gomes, C.P. (eds.) SAT 2006. LNCS, vol. 4121, pp. 339–352. Springer, Heidelberg (2006)
17. Lagniez, J.-M., Biere, A.: Factoring out assumptions to speed Up MUS extraction. In: Järvisalo, M., Van Gelder, A. (eds.) SAT 2013. LNCS, vol. 7962, pp. 276–292. Springer, Heidelberg (2013)

18. Letz, R.: Lemma and model caching in decision procedures for quantified boolean formulas. In: Egly, U., Fermüller, C. (eds.) TABLEAUX 2002. LNCS (LNAI), vol. 2381, p. 160. Springer, Heidelberg (2002)

19. Liffiton, M.H., Sakallah, K.A.: Algorithms for computing minimal unsatisfiable subsets of constraints. J. Autom. Reasoning **40**(1), 1–33 (2008)

20. Lonsing, F., Egly, U.: Incremental QBF solving. In: O'Sullivan, B. (ed.) CP 2014. LNCS, vol. 8656, pp. 514–530. Springer, Heidelberg (2014)

21. Lonsing, F., Egly, U.: Incremental QBF solving by DepQBF. In: Hong, H., Yap, C. (eds.) ICMS 2014. LNCS, vol. 8592, pp. 307–314. Springer, Heidelberg (2014)

22. Lonsing, F., Egly, U.: Incrementally Computing Minimal Unsatisfiable Cores of QBFs via a Clause Group Solver API. CoRR abs/1502.02484 (2015). http://arxiv.org/abs/1502.02484, SAT 2015 proceedings version (6-page tool paper) with appendix

23. Marin, P., Miller, C., Lewis, M.D.T., Becker, B.: Verification of Partial Designs using Incremental QBF Solving. In: Rosenstiel, W., Thiele, L. (eds.) DATE, pp. 623–628. IEEE (2012)

24. Marques-Silva, J.: Minimal unsatisfiability: models, algorithms and applications (Invited Paper). In: ISMVL, pp. 9–14. IEEE Computer Society (2010)

25. Marques-Silva, J., Lynce, I.: On improving MUS extraction algorithms. In: Sakallah, K.A., Simon, L. (eds.) SAT 2011. LNCS, vol. 6695, pp. 159–173. Springer, Heidelberg (2011)

26. Miller, C., Marin, P., Becker, B.: Verification of partial designs using incremental QBF. AI Commun. **28**(2), 283–307 (2015)

27. Moskewicz, M.W., Madigan, C.F., Zhao, Y., Zhang, L., Malik, S.: Chaff: Engineering an Efficient SAT Solver. In: DAC, pp. 530–535. ACM (2001)

28. Nadel, A.: Boosting Minimal Unsatisfiable Core Extraction. In: Bloem, R., Sharygina, N. (eds.) FMCAD, pp. 221–229. IEEE (2010)

29. Nadel, A., Ryvchin, V., Strichman, O.: Accelerated deletion-based extraction of minimal unsatisfiable cores. JSAT **9**, 27–51 (2014)

30. Nadel, A., Ryvchin, V., Strichman, O.: Ultimately incremental SAT. In: Sinz, C., Egly, U. (eds.) SAT 2014. LNCS, vol. 8561, pp. 206–218. Springer, Heidelberg (2014)

31. Yu, Y., Malik, S.: Validating the result of a Quantified Boolean Formula (QBF) solver: theory and practice. In: Tang, T. (ed.) ASP-DAC, pp. 1047–1051. ACM Press (2005)

32. Zhang, L., Malik, S.: Towards a symmetric treatment of satisfaction and conflicts in quantified boolean formula evaluation. In: Van Hentenryck, P. (ed.) CP 2002. LNCS, vol. 2470, p. 200. Springer, Heidelberg (2002)

On Compiling CNFs into Structured Deterministic DNNFs

Simone Bova[1], Florent Capelli[2], Stefan Mengel[3], and Friedrich Slivovsky[1]([⊠])

[1] Institute of Computer Graphics and Algorithms, TU Wien, Vienna, Austria
friedrich.slivovsky@tuwien.ac.at
[2] IMJ UMR 7586 - Logique, Université Paris Diderot, Paris, France
[3] LIX UMR 7161, Ecole Polytechnique, Université Paris-Saclay, Palaiseau, France

Abstract. We show that the traces of recently introduced dynamic programming algorithms for #SAT can be used to construct structured deterministic DNNF (decomposable negation normal form) representations of propositional formulas in CNF (conjunctive normal form). This allows us prove new upper bounds on the complexity of compiling CNF formulas into structured deterministic DNNFs in terms of parameters such as the treewidth and the clique-width of the incidence graph.

1 Introduction

The aim of knowledge compilation is to succinctly represent propositional knowledge bases in a format that allows for answering a number of queries in polynomial time [6]. Choosing a representation language generally involves a trade-off between succinctness and the range of queries that can be efficiently answered. Constraints arising in various domains can often be conveniently modeled by propositional formulas in conjunctive normal form (CNFs), but most queries of interest, such as model counting, are intractable for CNF formulas.

Decomposable Negation Normal Forms (DNNFs) are a restricted form of Boolean circuits in negation normal form (NNF) such that the subcircuits leading into an AND gate are defined on disjoint sets of variables [4]. DNNFs—which generalize variants of binary decision diagrams such as ordered binary decision diagrams (OBDDs)—are among the most succinct representation languages considered in knowledge compilation. Although CNFs do not have DNNF representations of polynomial size in general [1,6] they can be efficiently compiled into DNNFs when certain structural parameters are small, see [4,5,11,13–15].

Among the key properties of DNNFs is that they allow for clause entailment queries in polynomial time. By imposing further restrictions, one obtains languages that efficiently support a wider range of queries and operations. A DNNF is *deterministic* (a *d-DNNF*, for short) if the subcircuits leading into an OR gate do not have satisfying assignments in common, and *structured* if its variables can be associated with the leaves of a binary tree so that, for each AND gate, one can find a tree node whose principal subtrees contain the variables occurring in the subcircuits leading into that gate. Deterministic DNNFs support model counting in linear time [5], and structured DNNFs allow for an efficient conjoin operation [13].

M. Heule and S. Weaver (Eds.): SAT 2015, LNCS 9340, pp. 199–214, 2015.
DOI: 10.1007/978-3-319-24318-4_15

In this paper, we prove the following result (Theorem 1):

Theorem. A CNF formula with n variables, m clauses, and PS-width k can be compiled into a structured d-DNNF of size $O(k^3(n + m))$.

PS-width is a parameter that was introduced to characterize CNF formulas for which the model counting problem (#SAT) can be solved efficiently by means of recently developed dynamic programming algorithms [16,17]. We prove Theorem 1 by showing that the traces of these algorithms can be used to construct structured d-DNNF representations of CNF formulas.

Our rationale for stating and proving the above theorem in terms of PS-width is that this parameter generalizes most width measures of formulas commonly considered in the literature [16]. Accordingly, we are able to immediately derive a number of corollaries. For instance, a CNF formula with m clauses and an incidence graph[1] of clique-width k has PS-width at most m^k [16]. This allows us to state an upper bound in terms of incidence clique-width as follows (Corollary 2):

Corollary. A CNF formula with n variables, m clauses, and incidence clique-width k can be compiled into a structured d-DNNF of size $O(m^{3k}(n + m))$.

In particular, any class of formulas of bounded incidence clique-width admits compilation into structured d-DNNFs of polynomial size. Such classes can have unbounded incidence treewidth, effectively putting them out of reach of known compilation algorithms generating DNNFs of size exponential in the incidence treewidth [15].

One can further show that a formula with incidence treewidth k has PS-width at most 2^{k+1} (see Proposition 1). Accordingly, the upper bound of Theorem 1 translates into the following bound in terms of incidence treewidth (Corollary 1):

Corollary. A CNF formula with n variables, m clauses, and incidence treewidth k can be compiled into a structured d-DNNF of size $O(8^k(n + m))$.

This comes close to the best known upper bound of $O(3^k n)$ on the complexity of compiling CNFs with incidence treewidth k into structured DNNFs [11], while allowing us to compile into the more restrictive language of structured *deterministic* DNNFs.

As far as compilation of CNFs into d-DNNFs is concerned, the best known result using a structural parameter is an upper bound of $O(2^k n)$ for CNF formulas with n variables and *decision-width* k [12]. As the decision-width of a formula is no greater than the treewidth of its primal graph, this bound translates into an upper bound of $O(2^k n)$ for formulas with n variables and primal treewidth k. The incidence treewidth of a formula is at most its primal treewidth plus one, but there are classes of formulas with bounded incidence treewidth and unbounded primal treewidth, so Corollary 1 yields an improvement whenever the difference between primal treewidth and incidence treewidth is sufficiently large.

---

[1] The incidence graph of a formula is the bipartite graph whose vertex classes consist of variables and clauses, and a variable is adjacent to the clauses it occurs in.

The degree of the polynomial in the upper bound of Corollary 2 depends on the incidence clique-width k, and one may wonder whether this can be improved to a bound of the form, say, $2^{O(k)}(n + m)^c$ for some constant c. We show that such an improvement is impossible, subject to a complexity-theoretic assumption (Theorem 2).

The remainder of the paper is structured as follows. In Section 2 we introduce basic notation and terminology. Section 3 proves Theorem 1 by showing how ideas implemented in recently introduced dynamic programming algorithms for #SAT can be used for compilation into structured d-DNNFs. We present corollaries of this result in Section 4. Section 5 provides evidence that our upper bound on the DNNF size of formulas in terms of incidence clique-width (Corollary 2) cannot be substantially improved. We conclude in Section 6.

2 Preliminaries

Formulas. A *literal* is a variable x or a negated variable $\neg x$. A *clause* is a finite set of literals. A clause is *tautological* if it contains the same variable negated as well as unnegated. A *(CNF) formula* (or *CNF*, for short) is a finite set of non-tautological clauses. If x is a variable, we let $var(x) = var(\neg x) = x$. The set of variables occurring in a clause C is $var(C) = \{\, var(\ell) \mid \ell \in C \,\}$, and the set of variables occurring in a formula F is $var(F) = \bigcup_{C \in F} var(C)$. The *length* of a formula F is $\sum_{C \in F} |C|$. The *incidence graph* of a formula F is the bipartite graph $I(F) = (F, var(F), E)$ such that there is an edge $xC \in E$ joining a variable $x \in var(F)$ and a clause $C \in F$ if and only if $x \in var(C)$.

A *truth assignment* (*assignment,* for short) is a mapping $\tau : X \rightarrow \{0,1\}$, where X is a set of variables. Extending assignments to literals in the usual way, we say that an assignment τ *satisfies* a clause C if there is a literal $\ell \in C$ such that $\tau(\ell) = 1$. An assignment *satisfies* a formula F if it satisfies every clause $C \in F$.

DNNFs. A *(Boolean) circuit in negation normal form* (or *NNF*) is a directed acyclic graph (DAG) with a single sink node (outdegree 0) where each source node (indegree 0) is labelled by a constant (0 or 1) or by a literal, and each other node is labelled by $\wedge$ (AND) or $\vee$ (OR). If φ is an NNF and v is a vertex of φ, the *sub-NNF* of φ rooted at v is the NNF obtained from φ by deleting every vertex from which v cannot be reached along a directed path. We write $var(\varphi)$ for the set of variables occurring in an NNF φ. Let φ be an NNF and let τ be an assignment to $X \supseteq var(\varphi)$. Relative to τ, we associate each vertex v of φ with a value $val_\varphi(v, \tau) \in \{0,1\}$ as follows. If v is labelled with a constant $c \in \{0,1\}$ then $val_\varphi(v, \tau) = c$, and if v is labelled with a literal ℓ then $val_\varphi(v, \tau) = \tau(\ell)$. If v is an AND node then we let $val_\varphi(v, \tau) = \min\{\, val_\varphi(w, \tau) \mid w \text{ is a child of } v \,\}$, and if v is an OR node we define $val_\varphi(v, \tau) = \max\{\, val_\varphi(w, \tau) \mid w \text{ is a child of } v \,\}$. We say that τ *satisfies* φ if $val_\varphi(s, \tau) = 1$, where s denotes the (unique) sink of φ. An NNF φ is said to *compute* a CNF formula F if the satisfying assignments of φ and F coincide. Similarly, we say that two NNFs φ

and ψ are *equivalent*, in symbols $\varphi \equiv \psi$, if they have the same set of satisfying assignments. For convenience, we interpret propositional expressions over literals and $\{0, 1, \wedge, \vee\}$ as NNFs. We also use the names of NNFs in expressions involving logical connectives, writing, for instance, $\varphi \wedge \psi$ to denote the NNF constructed from φ and ψ by adding a new AND node as a sink that has incoming edges from the sinks of φ and ψ.

An NNF φ is *decomposable* (in short, a *DNNF*) if every AND node v of φ satisfies the following property: if v has incoming edges from v_1 and v_2, and φ_1 and φ_2 denote the sub-NNFs of φ rooted at v_1 and v_2, respectively, then $var(\varphi_1)$ and $var(\varphi_2)$ are disjoint. A DNNF φ is *deterministic* (a *d-DNNF*) if, for every pair of distinct children v_1 and v_2 of an OR node, the sub-NNFs rooted at v_1 and v_2 do not have satisfying assignments in common.

3   From Dynamic Programming to Structured d-DNNFs

In this section, we show how ideas implemented in #SAT algorithms by Slivovsky and Szeider [17] and Saether et. al. [16] can be used for compiling CNF formulas into structured d-DNNFs.

3.1   Branch Decompositions, Projections, and PS-width

Given a formula F, the algorithms of [16,17] perform dynamic programming on a *branch decomposition* of $F \cup var(F)$. Here, a branch decomposition of a finite set S is a binary tree whose leaves are in one-to-one correspondence with S (see the left-hand side of Figure 1 for an illustration). Formally, we will think of a branch decomposition as a pair (T, δ) consisting of a rooted binary tree T and a bijection δ from the set of leaves of T to the set S. Accordingly, if (T, δ) is a branch decomposition of the set $F \cup var(F)$ for some formula F, then δ bijectively maps each leaf of T to a variable or a clause of F.[2]

Partial solution counts computed by dynamic programming in [16,17] are stored in tables indexed by pairs of *projections*. Here, the projection of a truth assignment $\tau : X \rightarrow \{0, 1\}$ onto a formula F is the set $F(\tau)$ of clauses of F satisfied by τ. Observe that the projection of the union of two assignments σ and τ (that agree on the intersection of their domains) onto F satisfies $F(\sigma \cup \tau) = F(\sigma) \cup F(\tau)$, and that τ is a satisfying assignment of F if, and only if, $F(\tau) = F$. For a formula F and a set X of variables we write $proj(F, X)$ for the set of projections of truth assignments $\tau : X \rightarrow \{0, 1\}$ onto F, formally

$$proj(F, X) = \{\, F(\tau) \mid \tau : X \rightarrow \{0, 1\} \,\}.$$

Let F be a CNF formula and let $\mathbf{T} = (T, \delta)$ be a branch decomposition of $F \cup var(F)$. For a node v of T, let T_v denote the subtree of T rooted at v, and

---

[2] Such decompositions can be thought of as generalizations of *vtrees*, which are binary trees whose leaves are in one-to-one correspondence with a set of variables and that have been studied before in knowledge compilation [13].

let $L(T_v)$ denote the set of leaves of T_v. We write X_v^T for the set of variables in the image of $L(T_v)$ under δ, and F_v^T for the set of clauses in the image of $L(T_v)$ under δ. We write $\overline{X_v^T} = var(F) \setminus X_v^T$ for the set of variables and $\overline{F_v^T} = F \setminus F_v^T$ for the set of clauses outside the subtree rooted at v. When T is clear from the context (as will be the case) we will omit T from the superscript.

Our main result states that a CNF formula can be represented by a structured d-DNNF of size polynomial in the number of clauses and a parameter called PS-width, which is defined as follows [16]: let F be a formula and let $T = (T, \delta)$ be a branch decomposition of $F \cup var(F)$. The *PS-width* of T is defined

$$psw(T) = \max_{v \in V(T)} \max(|proj(\overline{F_v}, X_v)|, |proj(F_v, \overline{X_v})|).$$

That is, the PS-width of T is the maximum number of projections "across" one of the bipartitions of F and $var(F)$ induced by a node of T. The *PS-width* of a formula F is the minimum PS-width of a branch decomposition of $F \cup var(F)$.

3.2   Records and Dynamic Programming

We now describe the "records" used by the dynamic programming algorithms for #SAT [16,17].

Let F be a formula, let $T = (T, \delta)$ be a branch decomposition of $F \cup var(F)$, and let v be a node of T. A *shape* (for v, with respect to T) is a pair $S = (S, S')$ of subsets of F such that $S \in proj(\overline{F_v}, X_v)$ and $S' \in proj(F_v, \overline{X_v})$. We say that an assignment $\tau : X_v \to \{0, 1\}$ has shape S if

(A) $\overline{F_v}(\tau) = S$, and
(B) $F_v(\tau) \cup S' = F_v$.

We write $N_v^T(S)$ for the set of assignments of shape S (again, we drop T from the superscript if it is clear which branch decomposition we are using). The intermediate values for dynamic programming computed at node v are the cardinalities $|N_v(S)|$ for each shape S for v.

The reason for using shapes rather than just computing the number of assignments $\tau : X_v \to \{0, 1\}$ with projection $F(\tau) = S$ for each $S \in proj(F, X_v)$ is that, in some cases of interest (such as formulas of bounded clique-width [17]), the cardinality $|proj(F, X_v)|$ can be exponential in the number m of clauses while the number of shapes is bounded by a polynomial in m. This reduction in the amount of information required to represent partial solution counts is achieved by the use of an "expectation from the outside": by Condition (B), an assignment τ of shape (S, S') satisfies F_v when combined with an assignment $\sigma : \overline{X_v} \to \{0, 1\}$ such that $F_v(\sigma) = S'$. Since we are interested in satisfying assignments of F we expect τ to be paired with such an assignment σ and do not have to keep track of the projection $F_v(\tau)$.

We now explain how shapes for an inner node can be related to shapes for its child nodes in order to perform dynamic programming. Let F be a formula and let (T, δ) be a branch decomposition of $F \cup var(F)$. Let $S = (S, S')$ be a shape for an inner node v of T, and let $\boldsymbol{S}_1 = (S_1, S_1')$, $\boldsymbol{S}_2 = (S_2, S_2')$ be shapes for its children v_1 and v_2, respectively. We say that $\boldsymbol{S}_1$ and $\boldsymbol{S}_2$ *generate* $\boldsymbol{S}$ if

(a) $S = (S_1 \cup S_2) \cap \overline{F_v}$,
(b) $S_1' = (S' \cup S_2) \cap F_{v_1}$, and
(c) $S_2' = (S' \cup S_1) \cap F_{v_2}$.

The following result relates the shapes for an inner node to the generating shapes for its children (here, $\sqcup$ denotes the disjoint union).

Lemma 1. *Let F be a formula, let $\boldsymbol{T} = (T, \delta)$ be a branch decomposition of $F \cup var(F)$, and let v be an inner node of T with children v_1 and v_2. Let $\boldsymbol{S}$ be a shape for v, and let G denote the set of pairs of shapes $\boldsymbol{S}_1$ for v_1 and $\boldsymbol{S}_2$ for v_2 such that $\boldsymbol{S}_1$ and $\boldsymbol{S}_2$ generate $\boldsymbol{S}$. Then*

$$N_v(\boldsymbol{S}) = \bigsqcup_{(\boldsymbol{S}_1, \boldsymbol{S}_2) \in G} \{\, \tau_1 \cup \tau_2 \mid \tau_1 \in N_{v_1}(\boldsymbol{S}_1), \tau_2 \in N_{v_2}(\boldsymbol{S}_2) \,\}.$$

Lemma 1 is an easy consequence of the following two lemmas (cf. [17]).

Lemma 2. *Let v be a node of T with children v_1 and v_2. Let $\boldsymbol{S}_1 = (S_1, S_1')$ be a shape for v_1, let $\boldsymbol{S}_2 = (S_2, S_2')$ be a shape for v_2, and let $\boldsymbol{S} = (S, S')$ be a shape for v generated by $\boldsymbol{S}_1$ and $\boldsymbol{S}_2$. If $\tau_1 \in N_{v_1}(\boldsymbol{S}_1)$ and $\tau_2 \in N_{v_2}(\boldsymbol{S}_2)$ then $\tau_1 \cup \tau_2 \in N_v(\boldsymbol{S})$.*

Proof. Let $\tau_1 \in N_{v_1}(\boldsymbol{S}_1)$ and $\tau_2 \in N_{v_2}(\boldsymbol{S}_2)$. As $\overline{F_{v_1}}(\tau_1) = S_1$ and $\overline{F_{v_2}}(\tau_2) = S_2$, we get $\overline{F_v}(\tau_1 \cup \tau_2) = (S_1 \cup S_2) \cap \overline{F_v}$. This shows that Condition (A) is satisfied. Consider a clause $C \in F_v$ and assume without loss of generality that $C \in F_{v_1}$. Suppose $C \notin F_v(\tau_1 \cup \tau_2)$. Then τ_1 does not satisfy C and thus $C \in S_1'$ by Condition (B). But τ_2 does not satisfy C either, so $C \notin S_2$. The shapes $\boldsymbol{S}_1$ and $\boldsymbol{S}_2$ generate $\boldsymbol{S}$, so $S_1' \subseteq S' \cup S_2$ and thus $C \in S'$ by Condition (b). This proves that Condition (B) is satisfied. We conclude that $\tau_1 \cup \tau_2$ has shape $\boldsymbol{S}$ as claimed. □

Lemma 3. *Let v be a node of T with children v_1 and v_2, let $\boldsymbol{S} = (S, S')$ be a shape for v, and let $\tau \in N_v(\boldsymbol{S})$. Let τ_1 and τ_2 denote the restrictions of τ to X_{v_1} and X_{v_2}, respectively. There is a unique pair of shapes $\boldsymbol{S}_1$ for v_1 and $\boldsymbol{S}_2$ for v_2 generating $\boldsymbol{S}$ such that $\tau_1 \in N_{v_1}(\boldsymbol{S}_1)$ and $\tau_2 \in N_{v_2}(\boldsymbol{S}_2)$.*

Proof. Let $S_1 = \overline{F_{v_1}}(\tau_1)$ and $S_2 = \overline{F_{v_2}}(\tau_2)$. Let $\tau' : \overline{X_t} \to \{0, 1\}$ be the assignment such that $S' = F_v(\tau)$. Then the sets $S_1' = (S' \cup S_2) \cap F_{v_1}$ and $S_2' = (S' \cup S_1) \cap F_{v_2}$ are the projections of the assignments $\tau' \cup \tau_2$ and $\tau' \cup \tau_1$ onto F_{v_1} and F_{v_2}, respectively. It follows that $\boldsymbol{S}_1 = (S_1, S_1')$ is a shape for v_1 and that $\boldsymbol{S}_2 = (S_2, S_2')$ is a shape for v_2. We verify that τ_1 has shape $\boldsymbol{S}_1$. Condition (A) is satisfied by construction. To see that Condition (B) is satisfied as well, let $C \in F_{v_1}$ and suppose C is not satisfied by τ_1. There are two cases. If τ_2 does not satisfy C either then $\tau = \tau_1 \cup \tau_2$ does not satisfy C and $C \in S'$ since τ has shape $\boldsymbol{S}$. Otherwise we have $C \in \overline{F_{v_2}}(\tau_2)$, that is, $C \in S_2$. In either case we have $C \in S_1'$ by choice of S_1'. The proof that τ_2 has shape $\boldsymbol{S}_2$ is symmetric.

Let $R_1 = (R_1, R_1')$ and $R_2 = (R_2, R_2')$ be shapes for v_1 and v_2 such that R_1 and R_2 generate S and such that $\tau_1 \in N_{v_1}(R_1)$ and $\tau_2 \in N_{v_2}(R_2)$. We have $R_1 = \overline{F_{v_1}}(\tau_1) = S_1$ and $R_2 = \overline{F_{v_2}}(\tau_2) = S_2$ by Condition (A). As R_1 and R_2 generate S, we further have $R_1' = (S' \cup R_2) \cap F_{v_1}$ and $R_2' = (S' \cup R_1) \cap F_{v_2}$. That is, $R_1' = S_1'$ and $R_2' = S_2'$, so $R_1 = S_1$ and $R_2 = S_2$.   □

3.3   Constructing a Structured d-DNNF

Lemma 1 can be turned into a recurrence for determining the model count of F by dynamic programming [16,17]. It can also be used to construct a structured d-DNNF for F.

To simplify matters, for the remainder of this subsection let F be an arbitrary, but fixed, formula, and let $T = (T, \delta)$ be an arbitrary, but fixed, branch decomposition of $F \cup var(F)$. Starting at the leaves of T, we are going to construct a DNNF $\varphi_v(S)$ for each node v and each shape S for v. For a leaf node v of T, we have to consider two cases:

1. Suppose $\delta(v) = x$ for a variable x of F. For $\ell \in \{x, \neg x\}$, let τ_ℓ denote the assignment $\tau_\ell : \{x\} \to \{0, 1\}$ such that $\tau(\ell) = 1$. The pairs $S_x = (F(\tau_x), \emptyset)$ and $S_{\neg x} = (F(\tau_{\neg x}), \emptyset)$ are the only shapes for v, and $N_v(S_x) = \{\tau_x\}$ as well as $N_v(S_{\neg x}) = \{\tau_{\neg x}\}$. Accordingly, we let $\varphi_v(S_x) \equiv x$ and $\varphi_v(S_{\neg x}) \equiv \neg x$.
2. Let $\delta(v) = C$ for a clause $C \in F$. The pairs $S_\perp = (\emptyset, \emptyset)$ and $S_\top = (\emptyset, \{C\})$ are the only shapes for v. Since $X_v = \emptyset$ it suffices to determine whether the empty assignment $\varepsilon : \emptyset \to \{0, 1\}$ has one of these shapes. Because the empty assignment does not satisfy any clause we get $N_v(S_\top) = \{\varepsilon\}$ and $N_v(S_\perp) = \emptyset$, so we define $\varphi_v(S_\perp) \equiv 0$ and $\varphi_v(S_\top) \equiv 1$.

Let v be an inner node of T with children v_1 and v_2, and assume we have constructed $\varphi_{v_1}(S_1)$ for each shape S_1 for v_1 and $\varphi_{v_2}(S_2)$ for each shape S_2 for v_2. Let S be a shape for v and let G denote the set of pairs of shapes S_1 for v_1 and S_2 for v_2 that generate S. We construct $\varphi_v(S)$ as

$$\varphi_v(S) \equiv \bigvee_{(S_1, S_2) \in G} \varphi_{v_1}(S_1) \wedge \varphi_{v_2}(S_2). \tag{1}$$

That is, we create an AND node conjoining every pair $\varphi_{v_1}(S_1)$ and $\varphi_{v_2}(S_2)$ such that S_1 and S_2 generate S, and then add an OR node that has an incoming edge from each AND node thus created. We assume that the resulting DNNF has been simplified by propagating constants.

Lemma 4. *For each node v of T and shape S for v, $\varphi_v(S)$ is a d-DNNF such that $var(\varphi_v(S)) \subseteq X_v$ and such that an assignment $\tau : X_v \to \{0, 1\}$ satisfies $\varphi_v(S)$ if, and only if, $\tau \in N_v(S)$.*

Proof. It is easy to check that the statement holds for each leaf node v of T. Let v be an inner node and suppose the statement holds for its children v_1 and v_2. Let S be a shape for v. By assumption, $var(\varphi_{v_1}(S_1)) \subseteq X_{v_1}$ and $var(\varphi_{v_2}(S_2)) \subseteq X_{v_2}$

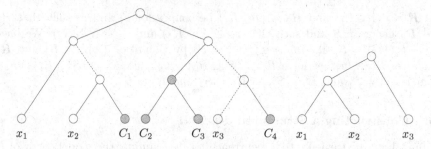

Fig. 1. The tree on the left is a branch decomposition of a formula $F = \{C_1, C_2, C_3, C_4\}$ with $var(F) = \{x_1, x_2, x_3\}$. To obtain the vtree on the right, we first delete each leaf node associated with a clause, as well inner nodes turned into leaf nodes by these deletions (the corresponding vertices are shown in grey). The resulting tree is turned into a binary tree by contracting edges incident to nodes of degree two (these edges are represented by dashed lines).

for every shape S_1 for v_1 and every shape S_2 for v_2. We have $X_v = X_{v_1} \cup X_{v_2}$ and since X_{v_1} and X_{v_2} are disjoint it follows that $\varphi_v(S)$ is a DNNF satisfying $var(\varphi_v(S)) \subseteq X_v$. Let $\tau : X_v \to \{0, 1\}$ be a satisfying assignment of $\varphi_v(S)$, and let τ_1 and τ_2 denote the restrictions of τ to X_{v_1} and X_{v_2}, respectively. There is a pair of shapes S_1 and S_2 generating S such that τ satisfies the disjunct $\varphi_{v_1}(S_1) \wedge \varphi_{v_2}(S_2)$. By assumption, the lemma holds for v_1 and v_2. In particular, $var(\varphi_{v_1}(S_1)) \subseteq X_{v_1}$ and $var(\varphi_{v_2}(S_2)) \subseteq X_{v_2}$, so τ_1 satisfies $\varphi_{v_1}(S_1)$ and τ_2 satisfies $\varphi_{v_2}(S_2)$, which in turn implies that $\tau_1 \in N_{v_1}(S_1)$ and $\tau_2 \in N_{v_2}(S_2)$. It now follows from Lemma 1 that τ has shape S. In addition to that, Lemma 1 tells us that (S_1, S_2) is the unique pair of shapes generating S such that τ_1 has shape S_1 and τ_2 has shape S_2. Thus $\varphi_{v_1}(S_1) \wedge \varphi_{v_2}(S_2)$ is the unique disjunct satisfied by τ. By assumption, $\varphi_{v_1}(S_1')$ and $\varphi_{v_2}(S_2')$ are deterministic DNNFs for each shape S_1' for v_1 and S_2' for v_2, so $\varphi_v(S)$ is deterministic as well. Now let $\tau : X_v \to \{0, 1\}$ be an assignment of shape S, and let τ_1 and τ_2 denote its restrictions to X_{v_1} and X_{v_2}, respectively. By Lemma 1, there has to be a pair (S_1, S_2) of shapes S_1 for v_1 and S_2 for v_2 generating S such that $\tau_1 \in N_{v_1}(S_1)$ and $\tau_2 \in N_{v_2}(S_2)$. It follows from our assumption that the lemma holds for v_1 and v_2 that τ_1 satisfies $\varphi_{v_1}(S_1)$ and that τ_2 satisfies $\varphi_{v_2}(S_2)$. Thus τ satisfies $\varphi_{v_1}(S_1) \wedge \varphi_{v_2}(S_2)$ and $\varphi_v(S)$. $\qquad\square$

To show that $\varphi_v(S)$ is a *structured* DNNF, we have to provide a vtree respected by $\varphi_v(S)$ [13]. A vtree is a binary tree whose leaves are in one-to-one correspondence with a set of variables. We will think of a vtree simply as a branch decomposition of a set X of variables. A DNNF φ *respects* a vtree (T, δ) if each AND node v of φ has exactly two children and furthermore satisfies the following property: let v_1 and v_2 be the children of v in T, and let φ_1 and φ_2 denote the sub-DNNFs of φ rooted at v_1 and v_2, respectively; then there is a node t of T with children t_1 and t_2 such that the sub-DNNFs satisfy $var(\varphi_1) \subseteq \delta(L(T_{t_1}))$ and $var(\varphi_2) \subseteq \delta(L(T_{t_2}))$. Here, $L(T_{t_i})$ denotes the set of leaves in the subtree T_{t_i}, for $i \in \{1, 2\}$.

For a node v of T, let $vtree(\boldsymbol{T}, v) = (T', \delta')$, where T' is the tree obtained from the subtree T_v by deleting all leaves w such that $\delta(w) \in F$, followed—if necessary—by a sequence of operations to make the resulting tree binary, and δ' is the restriction of δ to leaves of T'. Verify that $vtree(\boldsymbol{T}, v)$ is a branch decomposition of X_v and hence a vtree. We illustrate this construction in Figure 1.

Lemma 5. *For each node v of T and shape $\boldsymbol{S}$ for v, the DNNF $\varphi_v(\boldsymbol{S})$ respects $vtree(\boldsymbol{T}, v)$.*

Proof. The lemma trivially holds for each leaf node v of T and shape $\boldsymbol{S}$ for v, as $\varphi_v(\boldsymbol{S})$ does not contain any AND nodes. Let v be an inner node of T with children v_1 and v_2, and assume the lemma holds for v_1 and v_2 and their respective shapes. Let $\boldsymbol{S}$ be a shape for v. By construction, each AND node introduced in $\varphi_v(\boldsymbol{S})$ computes a conjunction $\varphi_{v_1}(\boldsymbol{S}_1) \wedge \varphi_{v_2}(\boldsymbol{S}_2)$, where $\boldsymbol{S}_1$ and $\boldsymbol{S}_2$ are shapes for v_1 and v_2, respectively, that generate $\boldsymbol{S}$. Since we assume $\varphi_v(\boldsymbol{S})$ to be simplified, both X_{v_1} and X_{v_2} have to be nonempty: otherwise, one of the conjuncts $\varphi_{v_i}(\boldsymbol{S}_i)$ for $i \in \{1, 2\}$ would satisfy $var(\varphi_{v_i}(\boldsymbol{S}_i)) = \emptyset$ by Lemma 4 and would have been simplified to a constant, which in turn would have been propagated through the AND node. Let $vtree(\boldsymbol{T}, v) = (T', \delta')$, let $vtree(\boldsymbol{T}, v_1) = (T_1, \delta_1)$, and let $vtree(\boldsymbol{T}, v_2) = (T_2, \delta_2)$. As both X_{v_1} and X_{v_2} are nonempty, T' is a binary tree whose principal subtrees are T_1 and T_2. By Lemma 4, the conjuncts satisfy $var(\varphi_{v_1}(\boldsymbol{S}_1)) \subseteq X_{v_1}$ and $var(\varphi_{v_2}(\boldsymbol{S}_2)) \subseteq X_{v_2}$. In combination with the assumption that the DNNF $\varphi_{v_i}(\boldsymbol{S}'_i)$ respects $vtree(\boldsymbol{T}, v_i)$ for each $i \in \{1, 2\}$ and shape $\boldsymbol{S}'_i$ for v_i, this implies that $\varphi_v(\boldsymbol{S})$ respects $vtree(\boldsymbol{T}, v)$. $\qquad\square$

Let r denote the root of T and let $\boldsymbol{\emptyset} = (\emptyset, \emptyset)$. We now prove that our construction yields a structured d-DNNF representation of F.

Lemma 6. *The pair $\boldsymbol{\emptyset}$ is the only shape for r and $\varphi_r(\boldsymbol{\emptyset})$ is a structured d-DNNF computing F.*

Proof. The first part follows from the fact that $X_r = var(F)$ and $F_r = F$, so that $\overline{X_r} = \emptyset$ and $\overline{F_r} = \emptyset$. By Lemma 4 and Lemma 5, $\varphi_r(\boldsymbol{\emptyset})$ is a structured d-DNNF such that an assignment $\tau : var(F) \to \{0, 1\}$ satisfies $\varphi_r(\boldsymbol{\emptyset})$ if, and only if, $\tau \in N_r(\boldsymbol{\emptyset})$. By Condition (B), an assignment $\tau : var(F) \to \{0, 1\}$ has shape $\boldsymbol{\emptyset}$ if, and only if, $F(\tau) \cup \emptyset = F$. That is, $N_r(\boldsymbol{\emptyset})$ is the set of satisfying assignments of F. $\qquad\square$

Let n be the number of variables of F, let m be the number of clauses in F, and let k denote the PS-width of $\boldsymbol{T}$. The size of the structured d-DNNF constructed for F can be bounded as follows.

Lemma 7. *The DNNF $\varphi_r(\boldsymbol{\emptyset})$ has size at most $7k^3(n + m)$.*

Proof. We can assume without loss of generality that T contains at least one inner node. Let v be an inner node of T with children v_1 and v_2. Consider the DNNFs $\varphi_{v_1}(\boldsymbol{S}_1)$ for shapes $\boldsymbol{S}_1$ for v_1 and $\varphi_{v_2}(\boldsymbol{S}_2)$ for shapes $\boldsymbol{S}_2$ for v_2. We claim that all DNNFs $\varphi_v(\boldsymbol{S})$ for shapes $\boldsymbol{S}$ of v can be constructed from these DNNFs by introducing at most $5k^3$ new nodes and edges. If S is a shape for v and S_1 and S_2 are shapes for v_1 and v_2 that generate S, we have to introduce an AND node and two edges to construct the DNNF computing $\varphi_{v_1}(S_1) \wedge \varphi_{v_2}(S_2)$, as well an edge from this AND node to the OR node that will eventually compute $\varphi_v(S)$. In the worst case, we have to create this OR node first. In total, we have to introduce at most 5 nodes and edges for each triple (S, S_1, S_2) of shapes such that S_1 and S_2 generate S. How many such triples are there? For any three projections $S_1 \in proj(\overline{F_{v_1}}, X_{v_1})$, $S_2 \in proj(\overline{F_{v_2}}, X_{v_2})$, and $S' \in proj(F_v, \overline{X_v})$, the projections $S_1' \in proj(F_{v_1}, \overline{X_{v_1}})$, $S_2' \in proj(F_{v_2}, \overline{X_{v_2}})$, and $S \in proj(\overline{F_v}, X_v)$ such that $S_1 = (S_1, S_1')$ and $S_2 = (S_2, S_2')$ generate $S = (S, S')$ is uniquely determined. As there are at most k^3 such projections, we have to introduce at most $5k^3$ nodes and edges. The tree T has exactly $n + m - 1$ inner nodes, so we need at most $5k^3(n + m)$ nodes and edges to construct the DNNF $\varphi_r(\emptyset)$ from the DNNFs constructed for leaves of T. For each leaf node there at most two DNNFs consisting of a single node and there are $n + m$ leaves, so we require at most $7k^3(n + m)$ nodes and edges in total.   □

Since we did not make any assumptions about the formula F and the branch decomposition T, Lemma 6 and Lemma 7 yield the following result.

Theorem 1. *A CNF formula with n variables, m clauses, and PS-width k can be compiled into a structured d-DNNF of size $O(k^3(n + m))$.*

The above construction leads to an algorithm which, given a formula F and a branch decomposition T of $F \cup var(F)$, computes a structured d-DNNF representation of F. The pseudocode listed as Algorithm 1 provides the outlines of this procedure.[3] Using an efficient method for computing the set of shapes for each node during the initialization phase (for details, see Saether et. al. [16]), this algorithm can be made to run in time $O(k^3 m(n + m))$, where n is the number of variables of F, m is the number of clauses of F, and k is the PS-width of T.

4   Corollaries

Theorem 1 allows us to derive compilation results for CNF formulas based on structural properties of their incidence graphs, namely *treewidth*, *directed clique-width*, and *clique-width* [8].

We first consider treewidth. A *tree decomposition* of a graph $G = (V, E)$ is a pair $(T, (B_t)_{t \in V(T)})$ where T is a tree and $(B_t)_{t \in V(T)}$ is a family of subsets of V (called "bags") such that:

1. For every vertex $v \in V$, the set $\{t \in V(T) \mid v \in B_t\}$ is non-empty and connected in T.
2. For every edge $uv \in E$, there is a $t \in V(T)$ such that $u, v \in B_t$.

---

[3] To enhance readability, we suppress double brackets around shapes, writing, for instance, $\varphi_v(S, S')$ instead of $\varphi_v((S, S'))$.

Algorithm 1. Compiling CNFs into structured d-DNNFs.

Input: a CNF F and a branch decomposition (T, δ) of $F \cup var(F)$
Output: a structured d-DNNF computing F
// initialization, precomputing shapes
1 **for** v in T
2   compute $proj(\overline{F_v}, X_v)$ and $proj(F_v, \overline{X_v})$
// compilation, leaf nodes
3 **for** v in $L(T)$
4   **if** $\delta(v)$ in $var(F)$
5         $x = \delta(v)$
6         $S_x = \{ C \in F \mid x \in C \}$
7         $S_{\neg x} = \{ C \in F \mid \neg x \in C \}$
8         $\varphi_v(S_x, \emptyset) = x$
9         $\varphi_v(S_{\neg x}, \emptyset) = \neg x$
10   **else**
11         $C = \delta(v)$
12         $\varphi_v(\emptyset, \{C\}) = 1$
13         $\varphi_v(\emptyset, \emptyset) = 0$
14   mark v as processed
// compilation, inner nodes
15 **while** T contains an unprocessed node
16   let v be an unprocessed node whose children v_1 and v_2 have been processed
17   **for** (S_1, S_2, S') in $proj(\overline{F_{v_1}}, X_{v_1}) \times proj(\overline{F_{v_2}}, X_{v_2}) \times proj(F_v, \overline{X_v})$
18         $S = S_1 \cup S_2$
19         $S'_1 = S' \cup S_2$
20         $S'_2 = S' \cup S_1$
21   **if** $\varphi_v(S, S')$ has not been created
            // initialize $\varphi_v(S, S')$
22             $\varphi_v(S, S') = 0$
23         $\varphi_v(S, S') = \varphi_v(S, S') \vee (\varphi_{v_1}(S_1, S'_1) \wedge \varphi_{v_2}(S_2, S'_2))$
24   propagate constants in $\varphi_v(S, S')$
25   mark v as processed
26 **return** $\varphi_r(\emptyset, \emptyset)$

The *width* of a tree decomposition $(T, (B_t)_{t \in V(T)})$ is the maximum size of a bag minus one, and the treewidth of G is the minimum of width attained over all tree decompositions of G. The incidence treewidth of a formula F defined as the treewidth of its incidence graph $I(F)$.

Proposition 1. *A formula of incidence treewidth k has PS-width at most 2^{k+1}.*

Proof (Sketch). Let F be a formula and let $\boldsymbol{T} = (T, (B_t)_{t \in V(T)})$ be a tree decomposition of its incidence graph such that $\boldsymbol{T}$ has width k. We can assume without loss of generality that T is binary (this can be achieved by copying nodes and bags of T). We construct a branch decomposition $\boldsymbol{T'} = (T', \delta)$ of $F \cup var(F)$ as follows: for every variable $x \in var(F)$ we introduce a vertex v_x and connect it to the node t of T such that t is closest to the root among nodes whose associated bags contain the variable x. For each clause $C \in F$ we add a vertex v_C in an analogous way. The result is a tree where every vertex has at most three neighbors. We obtain the desired branch decomposition T' by iteratively deleting all leaves not among the nodes v_x and v_C introduced in the first step and contracting paths to edges. We now claim the following.

- For every $v \in T'$, there are at most $k+1$ clauses in $\overline{F_v}$ that contain a variable from X_v. Each projection $\overline{F_v}(\tau)$ of an assignment $\tau : X_v \to \{0,1\}$ onto $\overline{F_v}$ is a subset of these clauses, so $|proj(\overline{F_v}, X_v)| \leq 2^{k+1}$.
- Symmetrically, for every $v \in T'$, there are at most $k+1$ variables in $\overline{X_v}$ that occur in a clause $C \in F_v$. It follows that $|proj(F_v, \overline{X_v})| \leq 2^{k+1}$ because there are at most 2^{k+1} assignments $\tau : \overline{X_v} \to \{0,1\}$.

That is, T' has PS-width at most 2^{k+1}.   □

Combining Proposition 1 and Theorem 1, we obtain the following result.

Corollary 1. *A formula with n variables, m clauses, and incidence treewidth k can be compiled into a structured deterministic DNNF of size $O(8^k(n+m))$.*

Clique-width is a generalization of treewidth defined as follows. A k-*graph* is a pair (G, λ) consisting of a graph $G = (V(G), E(G))$ and a mapping $\lambda : V(G) \to \{1, \ldots, k\}$. We call $\lambda(v)$ the *label* of vertex v. We define the following operations for constructing k-graphs:

(i) For $i \in \{1, \ldots, k\}$, we write $\bullet_i$ for the k-graph (G, λ) where G contains a single isolated vertex v and $\lambda(v) = i$.

(ii) Let $i, j \in \{1, \ldots, k\}$ such that $i \neq j$, and let $G = (G, \lambda)$ be a k-graph. Then $\rho_{i \to j}(G) = (G, \lambda')$, where $\lambda'(v) = \lambda(v)$ if $\lambda(v) \neq i$, and $\lambda'(v) = j$ if $\lambda(v) = i$, for each vertex $v \in V(G)$.

(iii) Let $i, j \in \{1, \ldots, k\}$ such that $i \neq j$, and let $G = (G, \lambda)$ be a k-graph. Then $\eta_{i,j}(G) = (G', \lambda)$, where G' is the graph such that $V(G') = V(G)$, and such that $E(G') = E(G) \cup \{ vw \mid \lambda(v) = i, \lambda(w) = j \}$. That is, G' is obtained from G by adding an edge between any two vertices v and w such that v is labelled i and w is labelled j.

(iv) We write $G \sqcup G'$ to denote the disjoint union of two k-graphs $G = (G, \lambda)$ and $G' = (G', \lambda')$, that is, $G \sqcup G' = (G \sqcup G', \lambda \cup \lambda')$.

A k-*expression* is a well-formed expression using the symbols $\bullet_i$ (constant), $\rho_{i \to j}$, $\eta_{i,j}$ (both unary), and $\sqcup$ (binary). The k-graph associated with a k-expression t (and any k-graph isomorphic to it) is called the *value* of t. If a k-expression t has the value (G, λ) we say that t is a k-*expression of* G. The *clique-width* of a graph G is the minimum k such that there is a k-expression of G.

A formula with m clauses and incidence clique-width k has PS-width at most m^k [16]. In combination with Theorem 1, this gives the following result.[4]

---

[4] By the same token, the statement could also be proved for other structural parameters, like Boolean width, rank-width, or MIM-width [16,18].

Corollary 2. *A formula with* n *variables, m clauses, and incidence clique-width* k *can be compiled into a structured d-DNNF of size* $O(m^{3k}(n+m))$.

The *directed clique-width* of a directed graph is defined analogously to clique-width. If F is a CNF formula with a directed incidence graph[5] of directed clique-width k, then F has PS-width at most 4^k [2]. By combining this fact and Theorem 1, we obtain the following.

Corollary 3. *A formula with* n *variables, m clauses, and directed incidence clique-width* k *can be compiled into a structured d-DNNF of size* $O(64^k(n+m))$.

5   A Lower Bound for Clique-Width

Note that there is a qualitative difference in the size bounds of Corollary 1 and Corollary 3 on the one hand, and Corollary 2 on the other hand. If k is the value of a structural parameter of a formula with n variables and m clauses, then the former bound has the shape $2^{O(k)}(n+m)$, whereas the latter bound has the shape $m^{O(k)}(n+m)$. For small values of k and large values of n and m, bounds of the form $2^{O(k)}(n+m)$ are preferable to bounds of the form $m^{O(k)}(n+m)$.

In this section we will give evidence that the size bound of Corollary 2 is optimal qualitatively, so that the qualitative difference discussed above is unavoidable. To this end, we introduce the following notions from parameterized complexity.

A *parameterized problem* is a pair (P, κ) where P is a decision problem and $\kappa : \{0,1\} \to \mathbb{N}$ is a computable function associating every instance of P with a *parameter*. A parameterized problem (P, κ) is in the complexity class FPT, or *fixed-parameter tractable*, if there is an algorithm solving P in time $f(\kappa(x))|x|^c$ for every instance x, where $f : \mathbb{N} \to \mathbb{N}$ is a computable function and c is a constant. A parameterized problem (P, κ) is in the complexity class FPT/ppoly if there is an algorithm that, given an instance x of P and $f'(\kappa(x))|x|^{c'}$ advice bits, correctly solves x in time $f(\kappa(x))|x|^c$ where $f : \mathbb{N} \to \mathbb{N}$ and $f' : \mathbb{N} \to \mathbb{N}$ are computable functions and c, c' are constants [3].[6] Clearly, FPT is contained in FPT/ppoly.

Theorem 2. *Assume that* $\mathrm{W}[1] \not\subseteq \mathrm{FPT/ppoly}$. *Then there is no computable function* $f : \mathbb{N} \to \mathbb{N}$ *and constant* c *such that for every CNF-formula* F *with* n *variables,* m *clauses and clique-width* k *there is a DNNF* D *such that* F *and* D *compute the same function and the size of* D *is at most* $f(k)(n+m)^c$.

The assumption $\mathrm{W}[1] \not\subseteq \mathrm{FPT/ppoly}$ is a parameterized analogue of the assumption $\mathrm{NP} \not\subseteq \mathrm{P/poly}$ in classical complexity; the latter is known to hold

---

[5] The directed incidence graph is an orientation of the incidence graph encoding positive and negative occurrences of variables.

[6] Note that the advice given to the algorithm may only depend on $\kappa(x)$ and $|x|$ but not directly on x. Thus for two instances x and x' with $\kappa(x) = \kappa(x')$ and $|x| = |x'|$ the algorithm is given *the same* advice string.

unless the Polynomial Hierarchy collapses to the second level [9]. Although $W[1] \not\subseteq FPT/poly$ is a stronger assumption than $W[1] \not\subseteq FPT$, which is standard in parameterized complexity, we still consider it plausible, since it is not clear how nonuniformity should help in solving $W[1]$-hard problems.

Proof (of Theorem 2). We use a reduction from partitioned clique to the satisfiability problem presented in [10]. The partitioned clique problem is to decide, given a k-partite graph G whose color classes all have the same size, whether G has a clique of size k, i.e. containing a vertex from every color class. Here, k is the parameter of the problem instance. The partitioned clique problem is $W[1]$-complete under fixed-parameter tractable many-one reductions; see [7] for more details.

In [10], it is shown (Theorem 4 and Corollary 1) that, given a k-partite graph $G = (V_1, \ldots, V_k, E)$ with the same number of vertices in each color class, one can construct a CNF formula F_G such that the incidence graph of F_G has clique-width at most $k+4$ and the size of F_G is polynomial in the size of G, and such that the formula F_G has a satisfying assignment if and only if G has a clique of size k. If $V_i = \{v_1^i, \ldots, v_n^i\}$, the formula contains the variables V_i for each $1 \leq i \leq k$. For each pair (u, v) such that $v \in V_i$, $u \in V_j$ $(i \neq j)$, and such that $uv \notin E$, the formula F_G contains the clause $C_{u,v} = \{\neg u, \neg v\} \cup \{w \mid w \in (V_i \cup V_j) \setminus \{u, v\}\}$. The idea is that the variables v_j^i mapped to 1 by a satisfying assignment of F_G correspond to a partitioned clique of G. The clauses $C_{u,v}$ are padded with the remaining variables in order to keep the clique-width of F_G's incidence graph small; to make sure that a clause $C_{u,v}$ cannot be satisfied by these extra variables when u and v are both assigned to 1, a "selection gadget" is attached to each color class V_i. This gadget (which we will not describe here) guarantees that each satisfying assignment of F_G maps exactly one of the variables in each color class V_i to 1.

We modify this construction in the following way. Let $G_n^k = (V_1, \ldots, V_k, \emptyset)$ denote the empty k-partite graph with n vertices in each color class. The formula $F_{G_n^k}$ contains a clause $C_{u,v}$ for each pair of variables $u \in V_i$, $v \in V_j$ $(i \neq j)$, as G_n^k does not contain any edges. Starting from $F_{G_n^k}$, we construct a new formula $F_{k,n}$ by adding a distinct relaxation variable $x_{u,v}$ to each clause $C_{u,v}$. These variables allow us to "switch clauses on and off" as needed. Adding the variable $x_{u,v}$ to the clause $C_{u,v}$ corresponds to adding a vertex $x_{u,v}$ and a "dangling edge" $\{x_{u,v}, C_{u,v}\}$ to the incidence graph of $F_{G_n^k}$. This can be done for each clause $C_{u,v}$ while increasing the clique-width of the incidence graph by at most 3, as can be seen from the following argument. Consider a $(k + 4)$-expression t of the incidence graph $I(F_{G_n^k})$ of $F_{G_n^k}$. For each clause $C_{u,v}$, the expression t contains a subexpression $\bullet_j$ that introduces the vertex $C_{u,v}$ with some label $j \in \{1, \ldots, k + 4\}$. Using fresh labels, we replace each such subexpression with the expression $\rho_{k+6 \to k+7}(\rho_{k+5 \to j}(\eta_{k+5,k+6}(\bullet_{k+5} \sqcup \bullet_{k+6})))$. That is, instead of introducing $C_{u,v}$ with label j, we first introduce it with label $k + 5$, along with the vertex $x_{u,v}$, which we label with $k + 6$. We then create the edge $\{C_{u,v}, x_{u,v}\}$ and before relabelling both vertices: the vertex $C_{u,v}$ gets its original label j, while vertex $x_{u,v}$ is assigned an auxiliary label $k + 7$ to make sure it does not become the endpoint of further edges. The resulting expression is a $(k + 7)$-expression of $I(F_{k,n})$.

Given a k-partite graph G with n vertices in each color class, the formula F_G can be obtained from $F_{k,n}$ by assigning the relaxation variables: simply set $x_{u,v}$ to 1 if uv is an edge of G, and to 0 if uv is not an edge of G.

Now assume, by way of contradiction, that there is a function $f : \mathbb{N} \to \mathbb{N}$ and a constant c such that for every CNF formula F with n variables, m clauses and clique-width k, there is a DNNF D such that F and D compute the same function and the size of D is at most $f(k)(n + m)^c$. Then in particular, there is a constant c' such that for every n and k, there is a DNNF $D_{k,n}$ of size $f(k)n^{c'}$ that computes $F_{k,n}$.

We now describe a non-uniform algorithm for the partitioned clique problem: Given a k-partite graph G with n vertices in each color class, the advice string is a desciption of $D_{k,n}$. The algorithm first sets the relaxation variables so as to get a DNNF D_G computing F_G. This can be done in linear time [4]. The graph G has a k-clique if and only if D_G is satisfiable. Since checking satisfiability of DNNF can be done in linear time [4], this gives the desired algorithm. It follows that the partitioned clique problem, and hence every problem in W[1], is in FPT/ppoly, which is a contradiction to the assumption of the lemma. We conclude that DNNFs of the desired size cannot exist if W[1] $\not\subseteq$ FPT/ppoly. □

6 Conclusion

We demonstrated how dynamic programming algorithms for #SAT [16,17] can be modified to construct structured d-DNNF representations of CNF formulas. This observation allowed us to prove an upper bound on the size of structured d-DNNF representations of CNFs in terms of a parameter called PS-width [16]. We showed that this bound translates into new upper bounds in terms of parameters such as the treewidth and the clique-width of the incidence graph. We also provided evidence that the upper bound in terms of incidence clique-width cannot be substantially improved, even for general DNNFs.

The d-DNNFs generated by our compilation algorithm do not necessarily fall into the more restricted subclass of *decision DNNFs*. We do not know whether this is an artifact of our methods or due to an inherent limitation of decision DNNFs. In particular, we would like to know if CNF formulas can be compiled into decision DNNFs of size exponential only in their incidence treewidth. Finally, it would be interesting to compare PS-width to other recently proposed width measures of CNF-formulas, such as CV-width [11] and decision-width [12].

Acknowledgments. The first and fourth author were supported by the FWF Austrian Science Fund (P26200). The second author was supported by the ANR Blanc AGGREG reference ANR-14-CE25-0017-01. The third author was supported by the ANR Blanc International ALCOCLAN.

References

1. Bova, S., Capelli, F., Mengel, S., Slivovsky, F.: Expander CNFs have exponential DNNF size. CoRR, abs/1411.1995 (2014)
2. Brault-Baron, J., Capelli, F., Mengel, S.: Understanding model counting for beta-acyclic CNF-formulas. In: Mayr, E.W., Ollinger, N. (eds.) 32nd International Symposium on Theoretical Aspects of Computer Science, STACS 2015. LIPIcs, vol. 30, pp. 143–156. Schloss Dagstuhl - Leibniz-Zentrum fuer Informatik (2015)
3. Chen, H.: Parameterized compilability. In: Proceedings of the Nineteenth International Joint Conference on Artificial Intelligence, IJCAI 2005, Edinburgh, Scotland, UK, July 30-August 5, 2005, pp. 412–417 (2005)
4. Darwiche, A.: Decomposable negation normal form. J. ACM **48**(4), 608–647 (2001)
5. Darwiche, A.: On the tractable counting of theory models and its application to truth maintenance and belief revision. Journal of Applied Non-Classical Logics **11**(1–2), 11–34 (2001)
6. Darwiche, A., Marquis, P.: A Knowledge Compilation Map. J. Artif. Intell. Res. (JAIR) **17**, 229–264 (2002)
7. Flum, J., Grohe, M.: Parameterized Complexity Theory. Springer-Verlag, New York (2006)
8. Hlinený, P., Oum, S., Seese, D., Gottlob, G.: Width parameters beyond tree-width and their applications. Comput. J. **51**(3), 326–362 (2008)
9. Karp, R.M., Lipton, R.J.: Some connections between nonuniform and uniform complexity classes. In: Proceedings of STOC 1980, pp. 302–309. ACM (1980)
10. Ordyniak, S., Paulusma, D., Szeider, S.: Satisfiability of acyclic and almost acyclic CNF formulas. Theoretical Computer Science **481**, 85–99 (2013)
11. Oztok, U., Darwiche, A.: CV-width: a new complexity parameter for CNFs. In: 21st European Conference on Artificial Intelligence, ECAI 2014, pp. 675–680 (2014)
12. Oztok, U., Darwiche, A.: On compiling CNF into decision-DNNF. In: O'Sullivan, B. (ed.) CP 2014. LNCS, vol. 8656, pp. 42–57. Springer, Heidelberg (2014)
13. Pipatsrisawat, K., Darwiche, A.: New compilation languages based on structured decomposability. In: Proceedings of the 23rd National Conference on Artificial Intelligence, AAAI 2008, vol. 1, pp. 517–522. AAAI Press (2008)
14. Pipatsrisawat, K., Darwiche, A.: Top-down algorithms for constructing structured DNNF: theoretical and practical implications. In: Proceedings of 19th European Conference on Artificial Intelligence, ECAI 2010, Lisbon, Portugal, August 16–20, 2010. Frontiers in Artificial Intelligence and Applications, vol. 215, pp. 3–8. IOS Press (2010)
15. Razgon, I., Petke, J.: Cliquewidth and knowledge compilation. In: Järvisalo, M., Van Gelder, A. (eds.) SAT 2013. LNCS, vol. 7962, pp. 335–350. Springer, Heidelberg (2013)
16. Sæther, S.H., Telle, J.A., Vatshelle, M.: Solving MaxSAT and #SAT on structured CNF formulas. In: Sinz, C., Egly, U. (eds.) SAT 2014. LNCS, vol. 8561, pp. 16–31. Springer, Heidelberg (2014)
17. Slivovsky, F., Szeider, S.: Model counting for formulas of bounded Clique-Width. In: Cai, L., Cheng, S.-W., Lam, T.-W. (eds.) Algorithms and Computation. LNCS, vol. 8283, pp. 677–687. Springer, Heidelberg (2013)
18. Vatshelle, M.: New Width Parameters of Graphs. PhD thesis, University of Bergen (2012)

SpySMAC: Automated Configuration and Performance Analysis of SAT Solvers

Stefan Falkner[✉], Marius Lindauer, and Frank Hutter

University of Freiburg, Freiburg Im Breisgau, Germany
{sfalkner,lindauer,fh}@cs.uni-freiburg.de

Abstract. Most modern SAT solvers expose a range of parameters to allow some customization for improving performance on specific types of instances. Performing this customization manually can be challenging and time-consuming, and as a consequence several automated algorithm configuration methods have been developed for this purpose. Although automatic algorithm configuration has already been applied successfully to many different SAT solvers, a comprehensive analysis of the configuration process is usually not readily available to users. Here, we present SpySMAC to address this gap by providing a lightweight and easy-to-use toolbox for (i) automatic configuration of SAT solvers in different settings, (ii) a thorough performance analysis comparing the best found configuration to the default one, and (iii) an assessment of each parameter's importance using the fANOVA framework. To showcase our tool, we apply it to Lingeling and probSAT, two state-of-the-art solvers with very different characteristics.

1 Introduction

Over the last decade, modern SAT solvers have become more and more sophisticated. With this sophistication, usually the number of parameters inside the algorithm increases, and the performance may crucially depend on the setting of these parameters. For example, in the case of the prominent competition-winning solver Lingeling [3], there are 323 parameters which give rise to approximately 10^{1341} possible settings. Exploring these parameter spaces manually is tedious and time-consuming at best. Consequently, automated methods for solving this so-called algorithm configuration problem have been developed to find parameter settings with good performance on a given class of instances [1,11,14,17].

Despite several success stories of automated configuration of SAT solvers [10, 16,19], the reasons why a configuration system chose a certain parameter setting often remain unclear to SAT solver developers. Especially, information about the importance of specific parameter settings is usually not provided. To give more insights into the configuration process of SAT solvers, we present SpySMAC, a lightweight toolbox that combines: (i) the state-of-the-art algorithm configuration system SMAC [11], (ii) automatic evaluation comparing the performance of the default and the optimized configuration across training and test

© Springer International Publishing Switzerland 2015
M. Heule and S. Weaver (Eds.): SAT 2015, LNCS 9340, pp. 215–222, 2015.
DOI: 10.1007/978-3-319-24318-4_16

instances, and (iii) an automatic method to quantify the importance of parameters, fANOVA [13]. In the end, SpySMAC generates a report with relevant tables and figures that summarize the results and reveal details about the configuration process. The standardized input and output of SAT solvers allowed us to design SpySMAC to be very easy to use for both developers and users of SAT solvers.

2   Algorithm Configuration and Analysis

The general task of algorithm configuration consists of determining a well-performing parameter configuration for a given instance set and a performance metric (e.g., runtime). To this end, an algorithm configuration system, or configurator for short, iteratively evaluates different configurations trying to improve the overall performance. After a given time budget is exhausted, the configuration process ends, and the configurator returns the best parameter setting found. The configurator can typically only explore a small fraction of the space of all possible configurations since that space is exponential in the number of parameters and evaluating a single configuration requires running it on multiple instances.

Several different approaches have been taken towards efficiently searching through the configuration space, among others: iterated local search (ParamILS [14]), genetic algorithms (GGA [1]), iterated racing procedures (irace [17]), and model-based Bayesian optimization (SMAC [11]). The Configurable SAT Solver Challenge (CSSC) [15] recently evaluated these configurators (except irace) and achieved significant speed-ups for various solvers and benchmarks. For example, in the CSSC 2014, the PAR10 score[1] of Lingeling [3], clasp [9], and probSAT [2] improved by up to a factor of 5, 108 and 1500, respectively. Across a wide range of solvers on a broad collection of benchmarks, SMAC consistently achieved the largest speedups in the challenge; therefore, we decided to use it in our tool.

SMAC is a sequential model-based algorithm configuration system: it models the performance metric based on finished runs (as a function of the parameter configuration used in each run and characteristics of the instance used), and uses this model to determine the next promising configuration to evaluate. It uses random forests as the underlying model [4], methods from Bayesian optimization [5], and applies mechanisms to evaluate poor configurations only on few instances terminating long runs early [11,14].

To give some insights into the configuration process, different complementary techniques have been developed towards identifying parameter importance. These include forward selection of parameters based on an empirical performance model [12], ablation paths between the default and optimized configuration to identify the important parameter value flips [8], and functional ANOVA (fANOVA) to quantify the importance of parameters in the entire configuration space based on random forests as empirical performance models [13].

---

[1] PAR10 is the penalized average runtime where timeouts are accounted for 10 time the runtime cutoff.

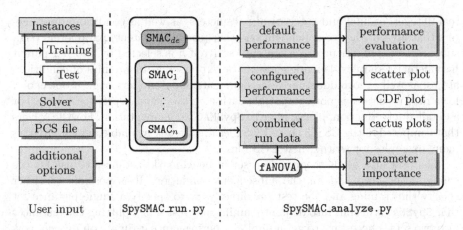

Fig. 1. Schematic of SpySMAC's workflow. The only significant user input (left) is needed to start the configuration phase via SpySMAC_run.py. The n independent SMAC runs search for a better configuration while SMAC$_{de}$ only evaluates the default performance. After all runs have finished, SpySMAC_analyze.py is called to prepare a report showing details about the configuration process and the final configuration.

Each of these methods has some advantages and disadvantages. The forward selection approach is the only of the three that can detect patterns relating instance characteristics to well-performing configurations, but forward selection can be computationally very demanding as it requires the fitting of hundreds of machine learning models. The ablation path is the only one that directly quantifies the performance difference of each changed parameter based on new experiments, but the drawback is that for long ablation paths these experiments can take even longer than the configuration step. Finally, the functional ANOVA approach is computationally efficient (it only requires fitting a single machine learning model and does not require any new algorithm runs) and does not only quantify which parameters are important but also how well each of the parameters' values perform. While we are ultimately planning to support all of these methods, SpySMAC's first version focuses on fANOVA to keep the computational cost of the analysis step low.

3   SpySMAC's Framework

The workflow in SpySMAC is as follows. First, the user provides information about the solver, its parameters and the instance set. Based on that, the configuration phase (running SMAC) and analysis phase (evaluating performance and parameter importance) are conducted. Figure 1 shows a schematic workflow.

The solver specifics provided by the user include the solver binary, and a specification of its parameters and their possible ranges. We use SMAC's *parameter configuration space* (PCS) file format. This format allows the declaration of real, integer, ordinal and categorical parameters, as well as conditional parameters that are only active dependent on other (so-called parent) parameters (e.g., subparameters of a heuristic h are only active if h is selected by another parent parameter). Complex dependencies can be expressed as hierarchies of conditionalities, as well as forbidden partial assignments of parameters (e.g., if one choice for a data structure is not compatible with a certain type of preprocessing). For a detailed introduction, please refer to SpySMAC's documentation. For the solvers that competed in the CSSC, these PCS files are already available, which provides many examples for writing new PCS files.

The user also needs to provide a set of benchmark instances to use for the configuration step and for the subsequent validation. It is possible to either specify the training and the test set directly, or to specify a single instance set that SpySMAC will split into disjoint training and test sets. Splitting the instances into two sets is necessary to get a unbiased performance estimate on unseen, new instances, to avoid over-tuning effects.

The configuration phase consists of multiple, independent SMAC runs (which should take place on the same type of hardware to yield comparable runtimes). Since the configuration of algorithms is a stochastic process and many local minima in the configuration space exist, multiple runs of SMAC can be used to improve the performance of the final configuration found. In principle, one very long run of SMAC would have the same effect, but multiple runs can be effectively parallelized on multi-core systems or compute clusters. We emphasize that we determine the best performing configurations among all SMAC runs based on the training set, not on the test set. This avoids over-tuning effects, again.

After all configuration runs have finished, the separate evaluation step can commence. The user simply executes the analyze script, SpySMAC_analyze.py, to automatically generate a report summarizing the results. This report includes a performance evaluation of the default configuration and the found configuration on the test and training instances[2], scatter plots to visualize the performance on each instance, as well as cumulative distribution function (CDF) and cactus plots to visualize the runtime distributions.

The analysis step can also run fANOVA based on the performance data collected during the configuration to compute parameter importance, producing a table with quantitative results for each parameter and plots to visualize the effect of different parameter values. The fANOVA step is based on a machine learning model fitted on the combined performance data of all solver runs performed in the configuration phase. For many solver runs (i.e., hundreds of thousands runs), even fANOVA's computations can take up to several hours and require several GB memory; therefore, fANOVA is an optional part of the analyzing step.

---

[2] Showing the training and test performance helps to identify over-tuning effects, i.e., the performance improved on the training set but not on test set.

Test Performance

	Default	Configured
Average Runtime	47.79	44.24
PAR10	316.00	276.69
Timeouts	30 / 302	26 / 302

Parameter	Importance
decolim	18.20
cce2wait	5.12
actvlim	4.24
rdpclslim	4.18
redoutvlim	3.92
lftmaxeff	3.30
phaseneginit	3.22
synclsglue	2.54
cardminlen	2.45
restartinit	2.34

Training Performance

	Default	Configured
Average Runtime	40.43	36.79
PAR10	248.13	199.33
Timeouts	23 / 299	18 / 299

Fig. 2. Performance overview for test and training data (left), and the parameter importance determined by fANOVA for Lingeling on CircuitFuzz (right)

4   Spying on Lingeling and probSAT

In this section, we apply SpySMAC to two solvers and three different benchmarks from the Configurable SAT Solver Challenge: Lingeling [3] on an instance set from circuit-based CNF fuzzing (CircuitFuzz [6]), and probSAT [2] on two collections of random satisfiable CNF formulas (7-SAT instances with 90 clauses, 7SAT90-SAT, and 3-SAT instances with 1000 clauses, 3SAT1k-SAT, see [19]).

Figure 2 shows tables generated for the report for the Lingeling example. By comparing the test and training performance, one can see that SMAC found a configuration improving over the standard parameter setting. Even though Lingeling's default already performed very well on this instance set, SMAC was still able to lower the average runtime further, and to reduce the number of timeouts. The table on the right shows the ten most important parameters of Lingeling on this set. The importance score quantifies the effect of varying a parameter across all instantiations of all other parameters. A high value corresponds to large variations in the performance meaning it is important to set this parameter to a specific value (see [13] for more detail).

As an example for probSAT, we used two other scenarios from the CSSC to show the differences our tool can reveal about the configuration on different instance sets. Figure 3 displays the kind of performance plots generated for the report for one of the sets. It clearly shows that configuration successfully improved the performance, reducing mean runtimes for training and test instances by more than a factor of four.

To demonstrate what insights can be gained from the analysis, Figure 4 shows the parameter importance plots for the parameter cb1, the constant probSAT uses to weight the break score in its scoring function. The fANOVA procedure reveals this parameter to be the most important one in both scenarios, but the values differ for the two sets: it should be set high for 7-SAT and low for 3-SAT. We note that this automatically-derived insight is aligned with expert practice for setting probSAT's cb1 parameter. By doing thorough sets of experiments, developers can use our tool to understand the impact of their parameters better, and to try to find ways to adapt parameters based on prior knowledge, such as the clause length in our example here.

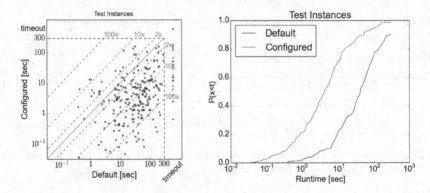

Fig. 3. Example scatter plot (left) and CDF plot (right) from applying `SpySMAC` to `probSAT` on `7SAT90-SAT`. Both show that parameter tuning significantly improves the overall performance across the whole range of runtimes.

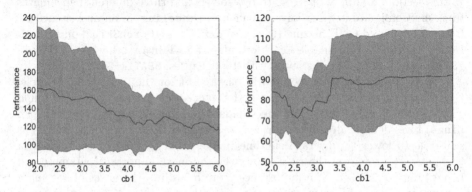

Fig. 4. Parameter importance plot for the most important parameter `cb1` on `7SAT90-SAT` (left) and `3SAT1k-SAT` (right). The plots show the mean performance (blue line) with confidence intervals (red area) as a function of `cb1`, marginalized over all other parameters. The best found configurations set it to 4.35 and 2.86 respectively.

5 Conclusion

We have presented `SpySMAC`: a tool for automatic SAT solver configuration using `SMAC` combined with extensive analysis allowing the user to "spy" into the configuration process of a solver on a given instance set. The report `SpySMAC` generates offers some insight into performance improvements and also quantifies parameter importance by applying `fANOVA`. We have shown for three examples how the framework works, re-running and analyzing three CSSC scenarios effortlessly. For the future, we plan to integrate more methods to evaluate parameter importance, including ablation [8] and forward selection to identify key parameters [12]. `SpySMAC` is available at www.ml4aad.org/spysmac under AGPL license with a long list of examples.

References

1. Ansótegui, C., Sellmann, M., Tierney, K.: A Gender-based genetic algorithm for the automatic configuration of algorithms. In: Gent, I.P. (ed.) CP 2009. LNCS, vol. 5732, pp. 142–157. Springer, Heidelberg (2009)
2. Balint, A., Schöning, U.: Choosing probability distributions for stochastic local search and the role of make versus break. In: Cimatti and Sebastiani [7], pp. 16–19
3. Biere, A.: Yet another local search solver and lingeling and friends entering the SAT competition 2014. In: Belov, A., Diepold, D., Heule, M., Järvisalo, M. (eds.) Proceedings of SAT Competition 2014: Solver and Benchmark Descriptions. Department of Computer Science Series of Publications B, vol. B-2014-2, pp. 39–40. University of Helsinki (2014)
4. Breimann, L.: Random forests. Machine Learning Journal **45**, 5–32 (2001)
5. Brochu, E., Cora, V., de Freitas, N.: A tutorial on Bayesian optimization of expensive cost functions, with application to active user modeling and hierarchical reinforcement learning. Computing Research Repository (2010). (CoRR) abs/1012.2599
6. Brummayer, R., Lonsing, F., Biere, A.: Automated testing and debugging of SAT and QBF solvers. In: Cimatti and Sebastiani [7], pp. 44–57
7. Cimatti, A., Sebastiani, R. (eds.): SAT 2012. LNCS, vol. 7317. Springer, Heidelberg (2012)
8. Fawcett, C., Hoos, H.H.: Analysing differences between algorithm configurations through ablation. Journal of Heuristics, 1–28 (2015)
9. Gebser, M., Kaufmann, B., Schaub, T.: Conflict-driven answer set solving: From theory to practice. Artificial Intelligence **187–188**, 52–89 (2012)
10. Hutter, F., Babić, D., Hoos, H.H., Hu, A.: Boosting verification by automatic tuning of decision procedures. In: O'Conner, L. (ed.) Formal Methods in Computer Aided Design (FMCAD 2007), pp. 27–34. IEEE Computer Society Press (2007)
11. Hutter, F., Hoos, H.H., Leyton-Brown, K.: Sequential model-based optimization for general algorithm configuration. In: Coello, C.A.C. (ed.) LION 2011. LNCS, vol. 6683, pp. 507–523. Springer, Heidelberg (2011)
12. Hutter, F., Hoos, H.H., Leyton-Brown, K.: Identifying key algorithm parameters and instance features using forward selection. In: Nicosia, G., Pardalos, P. (eds.) LION 7. LNCS, vol. 7997, pp. 364–381. Springer, Heidelberg (2013)
13. Hutter, F., Hoos, H.H., Leyton-Brown, K.: An efficient approach for assessing hyperparameter importance. In: Xing, E., Jebara, T. (eds.) Proceedings of the 31th International Conference on Machine Learning, (ICML 2014), vol. 32, pp. 754–762. Omniprdess (2014)
14. Hutter, F., Hoos, H.H., Leyton-Brown, K., Stützle, T.: ParamILS: An automatic algorithm configuration framework. Journal of Artificial Intelligence Research **36**, 267–306 (2009)
15. Hutter, F., Lindauer, M., Balint, A., Bayless, S., Hoos, H.H., Leyton-Brown, K.: The Configurable SAT Solver Challenge. Computing Research Repository (CoRR) (2015). http://arxiv.org/abs/1505.01221
16. KhudaBukhsh, A., Xu, L., Hoos, H.H., Leyton-Brown, K.: SATenstein: automatically building local search SAT solvers from components. In: Boutilier, C. (ed.) Proceedings of the 22th International Joint Conference on Artificial Intelligence (IJCAI 2009), pp. 517–524 (2009)

17. López-Ibáñez, M., Dubois-Lacoste, J., Stützle, T., Birattari, M.: The irace package, iterated race for automatic algorithm configuration. Tech. rep., IRIDIA, Université Libre de Bruxelles, Belgium (2011). http://iridia.ulb.ac.be/IridiaTrSeries/IridiaTr2011-004.pdf
18. Sakallah, K.A., Simon, L. (eds.): SAT 2011. LNCS, vol. 6695, pp. 134–144. Springer, Heidelberg (2011)
19. Tompkins, D.A.D., Balint, A., Hoos, H.H.: Captain jack: new variable selection heuristics in local search for SAT. In: Sakallah and Simon [18], pp. 302–316

Community Structure Inspired Algorithms for SAT and #SAT

Robert Ganian[✉] and Stefan Szeider

Algorithms and Complexity Group, TU Wien, Vienna, Austria
rganian@gmail.com, stefan@szeider.net

Abstract. We introduce h-modularity, a structural parameter of CNF formulas, and present algorithms that render the decision problem SAT and the model counting problem #SAT fixed-parameter tractable when parameterized by h-modularity. The new parameter is defined in terms of a partition of clauses of the given CNF formula into strongly interconnected communities which are sparsely interconnected with each other. Each community forms a hitting formula, whereas the interconnections between communities form a graph of small treewidth. Our algorithms first identify the community structure and then use it for an efficient solution of SAT and #SAT, respectively. We further show that h-modularity is incomparable with known parameters under which SAT or #SAT is fixed-parameter tractable.

1 Introduction

Large networks often exhibit a certain structure, where nodes form strongly interconnected communities which are sparsely connected with each other; to what extent a network exhibits such a structure can be measured by its *modularity* [17–19,31]. Recently the community structure and modularity of practical SAT instances has been empirically studied, revealing an interesting correlation between the modularity and the solving time of state-of-the art SAT solvers. Interestingly, learnt clauses tend to lie within communities and learnt clauses of low Literal Block Distance (LBD) are shared by few communities [1,20]. These findings contribute towards a better understanding of the spectacular performance of today's SAT solvers on practical instances, which is generally not well understood and remains a challenge for the research community [29].

However, the presence of a community structure with low modularity is not a *guarantee* for an instance to be easy; instead, the correlation between modularity and solving time is of statistical nature. In fact, it is not difficult to show that SAT remains NP-hard for highly modular instances. More specifically, given any SAT formula F, one can use a padding process (i.e., the addition of multiple variable-disjoint dense satisfiable subformulas) to create an equisatisfiable formula F' whose size is linear in F and whose modularity can be better than any arbitrarily fixed threshold.

Supported by the Austrian Science Fund (FWF), project P26696.

M. Heule and S. Weaver (Eds.): SAT 2015, LNCS 9340, pp. 223–237, 2015.
DOI: 10.1007/978-3-319-24318-4_17

In this paper we propose the notion of *h-modularity* for SAT instances that provides a worst-case performance guarantee for SAT decision. The h-modularity of a SAT instance is an integer-valued parameter, where instances with small h-modularity can provably be solved quickly. More precisely, we propose an algorithm that, given a SAT instance F of input length ℓ and h-modularity k, decides the satisfiability of F in time $f(k)\ell^2$, where f is singly exponential function in the parameter k. In other words, SAT is *fixed-parameter tractably* (FPT) in the parameter h-modularity. We also provide an FPT algorithm for propositional model counting (i.e., #SAT) parameterized by h-modularity. The parameter dependency is single-exponential for SAT and double-exponential for #SAT.

Our parameter is defined based on the partition of the set of clauses into subsets, which we call *h-communities*. Each h-community forms a strongly interconnected set of clauses. This is ensured by the requirement that any two clauses of an h-community clash in at least one variable (i.e., h-communities are so-called "hitting formulas" [11–13,22]). Furthermore, the h-communities are sparsely interconnected with each other, which is ensured by the requirement that a certain graph which represents the interaction between h-communities has small treewidth as well as h-communities are of small degree (graphs of small treewidth are sparse [14,24]). A formal definition of h-modularity is given in Section 3. We show that h-modularity is incomparable with the parameters signed clique-width and clustering-width, hence h-modularity is not dominated by well-known parameters that admit fixed-parameter tractability of SAT or #SAT. As a consequence, our parameter pushes the frontiers of tractability for SAT and exploits a type of structure not accessible to known FPT algorithms.

2   Preliminaries

2.1   SAT and #SAT

We consider propositional formulas in conjunctive normal form (CNF), represented as sets of clauses. That is, a *literal* is a (propositional) variable x or a negated variable $\overline{x}$; a *clause* is a finite set of literals not containing a complementary pair x and $\overline{x}$; a *formula* is a finite set of clauses. For a literal $l = \overline{x}$ we write $\overline{l} = x$; for a clause C we set $\overline{C} = \{\overline{l} \mid l \in C\}$. For a clause C, $var(C)$ denotes the set of variables x with $x \in C$ or $\overline{x} \in C$. Similarly, for a formula F we write $var(F) = \bigcup_{C \in F} var(C)$. The *length* of a formula F is defined as $\sum_{C \in F} |C|$.

We say that two clauses C, D *overlap* if $C \cap D \neq \emptyset$; we say that C and D *clash* if C and $\overline{D}$ overlap. Note that two clauses can clash and overlap at the same time. Two clauses C, D are *adjacent* if $var(C) \cap var(D) \neq \emptyset$ (i.e., if C and D clash or overlap), and the degree $\mathbf{deg}(C)$ of C in a formula F is the number of clauses $D \in F$ adjacent to C. The *dual graph* of a formula F is the graph whose vertices are clauses of F and whose edges are defined by the adjacency relation of clauses. The dual graph allows us to use standard graph terminology, such as *neighborhood* and *edge-disjoint paths*, when speaking about a formula.

We will also use the *primal graph* of a formula F, specifically in the proof of Theorem 3. The primal graph of F is the graph whose vertices are variables of

F and where two variables a, b are adjacent iff there exists a clause C such that $a, b \in C$.

A *truth assignment* (or *assignment*, for short) is a mapping $\tau : X \to \{0, 1\}$ defined on some set X of variables. We extend τ to literals by setting $\tau(\overline{x}) = 1 - \tau(x)$ for $x \in X$. $F[\tau]$ denotes the formula obtained from F by removing all clauses that contain a literal x with $\tau(x) = 1$ and by removing from the remaining clauses all literals y with $\tau(y) = 0$; $F[\tau]$ is the *restriction* of F to τ. Note that $var(F[\tau]) \cap X = \emptyset$ holds for every assignment $\tau : X \to \{0, 1\}$ and every formula F. A truth assignment $\tau : X \to \{0, 1\}$ *satisfies* a formula F if $F[\tau] = \emptyset$. A truth assignment $\tau : var(F) \to \{0, 1\}$ that satisfies F is a *model* of F. We denote by $\#(F)$ the number of models of F. A formula F is *satisfiable* if $\#(F) > 0$.

2.2 Parameterized Complexity

Next we give a brief and rather informal review of the most important concepts of parameterized complexity. For an in-depth treatment of the subject we refer the reader to other sources [7,21].

The instances of a parameterized problem can be considered as pairs (I, k) where I is the *main part* of the instance and k is the *parameter* of the instance; the latter is usually a non-negative integer. A parameterized problem is *fixed-parameter tractable* (FPT) if instances (I, k) of size n (with respect to some reasonable encoding) can be solved in time $O(f(k)n^c)$ where f is a computable function and c is a constant independent of k. The function f is called the *parameter dependence*.

2.3 Hitting Formulas

A *hitting formula* is a CNF formula with the property that any two of its clauses clash (see [11,12,22]). The same notion for DNF formulas is termed orthogonality [5]. The following result makes hitting formulas particularly attractive in the context of SAT and #SAT.

Fact 1 ([10]). *A hitting formula F with n variables has exactly $2^n - \sum_{C \in F} 2^{n-|C|}$ models.*

The following observation will be implicitly used in several of our proofs.

Fact 2. *Let F be a hitting formula, and let F' be obtained from F by an arbitrary sequence of clause deletions and restrictions under truth assignments. Then F' is also a hitting formula.*

2.4 Treewidth

Let G be a simple, undirected, finite graph with vertex set $V = V(G)$ and edge set $E = E(G)$. For standard graph-theoretic notions not defined here, we refer to [6]. A *tree decomposition* of G is a pair $(\{X_i : i \in I\}, T)$ where $X_i \subseteq V$, $i \in I$, and T is a tree with elements of I as nodes such that:

1. for each edge $uv \in E$, there is an $i \in I$ such that $\{u, v\} \subseteq X_i$, and
2. for each vertex $v \in V$, the set $\{ i \in I \mid v \in X_i \}$ induces a (connected) subtree in T with at least one node.

The *width* of a tree decomposition is $\max_{i \in I} |X_i| - 1$. The *treewidth* [14,23] of G is the minimum width taken over all tree decompositions of G and it is denoted by $\mathbf{tw}(G)$.

Fact 3 ([3]). *There exists an algorithm which, given a graph G and an integer k, runs in time $2^{k^{\mathcal{O}(1)}} \cdot (|V(G)| + |E(G)|)$, and either outputs a tree decomposition of G of width at most k or correctly determines that $\mathbf{tw}(G) > k$.*

It is well known that, for every clique over $Z \subseteq V(G)$ in G, it holds that every tree decomposition of G contains an element X_i such that $Z \subseteq X_i$ [14]. Furthermore, an n-vertex graph of treewidth k is sparse and has $\mathcal{O}(nk)$ edges [14,24].

3 h-Communities and h-Modularity

Let F be a formula. We call a hitting formula $H \subseteq F$ a *hitting community* (or *h-community* in brief) in F. The *degree* $\mathbf{deg}(H)$ of an h-community H is the number of edges in the dual graph of F between a clause in H and a clause outside of H. A *hitting community structure* (or *h-structure* in brief) $\mathcal{P}$ is a partitioning of F into h-communities, and the degree $\mathbf{deg}(\mathcal{P})$ of $\mathcal{P}$ is $\max\{ \mathbf{deg}(H) \mid H \in \mathcal{P} \}$.

To measure the treewidth of an h-structure $\mathcal{P}$, we construct a *community graph* G as follows. The vertices of G are the h-communities in $\mathcal{P}$, and two vertices A, B in G are joined by an edge if and only if there exist clauses $C \in A$ and $D \in B$ which are adjacent. Then we let $\mathbf{tw}(\mathcal{P}) = \mathbf{tw}(G)$.

We define the *h-modularity* of an h-structure $\mathcal{P}$ as the maximum over $\mathbf{deg}(\mathcal{P})$ and $\mathbf{tw}(\mathcal{P})$. The h-modularity $\mathbf{h\text{-}mod}(F)$ of a formula F is then defined as the minimum $\mathbf{h\text{-}mod}(\mathcal{P})$ over all h-structures $\mathcal{P}$ of F.

Observe that this definition ensures that clauses in individual h-communities are strongly interconnected (since they form hitting formulas), but each h-community is only sparsely connected to other h-communities (due to the community graph having small treewidth and degree). At the same time we will prove that, unlike modularity, h-modularity is a parameter that guarantees the existence of structure which can be algorithmically exploited to establish the fixed-parameter tractability of SAT and #SAT.

Example: Consider the formula $F = \{xy\bar{a}, \bar{x}ya, x\bar{y}, \overline{xy}, abc, \bar{b}, c\bar{d}ef, d\bar{e}, f\bar{g}\bar{h}, h\bar{i}, i\bar{j}, jkl\overline{mn}, u\bar{v}g\bar{k}lm\bar{n}, uv\bar{l}, \bar{u}v\}$. Figure 1 (left) then illustrates the dual graph of F with the indicated partition $\mathcal{P} = \{H_1, \ldots, H_6\}$ of F into h-communities $H_1 = \{xy\bar{a}, \bar{x}ya, x\bar{y}, \overline{xy}\}$, $H_2 = \{abc, \bar{b}\}$, $H_3 = \{c\bar{d}ef, d\bar{e}\}$, $H_4 = \{f\bar{g}\bar{h}, h\bar{i}\}$, $H_5 = \{i\bar{j}, jkl\overline{mn}\}$, and $H_6 = \{u\bar{v}g\bar{k}lm\bar{n}, uv\bar{l}, \bar{u}v\}$. Figure 1 (right) shows the community graph of $\mathcal{P}$; it is easy to verify that this graph has treewidth 2 [14] (observe, for instance, that the deletion of a single vertex turns it into a tree).

The h-communities H_1 and H_3 have degree 2, and all other h-communities have degree 3. Therefore the h-modularity of F is at most $\max(3, 2) = 3$.

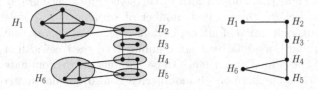

Fig. 1. The dual graph (left) and community graph (right) of the formula F and the h-structure $\mathcal{P}$.

An h-structure $\mathcal{P}$ of F is called a *witness* of **h-mod**$(F) \leq k$ if **h-mod**$(\mathcal{P}) \leq k$. Given an h-structure $\mathcal{P}$ of F and a subformula $F' \subseteq F$, we denote by $\mathcal{P}[F']$ the h-structure induced by $\mathcal{P}$ on F'; observe that **h-mod**$(\mathcal{P}[F']) \leq$ **h-mod**$(\mathcal{P})$.

We introduce some additional notation which will be useful later, always w.r.t. a fixed h-structure. A clause $C \in H$ is a *bridge clause* if there exists a clause outside of H adjacent to C. A variable x is a *bridge variable* if it occurs in a clause in one h-community and at least one other clause in another h-community. Notice that every clause containing a bridge variable is a bridge clause, and that h-structures of low h-modularity can still contain a large number of bridge variables, even in a single h-community.

We can now formalize the parameterized problems we are solving and present our main results.

#SAT[h-mod]
Instance: A formula F of length ℓ and an integer $k \geq 0$.
Task: Either compute the number of models of F, or correctly determine that **h-mod**$(F) > k$.
Parameter: k.

The problem SAT[**h-mod**] is then defined analogously to #SAT[**h-mod**], with the distinction that the task is only to determine whether the number of models is non-zero (in which case we say that F is satisfiable).

Theorem 1. *#SAT[**h-mod**] and SAT[**h-mod**] are fixed parameter tractable.*

Our approach for proving Theorem 1 can be separated into two main tasks: first, we compute an h-structure $\mathcal{P}$ of small h-modularity, and then we use $\mathcal{P}$ to solve the problem. Our techniques to achieve this are discussed in detail in the following two sections. We remark that the parameter dependence is single-exponential for our SAT algorithm and double-exponential for our #SAT algorithm.

Before proceeding, we make a short digression comparing the new notion of h-modularity to established parameters for SAT. We say that parameter X *dominates* parameter Y if there exists a computable function f such that for each formula F we have $X(F) \leq f(Y(F))$ [25]. In particular, if X dominates Y and SAT is FPT parameterized by X, then SAT is FPT parameterized by

Y [25]. We say that two parameters are *incomparable* if neither dominates the other. In the following, we show that h-modularity is incomparable with the *signed clique-width* (the clique-width of the signed incidence graph [4, 28]) and with *clustering-width* (the smallest number of variables whose deletion results in a variable-disjoint union of hitting formulas) [22]. We remark that the former claim implies that h-modularity is not dominated by the treewidth of neither the incidence nor the primal graph, since these parameters are dominated by signed clique-width [28]. Furthermore, h-modularity is also not dominated by signed rank-width [9], which both dominates and is dominated by signed clique-width.

Proposition 1. *The following claims hold.*

1. *Signed clique-width and h-modularity are incomparable.*
2. *Clustering-width and h-modularity are incomparable.*

Proof. We prove both claims by showing that there exist classes of formulas such that each formula in the class has one parameter bounded while the other parameter can grow arbitrarily. For a formula F, let $\mathbf{scw}(F)$ and $\mathbf{clu}(F)$ denote its signed clique-width and clustering width, respectively. Our proof does not require a formal definition of these parameters, as we refer to known properties of these notions.

Let $\mathbb{N}$ be the set of positive integers, and let us choose an arbitrary $i \in \mathbb{N}$. For the first claim, it is known that already the class of all hitting formulas has unbounded $\mathbf{scw}$ [22]. In particular, this means that there exists a hitting formula F_1 such that $\mathbf{scw}(F_1) \geq i$. Recall that, since F_1 is a hitting formula, clearly $\mathbf{h\text{-}mod}(F_1) = 0$.

Conversely, consider the following formula $F_2 = \{C, C_1, \ldots, C_{i+2}\}$. The formula contains variables $x_1, \ldots x_{i+2}$, and each variable x_j occurs (either positively or negatively) in clause C and C_j. Then the incidence graph of F_2 is a tree and hence has treewidth 1. Since signed clique-width dominates the treewidth of the incidence graph, it follows that there exists a constant c independent of i such that $\mathbf{scw}(F_2) \leq c$ (in particular, one can check from the definition of $\mathbf{scw}$ that $c \leq 2$). On the other hand, the degree of any h-community H containing C is at least $i + 1$, and hence $\mathbf{h\text{-}mod}(F_2) \geq i + 1$.

We proceed similarly for the second claim; let $i \in \mathbb{N}$. Let F_1'' be a hitting formula, let F_1' be constructed by adding a new variable z into an arbitrary clause in F_1'' and adding a clause Z containing only z (both occurrences can either be positive or negative). Observe that $\mathbf{clu}(F_1') = \mathbf{h\text{-}mod}(F_1') = 1$. Let F_1 then contain $i + 2$ disjoint copies of F_1'; clearly, $\mathbf{clu}(F_1) = i + 2$. However, since the h-modularity of a formula is equal to the maximum h-modularity over all of its connected components, it holds that $\mathbf{h\text{-}mod}(F_1) = 1$.

Conversely, let F_2' and F_2'' be variable-disjoint hitting formulas containing at least $i + 2$ clauses each, and let F_2 be obtained from a disjoint union of F_2' and F_2'' by adding a variable z which occurs (either positively or negatively) in $\lfloor i/2 \rfloor$ clauses in F_2' and in $\lfloor i/2 \rfloor$ clauses in F_2''. While F_2 is not a hitting formula, deleting z results in two variable-disjoint hitting formulas and hence $\mathbf{clu}(F_2) = 1$. On the other hand, the three inclusion-maximal h-communities in F_2 are F_2', F_2''

and possibly the set of clauses where z occurs; each of these have a degree which is greater than i. Consequently, it holds that $\textbf{h-mod}(F_2) \geq i + 1$.   □

4   Finding h-Structures

Our approach for finding h-structures of small h-modularity consists of two steps. Generally speaking, we introduce a preprocessing procedure which we exhaustively apply until all clauses have a sufficiently small degree (Lemma 1), and once the degree of all clauses is sufficiently small we compute a tree decomposition of the dual graph and use it to find a suitable h-structure (Lemma 2). The result is an FPT-approximation algorithm [16]. One of the technical obstacles we have to overcome is that the preprocessing procedure given by Lemma 1 only guarantees the preservation of h-modularity up to a certain bound. This bound then represents an additional constraint on the approximation algorithm presented in Lemma 2.

Lemma 1. *There exists an algorithm which, given $q \in \mathbb{N}$ and a formula F of length ℓ containing a clause C such that $\textbf{deg}(C) > 3q + 2$, runs in time $\mathcal{O}(\ell^2)$ and either correctly determines that $\textbf{h-mod}(F) > q$, or outputs a strictly smaller subformula F' with the following property: if $\textbf{h-mod}(F') \leq q$, then $\textbf{h-mod}(F) = \textbf{h-mod}(F')$. Furthermore, a witness $\mathcal{P}$ of $\textbf{h-mod}(F)$ can be computed from F, F' and a witness $\mathcal{P}'$ of $\textbf{h-mod}(F') \leq q$ in linear time.*

Proof. Let Z_0 be the set containing C and all clauses which are neighbors of C, let Z_1 be the subset of Z_0 containing clauses which have a neighbor outside of Z_0, and let $Z = Z_0 \setminus Z_1$. Let W be the subset of Z containing clauses which have at least $q + 2$ neighbors in Z. We now make a series of tests:

1. if $Z_1 > q$, then $\textbf{h-mod}(F) > q$;
2. if $|W| < q + 3$, then $\textbf{h-mod}(F) > q$;
3. if W is not a hitting formula, then $\textbf{h-mod}(F) > q$;
4. if Z contains a clause which clashes with exactly $|W| - 1$ clauses in W, then $\textbf{h-mod}(F) > q$;
5. let $B \in W$ be a clause with no neighbors outside W; if no such B exists, then $\textbf{h-mod}(F) > q$.

Otherwise we set $F' = F \setminus B$.

We prove correctness. Observe that if $|Z_1| > q$ then there exists no $\mathcal{P}$ of h-modularity at most q. Indeed, for each neighbor D of Z_1 outside of Z_0, it holds that D and C cannot be in the same h-community, since they are not adjacent. Hence each element of Z_1 increases the degree of the h-community containing C by at least 1; either due to the edge between C and that element, or the edge between D and that element. Hence we can assume that $|Z| \geq 2q + 3$.

For the second test, observe that if $|W| < q + 3$ then there exists no $\mathcal{P}$ of h-modularity at most q. Indeed, since the number of neighbors of C in Z is at least $2q + 2$, at least $q + 2$ of these neighbors must be in the same h-community as C if $\textbf{h-mod}(\mathcal{P}) \leq q$. This implies that at least $q+2$ of these neighbors would have to be pairwise-adjacent, and in particular would each have at least $q+2$ neighbors in Z. Then W necessarily must contain C and at least $q+2$ neighbors of C.

For the third test, if W is not a hitting formula, then any h-structure $\mathcal{P}$ of h-modularity at most q would need to partition W into (subsets of) at least two h-structures; let H_C be the hypothetical h-community containing C, and let $D \in W \setminus H_C$. Since D has $q+2$ neighbors in Z, there are at least $q+2$ edge-disjoint paths between D and C, and each of these paths contributes at least 1 to the degree of H_C. But then it follows that $\textbf{deg}(H_C) \geq q+2$, which would contradict $\textbf{h-mod}(\mathcal{P}) \leq q$, and hence W must be a hitting formula. Observe that this argument also implies that every clause in W is in fact adjacent to every other clause in W, and that every $\mathcal{P}$ of h-modularity at most q must contain an h-community H_C which contains W.

For the fourth test, assume there exists a clause D which clashes with exactly $|W| - 1$ clauses in W. Consider any witness $\mathcal{P}$ of $\textbf{h-mod}(F) \leq q$, and let H_C be the h-community containing C. Since $D \notin H_C$ and there are at least $q+1$ edge-disjoint paths between D and C, the existence of D would imply that $\textbf{deg}(H_C) \geq q+1$.

For the fifth test, recall that for any clause $Q \in Z \setminus W$ it holds that $W \cup \{Q\}$ cannot be a hitting formula because Q cannot be adjacent to every clause in W. Hence every clause in W with a neighbor outside of W contributes at least 1 to the degree of any h-community containing W. Together with $|W| > q+2$ this implies that if no clause B exists, then $\textbf{h-mod}(F) > q$.

Finally, assume there exists a clause $B \in W$ with no neighbors outside of W and let $F' = F \setminus B$. If $\textbf{h-mod}(F') > q$ then the lemma already holds, so assume there exists a witness $\mathcal{P}'$ of $\textbf{h-mod}(F') \leq q$. Let $W' = W \setminus B$. Observe that W' must be contained in a single h-community $H' \in \mathcal{P}'$, since otherwise the fact that each clause of W' is adjacent to every other clause of W' would contradict the degree bound given by $\textbf{h-mod}(\mathcal{P}') \leq q$. Then let $\mathcal{P}$ be obtained from $\mathcal{P}'$ by adding B to H'. Observe that there cannot exist a clause $D \in H'$ such that D and B do not clash; since D clashes with every other clause in W, it follows that D would clash with $|W| - 1$ clauses in W. Hence $B \cup H'$ is still an h-community. Furthermore, by our choice of B it holds that B contains no neighbors outside of W', and hence $\textbf{deg}(H') = \textbf{deg}(H' \cup \{B\})$ and in turn $\textbf{deg}(\mathcal{P}') = \textbf{deg}(\mathcal{P})$.

Finally, observe that, if we are given a witness $\mathcal{P}'$ of $\textbf{h-mod}(F') \leq q$, we can construct a witness of $\textbf{h-mod}(F)$ by adding B back into the unique h-community in $\mathcal{P}$ containing the neighbors of B (i.e., W'). □

Lemma 2. *There exists an algorithm which, given $k \in \mathbb{N}$ and a formula F of length ℓ such that $\textbf{deg}(F) \leq 12k^2 + 2$, runs in time $2^{k^{\mathcal{O}(1)}} \cdot \ell$, and either outputs an h-structure $\mathcal{P}$ of F such that $\textbf{h-mod}(\mathcal{P}) \leq k^2 + k$, or correctly determines that $\textbf{h-mod}(F) > k$.*

Proof. We first test whether the treewidth of the dual graph G of F is at most $k \cdot (12k^2 + 3)$; if not, then $\textbf{h-mod}(F) > k$, and if yes, we compute a tree decomposition of F. This can be achieved in time at most $2^{k^{\mathcal{O}(1)}} \cdot \ell$ by Fact 3. Next, we enumerate every inclusion-maximal clique in G of cardinality at least $k + 2$ in time $\mathcal{O}(k^3) \cdot \ell$ by a simple traversal of the tree decomposition. Let L be the set of all such cliques. For each clique $K \in L$ we test whether K is a hitting formula and whether $\mathbf{deg}(K) \leq k$; if not, then $\mathbf{h\text{-}mod}(F) > k$. For each pair of cliques $K_1, K_2 \in L$ we test that they are pairwise disjoint; if not, then $\mathbf{h\text{-}mod}(F) > k$. Let G' be the graph obtained from G by contracting each clique in L into a single vertex; that is, each $K \in L$ is replaced by a vertex adjacent to all neighbors of K. We test that $\mathbf{deg}(G') \leq 2k$ and $\mathbf{tw}(G') \leq k^2 + k$; if not, then $\mathbf{h\text{-}mod}(F) > k$. Finally, let $\mathcal{P}'$ be the vertex set of G'. Then $\mathcal{P}'$ is an h-structure witnessing $\mathbf{h\text{-}mod}(F) \leq k^2 + k$.

We prove correctness. First, assume for a contradiction that $\mathbf{tw}(G) > k \cdot (12k^2 + 3)$ and that there exists a witness $\mathcal{P}'$ of $\mathbf{h\text{-}mod}(F) \leq k$. Since $\mathbf{deg}(F) \leq 12k^2 + 2$, every h-community in $\mathcal{P}'$ must have size at most $12k^2 + 3$. Let (β, T) be a width-k tree decomposition of the community graph of $\mathcal{P}'$, and let β' be obtained by replacing each h-community $H \in \mathcal{P}'$ with $\bigcup_{C \in H} C$. Then (β', T) is a tree decomposition of G of width at most $k \cdot (12k^2 + 3)$, contradicting our assumptions.

Next, assume that there exists a clique $K \in L$ which is not a hitting formula. Then any hypothetical h-structure $\mathcal{P}$ of F must partition K into several h-communities. Let $C, D \in K$ and $H \in \mathcal{P}$ be such that $C \in H$ and $D \notin H$. Since there exist $k + 1$ edge-disjoint paths between C and D, this implies that $\mathbf{deg}(H) \geq k + 1$ and hence $\mathbf{h\text{-}mod}(\mathcal{P}) > k$.

Similarly, assume that there exist inclusion-maximal cliques $K_1, K_2 \in L$ which intersect in some clause C. Then any hypothetical h-structure $\mathcal{P}$ must contain an h-community H containing C, and there must exist a clause $D \in K_1 \cup K_2$ such that $D \notin H$. As in the previous case, this gives rise to at least $k + 1$ edge-disjoint paths between C and D and hence $\mathbf{h\text{-}mod}(\mathcal{P}) > k$. In particular, we conclude that each element of L must form an h-community in any hypothetical witness of $\mathbf{h\text{-}mod}(F) \leq k$. This in turn implies that if there exists an h-community $K \in L$ of degree at least $k + 1$, then $\mathbf{h\text{-}mod}(F) > k$.

We proceed by considering the graph G'. Assume it contains a vertex v of degree at least $2k + 1$. If v is a clause in F, then at most k neighbors of v can form an h-community with v (since we have contracted all cliques of cardinality at least $k + 2$). This means that at least $k + 1$ neighbors of v would contribute to the degree of the h-community containing v, which guarantees $\mathbf{h\text{-}mod}(F) > k$. On the other hand, if v is an element of L, then we already know that v itself must be an h-community in any witness of $\mathbf{h\text{-}mod}(F) \leq k$, and hence v having more than k neighbors also implies $\mathbf{h\text{-}mod}(F) > k$.

Next, consider the case $\mathbf{tw}(G') > k^2 + k$. Observe that each hitting subformula of F not contained in L contains at most $k + 1$ clauses. Consider a width-k tree decomposition (β, T) of the community graph Q of a hypothetical witness of $\mathbf{h\text{-}mod}(F) \leq k$. By replacing, in β, each h-community $H \in V(Q) \setminus L$ with the set of clauses contained in H, we would obtain a tree decomposition of G' of width at most $k \cdot (k+1)$, contradicting our assumption. Hence we conclude that $\textbf{h-mod}(F) > k$.

Finally, we summarize why $\mathcal{P}'$ is indeed an h-structure of G such that $\textbf{h-mod}(\mathcal{P}') \leq k^2 + k$. The fact that $\mathcal{P}'$ is an h-structure follows by construction; indeed, each element in $\mathcal{P}'$ is either a single clause, or an element of L which is guaranteed to be a hitting formula. Regarding the h-modularity of $\mathcal{P}'$, recall that G' is the community graph of $\mathcal{P}'$ and that $\textbf{tw}(G') \leq k^2 + k$. As for the degree bound, each vertex v in G' is either a clause C in F, which means that $\deg(v) \leq 2k$, or an element of K, in which case we have already tested that $\deg(v) \leq k$.   □

Theorem 2. *There exists an algorithm which, given $k \in \mathbb{N}$ and a formula F of length ℓ, runs in time $\mathcal{O}(\ell^3) + 2^{k^{\mathcal{O}(1)}} \cdot \ell$, and either outputs an h-structure $\mathcal{P}$ of F such that $\textbf{h-mod}(\mathcal{P}) \leq k^2 + k$, or correctly determines that $\textbf{h-mod}(F) > k$.*

Proof. We begin by exhaustively applying Lemma 1 on F for $q = 4k^2$; let us denote the resulting formula F'. Then we apply Lemma 2 on F' to find an h-structure $\mathcal{P}'$ of F' such that $\textbf{h-mod}(\mathcal{P}') \leq k^2 + k \leq q$. Finally, we use Lemma 1 to convert $\mathcal{P}'$ into an h-structure $\mathcal{P}$ of F. Correctness follows from the correctness of Lemmas 1 and 2.   □

5   Using h-Structures

With Theorem 2 in hand, we proceed to show how the identified h-structure of small h-modularity can be used to obtain fixed-parameter tractability of SAT and #SAT. The general strategy is to replace each h-community by a suitable object that represents all the satisfying assignments of this h-community. This way, variables only appearing in a single h-community are eliminated. In case of SAT, we represent an h-community by a set of clauses over the bridge variables of the h-community, and in the case of #SAT, we represent an h-community by a so-called valued constraint. This way, we reduce the problems SAT and #SAT parameterized by h-modularity to certain problems (SAT and SumProd, respectively) parameterized by primal treewidth. For solving the latter problems we can use known algorithms.

For making this general strategy work, we have to overcome the difficulty that the number of bridge variables of a single h-community can be arbitrarily large even when the input formula has small h-modularity. In the case of SAT we can handle this by replacing the input formula with a satisfiability-equivalent subformula using a known construction. This approach does not work for #SAT since this replacement does not preserve the number of models. However, by replacing equivalence classes of variables that appear in the same way in all clauses by 3-valued variables (which represent the three possibilities that all variables in the module are set to true, all are set to false, or some are set to true and some to false, respectively), we can reduce the number of variables for a single valued constraint so that we can make our overall strategy work.

We begin with the conceptually simpler case of SAT. Our solution relies on the following folklore result.

Fact 4 ([26]). *There exists an algorithm which takes as input a formula F of length ℓ and a tree decomposition of the primal graph of F of width k, runs in time $2^{\mathcal{O}(k)} \cdot \ell^2$, and determines whether F is satisfiable.*

Theorem 3. *Given a formula F' of length ℓ and an h-structure $\mathcal{P}'$ of F', we can decide whether F' is satisfiable in time $2^{\mathcal{O}(\mathbf{h\text{-}mod}(\mathcal{P}')^2)} \cdot \ell^2$.*

Proof. Our algorithm has three steps. First, we compute an equisatisfiable sub-formula F of F' where F has the following property: for every nonempty set X of variables of F there are at least $|X| + 1$ clauses C of F such that some variable in X occurs in C. Formulas with this property are called 1-*expanding* or *matching-lean*, and it is known that for any formula F' of length ℓ, an equisatis-fiable 1-expanding subformula F can be computed in time $\mathcal{O}(\ell^{3/2})$ [8,15,27]. We set $\mathcal{P} = \mathcal{P}'[F]$ and $k = \mathbf{h\text{-}mod}(\mathcal{P}')$; note that $\mathbf{h\text{-}mod}(\mathcal{P}) = \mathbf{h\text{-}mod}(\mathcal{P}'[F]) \le \mathbf{h\text{-}mod}(\mathcal{P}') = k$. Observe that since each $H \in \mathcal{P}$ satisfies $\deg(H) \le k$, it fol-lows that the number of bridge variables which occur in any clause in H is upper-bounded by k.

For the second step, we construct a formula I as follows. The variable set of I consists of all the bridge variables of $\mathcal{P}$. For each h-community $H \in \mathcal{P}$ containing bridge variables $X_H = \{x_1, \ldots, x_p\}$ and for each assignment α of variables in X_H, we test whether α satisfies H; if it does not, we add the clause C_α over X_α into I, where C_α is the unique clause which is not satisfied by α.

For the final third step, we compute a tree decomposition of the primal graph of I with width at most $k^2 + k$ by Fact 3, and then decide whether I is satisfiable by Fact 4. If it is, we output "YES", and otherwise we output "NO". The rest of the proof is dedicated to verifying the bound on the treewidth of I and arguing correctness.

We argue that the treewidth of the primal graph of I at most $k^2 + k$. Let (β, T) be a tree decomposition of the community graph G of $\mathcal{P}$ of width at most k. Consider the tree decomposition (γ, T) obtained from (β, T) by replacing each h-community H in β by X_H. Since F is 1-expanding and the variables of X_H only appear in at most $k+1$ clauses of F due to the degree bound, the cardinality of each X_H is upper-bounded by $k + 1$. Consequently, the cardinality of each element in γ is at most $k^2 + k$.

Next, we show that (γ, T) is indeed a tree decomposition of the primal graph of I. For every edge ab in this graph, there exists at least one clause $C \in H$ which contains both variable a and variable b in its scope, and hence a, b are both bridge variables for H, which in turn means that a, b will both be present in every element of γ which used to contain H; this proves that the first property of tree decompositions is satisfied. For every bridge variable a, let $\mathcal{D}_a$ denote the set of h-communities which contain a. Since each pair of h-communities containing a are adjacent in the community graph of $\mathcal{P}$, $\mathcal{D}_a$ forms a clique in the community graph of $\mathcal{P}$ and hence there must exist an element θ_a of β which contains every h-community in $\mathcal{D}_a$. Since a occurs in an element of γ if and only if this originated from an element of β containing an h-community in $\mathcal{D}_a$, and since all h-communities in $\mathcal{D}_a$ occur in θ_a, we conclude that the nodes of

T containing a are connected in (γ, T); this proves that the second property of tree decompositions is satisfied.

Finally, we argue that I is satisfiable if and only if F is satisfiable. Let τ_I be a satisfying assignment for I, and consider the assignment τ which assigns each bridge variable in F based on τ_I. The resulting instance $F[\tau]$ consists of variable-disjoint h-communities. Furthermore, by the construction of each constraint in I, it holds that each h-community in $F[\tau]$ is satisfiable, and hence both $F[\tau]$ and F are satisfiable. On the other hand, let τ_F be a satisfying assignment for F, and consider the restriction τ of τ_F to the set of bridge variables. Then applying τ on F once again results in a satisfiable formula $F[\tau]$ consisting of variable-disjoint h-communities. Furthermore, since each such h-community is satisfiable, it follows that τ also satisfies every clause in I. □

Our next goal is to show how h-structures of low h-modularity can be used to solve #SAT. To this end, we will make use of a reduction to the SUMPROD (Sum of Products) problem [2], sometimes also called VALUED #CSP [30], which can be viewed as a generalization of the CONSTRAINT SATISFACTION problem. An instance I of SUMPROD is a triple $(V, D, \mathcal{C})$, where V is a finite set of *variables*, D is a finite set of *domain values*, and $\mathcal{C}$ is a finite set of *valued constraints*. Each valued constraint C in $\mathcal{C}$ is a tuple (S_C, f_C), where S_C, the *constraint scope*, is a non-empty sequence $s_1, s_2, \ldots, s_r$ of distinct variables of V, and f_C, the *cost function*, is a function from D^r to $\mathbb{N} \cup \{0\}$.

An *assignment* is a mapping $\psi : V \to D$. Each assignment ψ results in a cost, $f_C(\psi)$, being assigned to each constraint C, where $f_C(\psi) = f_C((\psi(s_1), \psi(s_2), \ldots, \psi(s_r)))$. The task in the SUMPROD problem is to compute the value $\mathbf{cost}(I)$, defined as the sum over all assignments of the products of cost functions for that assignment. In other words, $\mathbf{cost}(I) = \sum_{\psi:V \to D} \prod_{C \in \mathcal{C}} f_C(\psi)$.

The *primal graph* G of a SUMPROD instance I is defined as follows. The vertices of G are the variables of I, and two vertices a, b of G are adjacent if and only if there exists a constraint whose scope contains both a and b. The *primal treewidth* of I, denoted $\mathbf{ptw}(I)$, is the treewidth of the primal graph of I. The crucial property which we exploit is that primal treewidth allows a straightforward dynamic programming FPT algorithm for SUMPROD over a fixed and finite domain D. The following fact assumes that arithmetic operations can be carried out in polynomial time in the number of variables.

Fact 5 ([2]). *Let D be a fixed set. There exists an algorithm which takes as input an n-variable instance $I = (V, D, \mathcal{C})$ of SUMPROD and a tree decomposition of the primal graph of I of width k, runs in time $2^{\mathcal{O}(k)} \cdot n^{\mathcal{O}(1)}$, and correctly outputs $\mathbf{cost}(I)$.*

Lemma 3. *There exists an algorithm which, given a formula F of length ℓ and an h-structure $\mathcal{P}$ of F, runs in time $\mathcal{O}(3^{\mathbf{h\text{-}mod}(\mathcal{P})} \cdot \ell^{\mathcal{O}(1)})$, and computes an instance $I = (V, D, \mathcal{C})$ of SUMPROD such that $\mathbf{ptw}(I) \leq 2^{\mathcal{O}(\mathbf{h\text{-}mod}(\mathcal{P}))}$, $D = \{0, 1, mix\}$, $|V| \leq \ell$ and $\mathbf{cost}(I)$ is the number of models of F.*

Proof (Sketch). Our goal is to capture the contribution of an h-community H to the total number of models of F by using only a small number of variables in

I; specifically, the number of these variables should depend only on $\mathbf{h\text{-}mod}(\mathcal{P})$. Unlike in Theorem 3, here we cannot directly use 1-expanding subformulas, since these do not preserve the number of models. So instead we group bridge variables into equivalence classes, where two bridge variables are in the same equivalence class iff they occur in the same way in the same clauses; crucially, the number of equivalence classes which intersect with each H is bounded by a function of $\mathbf{h\text{-}mod}(\mathcal{P})$. Furthermore, every "mixed" assignment (mapping at least one variable to 0 and at least one to 1) of an equivalence class satisfies the same clauses as any other mixed assignment of that equivalence class, allowing us to aggregate all such assignments without loss of information. Then we construct our instance I so that each of its variables represents one equivalence class, and each constraint represents one h-community. An assignment ψ of I then corresponds to determining whether all bridge variables of F in each equivalence class are assigned to 0, to 1, or mix.

The cost function is then constructed so as to capture the contribution of each h-community to the total number of models. However, since many assignments in F can be aggregated into a single assignment in I due to the mix value, the cost function also needs to reflect this. To this end, each equivalence class is assigned (arbitrarily) to some valued constraint C and whenever that equivalence class is mapped to mix, f_C is increased by a factor corresponding to the number of assignments in F aggregated into this mixed assignment.

The desired running time follows by showing that equivalence classes can be computed in at most $\mathcal{O}(\ell^3)$ time and that the number of equivalence classes which occur in the same h-community is upper-bounded by $3^{\mathbf{h\text{-}mod}(\mathcal{P})+1}$. The lemma then follows from the following two claims, whose proofs are omitted in this version: (i) $\mathbf{ptw}(I) \leq 2^{\mathcal{O}(\mathbf{h\text{-}mod}(\mathcal{P}))}$, and (ii) $\mathbf{cost}(I)$ is the number of models of F. $\qquad\square$

Theorem 4. *Given a formula F of length ℓ and an h-structure $\mathcal{P}$ of F, we can count the number of models of F in time $2^{2^{\mathcal{O}(\mathbf{h\text{-}mod}(\mathcal{P}))}} \cdot \ell^{\mathcal{O}(1)}$.*

Proof. Let $k = \mathbf{h\text{-}mod}(\mathcal{P})$. We apply Lemma 3 to obtain an instance $I = (V, D, \mathcal{C})$ of SumProd such that $\mathbf{ptw}(I) \leq 2^{\mathcal{O}(k)}$ and $\mathbf{cost}(I)$ is the number of models of F. Next, we compute a tree decomposition of the primal graph of I of width $2^{\mathcal{O}(k)}$: either by observing that the algorithm of Lemma 3 implicitly also computes such a tree decomposition of I, or in time $2^{2^{\mathcal{O}(k)}} \cdot \ell$ by Fact 3. Finally, we use Fact 5 to solve I in time $2^{2^{\mathcal{O}(k)}} \cdot \ell^{\mathcal{O}(1)}$. $\qquad\square$

Proof (of Theorem 1). Let F be the given CNF formula and k the parameter. First we apply Theorem 2 to either find an h-structure $\mathcal{P}$ of F of h-modularity at most $k^2 + k$, or correctly determine that $\mathbf{h\text{-}mod}(F) > k$. To decide whether F is satisfiable, we now use Theorem 3. This establishes that SAT[$\mathbf{h\text{-}mod}(F)$] is fixed parameter tractable. To compute the number of models of F, we use Theorem 4. This establishes the fixed parameter tractability of #SAT[$\mathbf{h\text{-}mod}(F)$] and concludes the proof. $\qquad\square$

6   Concluding Notes

We have introduced the notion of an h-community structure in CNF formulas and the associated parameter h-modularity. Furthermore, we have shown that it is fixed-parameter tractable to find a suitable h-community structure and to use it to solve the problems SAT and #SAT, all parameterized by the h-modularity (Theorems 2, 3, and 4, respectively). Since the h-modularity is small for formulas where other known parameters can be arbitrarily large (Proposition 1), our FPT results provide worst-case performance guarantees for instances that are not accessible by known methods. Our results give rise to the question of how the notion of h-community structure can be further generalized, for example by using a suitably defined property for the communities that generalizes hitting formulas. This way, we hope that ultimately one can build bridges between empirically observed problem hardness and theoretical worst case upper bounds.

References

1. Ansótegui, C., Bonet, M.L., Giráldez-Cru, J., Levy, J.: The fractal dimension of SAT formulas. In: Demri, S., Kapur, D., Weidenbach, C. (eds.) IJCAR 2014. LNCS, vol. 8562, pp. 107–121. Springer, Heidelberg (2014)
2. Bacchus, F., Dalmao, S., Pitassi, T.: Solving #SAT and Bayesian inference with backtracking search. J. Artif. Intell. Res. **34**, 391–442 (2009)
3. Bodlaender, H.L.: A linear-time algorithm for finding tree-decompositions of small treewidth. SIAM J. Comput. **25**(6), 1305–1317 (1996)
4. Courcelle, B., Makowsky, J.A., Rotics, U.: On the fixed parameter complexity of graph enumeration problems definable in monadic second-order logic. Discr. Appl. Math. **108**(1–2), 23–52 (2001)
5. Crama, Y., Hammer, P.L.: Boolean functions. Encyclopedia of Mathematics and its Applications, vol. 142. Cambridge University Press, Cambridge (2011). Theory, algorithms, and applications
6. Diestel, R.: Graph Theory. Graduate Texts in Mathematics, vol. 173, 4th edn. Springer Verlag, New York (2010)
7. Downey, R.G., Fellows, M.R.: Parameterized Complexity. Monographs in Computer Science. Springer, New York (1999)
8. Fleischner, H., Kullmann, O., Szeider, S.: Polynomial-time recognition of minimal unsatisfiable formulas with fixed clause-variable difference. Theoretical Computer Science **289**(1), 503–516 (2002)
9. Ganian, R., Hliněný, P., Obdržálek, J.: Better algorithms for satisfiability problems for formulas of bounded rank-width. Fund. Inform. **123**(1), 59–76 (2013)
10. Iwama, K.: CNF-satisfiability test by counting and polynomial average time. SIAM J. Comput. **18**(2), 385–391 (1989)
11. Büning, H.K., Kullmann, O.: Minimal unsatisfiability and autarkies. In: Biere, A., Heule, M.J.H., van Maaren, H., Walsh, T. (eds) Handbook of Satisfiability. Frontiers in Artificial Intelligence and Applications, vol. 185, chapter 11, pp. 339–401. IOS Press (2009)
12. Büning, H.K., Zhao, X.: Satisfiable formulas closed under replacement. In: Kautz, H.,Selman, B. (eds.) Proceedings for the Workshop on Theory and Applications of Satisfiability. Electronic Notes in Discrete Mathematics, vol. 9. Elsevier Science Publishers, North-Holland (2001)

13. Büning, K.H., Zhao, X.: On the structure of some classes of minimal unsatisfiable formulas. Discr. Appl. Math. **130**(2), 185–207 (2003)
14. Kloks, T.: Treewidth: Computations and Approximations. Springer Verlag, Berlin (1994)
15. Kullmann, O.: Lean clause-sets: Generalizations of minimally unsatisfiable clause-sets. Discr. Appl. Math. **130**(2), 209–249 (2003)
16. Marx, D.: Parameterized complexity and approximation algorithms. The Computer Journal **51**(1), 60–78 (2008)
17. Newman, M.E.J.: The structure and function of complex networks. SIAM Review **45**(2), 167–256 (2003)
18. Newman, M.E.J.: Modularity and community structure in networks. Proceedings of the National Academy of Sciences **103**(23), 8577–8582 (2006)
19. Newman, M.E.J., Girvan, M.: Finding and evaluating community structure in networks. Phys. Rev. E **69**(2), 026113 (2004)
20. Newsham, Z., Ganesh, V., Fischmeister, S., Audemard, G., Simon, L.: Impact of community structure on SAT solver performance. In: Sinz, C., Egly, U. (eds.) SAT 2014. LNCS, vol. 8561, pp. 252–268. Springer, Heidelberg (2014)
21. Niedermeier, R.: Invitation to Fixed-Parameter Algorithms. Oxford Lecture Series in Mathematics and its Applications. Oxford University Press, Oxford (2006)
22. Nishimura, N., Ragde, P., Szeider, S.: Solving #SAT using vertex covers. Acta Informatica **44**(7–8), 509–523 (2007)
23. Robertson, N., Seymour, P.D.: Graph minors. II. Algorithmic aspects of tree-width. J. Algorithms **7**(3), 309–322 (1986)
24. Rose, D.J.: On simple characterizations of k-trees. Discrete Math. **7**, 317–322 (1974)
25. Samer, M., Szeider, S.: Fixed-parameter tractability. In: Biere, A., Heule, M., van Maaren, H., Walsh, T. (eds.) Handbook of Satisfiability, chapter 13, pp. 425–454. IOS Press (2009)
26. Samer, M., Szeider, S.: Algorithms for propositional model counting. J. Discrete Algorithms **8**(1), 50–64 (2010)
27. Szeider, S.: Minimal unsatisfiable formulas with bounded clause-variable difference are fixed-parameter tractable. J. of Computer and System Sciences **69**(4), 656–674 (2004)
28. Szeider, S.: On fixed-parameter tractable parameterizations of SAT. In: Giunchiglia, E., Tacchella, A. (eds.) SAT 2003. LNCS, vol. 2919, pp. 188–202. Springer, Heidelberg (2004)
29. Vardi, M.Y.: Boolean satisfiability: theory and engineering. Communications of the ACM **57**(3), 5 (2014)
30. Živný, S.: The Complexity of Valued Constraint Satisfaction Problems. Cognitive Technologies. Springer (2012)
31. Zhang, W., Pan, G., Wu, Z., Li, S.: Online community detection for large complex networks. In: Rossi, F. (eds.) Proceedings of the 23rd International Joint Conference on Artificial Intelligence, IJCAI 2013, Beijing, China, August 3–9, 2013. IJCAI/AAAI (2013)

Using Community Structure to Detect Relevant Learnt Clauses

Carlos Ansótegui[1], Jesús Giráldez-Cru[2(✉)], Jordi Levy[2],
and Laurent Simon[3]

[1] DIEI, Universitat de Lleida, Lleida, Spain
carlos@diei.udl.cat
[2] Artificial Intelligence Research Institute,
Spanish National Research Council (IIIA-CSIC), Barcelona, Spain
{jgiraldez,levy}@iiia.csic.es
[3] LaBRI, Université de Bordeaux, Talence, France
lsimon@labri.fr

Abstract. Nowadays, Conflict-Driven Clause Learning (CDCL) techniques are one of the key components of modern SAT solvers specialized in industrial instances. Last years, one of the focuses has been put on strategies to select which learnt clauses are removed during the search. Originally, one need for removing clauses was motivated by the finiteness of memory. Recently, it has been shown that more aggressive clause deletion policies may improve solvers performance, even when memory is sufficient. Also, the utility of learnt clauses has been related to the modular structure of industrial SAT instances.

In this paper, we show that augmenting SAT instances with learnt clauses does not always make them easier for the SAT solver. In fact, it makes worse the solver performance in many cases. However, we identify a set of highly useful learnt clauses, and we show that augmenting SAT instances with this set of clauses contributes to improve the solver performance in many cases, especially in satisfiable formulas. These clauses are related to the community structure of the formula, and they can be computed in a fast preprocessing step. This would suggest that the community structure may play an important role in clause deletion policies.

1 Introduction

Modern CDCL SAT solvers have been shown to be very efficient at solving industrial, or real-world, SAT instances. They integrate four major components: conflict-driven clause learning [17], activity-based variable branching heuristics [13], lazy data structures [13], and restarts [8]. In [10], it is empirically shown that these four components contribute to such success, but clause learning is the most important. Most CDCL solvers learn just one clause each time a conflict is found for the partially computed assignment. It has been observed

This work is partially supported by the CSIC project 201450E045, and the Ministerio de Economía y Competividad research project TASSAT2: TIN2013-48031-C4-4-P.

M. Heule and S. Weaver (Eds.): SAT 2015, LNCS 9340, pp. 238–254, 2015.
DOI: 10.1007/978-3-319-24318-4_18

that not all learnt clauses have the same usefulness or relevance. Moreover, a clause may be relevant at a certain instant of the search, but it may become useless later. Clause removal policies were initially proposed with the objective of saving memory and speed up propagations by the solver [7,13]. But the picture is more complex now. Since Glucose [2], aggressive clause removal policies are essential ingredients of CDCL solvers (more than 95% of the learnt clauses can be removed) and the initial arguments for clause database managements (unit propagation speed and memory issues) do not completely hold anymore. The intriguing question on how to predict efficiently and effectively the relevance of new learnt clauses is still open.

The structure of a SAT instance may be modeled as a graph with variables as nodes and clauses as edges. In [1], authors use this model to show that industrial SAT instances usually exhibit a clear community structure, i.e., high modularity. This means that we can find a partition of this graph into communities, with many edges between nodes of the same community and few edges connecting distinct communities. In [15], it is shown that the measure proposed in Glucose (i.e., the Literal Block Distance or *LBD*) can be strongly related to the community structure of the initial formula. However, this last result was just a one-way observation of the CDCL SAT solvers behavior: while LBD seems related to the number of communities in a learnt clause, it was not possible until now to exploit this correlation the other way, i.e., by using the community structure to guide the search in a CDCL SAT solver.

In this work, we show that community structure can be used to detect relevant learnt clauses. In particular, we present a technique that uses this structure to transform the formula adding learnt clauses, and hence guiding the search. This causality is much stronger than the previous observed correlation. Although we present our technique as a preprocessor for readability, our contribution is to give empirical evidence that the community structure can be used to generate relevant clauses, which is much stronger than identifying them (e.g., LBD is used to rank existing clauses). This would suggest that the community structure may play an important role in clause deletion policies.

Our preprocessor uses the community structure to split the instance into disjoint subformulas, and augments it with the learnt clauses of solving pairs of such subformulas. Intuitively, these clauses could be related to the notion of *glue clauses* used in Glucose. Our inspiration comes from the observation that clause learning destroys the (original) community structure of the instance. We give empirical evidence about the commonly accepted claim that having more learnt clauses does not always speed up the solving process. However, we show that augmenting the instance with our technique works experimentally. This is the case in several sets of industrial benchmarks and several CDCL SAT solvers. Notice that augmenting a formula with learnt clauses is against the common idea of preprocessing, which generally tries to reduce the instance.

The rest of this paper is structured as follows. After some preliminaries in Section 2, we review in Section 3 some observations about the effect of clause learning on the community structure of SAT instances. In Section 4, we provide some insights on the relevance of clauses learnt by a CDCL solver. In Section 5, we propose an algorithm that exploits the community structure to detect relevant clauses, and evaluate its performance in Section 6. We review some related works in Section 7, and we conclude in Section 8.

2   Preliminaries

The Boolean Satisfiability Problem (SAT) is the problem of determining if the variables of a propositional formula can be assigned in such a way that the formula is evaluated as `true`. A *literal* is either a Boolean variable x or its negation $\neg x$, a *clause* is a disjunction of literals, and a conjunctive normal form (*CNF*) instance is a conjunction of clauses.

An undirected weighted graph G is a pair $G = (V, w)$, where V is the set of nodes, and $w : V \times V \to \mathbb{R}^+$ is the edge-weight function that satisfies $w(x, y) = w(y, x)$.

The Variable Incidence Graph (VIG) of a SAT instance Γ is the graph whose nodes represent the variables of Γ, and there exists an edge between two variables if they both appear in a clause c. A clause with l literals results into $\binom{l}{2}$ edges. Thus, to give the same relevance to all clauses, edges have a weight $w(x, y) = \sum_{\substack{c \in \Gamma \\ x, y \in c}} 1/\binom{|c|}{2}$, where $|c| = l$ is the length of the clause c.

The *community structure* of a graph is usually measured using the notion of *modularity* [14]. Defined for a graph G and a partition P of its vertexes into communities, the modularity Q (see Eq. 1) measures the fraction of internal edges (edges connecting vertexes of the same community) w.r.t. a random graph with same number of vertexes and same degree. This avoids that the best partition is the one made up by an only community containing all vertexes.

$$Q(G, P) = \sum_{P_i \in P} \frac{\sum_{x, y \in P_i} w(x, y)}{\sum_{x, y \in V} w(x, y)} - \left(\frac{\sum_{x \in P_i} deg(x)}{\sum_{x \in V} deg(x)} \right)^2 \tag{1}$$

The modularity of a graph is the maximal modularity for any possible partition: $Q(G) = \max\{Q(G, P) \mid P\}$. This optimal modularity will be in the range $[0, 1]$. Computing the modularity of a graph is NP-hard [5]. Due to its complexity, instead of computing the (exact) modularity, most of methods in the literature approximate a lower-bound in the value of Q, trying to find a partition that maximizes this value. One of the most accurate and fastest algorithms is the Louvain method [4], extensively used to compute the modularity of large real-world networks.

In this work, we use the Louvain method to compute a partition of the formula into disjoint subformulas (i.e., sets of clauses). The cost of this algorithm depends on the number of nodes of the graph. We run this algorithm on the VIG, which is one of the graph representation of the formula with smallest number of

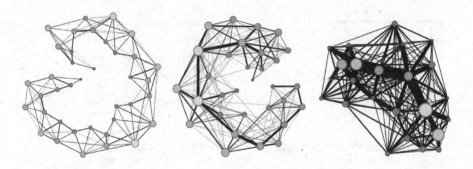

Fig. 1. Graph of communities of the instance `ibm-2002-22r-k60`: original formula (left), solved formula considering *small* learnt clauses (center), and solved formula considering *small* and *medium-sized* learnt clauses (right). Nodes and edges are accordingly scaled by community size and weight, respectively.

nodes[1], and we assign each clause to the most frequent community among its variables (randomly assigned in case of ties). We have observed that this formula partitioning (using the VIG) is similar to the one obtained using other graph models, but its computation is much faster.

3   Clause Learning Destroys the Community Structure

In this section, we review some observations about the community structure of real-world SAT formulas, clause learning, and the relation between them.

Industrial SAT instances have been shown to have a very clear community structure, with modularity Q in the VIG higher to 0.7 in most of the cases. Recall that the maximum value of Q is 1. This means that we can find a partition of their variables into communities, such that clauses mainly constraint variables of the same community. However, this partition is destroyed by the addition of learnt clauses [1], as we will see in this section.

In order to represent how this (initial) community structure is destroyed by the effects of clause learning, we can use the graph of communities[2]. This graph is built as follows: all nodes of the VIG (variables) that belong to the same community are merged into a single node in the graph of communities, and weighted edges are updated accordingly.[3] In Fig. 1 (*left*), we represent the graph of communities of the industrial formula `ibm-2002-22r-k60`. This instance has a modularity $Q = 0.91$ and 35 communities. Glucose solved this formula keeping a total of 504964 learnt clauses. We can recompute the graph of communities after adding some of these learnt clauses to the original instance. In Fig. 1 (*center* and

---

[1] In other models, clauses are represented as nodes in the graph.

[2] We cannot directly represent the VIG due to its large number of nodes (variables).

[3] The weight of the edge connecting communities A and B is the addition of the weights of the edges connecting one node from A and one node from B.

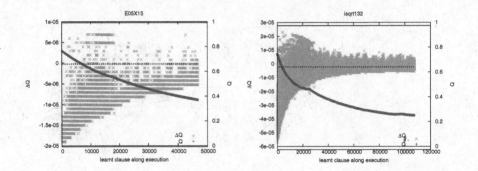

Fig. 2. Impact of adding learnt clauses on modularity, in instances E05X15 (left) and isqrt1_32 (right). Each point (x, y), with y measured in the left Y axis, represents a clause learnt at instant x and increasing Q on y. We also represent the evolution of the modularity Q (using the right Y axis).

right), we represent the graph of communities after adding *small* learnt clauses (up to 10 literals), and *medium-sized* learnt clauses (up to 50 literals), respectively.[4] In these graphs of communities, the node size is scaled according to the number of variables that belong to each community. Also, edges are scaled by their weights. Notice that edges weights are computed using the weights of the VIG (i.e., taking into account the length of the clauses). As it is stated in [1], the community structure is clear in all of these three graphs. However, as we consider more learnt clauses, we can observe two phenomena. First, the number of communities (number of nodes in the graph of communities) decreases. This means that variables that originally belonged to distinct communities are now grouped into the same community. Second, the weight of the inter-communities edges increases. Therefore, from the two previous effects, we observe that the solver prefers to learn clauses containing variables of distinct (original) communities (also stated in [1]). This means that, in general, clause learning contributes to decrease the modularity.

A question now is: are there some learnt clauses that contribute to increase the modularity even when most of them do not? In order to answer this question, we can measure the increase of the modularity ΔQ that each learnt clause produces. Notice that ΔQ is positive when most of the new edges generated by such clause connect nodes (variables) of the same community. Otherwise, ΔQ is negative. After an extensive experimentation, we see that, in general, learnt clauses produce a very small decrease of the modularity (i.e., $\Delta Q < 0$, in most cases). In Fig. 2, we represent this analysis for the industrial instances E05X15 and isqrt1_32. Each point (x, y), with y measured in the left Y axis, represents a clause learnt at instant x and increasing Q on y. We also represent (using the right Y axis) the value of the modularity Q using the original partition of

---

[4] As each clause of length l generates $\binom{l}{2}$ edges, it is hard to compute these graphs using *long* clauses.

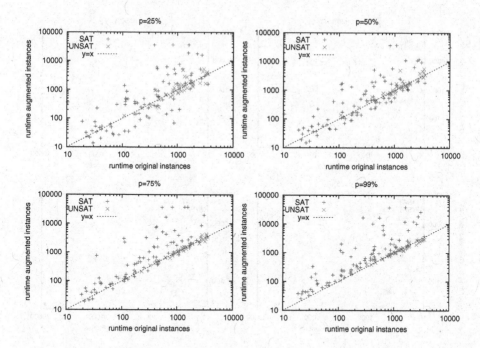

Fig. 3. Scatter plots of solving original instances (first step) versus generating and solving formulas augmented with learnt clauses (second plus third steps), at $p = 25\%, 50\%, 75\%$ and 99%

variables, along the execution. We can see that, even when some learnt clauses contribute to increase the value of Q, most of them do not (i.e., $\Delta Q < 0$), and thus Q tends to decrease. Due to space limitations, we only represent this analysis in two benchmarks. However, we observed similar results in most industrial SAT instances studied. Therefore, we can conclude that, in general, learnt clauses contribute to destroy the (original) community structure of the formula. It is not due to some particular clauses but rather a general phenomenon of the learning mechanism.

4   On the Relevance of Learnt Clauses

In this section, we try to answer the following question: if we augment the original formula with a set of learnt clauses obtained from some CDCL solver, will this contribute to solve the formula faster? In order to answer this question, we first introduce the notion of *relevant* clauses.

Definition 1. *Given a SAT solver S, a formula Γ, and a set of clauses φ, we say that φ is relevant for Γ and S, if φ is a logical consequence of Γ and $\Gamma \cup \varphi$ is easier to solve for S than Γ.*

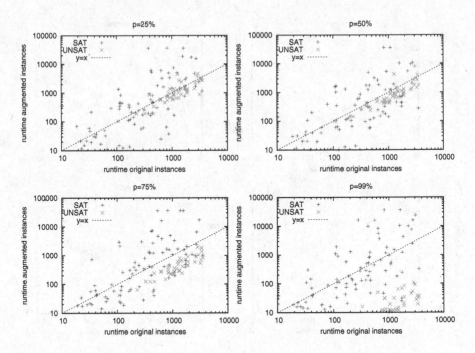

Fig. 4. Scatter plots of solving original instances (first step) versus solving formulas augmented with learnt clauses (third step), at $p = 25\%, 50\%, 75\%$ and 99%

Notice that in this definition we neglect the time needed to compute φ. Obviously, previous definition is informal. In order to experimentally validate if a set of clauses is relevant, we have considered a significant set of industrial instances.

In a first experiment, we select the set of instances of the application track of the SAT Competition 2013 solved in less than one hour. Notice that this set contains both satisfiable and unsatisfiable instances. This experiment is divided in three steps. In all of them, we use the CDCL SAT solver MiniSAT [7].

First step: we compute the number of conflicts c needed to solve the formula in an arbitrary run.

Second step: we repeat the same execution stopping the search after a certain number of conflicts $p \cdot c$ (where $0 < p < 1$), and we generate a new instance augmenting the original formula with the learnt clauses stored in the solver at that instant.

Third step: we solve the *augmented formula* generated in the previous step.

We could think that the third step is just the continuation of the second step due to a restart after $p \cdot c$ conflicts. But this is far from being true. First, CDCL SAT solvers do have more contextual information than learnt clauses, such as the activity counters, status of restarts, etc. It is also interesting to notice that the phase caching scheme [16] is not saved in the third step: a learnt clause could have been responsible for a propagation, and thus responsible for setting the phase caching scheme when backtracking, but this learnt clause could have been removed. Second, the learnt clauses used to generate the augmented formula will be treated as original clauses in the third step, i.e., they cannot be removed by the solver.

Since we limited the number of conflicts to $p \cdot c$ in the second step, you could expect to need around $(1-p) \cdot c$ conflicts to complete the search in the third step. Surprisingly, in our experiments, this is true when the instance is unsatisfiable, but not when it is satisfiable. If the formula is satisfiable, the aggregated runtime of generating the augmented formula (second step) and solving it (third step) is usually higher than the runtime required to solve the original instance (first step).

Let us present these observations in detail. In Fig. 3, we present the scatter plot of the runtime of solving the original formula (first step) versus generating and solving the augmented formula (second plus third steps), with $p = 25\%, 50\%, 75\%$ and 99%, and distinguishing SAT and UNSAT instances. In unsatisfiable instances, there is almost no difference (i.e., almost all points are on the diagonal). On the contrary, in satisfiable formulas the differences are much bigger (almost all points are far from the diagonal). Moreover, as we increase p, solving original instances is faster than generating and solving their corresponding augmented formulas (almost all points are above the diagonal). In Fig. 4, we present the scatter plots of solving the original formula (first step) versus just solving the augmented formula (third step). Notice that in this case, we do not take into account the runtime needed to generate these augmented instances. However, even in this case, solving some satisfiable augmented instances takes more time than solving their corresponding original formulas.

We have observed that augmenting an instance with learnt clauses does not always contribute to make it easier, when the formula is satisfiable. Let us conjecture why. First, although adding learnt clauses helps to reduce the search space, there are other key components, such as the activity counters and the phase component caching. These heuristics are set to their *optimal*[5] values after a certain number of conflicts. The phase component caching may play a crucial role here, since the solver may use this information to keep the solution to a subproblem. Therefore, even if we have an oracle providing a set of learnt clauses, this does not mean that you will find a satisfying assignment faster. Also notice that the status of the activity counters cannot be reproduced from this set of learnt clauses. These counters depend on all clauses learnt during the execution of a solver, but some of them may have been removed, and therefore they do not belong to the provided set anymore. Second, in [18] it was shown that the runtime of solving unsatisfiable formulas is much more robust than for satisfiable ones. Shuffling the instance may have an important impact on satisfiable problems, but not on unsatisfiable ones: the effort to find the UNSAT answer (and the size its proof) are always of the same order. If we try to link our result to

---

[5] In order to guide the search to a satisfying assignment.

Algorithm 1. Modularity-based SAT Instance Preprocessor (**modprep**)

Input: SAT Instance Γ
Output: SAT Instance Γ'

1 $\Gamma' := \Gamma$;
2 $C := communityStructure(\Gamma)$;
3 **foreach** *pair* (c_i, c_j) *of connected communities of* C **do**
4   Solver s;
5      $s.solve(c_i \cup c_j)$;
6   **if** $s == UNSAT$ **then**
7   **return** $\varnothing$;
8      $\Gamma' := \Gamma' \cup s.learntClauses$
9 **return** Γ';

this work, we think a reasonable explanation is the following one. For satisfiable instances, the solver is mostly starting again the whole search, trying to *learn* the correct phase component caching values. In this case, adding learnt clauses can slightly help, but the overall process is dominated by the high discrepancy of CPU time needed for satisfiable problems when shuffling the instance. For unsatisfiable instances, this shows that the solver is *continuing* the same proof.

Therefore, even when adding learnt clauses does not always help in satisfiable instances, is it possible to find a set of highly useful clauses that makes these formulas easier? In the next section, we will show that we can use the community structure to identify some clauses that are indeed relevant for those instances, i.e., they help to solve satisfiable instances faster.

5 Detecting Relevant Learnt Clauses

Learnt clauses are redundant by definition, hence not strictly necessary. However, they can help to prevent exploring the same unsatisfiable subspaces during the search. Moreover, their role could be to *guide* the solver in building the UNSAT proof by resolution. It is essential here to see CDCL SAT solvers as a combination of backtrack search algorithms (where learnt clauses are used to prevent exploring the same search space) and resolution proof engines (where learnt clauses are used to derive new learnt clauses).

In the early versions of CDCL solvers, memory was an important issue [7,13]. Therefore, some heuristics were proposed to remove useless clauses. Moreover, it is important to correctly manage the learnt clauses database in order to maintain a good unit propagation speed. More recently [2], some clause removal policies have been proposed. They aggressively remove most of the learnt clauses (95% of the learnt clauses can be removed). The proposed strategy is now one of the standards in CDCL engines. Thus, this policy is not only about maintaining good unit propagation rates, but also to guide the solver to some *easier* proofs. In Glucose, it was proposed to consider the number of decision levels occurring in a learnt clause as a measure of its quality (this was called Literal Block Distance, *LBD*, lower is better). The idea was that literals propagated at the same decision level were tightly connected and may often be propagated again and again together. Clauses of LBD 2 (called *glue clauses*) are kept forever in Glucose. Recently [15], it was shown that the LBD value was correlated to the number of communities of the clause. In this section, we show that community structure can be used to detect relevant learnt clauses.

In Alg. 1, we propose a technique presented as a preprocessing step, called *Modularity-based SAT Instance Preprocessor* (**modprec**). It augments the original formula with some learnt clauses based on its community structure. This algorithm proceeds as follows. First, it computes the community structure of the original formula (line 2), as described in Section 2. Recall that each community represents a set of clauses of the original instance. Then, for each pair of connected communities[6], it creates a subformula containing both communities, and solves it (line 5). If this subformula is UNSAT, it returns the empty clause. Otherwise, the original instance is augmented with the clauses the solver learnt for solving such subformula (line 8). Finally, it returns the augmented instance.

Notice that the previous algorithm imposes a very strong condition, which is solving *all* subformulas between two connected communities and keeping *all* learnt clauses found in this process. This could be further refined. Moreover, this preprocessing step could be heuristically applied during the search in the flavor of inprocessing approaches [9].

Although we will show in next section that this approach works experimentally, we may wonder why these learnt clauses indeed improve the performance of the solver. It is worth noticing that, by construction, these learnt clauses are usually composed of at most 2 communities, and thus are clearly related to the notion of *glue clauses* aforementioned. In addition, as we showed in Section 3, learnt clauses contribute to destroy the original community structure. In order to do this, we first need to connect pairs of communities, then triples of communities, and so forth; since we learn clauses that connect all communities (i.e., the whole formula) and we derive the empty clause. Therefore, we do not want to erase the base of this process (clauses connecting pairs of communities). Notice that a solver not aware of the community structure may remove them, unless, as we do, these clauses are added in a preprocessing step as original clauses. i.e., the solver will not remove them.

In this work, we only consider learnt clauses connecting pairs of communities at the preprocessing step, and not triples or higher arities. This is because the combinatorial space for pairs can be managed efficiently by the SAT solver. For bigger arities, we would need some additional filtering criterion, or working on a parallel solver (discussed in Section 8).

---

[6] Two communities are connected if there exists at least one variable appearing in both of them.

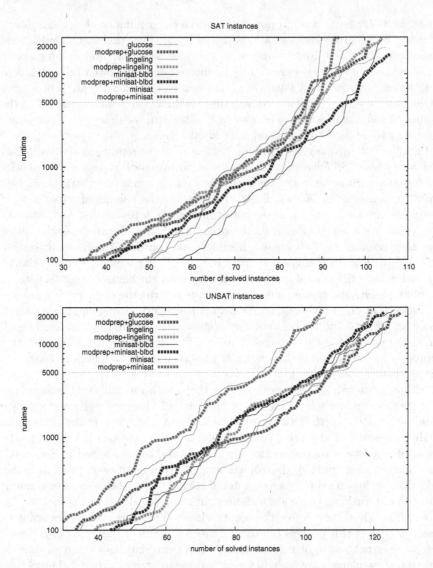

Fig. 5. Evaluation of application instances of the SAT Competition 2011, distinguishing satisfiable instances (top) and unsatisfiable instances (bottom), for Glucose, Lingering, MiniSAT-blbd, and MiniSAT; with and without using our preprocessor

6 Experimental Evaluation

In this section, we present an experimental evaluation of the modularity-based preprocessor presented in the previous section. All experiments were run in a cluster of 9 nodes IBM dx360 M2, each of them with 32GB of RAM and 2 processors Intel(R) Xeon(R) CPU L5520 2.27 GHz, limiting all experiments to a single core and to a maximum of 4GB of RAM. We use four representative CDCL SAT solvers: MiniSAT [7], Lingeling [3], Glucose [2], and MiniSAT-blbd [6]. MiniSAT is one of the most popular CDCL SAT solvers, while the three others were the best ranked solvers in the application track of the last SAT Competition 2014: Lingeling won both the UNSAT and the SAT+UNSAT tracks, MiniSAT-blbd won the SAT track, and Glucose was the second classified in the UNSAT track.

First, we evaluate how expensive is running the preprocessor described in Alg. 1 on a set of industrial SAT instances. We use the 300 application instances of the SAT Competition 2011. Notice that Alg. 1 can be split into two steps: i) partitioning the input formula into subformulas; and ii) solving them.

We compute the community structure as described in [1][7]. For this set of 300 application instances, this tool is able to correctly compute the community structure of 298 instances. This process is, in general, very fast. The average, median and maximum runtimes are respectively 12.6, 4.3 and 294.5 seconds.

Then, we solve all subformulas using MiniSAT. This step is performed on the 298 industrial formulas, with an average, median and maximum runtime of 78.0, 21.8 and 975.8 seconds, respectively. The average, median, maximum and minimum number of clauses that our preprocessor learnt is 11243.9, 512, 794950 and 1 clauses, respectively. A natural question now is if the number of clauses learnt with this preprocessor depends on the solver used to solve such subformulas. We run again this step using Glucose instead that MiniSAT. Notice that Glucose uses a more aggressive clause removal policy. However, we observe that this solver learns, in general, a similar number of clauses as MiniSAT, and needs a similar runtime to solve these subformulas. This is because the input subformulas are, in general, very easy.

In the next experiment, we evaluate the performance of the mentioned solvers, with and without using the presented preprocessor (referred in the plots as *<solver>* and *modprep+<solver>*, respectively). In Fig. 5, we represent the plots of this evaluation (solvers with and without using the preprocessor) for the industrial instances of the SAT Competition 2011, distinguishing between satisfiable and unsatisfiable instances. We represent a cactus plot (i.e., the maximum runtime of solving a set of instances) with logarithmic Y axis. The timeout is set to 25000 seconds (the timeout usually used in competitions is 5000 seconds). We remark that the reported runtime when the preprocessor is used include the runtime of computing the community structure and the runtime of solving all subformulas. We observe that using our preprocessor with MiniSAT, Glucose or MiniSAT-blbd improves their performance in satisfiable instances. Moreover, in unsatisfiable instances, Glucose also improves its performance. Interestingly, for this timeout of 25000 seconds, enhancing a solver with our preprocessor results into the best choice for solving satisfiable instances (using MiniSAT-blbd) and unsatisfiable instances (using Glucose). More interestingly, the solver MiniSAT-blbd enhanced with our preprocessor also results into the best technique to solve satisfiable instances when a timeout of 5000 seconds is considered (similar to the

---

[7] Tool available in http://www.iiia.csic.es/~jgiraldez/software.

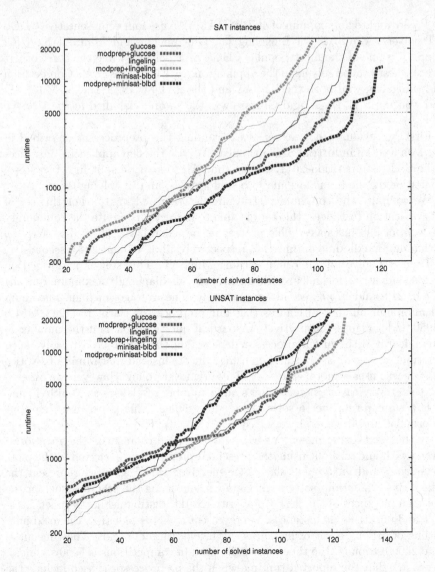

Fig. 6. Evaluation of application instances of the SAT Competition 2014, distinguishing satisfiable instances (top) and unsatisfiable instances (bottom), for Glucose, Lingering, and MiniSAT-blbd; with and without using our preprocessor

timeout used in the competition). It is worth noting that, for very easy instances, the overhead of the preprocessor (i.e., computing the community structure and solving all subformulas) does not compensate.

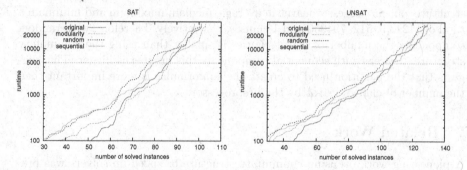

Fig. 7. Evaluation of random and sequential partitions, distinguishing between satisfiable (left) and unsatisfiable formulas (right), using the set of industrial instances of the SAT Competition 2011, and solved by Glucose

We want to validate if the previous results also hold in a different set of industrial instances. We repeat the same experiment[8] for the set of 300 application instances of the SAT Competition 2014. In Fig. 6, we represent the cactus plot of this experiment, distinguishing between satisfiable and unsatisfiable instances. Again, we observe that Glucose and MiniSAT-blbd improve their performance in both satisfiable and unsatisfiable instances when the preprocessor is used. In fact, MiniSAT-blbd enhanced with our technique is the best solver in satisfiable instances. Interestingly, these solvers also improve their performance using a shorter timeout of 5000 seconds. For instance, in our cluster MiniSAT-blbd solves 97 SAT instances, while this solver enhanced with our preprocessor solves 111. This difference is significant in the context of competitions. Also, Glucose solves 194 SAT+UNSAT instances, while using our technique with this solver results into a total of 206 SAT+UNSAT solved instances. Again, this difference is significant. However, our preprocessor does not improve the performance of Lingeling.

Finally, we want to check if a random partition of the formula would have the same effect as the partition provided by the community structure. For every instance, we compute a random partition of the formula with the same number of components as in the community structure. Also, we compute a sequential partition, where all variables of a component have sequential indexes. Then, we repeat all the experimentation with these random and sequential partitions. In Fig. 7, we show the cactus plot of the results on the set of industrial instances of the SAT Competition 2011. As expected, none of these methods performs better than either our proposed technique or solving the original instances.

Notice that in the previous experiment, the average, median, maximum and minimum number of clauses learnt by our preprocessor was respectively 4015.12, 28, 209085 and 0 clauses using the random partition, and 35360, 987, 951839 and 0 clauses using the sequential partition. Recall that using the community

---

[8] Excluding MiniSAT.

structure, our preprocessor learnt in average, median, maximum and minimum a total of 11243.9, 512, 794950 and 1 clauses, respectively. Therefore, with random components the number of learnt clauses is smaller than using the community structure, whereas with sequential components this number is bigger. This suggests that the partition used to create the subformulas is more important than the number of clauses learnt by the preprocessor.

7   Related Work

A pioneering work on using community structure to speed-up solvers was presented in [12]. In particular, they proposed to solve Maximum Satisfiability formulas by partitioning them according to the community structure and adding incrementally to the MaxSAT solver the sets of clauses related to communities.

In [11], it is shown that learnt clauses are most likely to be composed by variables on the fringes between communities. Interestingly enough, this confirms that the learning scheme tends to destroy the community structure: adding clauses with internal variables of communities would increase the clustering into communities. However, adding links between clusters by linking variables on their fringes seems to be more efficient.

As already mentioned, our work is also related to the work in [15]. Our current work contributes to confirm this by suggesting that good clauses are composed of variables from a few communities but, for the first time, it was possible to guide a CDCL SAT solver by the community structure of the formula. In particular, we think we were able to guide the solver to learn a set of initial *glue clauses*.

8   Conclusions and Future Work

In this paper, we use the community structure of industrial SAT instances to identify a set of highly useful learnt clauses. We show that augmenting a SAT instance with clauses learnt by the solver during its execution does not always mean to make the instance easier, especially in the case of satisfiable instances. However, we also show that augmenting the formula with a set of clauses based on the community structure of the formula improves the performance of the solver in many cases. Interestingly, this improvement is especially relevant in satisfiable instances. In particular, we use the set of clauses learnt from solving all subformulas consisting in pairs of connected communities.

We implement this approach as a preprocessor, and we show that it works experimentally on some representative sets of industrial instances, especially in satisfiable formulas. Interestingly, the SAT solver MiniSAT-blbd, which was the winner of the satisfiable track of the last SAT Competition 2014, enhanced with our technique improves its performance. It is also the case of Glucose, which improves its performance when it is enhanced with our technique in both satisfiable and unsatisfiable instances. To the best of our knowledge, this is the first time that community structure has been used to improve the performance of a CDCL SAT solver.

An important development of our work could be the design of a parallel solver. Each core could work only on a subset of the initial clauses, without communications. This could also allow us to extend our approach to tuples of communities instead of pairs of communities.

Our approach can also be improved by trying to guess which pairs of communities are important to work on. We are currently investigating this. At last, it is also important to link the community structure of formulas with their initial problem and generation. Linking the original problem with the detected communities is also an ongoing work.

References

1. Ansótegui, C., Giráldez-Cru, J., Levy, J.: The community structure of SAT formulas. In: Cimatti, A., Sebastiani, R. (eds.) SAT 2012. LNCS, vol. 7317, pp. 410–423. Springer, Heidelberg (2012)
2. Audemard, G., Simon, L.: Predicting learnt clauses quality in modern SAT solvers. In: Proc. of IJCAI 2009, pp. 399–404 (2009)
3. Biere, A.: Lingeling essentials, A tutorial on design and implementation aspects of the the SAT solver Lingeling. In: Proc. of POS 2014 (2014)
4. Blondel, V.D., Guillaume, J.L., Lambiotte, R., Lefebvre, E.: Fast unfolding of communities in large networks. Journal of Statistical Mechanics: Theory and Experiment 2008(10), P10008 (2008)
5. Brandes, U., Delling, D., Gaertler, M., Görke, R., Hoefer, M., Nikoloski, Z., Wagner, D.: On modularity clustering. IEEE Trans. on Knowledge and Data Engineering 20(2), 172–188 (2008)
6. Chen, J.: A bit-encoding phase selection strategy for satisfiability solvers. In: Gopal, T.V., Agrawal, M., Li, A., Cooper, S.B. (eds.) TAMC 2014. LNCS, vol. 8402, pp. 158–167. Springer, Heidelberg (2014)
7. Eén, N., Sörensson, N.: An extensible SAT-solver. In: Giunchiglia, E., Tacchella, A. (eds.) SAT 2003. LNCS, vol. 2919, pp. 502–518. Springer, Heidelberg (2004)
8. Gomes, C.P., Selman, B., Kautz, H.A.: Boosting combinatorial search through randomization. In: Proc. of the Fifteenth National Conf. on Artificial Intelligence, AAAI 1998, pp. 431–437 (1998)
9. Järvisalo, M., Heule, M.J.H., Biere, A.: Inprocessing rules. In: Gramlich, B., Miller, D., Sattler, U. (eds.) IJCAR 2012. LNCS, vol. 7364, pp. 355–370. Springer, Heidelberg (2012)
10. Katebi, H., Sakallah, K.A., Marques-Silva, J.P.: Empirical study of the anatomy of modern sat solvers. In: Sakallah, K.A., Simon, L. (eds.) SAT 2011. LNCS, vol. 6695, pp. 343–356. Springer, Heidelberg (2011)
11. Katsirelos, G., Simon, L.: Eigenvector centrality in industrial SAT instances. In: Milano, M. (ed.) CP 2012. LNCS, pp. 348–356. Springer, Heidelberg (2012)
12. Martins, R., Manquinho, V., Lynce, I.: Community-based partitioning for MaxSAT solving. In: Järvisalo, M., Van Gelder, A. (eds.) SAT 2013. LNCS, vol. 7962, pp. 182–191. Springer, Heidelberg (2013)
13. Moskewicz, M.W., Madigan, C.F., Zhao, Y., Zhang, L., Malik, S.: Chaff: Engineering an efficient SAT solver. In: Proc. of DAC 2001, pp. 530–535 (2001)
14. Newman, M.E.J., Girvan, M.: Finding and evaluating community structure in networks. Phys. Rev. E 69(2), 026113 (2004)

15. Newsham, Z., Ganesh, V., Fischmeister, S., Audemard, G., Simon, L.: Impact of community structure on SAT solver performance. In: Sinz, C., Egly, U. (eds.) SAT 2014. LNCS, vol. 8561, pp. 252–268. Springer, Heidelberg (2014)
16. Pipatsrisawat, K., Darwiche, A.: A lightweight component caching scheme for satisfiability solvers. In: Marques-Silva, J., Sakallah, K.A. (eds.) SAT 2007. LNCS, vol. 4501, pp. 294–299. Springer, Heidelberg (2007)
17. Silva, J.P.M., Sakallah, K.A.: GRASP: A search algorithm for propositional satisfiability. IEEE Trans. Computers 48(5), 506–521 (1999)
18. Simon, L.: Post mortem analysis of SAT solver proofs. In: Proc. of POS 2014, pp. 26–40 (2014)

Recognition of Nested Gates in CNF Formulas

Markus Iser[1], Norbert Manthey[2], and Carsten Sinz[1]($\boxtimes$)

[1] Karlsruhe Institute of Technology (KIT), Karlsruhe, Germany
{markus.iser,carsten.sinz}@kit.edu
[2] Dresden University of Technology (TUD), Dresden, Germany
norbert.manthey@tu-dresden.de

Abstract. We present a new algorithm to efficiently extract information about nested functional dependencies between variables of a formula in CNF. Our algorithm uses the relation between gate encodings and blocked sets in CNF formulas. Our notion of "gate" emphasizes this relation. The presented algorithm is central to our new tool, cnf2aig, that produces equisatisfiable and-inverter-graphs (AIGs) from CNF formulas. We compare the novel algorithm to earlier approaches and show that the produced AIG are generally more succinct and use less input variables. As the gate-detection is related to the structure of input formulas, we furthermore analyze the gate-detection before and after applying preprocessing techniques.

1 Introduction

The problem to automatically decide the satisfiability of propositional formulas (SAT) is important in numerous areas, from verification domains [9] to hardware layout [21] or AI planning [24]. Recent SAT solvers show a very good performance in solving large application instances with millions of variables. The reason is often thought to be their direct and indirect exploitation of problem-structure in application instances [25]. There are several notions of structure [4,7,29]. In this paper we focus on functional and partially functional relations between variables. Usually, application instances in CNF originate from more structured representations in full propositional logic, or are projections of formulations in higher order logics [26]. Tseitin-based CNF encodings are widely used to encode nested gate-structures to CNF [23,27,28]. Although the functional relations between variables are "hidden" in the CNF, they are still present and can be detected. We show that even random instances contain up to 10% variables that our approach recognizes as outputs of a gate-like structure. Furthermore we show that our approach can be used to improve SAT solver performance.

This work was partially supported by the SAT Association's Short-Term Mission (STM) program.

© Springer International Publishing Switzerland 2015
M. Heule and S. Weaver (Eds.): SAT 2015, LNCS 9340, pp. 255–271, 2015.
DOI: 10.1007/978-3-319-24318-4_19

1.1 Related Work

Efficient encodings of Boolean formulas to CNF rely on the introduction of new variable symbols for each connective in the formula. The well-known Tseitin [28] encoding and optimizations like the one of Plaisted and Greenbaum [23] are of this kind. Direct access to information about the original formula structure of the CNFs that are encoded that way is lost, however, yet it would be beneficial in many applications. For example, structural information can be used to minimize models [13] or to increase SAT solver performance [3,11,14]. Preprocessing techniques are also directly or indirectly related to the structure of CNFs that originate from circuits [16].

There are approaches that explicitly recover and extract such functional structure from existing CNFs. These are mostly based on fixed clausal patterns (as in [10,19,22]) or on detection of functional dependencies for variable elimination [8,18]. Detection is often limited to specific clausal patterns that arise from encoding basic Boolean functions such as AND or OR. For cardinality constraints, a semantic detection based on unit propagation has also been presented [3]. Blocked Set Decomposition has been successfully used to generate and-inverter-graphs (AIGs) from CNF; Balyo et al. [2] present an AIG encoding that is based on the solution algorithm for blocked sets that we explain in Section 2.3.

1.2 Contributions

We present a general notion of gates that encompasses arbitrary functional relations between one output and several input variables in CNF. We also investigate the relation between *blocked clauses* and functional relations. In our opinion, this has not been made clear enough in the past. We also see great potential in connecting CNF encodings and pre-processing techniques more closely.

We developed the tool cnf2aig, which implements a new algorithm for gate detection and generates AIGs from arbitrary CNFs. We show that the AIGs generated by our tool are more compact and use fewer variables than those produced by the tool presented in [2]. Furthermore we compared SAT solver performance in an experimental tool-chain that was first presented in [2] and show that using our gate recognition approach results in faster runtimes.

We furthermore experimentally show that some pre-processing techniques, such as bounded variable elimination, are detrimental to our gate detection approach, whereas especially blocked clause decomposition (BCD) but also bounded variable addition (BVA) have a rather positive effect on recognition rates.

2 Theoretical Background

We use a Boolean algebra with variables V and the common operators $\wedge, \vee, \neg$ with their usual semantics. A literal l is either a variable or its complement. Given a Boolean formula F, $\mathsf{vars}(F)$ denotes the set of variables and $\mathsf{lits}(F)$ the set of literals that occur in F. A formula is in Conjunctive Normal Form (CNF) if it is a conjunction of disjunctions of literals. A CNF C is represented by a set of clauses, where a clause is a set of literals. In the following, C_l denotes the subset of clauses in C that contain literal l, i.e. $C_l = \{c \in C \mid l \in c\}$. Given a CNF C, a literal $l \in \mathsf{lits}(C)$ is *pure* in F iff $C_{\bar{l}} = \emptyset$. The restriction $C|_{l=v}$ is derived from C by assigning v to literal l and subsequent simplification. A variable assignment $a : V \rightarrow \{0,1\}$ is represented by a set M_a of literals such that $v \in M_a$ iff $a(v) = 1$ and $\neg v \in M_a$ iff $a(v) = 0$.

Given two clauses c_1 and c_2 and a literal l, such that $l \in c_1$ and $\bar{l} \in c_2$, the *resolvent* $c_1 \otimes_l c_2$ is the clause $(c_1 \cup c_2) \setminus \{l, \bar{l}\}$ It holds that $c_1, c_2 \models c_1 \otimes_l c_2$. For sets of clauses C_1 and C_2, the set $C_1 \otimes_l C_2$ denotes the set of all resolvents between clauses in C_1 and C_2 on l.

Given a CNF C, a clause $c \in C$ is *blocked* in C if there exists a literal $l \in c$ such that for every clause $d \in C_{\bar{l}}$ the resolvent $c \otimes_l d$ is tautological. The literal l is also called the *blocking literal* of c. A set of clauses $D \subseteq C$ is blocked iff each clause D is blocked in C.

Given a formula F in CNF we can construct an equivalent formula G via *unit propagation*, denoted by $F \vdash_{\mathsf{UP}} G$, by removing for each unit clause $\{l\} \in F$ all clauses in F_l and replacing each clause $c \in F_{\bar{l}}$ by $c \setminus \{\bar{l}\}$, repeating the whole process until all unit clauses are processed.

2.1 Gates and Monotonicity

In this section we introduce gates as relations and show some interesting properties and how they correspond to their propositional encodings. By $\mathbb{B} = \{0, 1\}$ we denote the set of Boolean constants.

Definition 1 (Gate). *An n-ary gate G is a functional relation $G \subseteq \mathbb{B}^n \times \mathbb{B}$ of n input variables $P = (p_1, \ldots, p_n)$ and one output variable o.*

Functionality breaks down into the two properties *left-totality* and *right-uniqueness*, i.e. a relation is functional iff it is left-total and right-unique. A $n{+}1$-ary relation G with inputs $P = (p_1, \ldots, p_n)$ and output o is

- **left-total**   if $\forall P \in \mathbb{B}^n. \exists o \in \mathbb{B}.(p_1, \ldots, p_n, o) \in G$
- **right-unique** if $\forall P \in \mathbb{B}^n. \exists o \in \mathbb{B}.(p_1, \ldots, p_n, o) \notin G$

From left-totality follows that for every 2^n combinations of inputs there exists an output such that the tuple is in the gate and from right-uniqueness follows that there is *exactly* one output such that the tuple is in the gate.

Given a gate $G(p_1, \ldots, p_n, o)$ there exists a corresponding Boolean function g and a propositional encoding Γ of $o \leftrightarrow g(p_1, \ldots, p_n)$ such that for every tuple $T \in \mathbb{B}^n \times \mathbb{B}$ there exists a corresponding model $M_T \models \Gamma$ iff $T \in G$. Given a tuple $T = (t_1, \ldots, t_{n+1})$ and corresponding propositional variables τ_i for every t_i, then the corresponding model M is constructed such that $\tau_i \in M$ iff $t_i = 1$ and $\neg\tau_i \in M$ iff $t_i = 0$. In the following, for simplicity, we assume Γ to be directly encoded in CNF.

From *left-totality* follows that an encoding Γ of an n-ary gate G can be satisfied for all 2^n assignments to its input variables by picking the proper output assignment. This basic insight leads us to Proposition 1.

Proposition 1 (Left-Totality and CNF Encodings of Gates). *For any direct CNF encoding Γ of a gate G with output variable ω, each non-tautological clause $c \in \Gamma$ contains either ω or $\neg\omega$. Furthermore all resolvents $r \in \Gamma_\omega \otimes_\omega \Gamma_{\overline{\omega}}$ are necessarily tautologic.*

Proof. Let G be a gate with encoding Γ and let $c \in \Gamma$ be a non-tautological clause. Now assume that $\omega \notin c$ and also $\neg\omega \notin c$. This contradicts *left-totality* as such a clause c imposes a restriction on the input variables such that there is no possible output (0 or 1) in the corresponding gate G. The same argument can be applied to resolvents $r \in \Gamma_\omega \otimes_\omega \Gamma_{\overline{\omega}}$. By definition of resolution it holds that $\omega \notin \mathsf{vars}(r)$. As $\Gamma \models r$ the assumption that r is non-tautological also contradicts *left-totality* as such a resolvent also imposes a restriction solely on the input variables. So each resolvent $r \in \Gamma_\omega \otimes_\omega \Gamma_{\overline{\omega}}$ is necessarily tautologic. □

Usually, in CNF encodings a gate's output can be input to other gates and can even be inverted such that in compressed encodings its output is used in several polarities as input to other gates. We address these issues in the following definitions of *nesting* and *monotonicity*.

Definition 2 (Nesting of Gates). *A gate G with output o is directly nested in another gate H if output o is input to H. We denote this by $G < H$. Moreover, we use the symbol $<^+$ for the transitive closure of $<$, and say that G is nested in H if $G <^+ H$. Nesting is a transitive, irreflexive and asymmetric relation on gates, i.e. it imposes a strict partial order.*[1]

Assume a formula F in CNF and a set of gates $\mathbb{G} = \{G_1, \ldots, G_n\}$ contained in F, i.e. the clauses of their CNF encodings are contained in F. Then F can be partitioned into a "gate part", $F_\mathbb{G}$ and a remaining part $F_R = F \setminus F_\mathbb{G}$. In $\mathbb{G}$, some of the gates are "output gates" which are not nested in other gates. These are the maximal elements of $<$ in $\mathbb{G}$. We assume in the following that there is a unique maximal gate in $\mathbb{G}$. If this is not the case, we add an additional AND gate on top (with output o_F), connecting all outputs of the maximal gates in $\mathbb{G}$. We also integrate the remainder F_R into this gate by adding $\overline{o}_F$ as additional literal to each clause in F_R. Moreover, we add the unit clause $\{o_F\}$ to F. The resulting formula F' is equisatisfiable to F.

Thereby we can more easily detect partially encoded gates that provide a sufficient encoding if the gate is monotonic as we argue as follows.

A Boolean function $g(p_1, \ldots, p_i, \ldots, p_n)$ is monotonically increasing in argument i, if $a \leq b$ implies $g(p_1, \ldots, a, \ldots, p_n) \leq g(p_1, \ldots, b, \ldots, p_n)$, and monotonically decreasing if $a \leq b$ implies $g(p_1, \ldots, a, \ldots, p_n) \geq g(p_1, \ldots, b, \ldots, p_n)$. A Boolean function is monotonic iff it is monotonic in every argument. Common examples for monotonic functions are AND and OR, common examples for

---

[1] We assume combinational circuits here (as opposed to sequential circuits).

non-monotonic functions are XOR and EQIV. A gate is monotonic iff its output variable is determined by a monotonic function.

To detect partially encoded gates, we define in definition 3 the notion of a nesting polarity np for each gate. $np(G) \in \{p, n, 0\}$, where p indicates positive polarity, n negative polarity, and 0 indicates that the gate has no polarity and thus has to be encoded fully.

Definition 3 (Nesting Polarity, Monotonic Nesting). *Given a set of nested gates* $\mathbb{G}$ *with maximum gate* M, *then* $np(M) = p$, *and for each other gate the polarity is defined as follows: Given a gate* $G \neq M$, *we define the successor set* $S(G) = \{G' \in \mathbb{G} \mid G < G'\}$. *Now the polarity of* G *is determined by:*

- $np(G) = p$ *if for all elements* $G' \in S(G)$ *the following holds: if* $np(G') = p$ *then* G *is a monotonically increasing argument of* G', *or, if* $np(G') = n$ *then* G *is a monotonically decreasing argument of* G'.
- $np(G) = n$ *if for all elements* $G' \in S(G)$ *the following holds: if* $np(G') = n$ *then* G *is a monotonically increasing argument of* G', *or, if* $np(G') = p$ *then* G *is a monotonically decreasing argument of* G'.
- *Otherwise* $np(G) = 0$.

We call a gate G *monotonically nested in* $\mathbb{G}$ *if* $np(G) \neq 0$.

Optimizations of encodings Γ of gates often skip *right-uniqueness* of monotonic nested gates by using *partial gate encodings* (i.e. either $\Gamma_{\bar{o}}$ or Γ_o) to reduce the number of clauses [23]. In a nested gate structure we may use a partial encoding for G without changing satisfiability iff it is monotonically nested. We may use $\Gamma_{\bar{o}}$ if $np(G) = p$ and Γ_o if $np(G) = n$.[2]

When we decode a nested gate structure from CNF we basically reduce the number of input variables. We will later use the remaining number of input variables as a quality measure of our applied recognition methods.

2.2 And-Inverter Graphs (AIG)

An And-Inverter Graph (AIG) encodes a nested gate structure by only using binary AND-gates and logical negation [17]. Our tool cnf2aig produces AIGs in the input format that is described on the website http://fmv.jku.at/aiger/.

2.3 Preprocessing

Preprocessing, or formula simplification, is to apply techniques to a given formula F, such that the resulting formula F' is equisatisfiable. Furthermore, these simplification techniques are assumed to make solving an application formula simpler [8,15]. Some of the most powerful simplification techniques are *bounded variable elimination*, *bounded variable addition* and *blocked clause elimination* [1,15].

---

[2] There "emerge" additional models M that are no model for the fully encoded formula F, i.e. $M \models F \setminus \Gamma_o$ but $M \not\models F$. These models can be "repaired" by flipping the output literal, i.e. if G is nested monotonic it is guaranteed [23] that $M_{\bar{o}/o} \models F$.

However, these preprocessing techniques are also assumed to destroy the structure of F, such that F' is "less structured". In the following paragraphs we briefly summarize the properties of the mentioned simplification techniques. Furthermore, we describe blocked clause decomposition, which is based on blocked clause elimination.

Bounded Variable Elimination. Bounded variable elimination (BVE) [6,8] eliminates variables from a formula F by resolution. Given F_x and $F_{\overline{x}}$, then the set of all non-tautological resolvents S is created. If the number of clauses in S is less or equal to the number of clauses that contain the variable x, then F_x and $F_{\overline{x}}$ are replaced with S.

The result of applying variable elimination to a formula F on the variable x is the formula $F' = (F \cup (F_x \otimes_v F_{\overline{x}})) \setminus (F_x \cup F_{\overline{x}})$. Usually, the clauses in S are used for *subsumption* and *strengthening*, before the next variable of F is eliminated [8]. In case $F_x \cup F_{\overline{x}}$ contain a functional dependency for the variable x, the set of resolvents can be reduced further [8].

Bounded Variable Addition. Bounded variable addition (BVA) [20] can be understood as the reverse operation of BVE, because BVA adds a partial definition of a fresh variable v to the formula: First, a fresh variable v is introduced like in extended resolution [28], resulting in the intermediate formula $G = F \cup \{\{v, \overline{x}, \overline{y}\}, \{\overline{v}, x\}, \{\overline{v}, y\}\}$, where $x, y \in \text{lits}(F)$. Next, all clauses $C, D \in F$, which have a common subclause E such that $C = E \cup \{x\}$ and $D = E \cup \{y\}$ are replaced by the new clause $(v \vee E)$, resulting in the formula H. Finally, the formula F', the result of applying bounded variable addition, is obtained from the formula H by removing the clause $\{v, \overline{x}, \overline{y}\}$, because this clause is blocked. Hence, the variable v represents the output of a newly introduced partial gate. In implemented variants this gate is an and-gate with an arbitrary number of inputs [20].

Blocked Clause Elimination. Blocked clause elimination (BCE) [15] removes *blocked clauses* from a formula. When C is a blocked clause in F, then F' is obtained as $F \setminus C$. This removal is usually repeated until F' does not contain any blocked clause any longer. BCE can also reduce a CNF with a gate encoding from a full encoding to a partial encoding, wherever the formula allows this transformation [15].

Blocked Clause Decomposition. A formula F can be decomposed by blocked clause decomposition (BCD) [12] into two disjoint formulas G and H. Furthermore, by applying BCE to G the empty formula is obtained and likewise BCE on H returns the empty formula. Hence, both G and H are satisfiable.

For a blocked set G a satisfying assignment can be computed in a polynomial number of steps [12]. Assume the order of removing blocked clauses from G is known. Then, the assignment I is modified to satisfy the current blocked clause $C \in G$ with the blocking literal $l \in C$, if $I \not\models C$ by flipping the assignment of the literal l. Hence, the literal l can also be seen as the output of a gate C, as its assignment is flipped according to the remaining literals of C.

Balyo et al. present several methods to perform blocked clause decomposition in [2] . Their goal is to create one blocked set that is as large as possible next to a smaller blocked set. For that they apply a post-processing step to an original decomposition in order to move clauses from the smaller to the larger blocked set as long as this set remains blocked. In the end they create a new variable v, add a unit-clause $\{v\}$, and append the literal $\neg v$ to every clause in the small set. They also present an algorithm where they generate AIG from the two blocked sets they previously generated. These AIG basically simulate the algorithm for solving blocked sets, such that it introduces *versions* of blocking variables for each possible flip that occurs in the process of solving the blocked set.

3   Recognition of Nested Gates

In the following we use the fact that we can always construct a *maximum* partial gate encoding $M_{\overline{F}}$ with output r and function F by adding $\neg r$ to every clause in F such that $r \wedge M_{\overline{F}} \models F$. Our method is presented by the two algorithms 1 and 2, where algorithm 1 iteratively selects a root literal o as possible output of a gate and then uses algorithm 2 to test for existence of a recursive gate-structure with the given output o. If algorithm 2 successfully decodes a gate with output o it recursively continues on the inputs to decode possibly nested gates.

3.1   Iterative Root Selection

Algorithm 1 displays the outer loop recognizeGates of our gate recognition method. The input is a CNF formula F and a constant max-tries that constrains the number of iterations. The output is a set of tuples $\{(o, \Gamma)\}$, where each tuple (o, Γ) represents a gate with output literal o and encoding clauses Γ.

First (in line 1) we initialize the output set, and introduce two more sets S and C to keep track of the already processed clauses. For the given number of tries (line 2) we select a clause from F (line 3) and keep track of the such selected clauses in the set S (line 4) as we use them later to create the *maximum* partial gate.

As long as the formula contains unprocessed unit-clauses selectClause returns a unit-clause. If there is no unit-clause left we choose the literal with the least number of occurrences and randomly select one clause which contains that literal.

Then in line 6 we globally mark each literal l in the selected clause as *input*. This flag is used in algorithm 2 to check for monotonic nesting of the candidate gate encoding.

For each literal l in the currently selected clause we descent into recursive gate extraction via extractGate (line 8). The method extractGate (algorithm 2) checks if there exists an encoding $\Gamma \subseteq F$ with output literal l and if this is the case it recursively extracts all the nested gates. The method extractGate returns

---

Algorithm 1. Gate Recognition: recognizeGates(F, max-tries)

Data: F : CNF formula, max-tries: maximum number of extraction attempts
Result: G : a set of tuples (o, Γ) where Γ is the CNF of a gate encoding with
output literal o

1 $G \leftarrow C \leftarrow S \leftarrow \emptyset$
2 **for** $i \leftarrow 1$ **to** *max-tries* **do**
3      $c \leftarrow \text{selectClause}(F \setminus (C \cup S))$
4      $S \leftarrow S \cup \{c\}$
       // Globally mark inputs literals:
5   **for** $l \in c$ **do**
6         $\lfloor$ setAsInput (l)
       // Recursively extract nested gates:
7   **for** $l \in c$ **do**
8         $G' \leftarrow \text{extractGates}(l, (F \setminus (C \cup S))$
9         $G \leftarrow G \cup G'$
10   **for** $(o, \Gamma) \in G'$ **do**
11           $\lfloor$ $C \leftarrow C \cup \Gamma$

   // Create unique maximal gate:
12 $o \leftarrow newVar$
13 $\Gamma \leftarrow \emptyset$
14 **for** $c \in F \setminus C$ **do**
15   $\lfloor$ $\Gamma \leftarrow \Gamma \cup \{c \cup \{\bar{o}\}\}$
16 **return** $G \cup \{(o, \Gamma)\}$

---

a set of tuples with one tuple per recognized gate. We keep track of the encoding clauses of already recognized gates in C (line 11).

In the end (lines 12-15) we create a partial gate encoding for the remaining clauses including the previously selected clauses. This partial gate is then the maximum of our nested gate structure.

3.2 Recursive Gate Extraction

Algorithm 2 displays the inner recursive function extractGate that is used in the outer loop of algorithm 1 do the actual gate recognition. The algorithm's input is a CNF that contains the yet unprocessed clauses and an output literal. The algorithm's output like in algorithm 1 is the set of recognized gates represented by tuples (o, Γ) of an output literal o and encoding clauses Γ.

If for the given literal o a gate encoding $\Gamma \subseteq F$ is recognized, the algorithm descends recursively for each input of the extracted gate. At the end of the recursion the union of all extracted gates is created (line 17) and returned.

Note that we always extract all the clauses that contain the output literal in question. The first test in line 2 checks if these clauses block each other on the output literal. As we have seen in Proposition 1 this is a mandatory property to

---

Algorithm 2. Gate Extraction: extractGates(o,F)

Data: o : output literal, F : CNF formula
Result: G : a set of tuples (o, Γ) where Γ is the CNF of a gate encoding with output literal o

1 $G \leftarrow C \leftarrow \emptyset$
2 **if** $F_{\bar{o}}$ *block* F_o **then**
3      $C \leftarrow C \cup \{F_{\bar{o}} \cup F_o\}$
    // Check monotonicity of nesting:
4   **if** isSetAsInput($\bar{o}$) **then**
5   monotonic $\leftarrow$ false
6   **else**
7   monotonic $\leftarrow$ true
8   **if** monotonic $\vee$ isFullEncoding($o, F_{\bar{o}} \cup F_o$) **then**
9          $G \leftarrow G \cup \{(o, F_{\bar{o}} \cup F_o)\}$
10   inputs $\leftarrow$ lits($F_{\bar{o}}$) $\setminus \{\bar{o}\}$
        // Globally mark inputs literals:
11   **for** $l \in$ inputs **do**
12   setAsInput (l)
13   **if** $\neg$monotonic **then**
14   setAsInput ($\neg l$)

        // Recursively extract nested gates:
15   **for** $l \in$ inputs **do**
16             $G' \leftarrow$ extractGates($l, F \setminus C$)
17             $G \leftarrow G \cup G'$
18   **for** $(o, \Gamma) \in G'$ **do**
19                 $C \leftarrow C \cup \Gamma$

20 **return** G

---

ensure *left-totality*. If we are in a monotonic branch (line 7) we can then eagerly continue as we may skip *right-uniqueness* in the monotonic case (see Section 2.1).

Otherwise (line 8) we have to test if the clauses realize a full gate encoding, which includes an equivalence check. Details about the different methods of equivalence detection that we have implemented and evaluated can be found in Section 3.3.

In lines 12 and 14 we globally mark the input literals for follow-up monotonicity detection. Note that by using literals instead of variables as gate outputs we implicitly keep track of the nesting polarity. Also note that in line 14 we pass-on non-monotonicity to subsequently nested gates.

3.3   Equivalence Detection Methods

Given the candidate clauses Γ_p and $\Gamma_{\bar{p}}$ such that all clauses from $\Gamma_{\bar{p}} \equiv p \rightarrow g$ and $\Gamma_p \equiv p \leftarrow g'$ are blocked on p, we have previously seen that *left-totality* follows from the *blocked*-property. For non-monotonically nested gates we still have to show *right-uniqueness* to obtain equivalence. We implemented three methods to deal with that situation.

Skip. In this variant we just stop gate-recognition at non-monotonic inputs. This typically results in a larger remainder at the end of the recognition process. In the following, we refer to this method as SKIP.

Clausal Patterns. Our second method compares the candidate clauses Γ_p and $\Gamma_{\overline{p}}$ with a fixed set of known clausal patterns to ensure that both encode the same function. We implemented a simple check for standard AND- and OR-gate encodings. In the following we refer to this method as PATTERNS.

Semantic Analysis. Given the candidate clauses Γ_p and $\Gamma_{\overline{p}}$ such that $\Gamma_{\overline{p}} \equiv p \rightarrow g$ and $\Gamma_p \equiv p \leftarrow g'$ our third and most advanced method indirectly proves right-uniqueness. From $\Gamma_{\overline{p}}$ we extract g by $g = \Gamma_{\overline{p}}|_{p=1}$. Then we prove that $g \wedge \overline{p} \wedge \Gamma_p$ is UNSAT. In our current implementation we use a simplified, but incomplete method that uses unit-propagation and Minisat's *assumptions*. We start by initializing Minisat with the complete input formula F, and then try to prove by unit-propagation the unsatisfiability of $M \wedge \overline{p} \wedge F$ for every model M of g (note that $F \models \Gamma_p$) using the assumptions $M \wedge \overline{p}$.

As these checks are run quite often (eventually more than once per variable) and as the construction of all models of g can be exponential, we have bounded the number and the size of the clauses in g; i.e. we start semantic analysis only if $|\Gamma_{\overline{p}}| \leq k_1$ and for each $c \in \Gamma_{\overline{p}}, |c| \leq k_2$. We used the values $k_1 = 3$ and $k_2 = 4$. In the following we refer to this method as SEMANTIC.

3.4   AIG Construction

The output of our gate-recognition algorithm is a set of tuples $\{(o, \Gamma)\}$ of output literals o and encoding clauses Γ. These tuples can be further processed to produce an AIG. The following paragraphs describe the method that is also applied in our tool cnf2aig that can be downloaded at https://github.com/IserMk/cnf2aig.

For each tuple (o, Γ) with output literal o and clauses $\Gamma = \{c_1, \ldots, c_n\}$ we create a set of and-gates in which we reuse the given variables. For each clause $c_i \in \Gamma$ we create a new variable γ_i that is output of a new and-gate $\gamma_i = \text{and}(\overline{l}_1, \overline{l}_2, \ldots, \overline{l}_{|c_i|})$ with distinct $l_j \in c_i$. Then we use the output literals γ_i to construct the outer and-gate $o = \text{and}(\overline{\gamma}_1, \ldots, \overline{\gamma}_n)$.

The result of the first step is a directed acyclic graph of n-ary and-gates with negation. In the next step each *n*-ary and-gate is converted to a set of binary and-gates via introduction of new output-variables for each newly created binary and. The result can be directly converted to the AIG format.[3]

---

[3] Currently we do *not* check for duplicate binary and-gates.

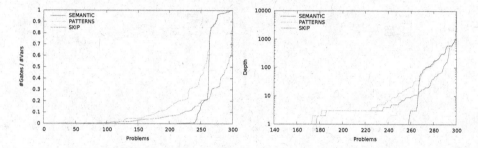

Fig. 1. Relative number of recognized gates (left) and maximum nesting-depth (right) by equivalence detection method on the application instance set

4 Experimental Results

We experimented with diverse configurations of our algorithm using the benchmark set from `SAT-Competition 2014`. We also used preprocessed CNF of the chosen benchmark set as input to our algorithm.

Hardware. For our experiments we used a compute cluster where each node is equipped with 2 Intel Xeon E5430 CPUs running at 2.66 GHz and 32 GB of RAM. The operating system is OpenSuSE 11.1 Linux 64 bit. We ran each process with a CPU time limit of 1 hour and a memory limit of 4 GB.

Software. We used existing tools for CNF preprocessing with the methods *BVA*, *BVE* and *BCD*. For *BVA* and *BVE* we used version 4.27 of Coprocessor [18].[4] For BCD we used the tool mvSAT[5] and applied the method *solitaire decomposition* as described in [2].

Evaluation. We compare the quality of our recognition method in terms of number of *recognized gates* (`#Gates`). As some preprocessing methods change the total *number of variables* (`#Vars`) in a problem, we additionally use the number of recognized gates relative to the total number of variables. Note that the *number of input variables* of the recognized nested structure is exactly the difference of `#Vars` and `#Gates`. The cactus plots that we used to compare the methods either show for each problem the total or relative amount of recognized gates, for each method sorted by that amount.

4.1 Equivalence Detection Methods

We compared the quality of our recognition algorithm using the three equivalence detection methods that are described in Section 3.3. Figure 1 shows that while the pattern-based equivalence detection PATTERNS gives us almost unrecognizable runtime overhead it has an immense effect on the quantity of recognized gates (compared to SKIP).

---

[4] available at tools.computational-logic.org
[5] available at http://ktiml.mff.cuni.cz/~balyo/bcd/

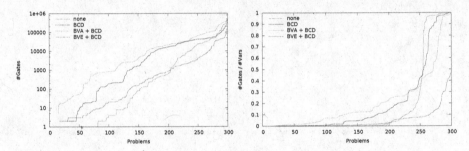

Fig. 2. Total and relative number of gates for the application instance set.

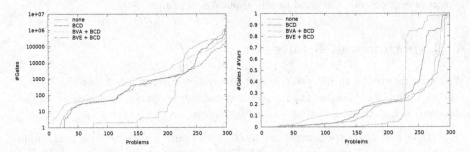

Fig. 3. Total and relative number of gates for the crafted instance set.

Our implementation of the semantic equivalence detection method SEMAN-TIC is implemented as a fallback when the PATTERNS method fails. As the construction of models of g is exponential we imposed a bound on the number and sizes of clauses in g for which we run the method. However it still has such a bad runtime (many time- and memory-outs) that the methods can hardly be compared. In the best cases of our experiments it still has some slightly improved recognition rates.

4.2   Preprocessing and Gate Recognition

We experimented with several combinations of preprocessors before recognition. Figures 2, 3 and 4 show the absolute and relative numbers of recognized gates per preprocessing method for the application, crafted and random instance set, respectively.

The plots display the quantitative recognition results on the unprocessed CNF (none) on the CNF after BCD (BCD) on the CNF after BVA with subsequent BCD (BVA + BCD) and after BVE with subsequent BCD (BVE + BCD).

Unprocessed CNF. On the crafted and application instance set the recognition results are best for the unprocessed CNF only if there are a lot of gates to recognize (i.e. higher rates at the "head").

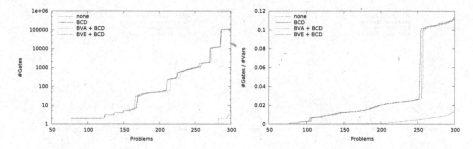

Fig. 4. Total and relative number of gates for the random instance set.

Blocked Clause Decomposition (BCD). Running BCD before recognition causes our algorithm to always start recursion on the inputs of the partial gate that mvSAT constructs from the small blocked set that is generated by the BCD algorithm [2]. This improves recognition rate at the tail of the distribution for the crafted and for the application instance set (Figures 2 and 3). For random instances BCD is responsible for getting a significant recognition rate of up to 10% of the variables (Figure 4).

Bounded Variable Elimination (BVE). BVE seems to "destroy" most of the gates that our algorithm is capable of recognizing. Even BVE with a follow-up BCD (this can be seen in the figures) does not help to improve recognition rates.

Bounded Variable Addition (BVA). Recognition rates deteriorate after application of BVA compared to the unprocessed CNF. However, after application of BCD (this can be seen in the figures) BVA sometimes has a positive effect on the recognition rates. Running BVA and then BCD improves recognition rates at the tail of the distribution even more than plain BCD does. On the other hand recognition rates on the head of the distribution rates deteriorate even further.

4.3   Maximum Number of Recognition Tries

We investigated the impact of changing the maximum number of recognition tries and ran gate-detection with different values of maximum tries on the unprocessed and on the BCD-preprocessed instance set. Figure 5 shows that an increased number of recognition tries typically does not improve recognition rates (especially when many gates can be detected) while making the algorithm slower.

4.4   And-Inverter Graphs

We used our gate-recognition approach on top of BCD to generate AIGs and compared them to the AIGs generated by the likewise BCD-based approach presented in [2].

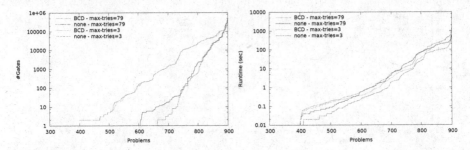

Fig. 5. Total number of gates and runtime for different numbers of recognition tries for the complete benchmark set with and without BCD.

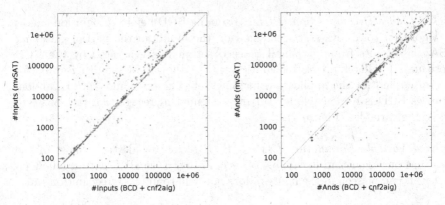

Fig. 6. Our new approach generates AIG with *less* input variables and *more* gates than the approach implemented in mvSAT.

Quality. The number of input variables controls the degrees of freedom of a functional structure. Figure 6 shows that our approach produces AIG with less input variables and but slightly more gates than the reference approach in [2].

Performance. We have run some initial experiments to check whether better structure recognition can help to elevate SAT solver performance. We tried to recreate the setup that was used in [2]. So we used an older version (ats) of lingeling on the benchmark set of 2013 and compared its runtime on the plain CNF with the runtime on the reencoded CNF with both the old mvSAT and the new cnf2aig approach. The reencoding was done as described in [2] with AIG construction (using mvSAT or cnf2aig), followed by time-bounded circuit-level simplifications with the tool abc (with option -dc2) and a subsequent translation of the resulting AIG back to CNF.

Figure 7 shows the results for the 138 instances that have a high quality BCD (at least 90% of the clauses are blocked). The experiments show that structure-based reencoding can improve the performance of lingeling in some cases. It also

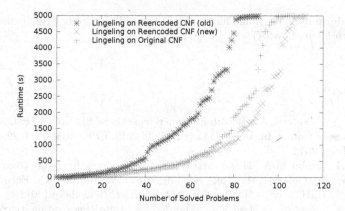

Fig. 7. Comparison of the runtimes of lingeling on the application benchmark set of SAT Competition 2013 with the re-encoded CNF using mvSAT (old) and cnf2aig (new).

shows the superiority of our gate recognition method with respect to the previous approach.[6]

5   Conclusion and Future Work

We have seen that our algorithm recognizes nested gates in CNF more effectively than previous approaches. Preliminary results show that our algorithm can be used to increase SAT solver performance.

From all preprocessing techniques *blocked clause decomposition* (BCD) has an outstanding effect on the recognition rate of our algorithm. Also *bounded variable addition* (BVA) leads to a better recognition rate for several problems. In the future it could be possible to integrate BCD and BVA into the gate recognition process to achieve even better results with special tradeoffs.

When it comes to *semantic analysis* for detection of full gate-encodings the performance of our algorithm can still be improved, e.g. with better data-structures or by using easy to check necessary criteria for equivalence beforehand in addition to the currently used bounds.

Knowledge about the gate structure of propositional formulas has been used in the past to elevate SAT solver performance [11,14]. Equipped with our improved gate-recognition some of these methods could become more effective and useful, e.g. Counterexample-Guided Abstraction Refinement (CEGAR) [5]. Structure recognition might be a way to make CEGAR applicable even inside a SAT solver.

---

[6] Deviations in the relative performance of lingeling and mvSAT in our experiments compared to those presented in [2] we discussed with one of the authors of [2]. Differing results might come from another version of abc we used or from differences in the computing infrastructure.

Acknowledgments. We would like to thank Tomáš Balyo for helpful discussions and support with his tool mvSAT.

References

1. Balint, A., Manthey, N.: Boosting the performance of SLS and CDCL solvers by preprocessor tuning. In: Berre, D.L. (ed.) POS 2013. EPiC Series, vol. 29, pp. 1–14. EasyChair (2014)
2. Balyo, T., Fröhlich, A., Heule, M.J.H., Biere, A.: Everything you always wanted to know about blocked sets (but were afraid to ask). In: Sinz, C., Egly, U. (eds.) SAT 2014. LNCS, vol. 8561, pp. 317–332. Springer, Heidelberg (2014)
3. Biere, A., Le Berre, D., Lonca, E., Manthey, N.: Detecting cardinality constraints in CNF. In: Sinz, C., Egly, U. (eds.) SAT 2014. LNCS, vol. 8561, pp. 285–301. Springer, Heidelberg (2014). http://dx.doi.org/10.1007/978-3-319-09284-3_22
4. Bjesse, P., Kukula, J.H., Damiano, R., Stanion, T., Zhu, Y.: Guiding SAT diagnosis with tree decompositions. In: Giunchiglia, E., Tacchella, A. (eds.) SAT 2003. LNCS, vol. 2919, pp. 315–329. Springer, Heidelberg (2004)
5. Clarke, E., Grumberg, O., Jha, S., Lu, Y., Veith, H.: Counterexample-guided abstraction refinement for symbolic model checking. J. ACM **50**(5), 752–794 (2003)
6. Davis, M., Putnam, H.: A computing procedure for quantification theory. Journal of the ACM **7**(3), 201–215 (1960). http://doi.acm.org/10.1145/321033.321034
7. Dixon, H.E., Ginsberg, M.L., Luks, E.M., Parkes, A.J.: Generalizing Boolean satisfiability II: Theory (2011). CoRR abs/1109.2134
8. Eén, N., Biere, A.: Effective preprocessing in SAT through variable and clause elimination. In: Bacchus, F., Walsh, T. (eds.) SAT 2005. LNCS, vol. 3569, pp. 61–75. Springer, Heidelberg (2005). http://dx.doi.org/10.1007/11499107_5
9. Falke, S., Merz, F., Sinz, C.: LLBMC: improved bounded model checking of C programs using LLVM (competition contribution). In: Piterman, N., Smolka, S.A. (eds.) TACAS 2013 (ETAPS 2013). LNCS, vol. 7795, pp. 623–626. Springer, Heidelberg (2013)
10. Fu, Z., Malik, S.: Extracting logic circuit structure from conjunctive normal form descriptions. In: 20th International Conference on VLSI Design (VLSI Design 2007), Sixth International Conference on Embedded Systems (ICES 2007), January 6–10, 2007, Bangalore, India. pp. 37–42 (2007)
11. Fu, Z., Yu, Y., Malik, S.: Considering circuit observability don't cares in CNF satisfiability. In: 2005 Design, Automation and Test in Europe Conference and Exposition (DATE 2005), March 7–11, 2005, Munich, Germany, pp. 1108–1113 (2005)
12. Heule, M.J.H., Biere, A.: Blocked clause decomposition. In: McMillan, K., Middeldorp, A., Voronkov, A. (eds.) LPAR-19 2013. LNCS, vol. 8312, pp. 423–438. Springer, Heidelberg (2013). http://dx.doi.org/10.1007/978-3-642-45221-5_29
13. Iser, M., Sinz, C., Taghdiri, M.: Minimizing models for tseitin-encoded SAT instances. In: Järvisalo, M., Van Gelder, A. (eds.) SAT 2013. LNCS, vol. 7962, pp. 224–232. Springer, Heidelberg (2013)
14. Iser, M., Taghdiri, M., Sinz, C.: Optimizing MiniSAT variable orderings for the relational model finder kodkod (poster presentation). In: Cimatti, A., Sebastiani, R. (eds.) SAT 2012. LNCS, vol. 7317, pp. 483–484. Springer, Heidelberg (2012)

15. Järvisalo, M., Biere, A., Heule, M.J.H.: Blocked clause elimination. In: Esparza, J., Majumdar, R. (eds.) TACAS 2010. LNCS, vol. 6015, pp. 129–144. Springer, Heidelberg (2010)
16. Järvisalo, M., Biere, A., Heule, M.J.H.: Simulating circuit-level simplifications on cnf. Journal of Automated Reasoning 49(4), 583–619 (2012)
17. Kuehlmann, A., Paruthi, V., Krohm, F., Ganai, M.K.: Robust boolean reasoning for equivalence checking and functional property verification. IEEE Trans. on CAD of Integrated Circuits and Systems 21(12), 1377–1394 (2002)
18. Manthey, N.: Coprocessor 2.0 – a flexible CNF simplifier. In: Cimatti, A., Sebastiani, R. (eds.) SAT 2012. LNCS, vol. 7317, pp. 436–441. Springer, Heidelberg (2012). http://dx.doi.org/10.1007/978-3-642-31612-8_34
19. Manthey, N.: Extended resolution in modern SAT solving. In: Joint Automated Reasoning Workshop and Deduktionstreffen (2014)
20. Manthey, N., Heule, M.J.H., Biere, A.: Automated reencoding of boolean formulas. In: Biere, A., Nahir, A., Vos, T. (eds.) HVC. LNCS, vol. 7857, pp. 102–117. Springer, Heidelberg (2013). http://dx.doi.org/10.1007/978-3-642-39611-3_14
21. Mihal, A., Teig, S.: A constraint satisfaction approach for programmable logic detailed placement. In: Järvisalo, M., Van Gelder, A. (eds.) SAT 2013. LNCS, vol. 7962, pp. 208–223. Springer, Heidelberg (2013)
22. Grégoire, R.O.É., Saïs, L.: Recovering and exploiting structural knowledge from CNF formulas. In: Van Hentenryck, P. (ed.) CP 2002. LNCS, vol. 2470, p. 185. Springer, Heidelberg (2002)
23. Plaisted, D.A., Greenbaum, S.: A structure-preserving clause form translation. J. Symb. Comput. 2(3), 293–304 (1986)
24. Rintanen, J.: Engineering efficient planners with SAT. In: ECAI, pp. 684–689 (2012)
25. Sinz, C.: Visualizing SAT instances and runs of the DPLL algorithm. J. Autom. Reasoning 39(2), 219–243 (2007)
26. Torlak, E., Jackson, D.: Kodkod: a relational model finder. In: Grumberg, O., Huth, M. (eds.) TACAS 2007. LNCS, vol. 4424, pp. 632–647. Springer, Heidelberg (2007)
27. Boy de la Tour, T.: An optimality result for clause form translation. J. Symb. Comput. 14(4), 283–301 (1992)
28. Tseitin, G.S.: On the complexity of derivation in propositional calculus. In: Slisenko, A.O. (ed.) Studies in Constructive Mathematics and Mathematical Logic, pp. 115–125 (1970)
29. Williams, R., Gomes, C.P., Selman, B.: Backdoors to typical case complexity. In: IJCAI, pp. 1173–1178 (2003)

Exploiting Resolution-Based Representations for MaxSAT Solving

Miguel Neves[1], Ruben Martins[2,3]([✉]), Mikoláš Janota[1], Inês Lynce[1], and Vasco Manquinho[1]

[1] INESC-ID / Instituto Superior Técnico,
Universidade de Lisboa, Lisbon, Portugal
{neves,mikolas,ines,vmm}@sat.inesc-id.pt
[2] University of Texas at Austin, Austin, USA
ruben.martins@cs.ox.ac.uk
[3] Department of Computer Science, University of Oxford, Oxford, UK

Abstract. Most recent MaxSAT algorithms rely on a succession of calls to a SAT solver in order to find an optimal solution. In particular, several algorithms take advantage of the ability of SAT solvers to identify unsatisfiable subformulas. Usually, these MaxSAT algorithms perform better when small unsatisfiable subformulas are found in early iterations of the algorithm. However, this is not the case in many problem instances, since the whole formula is given to the SAT solver in each call.

In this paper, we propose to partition the MaxSAT formula using a resolution-based graph representation. Partitions are then iteratively joined by using a proximity measure extracted from the graph representation of the formula. The algorithm ends when only one partition remains and the optimal solution is found. Experimental results show that this new approach further enhances a state of the art MaxSAT solver to optimally solve a larger set of industrial problem instances.

1 Introduction

The improvements of Maximum Satisfiability (MaxSAT) technology in recent years lead to a number of applications of MaxSAT. Many real-world problems in different areas such as fault localization in C programs, design debugging, upgradability of software systems, among others, can now be solved using MaxSAT [2,10,12,15,24]. In the last decade, several new techniques and algorithms have been proposed that improved on previous MaxSAT solvers by several orders of magnitude. Moreover, the developments in the underlying SAT technology, namely identification of unsatisfiable subformulas and incrementality have also been a factor in the improvements of MaxSAT solving.

MaxSAT solvers for industrial instances are usually based on iterative calls to a SAT solver. Moreover, most of these MaxSAT algorithms take advantage of the ability of SAT solvers to identify unsatisfiable subformulas. However, in most cases, algorithms deal with the whole formula at each call of the SAT solver. As a result, unnecessarily large unsatisfiable subformulas can be found at

© Springer International Publishing Switzerland 2015
M. Heule and S. Weaver (Eds.): SAT 2015, LNCS 9340, pp. 272–286, 2015.
DOI: 10.1007/978-3-319-24318-4_20

each SAT call, resulting in a slow down of the MaxSAT algorithm. In this work, we try to avoid this behavior by partitioning the formula and taking advantage of structural information obtained from a formula's graph representation.

In this paper, we improve on the current state of the art MaxSAT solving by proposing a new unsatisfiability-based algorithm for MaxSAT. The new algorithm integrates several new features, namely: (1) usage of resolution-based graphs to represent the MaxSAT formula, (2) partition of soft clauses in the MaxSAT formula using the referred representation, (3) usage of structural information obtained from the graph representation to drive the merge of partitions and, (4) integration of these features into a new fully incremental algorithm that improves on one of the best non-portfolio solvers from the last MaxSAT Solver Evaluation on several partial MaxSAT industrial benchmark sets.

The paper is organized as follows. Section 2 formally defines MaxSAT and briefly reviews the MaxSAT algorithms more closely related to the proposed approach. In section 3, graph representations of CNF formulas are described. Moreover, the adaptation of resolution-based graphs is proposed. The new MaxSAT algorithm is presented in section 4. Besides a detailed description, we show how to extract structural information from the graph representations and integrate it in the new algorithm. Section 5 presents the experimental results of the new MaxSAT solver on a large set of industrial benchmark sets used at MaxSAT evaluations. Finally, the paper concludes in section 6.

2   Preliminaries

A propositional formula in Conjunctive Normal Form (CNF), using n Boolean variables $x_1, x_2, \ldots, x_n$, is defined as a conjunction of clauses, where a clause is a disjunction of literals. A literal is either a variable x_i or its complement $\bar{x}_i$. The Propositional Satisfiability (SAT) problem consists of deciding whether there exists a total assignment to the variables such that the formula is satisfied.

The Maximum Satisfiability (MaxSAT) can be seen as an optimization version of the SAT problem. In MaxSAT, the objective is to find a total assignment to the variables of a CNF formula that minimizes the number of unsatisfied clauses. Notice that minimizing the number of unsatisfied clauses is equivalent to maximizing the number of satisfied clauses.

In a partial MaxSAT formula $\varphi = \varphi_h \cup \varphi_s$, some clauses are considered as hard (φ_h), while others are declared as soft (φ_s). The goal in partial MaxSAT is to find a total assignment to the formula variables such that all hard clauses in φ_h are satisfied, while minimizing the number of unsatisfied soft clauses in φ_s. There are also weighted variants of MaxSAT where soft clauses are associated with weights greater than or equal to 1. In this case, the objective is to satisfy all hard clauses and minimize the total weight of unsatisfied soft clauses. In this paper, we focus solely on partial MaxSAT, but the proposed approach can be generalized to its weighted variants. Furthermore, in all algorithms we assume that the set of hard clauses φ_h is satisfiable. Otherwise, the MaxSAT formula does not have a solution. This can easily be checked through a SAT call on φ_h.

Algorithm 1. Linear Search Unsat-Sat Algorithm

Input: $\varphi = \varphi_h \cup \varphi_s$
Output: satisfying assignment to φ

1   $(\varphi_W, V_R, \lambda) \leftarrow (\varphi_h, \emptyset, 0)$
2   **foreach** $c_i \in \varphi_s$ **do**
3      $V_R \leftarrow V_R \cup \{r_i\}$ // r_i is a new relaxation variable
4      $c_R \leftarrow c_i \cup \{r_i\}$
5      $\varphi_W \leftarrow \varphi_W \cup \{c_R\}$
6   **while** true **do**
7      $(st, \nu, \varphi_C) \leftarrow \mathtt{SAT}(\varphi_W \cup \{\mathtt{CNF}(\sum_{r_i \in V_R} r_i \leq \lambda)\})$
8   **if** $st = \mathsf{SAT}$ **then**
9   **return** ν // satisfying assignment to φ
10     $\lambda \leftarrow \lambda + 1$

The most recent state of the art MaxSAT solvers are based on iterative calls to a SAT solver. One of the most classic approaches is the linear Sat-Unsat algorithm that performs a linear search on the number of unsatisfied clauses. In this case, a new relaxation variable is initially added to each soft clause and the resulting formula is given to a SAT solver. Whenever a solution is found, a new cardinality constraint on the number of relaxation variables is added, such that solutions where a higher or equal number of relaxation variables assigned the value 1 are excluded. The cardinality constraint is encoded into a set of propositional clauses, which are added to the working formula [3,13,17]. The algorithm stops when the SAT call is unsatisfiable. As a result, the last solution found is an optimal solution of the MaxSAT formula.

A converse approach is the linear search Unsat-Sat presented in Algorithm 1. Here, a lower bound λ on the number of unsatisfied soft clauses is maintained between iterations of the algorithm. Initially, λ is assigned value 0. In each iteration, while the working formula given to the SAT solver (line 7) is unsatisfiable, λ is incremented (line 10). Otherwise, an optimal solution to the MaxSAT formula has been found (line 9).

Observe that a SAT solver call on a CNF formula φ_W returns a triple (st, ν, φ_C), where st denotes the status of the solver: satisfiable (SAT) or unsatisfiable (UNSAT). If φ_W is satisfiable, then ν stores the total assignment found for φ_W. Otherwise, φ_C contains an unsatisfiable subformula that explains a reason for the unsatisfiability of φ_W.

Even though the linear search Unsat-Sat algorithm does not take advantage of current SAT solvers being able to identify unsatisfiable subformulas, there are several more effective algorithms for MaxSAT that use this information to delay the relaxation of soft clauses. An example is the MSU3 algorithm [16] presented in Algorithm 2. Observe that this algorithm also performs an Unsat-Sat linear search, but soft clauses are only relaxed when they appear in an unsatisfiable subformula. The MSU3 algorithm takes as input a MaxSAT formula φ, a set of relaxation variables V_R, and a given lower bound λ. If no additional information

---

Algorithm 2. MSU3 Algorithm

    Input: (φ, V_R, λ)
    Output: satisfying assignment to φ
1 $\varphi_W \leftarrow \varphi$
2 **while** true **do**
3     $(\mathsf{st}, \nu, \varphi_C) \leftarrow \mathsf{SAT}(\varphi_W \cup \{\mathsf{CNF}(\sum_{r_i \in V_R} r_i \leq \lambda)\})$
4   **if** $\mathsf{st} = \mathsf{SAT}$ **then**
5         $\lfloor$ **return** ν // satisfying assignment to φ
6   **foreach** $c_i \in (\varphi_C \cap \varphi_s)$ **do**
7         $V_R \leftarrow V_R \cup \{r_i\}$ // r_i is a new variable
8         $c_R \leftarrow c_i \cup \{r_i\}$ // c_i was not previously relaxed
9         $\varphi_W \leftarrow (\varphi_W \setminus \{c_i\}) \cup \{c_R\}$
10   $\lfloor \lambda \leftarrow \lambda + 1$

---

is known about φ, then MSU3 is called with the default values $\varphi = \varphi_h \cup \varphi_s$, $V_R = \emptyset$ and $\lambda = 0$.

Although more sophisticated MaxSAT algorithms exist [20], an implementation of MSU3 algorithm on the Open-WBO framework was one of the best performing non-portfolio algorithms for industrial partial MaxSAT at the MaxSAT Solver Evaluation of 2014[1]. One of the crucial features for its success relies on the fact that only one SAT solver instance needs to be created [17]. Therefore, a proper implementation of MSU3 should take advantage of incrementality in SAT solver technology. In this paper, the MSU3 algorithm is further improved with structural information of the problem instance to solve.

3   Graph Representations

In order to extract structural properties of CNF formulas, different graph-based models have been previously proposed. For instance, graph representations have been used to characterize industrial SAT instances [1] and to improve on the performance of MaxSAT algorithms [19]. In this section, we briefly review the Clause-Variable Incidence Graph (CVIG) and adapt the use of Resolution-based Graphs (RES) [26] to model relations in CNF formulas. Although other models exist [1,19,25], in the context of our algorithm for MaxSAT solving, these were found to be the best suited.

In the CVIG model, a weighted undirected graph G is built such that a vertex is added for each variable x_j and for each clause c_i occurring in the CNF formula φ. Moreover, for each variable x_j occurring in clause c_i (either as literal x_j or $\bar{x}_j$), an edge (c_i, x_j) is added to graph G. The edge weight $w(c_i, x_j)$ is defined as:

$$w(c_i, x_j) = \frac{I(x_j)}{|c_i|} \tag{1}$$

---

[1] Results available at http://www.maxsat.udl.cat/

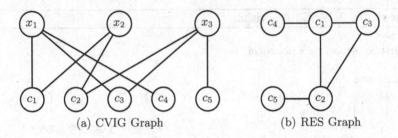

(a) CVIG Graph   (b) RES Graph

Fig. 1. Example of Graph Models

where $|c_i|$ denotes the number of literals in clause c_i and $I(x_j)$ is defined as the incidence function of x_j in soft clauses as:

$$I(x_j) = 1 + \sum_{x_j \in c \,\wedge\, c \in \varphi_s} \frac{1}{|c|} \tag{2}$$

As described in section 2, several MaxSAT solvers rely on the identification of unsatisfiable subformulas. In order to capture sets of clauses more closely related that would result in an unsatisfiable subformula, we propose to adapt Resolution Graphs (RES) to MaxSAT.

In the RES model, we have one vertex in graph G for each clause $c_i \in \varphi$. Let c_i and c_j denote two clauses such that $x_k \in c_i$ and $\bar{x}_k \in c_j$. Moreover, let c_{ij}^{res} be the resulting clause of applying the resolution operation on these clauses. In this case, if c_{ij}^{res} is not a tautology, then an edge (c_i, c_j) is added to G whose weight is defined as:

$$w(c_i, c_j) = \frac{1}{|c_{ij}^{res}|} \tag{3}$$

Notice that in the RES model, clauses are related if the application of the resolution operation results in a non-trivial resolvent. Moreover, observe that the weight of edges between pairs of clauses is greater when the size of the resolvent is smaller. The goal is to make tighter the relations between clauses that produce smaller clauses when resolution is applied.

Consider the following MaxSAT formula where $c_1 : (x_1 \vee x_2)$, $c_2 : (\bar{x}_2 \vee x_3)$ and $c_3 : (\bar{x}_1 \vee \bar{x}_3)$ are hard clauses and $c_4 : (\bar{x}_1)$, $c_5 : (\bar{x}_3)$ are soft clauses. Figures 1(a) and 1(b) illustrate the structure of the graph representation of this formula when using the CVIG and RES models. The weights of edges are not represented for simplicity but can be obtained via Equations (1) and (3). For example, for the CVIG model $w(c_1, x_1) = \frac{2}{2}$ and for the RES model $w(c_2, c_3) = \frac{1}{2}$. Observe that if the clause $c_6 : (\bar{x}_1 \vee \bar{x}_2)$ was added to the formula, it would not connect to any other clause in the RES graph because the only clause containing x_1 positively is $c_1 = (x_1 \vee x_2)$, but that does not connect to c_6 due to x_2 appearing negatively and positively in c_6 and c_1, respectively. A similar type of analysis is done in *blocked clause elimination* [11,14] — a technique commonly used in formula preprocessing.

Algorithm 3. Extended MSU3 Algorithm

    Input: $\varphi = \varphi_h \cup \varphi_s$
    Output: satisfying assignment to φ
1 $\gamma \leftarrow \langle \gamma_1, \ldots, \gamma_n \rangle \leftarrow \text{partitionSoft}(\varphi_s, \varphi_h)$
2 **foreach** $\gamma_i \in \gamma$ **do**
3      $(V_R^i, \lambda_i) \leftarrow (\emptyset, 0)$
4      $\nu \leftarrow \text{MSU3}(\varphi_h \cup \gamma_i, V_R^i, \lambda_i)$
5 **if** $|\gamma| = 1$ **then**
6   **return** ν `// no partitions were identified`
7 **while** true **do**
8      $(\gamma_i, \gamma_j) \leftarrow \text{selectPartitions}(\gamma)$
9      $\gamma \leftarrow \gamma \setminus \{\gamma_i, \gamma_j\}$
10     $(\gamma_k, V_R^k, \lambda_k) \leftarrow (\gamma_i \cup \gamma_j, V_R^i \cup V_R^j, \lambda_i + \lambda_j)$
11     $\nu \leftarrow \text{MSU3}(\varphi_h \cup \gamma_k, V_R^k, \lambda_k)$
12   **if** $\gamma = \emptyset$ **then**
13   **return** ν
14   **else**
15        $\gamma \leftarrow \gamma \cup \{\gamma_k\}$

Although resolution-based graphs are not novel [26] and have been used in other domains [25], in this paper we propose to enhance the resolution-based graph representation by adding weights to edges. Moreover, as far as we know, this representation has never been used for MaxSAT solving.

4 New Partition-Based Algorithm for MaxSAT

Despite its very good performance in industrial partial MaxSAT instances, the MSU3 algorithm (see Algorithm 2) may suffer from two issues: (1) identification of unnecessarily large unsatisfiable subformulas and, (2) a potentially large cardinality constraint to be maintained between iterations. In fact these issues are related. If an unsatisfiable subformula with an unnecessarily large number of soft clauses is encountered early, then an unnecessarily large cardinality constraint has to be dealt with through most of the algorithm's iterations.

Our approach to tackle these issues is to split the set of soft clauses. The goal is that, at each iteration, the algorithm should only consider part of the problem, instead of dealing with the whole problem instance in each iteration.

4.1 Algorithm Description

Algorithm 3 presents our enhancement of MSU3 with partitioning the soft clause set. The algorithm starts by partitioning φ_s into n disjoint sets of soft clauses $\gamma_1, \gamma_2 \ldots \gamma_n$ (line 1). Observe that several methods can be used to partition φ_s. Details of this procedure are discussed later.

For each set γ_i, we apply the MSU3 algorithm to the formula $\varphi_h \cup \gamma_i$ with starting values $V_R^i = \emptyset$ and $\lambda_i = 0$ (lines 2-4). As a result, we obtain a lower bound value λ_i associated with each set of soft clauses γ_i. If the partitioning procedure creates a single partition, then the algorithm terminates (line 6). Otherwise, it is necessary to build the solution of the MaxSAT instance by merging the different sets of soft clauses.

The merge process works as follows. At each iteration, two sets of soft clauses γ_i and γ_j are selected to be merged (line 8) and removed from γ. Let γ_k denote the union of γ_i and γ_j. Since γ_i and γ_j are disjoint, we necessarily have that $\lambda_i + \lambda_j$ is a lower bound for γ_k. Hence, we can safely initialize $\lambda_k = \lambda_i + \lambda_j$ (line 10). Next, the lower bound λ_k is refined by applying the MSU3 algorithm to $\varphi_h \cup \gamma_k$ with starting values $V_R^k = V_R^i \cup V_R^j$ and $\lambda_k = \lambda_i + \lambda_j$ (line 11). When set γ becomes empty, then all soft clauses were merged and the last solution found is an optimal solution (line 13). Otherwise, there are still more sets to be merged and γ_k is added to γ (line 15).

4.2 Partition and Merge of Soft Clauses

Algorithm 3 can be configured differently depending on two procedures: (1) how the set of soft clauses is partitioned (line 1) and (2) how to merge two sets of soft clauses (line 8).

In the partition procedure, our algorithm starts by representing the CNF formula as a graph using one of the models described in section 3. Next, we apply a community-finding algorithm on the graph representation that maximizes a modularity measure [4] in order to obtain a graph partitioning.

Recently, the use of modularity measures has become widespread when analyzing the structure of graphs, in particular for the identification of communities [7,23]. In fact, this has already been used in the analysis of SAT instances [1] and to improve the initial unsatisfiability-based approach proposed by Fu and Malik [6,19]. The purpose of the modularity measure is to evaluate the quality of the partitions, where vertices inside a partition should be densely connected and vertices assigned to different partitions should be loosely connected. However, finding a set of partitions with an optimal modularity value is computationally hard [5]. In our implementation, we use the approximation algorithm proposed by Blondel et al. [4].

At each iteration in Algorithm 3, two partitions are selected to be merged. One can devise several different criteria to select and merge the partitions of soft clauses. In early attempts, the merge process was sequential [19]. Given n partitions $\gamma_1, \gamma_2 \ldots \gamma_n$, at iteration i ($i < n$) of the algorithm, the first i partitions $\gamma_1, \gamma_2 \ldots \gamma_i$ were merged sequentially.

Figure 2(a) illustrates the sequential merging procedure. Observe that the sequential merging process is not balanced. This results in an early growth of the identified subformulas and, as a result, an early growth of the cardinality constraints to be maintained at each iteration of the algorithm.

In this paper, we propose a weighted balanced merge procedure that depends on the strength of the graph connections between partitions. The goal is to delay

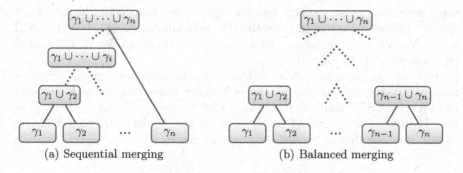

(a) Sequential merging   (b) Balanced merging

Fig. 2. Examples of merge processes

having to deal with a large number of soft clauses, until the latter iterations of the algorithm. Figure 2(b) illustrates the weighted balanced merging procedure.

Let $G = (V, E)$ denote an undirected weighted graph where V is the set of vertices and E the set of edges. Let $w : E \to \mathbb{R}$ be a weight function for each edge in the graph. The community-finding algorithm identifies a set of communities $C = \{C_1, C_2, \ldots, C_n\}$ where every vertex $u \in V$ is assigned to one and only one community in C. Hence, since in both CVIG and RES model there is a node for each propositional clause, one can build the partitions in a straightforward manner. For each community C_i with vertices representing soft clauses, there is a partition γ_i containing the respective soft clauses.

Based on the graph representation, one can define the strength of the connection between partitions. Let d_{ij} denote the strength between partition γ_i and γ_j. One can define d_{ij} based on the weight between the vertices of their respective communities C_i and C_j in the graph. Hence, d_{ij} can be defined as follows:

$$d_{ij} = \sum_{u \in C_i \land v \in C_j} w(u, v) \tag{4}$$

Considering that the graph is undirected, we necessarily have that $d_{ij} = d_{ji}$.

Given an initial set γ of n partitions $\gamma_1, \gamma_2 \ldots, \gamma_n$, our algorithm applies a greedy procedure that pairs all partitions γ_i and γ_j from γ to be merged, starting with the pair with largest d_{ij}. After pairing all partitions in the initial set, we perform the same procedure to the next $n/2$ partitions that result from the initial merging iterations. This is iteratively applied until we only have a single partition (see Figure 2(b)).

Observe that if partitions γ_i and γ_j are merged into a new partition γ_k, then the connectivity strength d_{kl} between γ_k to another partition γ_l is given by $d_{kl} = d_{il} + d_{jl}$. This follows from the fact that the communities in the graph are disjoint.

Finally, we would like to reference other solvers that split the set of soft clauses by identifying disjoint unsatisfiable subformulas [8,21]. However, there are major differences with regard to our proposed approach. First, our solver takes advantage of an explicit formula representation to split the set of soft clauses, instead of using the unsatisfiable subformulas provided by the SAT solver. Moreover, in our solver, the merge process is also guided by the explicit representation of the formula.

Furthermore, in solvers where disjoint unsatisfiable subformulas are identified [8,21], the split occurs on the cardinality constraints at each iteration. However, each SAT call still has to deal with the whole formula at each iteration. In Algorithm 3, the SAT solver does not have to deal with all soft clauses at each iteration, but only after the final merge step.

4.3   Algorithm Analysis

In this section a proof sketch of the correctness Algorithm 3, as well as an analysis on the number of SAT calls is presented.

Proof (Correctness of Algorithm 3). As mentioned in section 2, we assume the set of hard clauses φ_h is satisfiable. Otherwise, the MaxSAT formula is unsatisfiable. This can be verified by a single SAT call on φ_h before applying Algorithm 3.

For the proof we adopt the following notation. For some set γ_i processed in Algorithm 3, we write $\gamma_i^R \subseteq \varphi_s$ for the set of clauses that were relaxed in the algorithm (but clauses in γ_i^R do not contain the relaxation variables). We will prove by induction the invariant that $\varphi_h \cup \gamma_i^R$ cannot be satisfied unless at least λ_i clauses are removed from γ_i^R. The induction hypothesis is satisfied trivially at the beginning of the algorithm as each λ_i is initialized to 0.

Consider the case where λ_i is augmented by 1 when $\varphi_h \cup \gamma_i \cup \{\sum_{r \in V_R^i} r \leq \lambda_i\}$ is unsatisfiable. Let φ_C be the obtained unsatisfiable subformula from the SAT call, let $\varphi_C^R \subseteq \varphi_s$ be the soft clauses of φ_C that appear as relaxed in γ_i and let $\varphi_C^N = \varphi_s \cap \varphi_C$ be the rest of the soft clauses in the unsatisfiable subformula (not yet relaxed). From induction hypothesis $\varphi_h \cup \varphi_C^R$ cannot be satisfied unless at least λ_i clauses are removed from $\varphi_C^R \subseteq \gamma_i^R$. Since φ_C is an unsatisfiable subformula, it is *impossible* to satisfy $\varphi_h \cup \varphi_C^R \cup \varphi_C^N$ by removing λ_i clauses from φ_C^R. Now we need to also show that it is impossible to satisfy $\varphi_h \cup \gamma_i^R \cup \varphi_C^N$ by removing λ_i clauses from $\gamma_i^R \cup \varphi_C^N$ (this is the new set of relaxed clauses).

Let us assume for contradiction that it is possible to satisfy $\gamma_i^R \cup \varphi_C^N$ by removing some set of clauses ξ s.t. $|\xi| = \lambda_i$. To show the contradiction we consider two cases: (1) $\xi \subseteq \gamma_i$ and (2) $\xi \nsubseteq \gamma_i$. Case (1) yields an immediate contradiction as we would have not obtained unsatisfiability in the SAT call as it would be possible to satisfy $\varphi_h \cup \gamma_i^R$ by removing λ_i clauses from γ_i^R. For case (2) consider that there is a clause $c \in \xi$ s.t. c is not yet relaxed, i.e. $c \notin \gamma^R$. This means that $\varphi_h \cup \gamma_i^R$ is satisfiable after removing *less* than λ_i clauses, which is a contradiction with the induction hypothesis.

To show that the invariant is preserved by the merge operation, we observe that any merged γ_i and γ_j are disjoint and therefore so are γ_i^R and γ_j^R. In order to satisfy $\varphi_h \cup (\gamma_i^R \cup \gamma_j^R)$, both $\varphi_h \cup \gamma_i^R$, $\varphi_h \cup \gamma_j^R$ must be satisfied. Consequently, at least $\lambda_i + \lambda_j$ clauses must be removed from $(\gamma_i^R \cup \gamma_j^R)$. ☐

Table 1. Experimental evaluation of Open-WBO's MSU3 algorithm, Eva500a, MSCG and 4 different configurations of the partition-based algorithm

Instance Group	Total	MSU3	Eva500a	MSCG	S-CVIG	S-RES	W-CVIG	W-RES
aes	7	1	1	1	1	1	1	1
atcoss/mesat	18	11	11	4	11	1	11	11
atcoss/sugar	19	12	11	4	12	3	12	12
bcp/fir	59	59	55	59	56	44	51	51
bcp/hipp-yRa1/simp	17	16	16	16	16	16	16	16
bcp/hipp-yRa1/su	38	35	34	33	34	34	35	33
bcp/msp	64	26	37	29	23	41	27	42
bcp/mtg	40	40	40	40	40	40	40	40
bcp/syn	74	43	48	47	47	48	46	49
circuit-trace-compaction	4	4	4	4	4	3	4	4
close-solutions	50	48	48	46	40	32	40	45
des	50	42	41	41	49	48	50	48
haplotype-assembly	6	5	5	5	5	5	5	5
hs-timetabling	2	1	1	0	1	1	1	1
mbd	46	45	42	43	44	45	45	45
packup-pms	40	40	40	40	40	40	40	40
pbo/mqc/nencdr	84	84	84	84	84	84	84	84
pbo/mqc/nlogencdr	84	84	84	84	84	84	84	84
pbo/routing	15	15	15	15	14	15	15	15
protein_ins	12	12	8	12	12	12	12	12
tpr/Multiple_path	48	48	44	42	48	48	48	48
tpr/One_path	50	50	50	50	50	50	50	50
Total	827	721	719	699	715	695	717	736

Finally, we note that the number of SAT calls performed by Algorithm 3 is larger than the MSU3 algorithm. Observe that the number of unsatisfiable SAT calls is the same for both algorithms. Let λ be the number of unsatisfiable soft clauses at any optimal solution of the MaxSAT instance. In this case, both algorithms perform λ unsatisfiable SAT calls. However, while MSU3 performs only one satisfiable SAT call, Algorithm 3 performs $2n-1$, where n is the number of identified partitions (line 1).

5   Experimental Results

In this section we compare different configurations of Algorithm 3 with the top 3 non-portfolio solvers of the MaxSAT 2014 Evaluation's industrial partial MaxSAT category. The top 3 were Open-WBO's MSU3 incremental algorithm [17,18], Eva500a [22] and MSCG [9]. The new partition-based algorithm is also implemented using the Open-WBO framework[2].

The algorithms were evaluated running on the set union of the partial MaxSAT industrial instances of the MaxSAT evaluations of 2012, 2013 and 2014. For each instance, algorithms were executed with a timeout of 1800 seconds and a memory limit of 4 GB. Similar resource limitations were used during the last MaxSAT Evaluation of 2014. These tests were conducted on a machine with 4 AMD Opteron 6376 (2.3 GHz) and 128 GB of RAM, running Debian jessie.

Table 1 presents the number of instances solved by each algorithm, per instance set. Besides MSU3, Eva500a and MSCG, results for the best 4 configurations of the partition-based enhanced MSU3 algorithm are shown. S-CVIG

---

[2] Available at http://sat.inesc-id.pt/open-wbo/

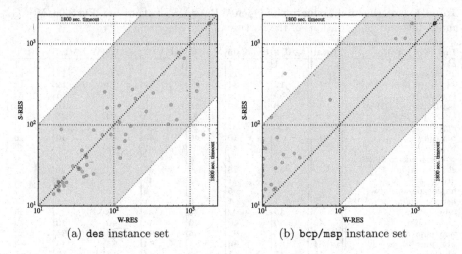

(a) des instance set   (b) bcp/msp instance set

Fig. 3. Comparison between run times of S-RES and W-RES on des and bcp/msp instance sets

applies the sequential merging of partitions using the CVIG graph model. S-RES also applies sequential merging, but using the RES graph model. W-CVIG and W-RES apply the weighted balanced merging of partitions, using the CVIG and RES graph models, respectively. Note that all our implementations are fully incremental, i.e. only one instance of the SAT solver is created throughout the execution of the proposed algorithm. As with the MSU3 implementation on Open-WBO, we take advantage of assumptions usage at each SAT call and incremental encoding of cardinality constraints [17].

Results from Table 1 show that all variants of the partition-based algorithm are competitive with the remaining state of the art algorithms. However, overall results clearly show that W-RES outperforms all remaining algorithms, since it is able to solve more instances in total. Moreover, results for the configurations of partition-based algorithm also show that weight-based balanced merging of partitions is preferable to sequential partitioning.

Considering that MSU3 is our base solver, most gains occur in instance sets bcp/msp, bcp/syn and des. While in the bcp/syn and des instance sets, all partition-based configurations perform better, in bcp/msp the resolution-based graph partitioning allowed a significant performance boost.

Figures 3(a) and 3(b) compare the results of S-RES and W-RES on the des and bcp/msp instance sets. In the des instances, the run time of sequential merging is slightly better, despite solving the same number of instances. Nevertheless, in the bcp/msp instance set the weight-based balanced merging used in W-RES clearly outperforms the sequential merging approach used in S-RES.

In Figures 4(a) and 4(b) we compare MSU3 and W-RES on the same benchmark sets. It can be observed that W-RES performs much better in these instances. In the des instance set, there are some instances where W-RES is

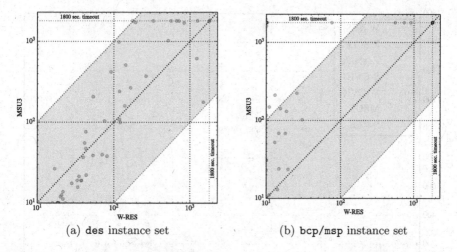

(a) des instance set   (b) bcp/msp instance set

Fig. 4. Comparison between run times of MSU3 and W-RES on des and bcp/msp instance sets

not as fast, since there is some time spent in finding partitions and additional SAT calls. We note that there is always some time spent in building the graph, applying the community finding algorithm and splitting the set of soft clauses. However, this partitioning step is usually not very time consuming. Nevertheless, W-RES is able to scale better and solve more instances. In the bcp/msp instances, the proposed techniques allow W-RES to be much better than MSU3, as well as all other algorithms tested.

Resolution-based graph models performed worst in the bcp/fir category. It was observed that the modularity values obtained for the resolution-based graphs were low in this particular instance set. As a result, the partitioning obtained for S-RES and W-RES in bcp/fir instances is not as meaningful as for other instance sets. When this occurs, it can deteriorate the algorithm's performance, since the partition-based algorithm performs more SAT calls than MSU3.

When considering all benchmark sets, W-CVIG and W-RES solve different instances and the Virtual Best Solver[3] (VBS) between them solves 747 instances (11 more than W-RES). Furthermore, there are a few instances which are only solved by MSU3 but not by W-CVIG nor W-RES. The VBS between MSU3, W-CVIG, W-RES can solve 752 (5 more than the VBS between W-CVIG and W-RES). Even though W-RES outperforms the remaining algorithms, this suggests that dynamically choosing the partition type could further improve the performance of the solver.

Finally, Figure 5 shows a cactus plot with the run times of all algorithms considered in the experimental evaluation. Here we can observe that S-RES is much slower than W-RES, clearly showing the effectiveness of the newly proposed

---

[3] The Virtual Best Solver between a set of solvers shows the total number of instances that can be solved by at least one of those solvers.

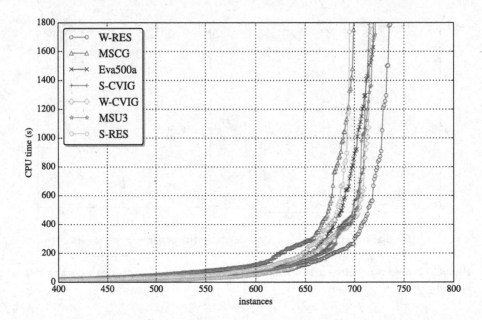

Fig. 5. Cactus plot with the run times for MSU3, Eva500a, MSCG, S-CVIG, S-RES, W-CVIG and W-RES.

weight-based merging. Overall, W-RES clearly outperforms the remaining algorithms, being able to solve 700 instances in 300 seconds or less.

6 Conclusions and Future Work

In this paper we exploit resolution-based graph representations of CNF formulas in order to develop a new state of the art algorithm for MaxSAT. In the proposed approach, soft clauses are initially partitioned in disjoint sets by analyzing the formula structure. The partitioning process is attained by applying a community-finding algorithm on weighted resolution-based graphs. Next, at each iteration of the algorithm, partitions are merged using structural information from the graph representation until an optimal solution is found.

The proposed approach is novel in many aspects. First, the use of a resolution-based graph representation allows to better model the interaction between clauses. Furthermore, instead of applying a sequential merging process, the graph representation is also used in a weight-based balanced merging procedure. Moreover, since the algorithm does not have to deal with the whole formula at each iteration, smaller unsatisfiable cores are identified. As a result from this process, smaller cardinality constraints are encoded into CNF at each iteration, thus improving the algorithm's performance.

Experimental results obtained in industrial partial MaxSAT instances clearly show the effectiveness of the proposed algorithm. As a result, our solver improves upon one of the best non-portfolio solvers on the industrial partial category from the 2014 MaxSAT solver evaluation.

The source code of the new solver will become available as part of the Open-WBO framework. This will allow the research community to build upon the current work to further improve MaxSAT solving.

As future work, we propose to extend the proposed approach for weighted MaxSAT solving. Moreover, different model representations of CNF formulas are to be tested, as well as new techniques for building and merging partitions of soft clauses in MaxSAT formulas. Furthermore, the proposed techniques are not exclusive to MSU3 and can also be integrated into other MaxSAT algorithms. Additionally, these techniques can also be applied to other extensions of SAT.

Acknowledgments. This work was partially supported by DARPA MUSE award #FA8750-14-2-0270, FCT grant POLARIS (PTDC/EIA-CCO/123051/2010), FCT grant AMOS (CMUP-EPB/TIC/0049/2013), and national funds through Fundação para a Ciência e a Tecnologia (FCT) with reference UID/CEC/50021/2013. The views, opinions, and/or findings contained in this article are those of the authors and should not be interpreted as representing the official views or policies of the Department of Defense or the U.S. Government.

References

1. Ansótegui, C., Giráldez-Cru, J., Levy, J.: The Community structure of SAT formulas. In: Cimatti, A., Sebastiani, R. (eds.) SAT 2012. LNCS, vol. 7317, pp. 410–423. Springer, Heidelberg (2012)
2. Asín, R., Nieuwenhuis, R.: Curriculum-based course timetabling with SAT and MaxSAT. Annals of Operations Research **218**(1), 71–91 (2014)
3. Bailleux, O., Boufkhad, Y.: Efficient CNF encoding of boolean cardinality constraints. In: Rossi, F. (ed.) CP 2003. LNCS, vol. 2833, pp. 108–122. Springer, Heidelberg (2003)
4. Blondel, V., Guillaume, J., Lambiotte, R., Lefebvre, E.: Fast unfolding of communities in large networks. Journal of Statistical Mechanics **2008**(10), P10008 (2008)
5. Brandes, U., Delling, D., Gaertler, M., Goerke, R., Hoefer, M., Nikoloski, Z., Wagner, D.: Maximizing modularity is hard (2006). arXiv: physics/0608255
6. Fu, Z., Malik, S.: On solving the partial MAX-SAT problem. In: Biere, A., Gomes, C.P. (eds.) SAT 2006. LNCS, vol. 4121, pp. 252–265. Springer, Heidelberg (2006)
7. Girvan, M., Newman, M.E.J.: Community structure in social and biological networks. Proceedings of the National Academy of Sciences **99**(12), 7821–7826 (2002)
8. Heras, F., Morgado, A., Marques-Silva, J.: Core-guided binary search algorithms for maximum satisfiability. In: AAAI Conference on Artificial Intelligence. AAAI Press (2011)
9. Ignatiev, A., Morgado, A., Manquinho, V., Lynce, I., Marques-Silva, J.: Progression in maximum satisfiability. In: European Conference on Artificial Intelligence, pp. 453–458. IOS Press (2014)
10. Janota, M., Lynce, I., Manquinho, V., Marques-Silva, J.: PackUp: Tools for Package Upgradability Solving. Journal on Satisfiability, Boolean Modeling and Computation **8**(1/2), 89–94 (2012)

11. Järvisalo, M., Biere, A., Heule, M.: Blocked clause elimination. In: Esparza, J., Majumdar, R. (eds.) TACAS 2010. LNCS, vol. 6015, pp. 129–144. Springer, Heidelberg (2010)

12. Jose, M., Majumdar, R.: Cause clue clauses: error localization using maximum satisfiability. In: Programming Language Design and Implementation, pp. 437–446. ACM (2011)

13. Koshimura, M., Zhang, T., Fujita, H., Hasegawa, R.: QMaxSAT: A Partial MaxSAT Solver. Journal on Satisfiability, Boolean Modeling and Computation $8(1/2)$, 95–100 (2012)

14. Kullmann, O.: On a generalization of extended resolution. Discrete Applied Mathematics **96–97**, 149–176 (1999)

15. Le Berre, D., Rapicault, P.: Dependency management for the eclipse ecosystem: an update. In: International Workshop on Logic and Search

16. Marques-Silva, J., Planes, J.: On Using Unsatisfiability for Solving Maximum Satisfiability (2007). CoRR

17. Martins, R., Joshi, S., Manquinho, V., Lynce, I.: Incremental cardinality constraints for MaxSAT. In: O'Sullivan, B. (ed.) CP 2014. LNCS, vol. 8656, pp. 531–548. Springer, Heidelberg (2014)

18. Martins, R., Manquinho, V., Lynce, I.: Open-WBO: a modular MaxSAT solver'. In: Sinz, C., Egly, U. (eds.) SAT 2014. LNCS, vol. 8561, pp. 438–445. Springer, Heidelberg (2014)

19. Martins, R., Manquinho, V., Lynce, I.: Community-based partitioning for MaxSAT solving. In: Järvisalo, M., Van Gelder, A. (eds.) SAT 2013. LNCS, vol. 7962, pp. 182–191. Springer, Heidelberg (2013)

20. Morgado, A., Heras, F., Liffiton, M., Planes, J., Marques-Silva, J.: Iterative and core-guided MaxSAT solving: A survey and assessment. Constraints $18(4)$, 478–534 (2013)

21. Morgado, A., Heras, F., Marques-Silva, J.: Improvements to core-guided binary search for MaxSAT. In: Cimatti, A., Sebastiani, R. (eds.) SAT 2012. LNCS, vol. 7317, pp. 284–297. Springer, Heidelberg (2012)

22. Narodytska, N., Bacchus, F.: Maximum satisfiability using core-guided MaxSAT resolution. In: AAAI Conference on Artificial Intelligence, pp. 2717–2723. AAAI Press (2014)

23. Newman, M.E.J., Girvan, M.: Finding and evaluating community structure in networks. Physical Review E **69**(026113) (2004)

24. Safarpour, S., Mangassarian, H., Veneris, A.G., Liffiton, M.H., Sakallah, K.A.: Improved design debugging using maximum satisfiability. In: Formal Methods in Computer-Aided Design, pp. 13–19. IEEE Computer Society (2007)

25. Van Gelder, A.: Variable independence and resolution paths for quantified boolean formulas. In: Lee, J. (ed.) CP 2011. LNCS, vol. 6876, pp. 789–803. Springer, Heidelberg (2011)

26. Yates, R.A., Raphael, B., Hart, T.P.: Resolution graphs. Artificial Intelligence **1**(4), 257–289 (1970)

SAT-Based Formula Simplification

Alexey Ignatiev[1]($\boxtimes$), Alessandro Previti[2], and Joao Marques-Silva[1,2]

[1] INESC-ID, IST, University of Lisbon, Lisbon, Portugal
[2] CASL, University College Dublin, Dublin, Ireland

Abstract. The problem of propositional formula minimization can be traced to the mid of the last century, to the seminal work of Quine and McCluskey, with a large body of work ensuing from this seminal work. Given a set of implicants (or implicates) of a formula, the goal for minimization is to find a smallest set of prime implicants (or implicates) equivalent to the original formula. This paper considers the more general problem of computing a smallest prime representation of a non-clausal propositional formula, which we refer to as formula simplification. Moreover, the paper proposes a novel, entirely SAT-based, approach for the formula simplification problem. The original problem addressed by the Quine-McCluskey procedure can thus be viewed as a special case of the problem addressed in this paper. Experimental results, obtained on well-known representative problem instances, demonstrate that a SAT-based approach for formula simplification is a viable alternative to existing implementations of the Quine-McCluskey procedure.

1 Introduction

The Quine-McCluskey [36,47,48] procedure for the minimization of clausal formulae (i.e. formulae either represented in Conjunctive Normal Form (CNF) or Disjunctive Normal Form (DNF)) is widely known, being a standard topic in a number of textbooks (e.g. [22]), with a number of publicly available implementations. This problem is referred to as *formula minimization* in this paper. Formula minimization finds a wide range of practical applications [5,8,11,14,17,19,28,46,49,54,55,58], ranging from security to biology. A typical implementation of Quine-McCluskey starts by computing all the prime implicates (or implicants) of a CNF (or DNF) formula, and then implements a set covering step, where a minimum number of prime implicates (implicants) is selected that is equivalent to the original function. A more general scenario is when the original formula is non-clausal. Clearly, one can still generate all the implicates (or implicants) of the formula, then generate all the prime implicates (or prime implicants), and then execute the set covering step. However, in practice the number of implicates may be much larger than the number of prime

---

This work is partially supported by SFI PI grant BEACON (09/IN.1/I2618), FCT grant POLARIS (PTDC/EIA-CCO/123051/2010) and national funds through Fundação para a Ciência e a Tecnologia (FCT) with reference UID/CEC/50021/2013.

M. Heule and S. Weaver (Eds.): SAT 2015, LNCS 9340, pp. 287–298, 2015.
DOI: 10.1007/978-3-319-24318-4_21

implicates. In contrast to the more restricted problem, this problem is referred to as *formula simplification* in this paper. Moreover, regarding the existing implementations of Quine-McCluskey, these are not only limited to clausal formula minimization but also usually restricted to a small number of variables. The latter is also the case for other formula simplification alternatives based on Binary Decision Diagrams (BDDs) [10].

This paper develops novel approaches for formula simplification as well as formula minimization, both of which are entirely SAT-based[1]. The proposed approaches exploit recent work on computing prime implicates (and implicants) with SAT solvers [27,45], but also recent work on solving MaxSAT [40] and on computing smallest minimal unsatisfiable subformulae (SMUS) [23,24,26,31]. For the formula minimization problem, the main technical contribution is a new way to compute the prime implicates (or implicants) of the formula. For the formula simplification problem, the main technical contribution is the integration of prime enumeration with smallest MUS extraction.

Throughout the paper, and similarly to the most common description of the Quine-McCluskey procedure, the focus will be to compute the prime implicants of a propositional formula (possibly represented in DNF) and then to select a minimum size set of prime implicants equivalent to the original formula. However, the algorithms described in the paper also apply when computing and minimizing the set of prime implicates, possibly starting from a CNF representation.

The paper is organized as follows. Section 2 introduces the notation and definitions used throughout the paper. Section 3 describes the novel approach to formula simplification proposed in the paper. Preliminary experimental results are analyzed in Section 4. Finally, Section 5 concludes the paper.

2   Preliminaries

Definitions standard in propositional satisfiability (SAT) and maximum satisfiability (MaxSAT) solving are assumed [4]. In what follows, $\mathcal{F}$ denotes an arbitrary propositional formula. A term t is a conjunction of literals and a clause c is a disjunction of literals, while a literal l is either a Boolean variable or its negation. Whenever convenient, terms and clauses are treated as sets of literals. A formula is said to be in *conjunctive* or *disjunctive normal form* (CNF or DNF, respectively) if it is a conjunction of clauses or disjunction of terms, respectively. Set theory notation will be also used with respect to CNF and DNF formulae when necessary. Moreover, the term *clausal* will be used to denote formulae represented as sets of sets of literals, i.e. either in CNF or DNF.

Definition 1. *A term* $\mathcal{I}_n$ *is called an implicant of* $\mathcal{F}$ *if* $\mathcal{I}_n \models \mathcal{F}$. *An implicant* $\mathcal{I}_n$ *of* $\mathcal{F}$ *is called* **prime** *if any subset* $\mathcal{I}'_n \subsetneq \mathcal{I}_n$ *is not an implicant of* $\mathcal{F}$.

Definition 2. *A clause* $\mathcal{I}_e$ *is called an implicate of* $\mathcal{F}$ *if* $\mathcal{F} \models \mathcal{I}_e$. *An implicate* $\mathcal{I}_e$ *of* $\mathcal{F}$ *is called* **prime** *if any subset* $\mathcal{I}'_e \subsetneq \mathcal{I}_e$ *is not an implicate of* $\mathcal{F}$.

---

[1] Earlier work [52] used SAT as part of the ESPRESSO algorithm [7].

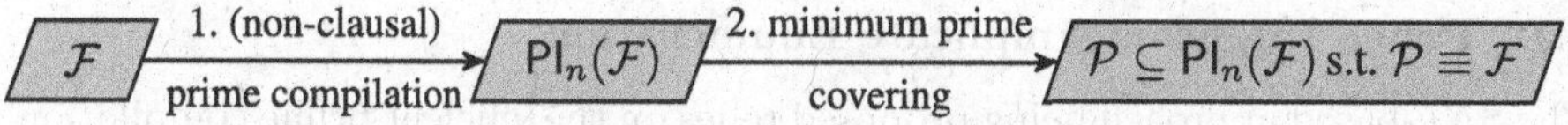

Fig. 1. General steps of the approach

The sets of all prime implicants and prime implicates of a Boolean formula $\mathcal{F}$ are denoted by $\mathsf{PI}_n(\mathcal{F})$ and $\mathsf{PI}_e(\mathcal{F})$, respectively. A subset $\mathcal{P}$ of $\mathsf{PI}_n(\mathcal{F})$ (or $\mathsf{PI}_e(\mathcal{F})$) such that $\mathcal{P} \equiv \mathcal{F}$ is said to be a *prime cover* of $\mathcal{F}$. Observe that given $\mathcal{F}$ and a prime implicant $\mathcal{I}_n \vDash \mathcal{F}$, the clause $\neg\mathcal{I}_n$ is a prime implicate of $\neg\mathcal{F}$, and the other way around. Moreover, a similar connection between $\mathsf{PI}_n(\mathcal{F})$ and $\mathsf{PI}_e(\neg\mathcal{F})$ also holds. Additionally, the concept of an *essential* prime implicant is exploited in the paper. A prime implicant is called essential if it is included in any set of prime implicants covering $\mathcal{F}$. With respect to CNF formulae, the following definitions related to MUSes and MCSes are also used:

Definition 3. *Given a CNF formula $\mathcal{F}$, a set of clauses $\mathcal{U} \subseteq \mathcal{F}$ is called a minimal unsatisfiable subset (MUS) if $\mathcal{U}$ is unsatisfiable and any subset $\mathcal{U}' \subset \mathcal{U}$ is satisfiable. A minimum size MUS of $\mathcal{F}$ is called a smallest MUS (SMUS).*

Definition 4. *A subset $\mathcal{C}$ of a CNF formula $\mathcal{F}$ is a minimal correction subset (MCS) if $\mathcal{F} \setminus \mathcal{C}$ is satisfiable and $\forall \mathcal{C}' \subseteq \mathcal{C} \wedge \mathcal{C}' \neq \emptyset$, $(\mathcal{F} \setminus \mathcal{C}) \cup \mathcal{C}'$ is unsatisfiable.*

These notions can be extended to the case of *group oriented* CNF formulae [33, 41]. A group oriented CNF formula contains groups of clauses instead of single clauses, i.e. $\mathcal{F} = \mathcal{D} \cup \mathcal{G}$, where $\mathcal{G} = \mathcal{G}_1 \cup \ldots \cup \mathcal{G}_k$ is a set of k groups while $\mathcal{D}$ is a *don't care* group. Accordingly, a *group MUS* of $\mathcal{F}$ is a subset of groups $\mathcal{G}' \subseteq \mathcal{G}$ such that formula $\mathcal{D} \cup \bigcup_{G \in \mathcal{G}'} G$ is unsatisfiable and $\forall \mathcal{G}'' \subset \mathcal{G}'$ formula $\mathcal{D} \cup \bigcup_{G \in \mathcal{G}''} G$ is satisfiable.

3   Formula Simplification with SAT

The approach proposed below follows the general steps of the original Quine-McCluskey algorithm [36,47,48] outlined in Figure 1. Given a propositional formula $\mathcal{F}$ in an non-clausal form, it (i) enumerates all prime implicants $\mathsf{PI}_n(\mathcal{F})$ (or prime implicates $\mathsf{PI}_e(\mathcal{F})$); and (ii) computes a minimum size subset $\mathcal{P} \subseteq \mathsf{PI}_n(\mathcal{F})$ (or $\mathcal{P} \subseteq \mathsf{PI}_e(\mathcal{F})$) such that $\mathcal{P} \equiv \mathcal{F}$. Hereinafter, the discussion is conducted with respect to computing a minimum size DNF representation of $\mathcal{F}$ (i.e. using prime implicants of $\mathcal{F}$). However, all of the proposed techniques can be easily adapted for the case of computing a minimum size CNF of $\mathcal{F}$ (i.e. with the use of prime implicates). Indeed, this results from the well-known connection between prime implicants of $\mathcal{F}$ and prime implicates of $\neg\mathcal{F}$ (see Section 2). For this reason and whenever convenient, some particular ideas are explained for implicate-based formula simplification.

3.1 Prime Implicant/Implicate Enumeration

The SAT-based approach being proposed relies on the efficient prime compilation of Boolean formulae. Although (and in contrast to [36,47,48]) the paper is mainly focused on non-clausal Boolean formulae, this section provides a description of the simplified version of the algorithm targeting clausal formulae. The reader is referred to [45] for further details and properties of the general algorithm.

Prime Compilation of Clausal Formulae. Although in general the extraction of a prime implicant requires a linear number of calls to a SAT solver, for the case of CNF formulae minimizing a model can be done in polynomial time. The algorithm used in this paper for the extraction of prime implicates is based on the algorithm *primer-b* recently introduced in [45]. When executed on a non-clausal formula, *primer-b* produces the complete set of prime implicates and (as a by-product) a prime implicant cover. At each step, *primer-b* identifies a new partial assignment to be tested. As highlighted in earlier work [45], when a partial assignment falsifies the formula, then its negation is guaranteed to be a prime implicate. Instead, if it satisfies the formula, the corresponding model has to be reduced to a prime implicant. However, when we deal with CNF formulae, the model can be reduced without employing a SAT solver by means of a procedure running in polynomial time. Suppose that m is a model for a CNF formula $\mathcal{F}$. Then we have to scan all the literals in m one at a time. Let l be the last picked literal. If when setting l to *don't care*, the implicant still satisfies the formula, then literal l is removed. Otherwise, it is a part of the prime implicant under construction. Note that in order to test if a literal is necessary, it is enough to check only the clauses containing it. This can be easily done by using an occurrence list, which for each literal stores the set of clauses where it appears. Additionally, more sophisticated techniques [13] can be also applied for improving the performance of the algorithm.

3.2 Computing a Smallest Prime Cover

This section describes the second phase of the proposed approach, which consists in the following. Given a complete set of prime implicants of a Boolean formula, it computes its subset of the smallest size such that the subset is equivalent to the original formula.

Prime Covering Non-Clausal Formulae. Let us assume that for a given non-clausal formula $\mathcal{F}$, the complete set of prime implicants $\mathsf{PI}_n(\mathcal{F})$ is computed as described in Section 3.1. Now one needs to find a minimum size subset $\mathcal{P} \subseteq \mathsf{PI}_n(\mathcal{F})$ such that $\mathcal{P} \equiv \mathcal{F}$. Clearly, by definition of a prime implicant, for any subset $\mathcal{P}$ (and, thus, for the smallest one) the following holds: $\mathcal{P} \vDash \mathcal{F}$. Therefore, it is enough to check whether $\mathcal{F} \vDash \mathcal{P}$, which can be done by testing if formula $\neg \mathcal{P} \wedge \mathcal{F}$ is unsatisfiable. Observe that $\mathsf{PI}_n(\mathcal{F}) \equiv \mathcal{F}$ and, thus, $\mathcal{F} \vDash \mathsf{PI}_n(\mathcal{F})$. Hence, formula

$$\neg \mathsf{PI}_n(\mathcal{F}) \wedge \mathcal{F} \tag{1}$$

is obviously unsatisfiable. This means that finding a minimum size cover $\mathcal{P}$ of $\mathcal{F}$ consists in computing a smallest size group MUS (e.g. see [33,41]) of formula (1) where subformula $\mathcal{F}$ is a *don't care group*, i.e. $\mathcal{F}$ is irrelevant for the size of the solution, and so only clauses $\neg I_n \in \neg \mathsf{PI}_n(\mathcal{F})$ are taken into account. This problem can be solved with an off-the-shelf SMUS extractor (e.g. [23,24,26,31, 34,38]).

Note that a smallest size group MUS of formula (1) corresponds to a minimum size prime cover $\mathcal{P}$ of $\mathcal{F}$ with respect to the number of prime implicants in $\mathcal{P}$. However, one might prefer to compute a minimum cover in terms of the *total number of occurrence of literals* in it. For this, a weighted group MUS formula can be considered, i.e. each clause $\neg I_n \in \neg \mathsf{PI}_n(\mathcal{F})$ is associated with a cost equal to $|I_n|$. Now, a smallest cost group MUS of (1) corresponds to a minimum cost prime cover of $\mathcal{F}$.

Observe that essential prime implicants of $\mathcal{F}$ can be identified by group MCS extraction (e.g. see [35,43] and references therein) on the considered formula (1). This is stated in the following proposition.

Proposition 1. *Any unit MCS (i.e. an MCS containing just one clause) of formula (1) corresponds to an essential prime implicant of formula $\mathcal{F}$.*

Proof. Due to the minimal hitting set duality between MCSes and MUSes of a (group) CNF formula [33,50], a clause of a unit MCS of the formula is included into any MUS of the formula. Since, by construction of (1), group unsatisfiable subformulae (hence, MUSes as well) define prime covers of $\mathcal{F}$, unit MCSes of (1) define prime implicants of $\mathcal{F}$ that must be included into *any* prime cover of $\mathcal{F}$. Thus, by definition of essential prime implicants, unit MCSes correspond to essential prime implicants of $\mathcal{F}$.   □

Unit MCSes (if any) can be identified with the use of MaxSAT (e.g. [23, 24,26,31]). This requires a SAT call for extracting an unsatisfiable core of (1), *relaxing* the corresponding clauses in the core, and enumerating models of the relaxed formula. Each unit MCS is defined by such a model and, thus, requires one SAT call per MCS. Thus, assuming that $\mathcal{F}$ has n essential primes, they can be enumerated with $n+1$ calls to a SAT oracle. Observe that this approach should be practically more efficient than the well-known alternative of separately checking each prime implicant for essentiality [22,51,53], especially if $|\mathsf{PI}_n(\mathcal{F})|$ is much larger than the number of essential primes.

Moreover, identification of the essential primes can be used for the further simplification of the group SMUS problem. Indeed, since essential prime implicants are included in any cover of $\mathcal{F}$, they can be excluded from $\neg \mathsf{PI}_n(\mathcal{F})$ and added to the don't care group. Let $\mathcal{E}$ denote the set of all essential primes of $\mathcal{F}$, and $\mathcal{Q} = \mathsf{PI}_n(\mathcal{F}) \setminus \mathcal{E}$. Then consider formula $\neg \mathcal{Q} \wedge (\mathcal{F} \wedge \mathcal{E})$ where $\mathcal{F} \wedge \mathcal{E}$ represents the don't care group. For any group SMUS $\mathcal{P}'$ of this formula, the corresponding group SMUS of (1) is $\mathcal{P}' \cup \mathcal{E}$.

Clausal Formulae Minimization. This section briefly explains how one can deal with a particular case of clausal formulae. Recall that given a clausal formula

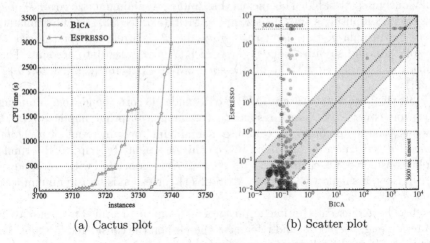

(a) Cactus plot   (b) Scatter plot

Fig. 2. Performance of BICA and ESPRESSO on PLA instances

$\mathcal{F}$, the approach being proposed is able to compute the exact minimum size representation of $\mathcal{F}$. Although the general technique described in Section 3.2 can be also applied to clausal formulae, a specialized MaxSAT-based approach to clausal formulae minimization can be proposed, which can be more efficient in practice.

Following the ideas of [36, 47, 48], one can formulate a set covering problem: given a set of terms $\mathcal{F} = \mathcal{T}_1 \cup \ldots \cup \mathcal{T}_m$ and a complete set of its prime implicants $\mathsf{PI}_n(\mathcal{F})$, one needs to compute a smallest size set of prime implicants $\mathcal{P} \subseteq \mathsf{PI}_n(\mathcal{F})$ such that for each $\mathcal{T}_i \in \mathcal{F}$ there is a prime implicant $\mathcal{I}_j \in \mathcal{P}$ covering term $\mathcal{T}_i$, i.e. $\mathcal{I}_j \subseteq \mathcal{T}_i$. The relation between the set covering problem and MaxSAT was originally put forward in [18, 44]. The translation from the set covering problem to MaxSAT is well-known and has been studied elsewhere (e.g. see [2, 56]). Note that compared to the general case SMUS-based approach, using this MaxSAT formulation of the problem is preferred for clausal formulae due to a better complexity characterization (decision versions of MaxSAT and SMUS are complete for NP [18, 44] and Σ_2^P [21, 30], respectively).

Approximated Solutions. Once the SMUS and MaxSAT formulations of the simplification phase of the approach are introduced, one can immediately notice that various techniques can be applied in order to get approximate solutions of the considered problems. For SMUS, these include MUS extraction (e.g. see [3, 42]) and MUS enumeration (see [32]) algorithms. As for MaxSAT, a number of MCS enumeration techniques approximating MaxSAT solutions were proposed in the past (e.g. [20, 35, 43]).

4 Preliminary Results

This section evaluates the proposed approach to Boolean formula simplification. The experiments were performed in Ubuntu Linux on an Intel Xeon E5-2630 2.60GHz processor with 64GByte of memory. The time limit was set to 3600s and the memory limit to 10GByte. The approach proposed above was implemented in a prototype called BICA (*Boolean simplifier for non-clausal formulae*). The BICA Boolean formula simplifier is written as a Python script, which instruments the flow of the proposed approach and calls the existing binaries both for doing the prime compilation phase and the minimum covering phase. Prime implicate enumeration is done by calling PRIMER [45], while minimum covering is done with the FORQES SMUS extractor [26] for non-clausal formulae, and with the MSCG MaxSAT solver [25,39] if the formulae are clausal. Also note that PRIMER is implemented on top of the MINISAT[2] SAT solver [15] while the underlying SAT solver of MSCG and FORQES is Glucose 3.0[3] [1]. Further details on the experimental evaluation including the chosen benchmark sets are presented below.

4.1 PLA Benchmarks

In order to assess the efficiency of the new approach applied to clausal Boolean formulae, two sets of PLA circuit benchmark sets were considered. The first set was originally described in [7] and includes 123 easy and 19 hard instances [16]. The second benchmark set called *MCNC91 suite* was proposed in [57] and comprises 41 PLA circuits. Since the approach being proposed currently cannot be applied to multi-output Boolean circuits and in order to compare it with the well-known implementation of the Quine-McCluskey procedure called Espresso [7,16], each of the considered instances was split in the following way. Given a PLA circuit with n inputs and m outputs, m single-output PLA circuits were created, each having n inputs. The total number of resulting PLA circuits constructed this way and considered in the evaluation is 3744.

The new approach was compared to the *exact* version of Espresso [7,16], which is referred to as ESPRESSO and implements the Quine-McCluskey algorithm. Figure 2 shows the performance of ESPRESSO compared to BICA for the considered set of clausal instances. As one can see in Figure 2a, both solvers can minimize most of the circuits. BICA is able to solve 3740 instances (out of 3744), ESPRESSO is not far with 3731 formulae minimized. However, the detailed scatter plot shown in Figure 2b indicates that BICA generally performs better than ESPRESSO (up to 4 orders of magnitude).

4.2 Bi-decomposition Interpolation Benchmarks

The following benchmark set comes from the area of bi-decomposition of a Boolean function (e.g. see [9]). An earlier work on using interpolants for Boolean

---

[2] https://github.com/niklasso/minisat
[3] http://www.labri.fr/perso/lsimon/glucose

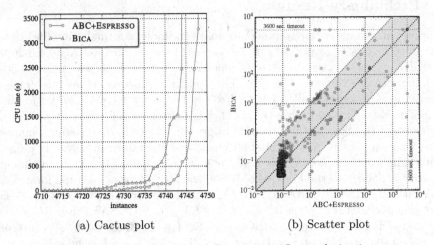

(a) Cactus plot   (b) Scatter plot

Fig. 3. Performance of BICA and ESPRESSO on Interpolation instances

Table 1. Performance of BICA and ABC+ESPRESSO on QG6 instances

	# solved	max. time (s)	min. time (s)	avg. time (s)
BICA	63	3600	0.56	1592.65
ABC+ESPRESSO	0	3600	3600	3600

function decomposition is for example [29], where the function's components are computed through *Craig's interpolation* [12]. Thus, given such interpolants representing the function's components, one can try to simplify them in order to get a simpler decomposed representation of the original Boolean function. The interpolant formulae were generated for the standard ISCAS, ITC, and LGSynth benchmark suites. The total number of the considered interpolant formulae is 4815.

Note that the interpolants are given in a non-clausal form. In this case, one cannot use ESPRESSO directly. First, the formulae need to be translated into a clausal form. For this purpose, the well-known logic synthesis system ABC [6] was used, namely its ability to *collapse* a circuit with the use of BDDs. Figure 3a shows a cactus plot illustrating the performance of both BICA and ABC+ESPRESSO for the considered interpolation benchmarks. Analogously to the PLA benchmarks, both competitors perform quite well being able to solve almost all the instances. BICA simplifies 4744 formulae while ABC+ESPRESSO solves 4748 instances. Figure 3b indicates that there is no clear winner in this case even though ABC+ESPRESSO has some advantage over BICA. A reason for this can be that the CUDD BDD package[4] used in ABC is usually able to *clausify* the considered circuits within a very short time (less than a second). Also, the number of terms reported by CUDD is usually very close to the optimum, which simplifies the Quine-McCluskey procedure performed by ESPRESSO.

---

[4] http://vlsi.colorado.edu/~fabio/CUDD/

4.3   Quasigroup Classification Benchmarks

This set of non-clausal benchmarks called QG6 was proposed in [37] when encoding classification theorems for quasigroups. Out of 256 formulae we chose 83 that are satisfiable. Note that these 83 benchmark instances have either 252 or 360 variables, which is larger than the number of inputs in all circuits considered in Section 4.1 and Section 4.2. Similarly to the interpolation benchmarks, ABC+ESPRESSO was used as an alternative to BICA. However, it was not able to simplify any of these formulae, which is not surprising because these instances are hard for BDDs (this may be caused by the number of variables). (For this reason, no plots are presented for QG6 benchmarks and Table 1 is shown instead). In contrast, BICA is able to simplify 63 (out of 83) formulae.

In summary, the experimental results indicate that the proposed approach is a viable alternative to the existing implementations of the Quine-McCluskey procedure for the case of clausal Boolean formulae. Moreover and as stated in Section 4.3, being focused on non-clausal formulae and based on the state-of-the-art SAT technology, the new approach performs reasonably well for non-clausal formulae with a large number of variables, which can be out of reach for the alternative approaches, e.g. the ones based on BDDs, or ABC and Espresso.

5   Conclusions

This paper develops entirely SAT-based solutions for propositional formula minimization and simplification. In both cases the set of prime implicates (or implicants) is computed using recent work on prime implicate (or implicant) enumeration. For the clausal formula minimization problem, a minimum-size subset of the prime implicates that covers an initial set of implicates is obtained with a set covering approach, which is done with MaxSAT. For non-clausal formula simplification, the problem is more challenging, and the problem is shown to be solved by computing a smallest MUS.

The experimental results are encouraging. For two classes of problem instances, the new approach outperforms a well-known implementation of Quine-McCluskey, whereas for another class of problem instances it loses to the Quine-McCluskey procedure. Future work will investigate settings in which SAT-based formula minimization and simplification can be shown to be the preferred option.

References

1. Audemard, G., Lagniez, J.-M., Simon, L.: Improving glucose for incremental SAT solving with assumptions: application to MUS extraction. In: Järvisalo, M., Van Gelder, A. (eds.) SAT 2013. LNCS, vol. 7962, pp. 309–317. Springer, Heidelberg (2013)
2. Bautista, J., Pereira, J.: A GRASP algorithm to solve the unicost set covering problem. Computers & OR **34**(10), 3162–3173 (2007)

3. Belov, A., Lynce, I., Marques-Silva, J.: Towards efficient MUS extraction. AI Commun. **25**(2), 97–116 (2012)
4. Biere, A., Heule, M., van Maaren, H., Walsh, T. (eds.) Handbook of Satisfiability, Frontiers in Artificial Intelligence and Applications, vol. 185. IOS Press (2009)
5. Boyar, J., Matthews, P., Peralta, R.: Logic minimization techniques with applications to cryptology. J. Cryptology **26**(2), 280–312 (2013)
6. Brayton, R., Mishchenko, A.: ABC: an academic industrial-strength verification tool. In: Touili, T., Cook, B., Jackson, P. (eds.) CAV 2010. LNCS, vol. 6174, pp. 24–40. Springer, Heidelberg (2010)
7. Brayton, R.K., Sangiovanni-Vincentelli, A.L., McMullen, C.T., Hachtel, G.D.: Logic Minimization Algorithms for VLSI Synthesis. Kluwer Academic Publishers, Norwell (1984)
8. Cabalar, P., Pearce, D.J., Valverde, A.: Minimal logic programs. In: Dahl, V., Niemelä, I. (eds.) ICLP 2007. LNCS, vol. 4670, pp. 104–118. Springer, Heidelberg (2007)
9. Chen, H., Janota, M., Marques-Silva, J.: QBF-based Boolean function bi-decomposition. In: DATE, pp. 816–819 (2012)
10. Coudert, O.: Two-level logic minimization: an overview. Integration **17**(2), 97–140 (1994)
11. Courtois, N., Hulme, D., Mourouzis, T.: Solving circuit optimisation problems in cryptography and cryptanalysis. IACR Cryptology ePrint Archive **2011**, 475 (2011)
12. Craig, W.: Linear reasoning. A new form of the Herbrand-Gentzen theorem. J. Symb. Log. **22**(3), 250–268 (1957)
13. Déharbe, D., Fontaine, P., Berre, D.L., Mazure, B.: Computing prime implicants. In: FMCAD, pp. 46–52 (2013) ·
14. Durzinsky, M., Wagler, A., Marwan, W.: Reconstruction of extended petri nets from time series data and its application to signal transduction and to gene regulatory networks. BMC Systems Biology **5**, 113 (2011)
15. Eén, N., Sörensson, N.: An extensible SAT-solver. In: Giunchiglia, E., Tacchella, A. (eds.) SAT 2003. LNCS, vol. 2919, pp. 502–518. Springer, Heidelberg (2004)
16. Espresso – multi-valued PLA minimization. http://embedded.eecs.berkeley.edu/pubs/downloads/espresso (accessed April 30, 2015)
17. Filiol, E.: Malware pattern scanning schemes secure against black-box analysis. Journal in Computer Virology **2**(1), 35–50 (2006)
18. Garey, M.R., Johnson, D.S.: Computers and Intractability: A Guide to the Theory of NP-Completeness. W. H. Freeman (1979)
19. Ghosh, J., Philip, S.J., Qiao, C.: Sociological orbit aware location approximation and routing (SOLAR) in MANET. Ad Hoc Networks **5**(2), 189–209 (2007)
20. Grégoire, É., Lagniez, J., Mazure, B.: An experimentally efficient method for (MSS, CoMSS) partitioning. In: AAAI, pp. 2666–2673 (2014)
21. Gupta, A.: Learning Abstractions for Model Checking. PhD thesis, Carnegie Mellon University, June 2006
22. Hachtel, G.D., Somenzi, F.: Logic synthesis and verification algorithms. Kluwer (1996)
23. Ignatiev, A., Janota, M., Marques-Silva, J.: Quantified maximum satisfiability: a coreguided approach. In: Järvisalo, M., Van Gelder, A. (eds.) SAT 2013. LNCS, vol. 7962, pp. 250–266. Springer, Heidelberg (2013)
24. Ignatiev, A., Janota, M., Marques-Silva, J.: Quantified maximum satisfiability. Constraints (2015). http://dx.doi.org/10.1007/s10601-015-9195-9

25. Ignatiev, A., Morgado, A., Manquinho, V.M., Lynce, I., Marques-Silva, J.: Progression in maximum satisfiability. In: ECAI, pp. 453–458 (2014)
26. Ignatiev, A., Previti, A., Liffiton, M., Marques-Silva, J.: Smallest MUS extraction with minimal hitting set dualization. In: Pesant, G. (ed.) CP 2015. LNCS, vol. 9255, pp. 173–182. Springer, Heidelberg (2015)
27. Jabbour, S., Marques-Silva, J., Sais, L., Salhi, Y.: Enumerating prime implicants of propositional formulae in conjunctive normal form. In: Fermé, E., Leite, J. (eds.) JELIA 2014. LNCS, vol. 8761, pp. 152–165. Springer, Heidelberg (2014)
28. Lafond, D., Lacouture, Y., Mineau, G.: Complexity minimization in rule-based category learning: Revising the catalog of Boolean concepts and evidence for non-minimal rules. Journal of Mathematical Psychology **51**(2), 57–74 (2007)
29. Lee, R., Jiang, J.R., Hung, W.: Bi-decomposing large boolean functions via interpolation and satisfiability solving. In: DAC, pp. 636–641 (2008)
30. Liberatore, P.: Redundancy in logic I: CNF propositional formulae. Artif. Intell. **163**(2), 203–232 (2005)
31. Liffiton, M.H., Mneimneh, M.N., Lynce, I., Andraus, Z.S., Marques-Silva, J., Sakallah, K.A.: A branch and bound algorithm for extracting smallest minimal unsatisfiable subformulas. Constraints **14**(4), 415–442 (2009)
32. Liffiton, M.H., Previti, A., Malik, A., Marques-Silva, J.: Fast, flexible MUS enumeration. Constraints (2015). http://dx.doi.org/10.1007/s10601-015-9183-0
33. Liffiton, M.H., Sakallah, K.A.: Algorithms for computing minimal unsatisfiable subsets of constraints. J. Autom. Reasoning **40**(1), 1–33 (2008)
34. Lynce, I., Marques-Silva, J.: On computing minimum unsatisfiable cores. In: SAT (2004)
35. Marques-Silva, J., Heras, F., Janota, M., Previti, A., Belov, A.: On computing minimal correction subsets. In: IJCAI, pp. 615–622 (2013)
36. McCluskey, E.J.: Minimization of Boolean functions. Bell system technical Journal **35**(6), 1417–1444 (1956)
37. Meier, A., Sorge, V.: A new set of algebraic benchmark problems for SAT solvers. In: Bacchus, F., Walsh, T. (eds.) SAT 2005. LNCS, vol. 3569, pp. 459–466. Springer, Heidelberg (2005)
38. Mneimneh, M., Lynce, I., Andraus, Z.S., Marques-Silva, J., Sakallah, K.A.: A branch-and-bound algorithm for extracting smallest minimal unsatisfiable formulas. In: Bacchus, F., Walsh, T. (eds.) SAT 2005. LNCS, vol. 3569, pp. 467–474. Springer, Heidelberg (2005)
39. Morgado, A., Dodaro, C., Marques-Silva, J.: Core-guided maxSAT with soft cardinality constraints. In: O'Sullivan, B. (ed.) CP 2014. LNCS, vol. 8656, pp. 564–573. Springer, Heidelberg (2014)
40. Morgado, A., Heras, F., Liffiton, M.H., Planes, J., Marques-Silva, J.: Iterative and core-guided maxsat solving: A survey and assessment. Constraints **18**(4), 478–534 (2013)
41. Nadel, A.: Boosting minimal unsatisfiable core extraction. In: FMCAD, pp. 221–229 (2010)
42. Nadel, A., Ryvchin, V., Strichman, O.: Efficient MUS extraction with resolution. In: FMCAD, pp. 197–200 (2013)
43. Nöhrer, A., Biere, A., Egyed, A.: Managing SAT inconsistencies with HUMUS. In: VAMOS, pp. 83–91 (2012)
44. Papadimitriou, C.H.: Computational complexity. Addison-Wesley (1994)
45. Previti, A., Ignatiev, A., Morgado, A., Marques-Silva, J.: Prime compilation of non-clausal formulae. In: IJCAI, pp. 1980–1987 (2015)

46. Priesterjahn, C., Steenken, D., Tichy, M.: Timed hazard analysis of self-healing systems. In: Cámara, J., de Lemos, R., Ghezzi, C., Lopes, A. (eds.) Assurances for Self-Adaptive Systems. LNCS, vol. 7740, pp. 112–151. Springer, Heidelberg (2013)

47. Quine, W.V.: The problem of simplifying truth functions. American mathematical monthly, 521–531 (1952)

48. Quine, W.V.: A way to simplify truth functions. American mathematical monthly, 627–631 (1955)

49. Rehák, M., Staab, E., Fusenig, V., Stiborek, J., Grill, M., Bartos, K., Pechoucek, M., Engel, T.: Threat-model-driven runtime adaptation and evaluation of intrusion detection system. In: ICAC, pp. 65–66 (2009)

50. Reiter, R.: A theory of diagnosis from first principles. Artif. Intell. 32(1), 57–95 (1987)

51. Rudell, R.L., Sangiovanni-Vincentelli, A.L.: Multiple-valued minimization for PLA optimization. IEEE Trans on CAD of Integrated Circuits and Systems 6(5), 727–750 (1987)

52. Sapra, S., Theobald, M., Clarke, E.M.: SAT-based algorithms for logic minimization. In: ICCD, pp. 510–517 (2003)

53. Sasao, T.: Input variable assignment and output phase optimization of PLA's. IEEE Trans. Computers 33(10), 879–894 (1984)

54. Vigo, R.: A note on the complexity of Boolean concepts. Journal of Mathematical Psychology 50(5), 501–510 (2006)

55. Wu, M.: Query optimization for selections using bitmaps. In: SIGMOD, pp. 227–238 (1999)

56. Yagiura, M., Ibaraki, T.: Efficient 2 and 3-flip neighborhood search algorithms for the MAX SAT: experimental evaluation. J. Heuristics 7(5), 423–442 (2001)

57. Yang, S.: Logic synthesis and optimization benchmarks user guide: Version 3.0. Technical Report 1991-IWLS-UG-Saeyang, MCNC, Research Triangle Park, January 1991

58. Zhou, Z., Huang, D., Wang, Z.: Efficient privacy-preserving ciphertext-policy attribute based-encryption and broadcast encryption. IEEE Trans. Computers 64(1), 126–138 (2015)

Volt: A Lazy Grounding Framework for Solving Very Large MaxSAT Instances

Ravi Mangal[1]([⊠]), Xin Zhang[1], Aditya V. Nori[2], and Mayur Naik[1]

[1] Georgia Institute of Technology, Atlanta, USA
ravi.mangal@gatech.edu
[2] Microsoft Research, Bangalore, India

Abstract. Very large MaxSAT instances, comprising 10^{20} clauses and beyond, commonly arise in a variety of domains. We present VOLT, a framework for solving such instances, using an iterative, lazy grounding approach. In each iteration, VOLT grounds a subset of clauses in the MaxSAT problem, and solves it using an off-the-shelf MaxSAT solver. VOLT provides a common ground to compare and contrast different lazy grounding approaches for solving large MaxSAT instances. We cast four diverse approaches from the literature on information retrieval and program analysis as instances of VOLT. We have implemented VOLT and evaluate its performance under different state-of-the-art MaxSAT solvers.

1 Introduction

MaxSAT solvers have made remarkable progress in performance over the last decade. Annual evaluations to assess the state-of-the-art in MaxSAT solvers began in 2006. These evaluations primarily focus on efficiently solving difficult MaxSAT instances. Due to several advances in solving such instances, many emerging problems in a variety of application domains are being cast as large MaxSAT instances, comprising 10^{20} clauses and beyond.[1]

Large MaxSAT instances pose scalability challenges to existing solvers. Researchers in other communities, notably statistical relational learning and program analysis, have proposed various *lazy grounding* techniques to solve such instances that arise in their application domains [4,9,15–17,19]. The high-level idea underlying these techniques is to use an iterative counterexample-guided approach that, in each iteration, poses a subset of clauses in the original large MaxSAT instance to an off-the-shelf MaxSAT solver. The construction of this subset of clauses is guided by means of *counterexamples*—these are clauses in the original problem that are unsatisfied by the current solution.

This paper presents a formal framework VOLT for systematically studying the class of lazy grounding techniques. We show how diverse existing techniques in the literature are instances of our framework (Table 1 in Section 2). In doing so,

---

[1] Throughout the paper, we slightly abuse terminology by using MaxSAT to refer to the *weighted partial maximum satisfiability* problem, which asks for a solution that satisfies all hard clauses and maximizes the sum of weights of satisfied soft clauses.

© Springer International Publishing Switzerland 2015
M. Heule and S. Weaver (Eds.): SAT 2015, LNCS 9340, pp. 299–306, 2015.
DOI: 10.1007/978-3-319-24318-4_22

$$
\begin{array}{ll}
\text{(relation)} \quad r \in \mathbf{R} & \text{(argument)} \quad a \in \mathbf{A} = \mathbf{V} \cup \mathbf{C} \\
\text{(constant)} \quad c \in \mathbf{C} & \text{(fact)} \quad t \in \mathbf{T} = \mathbf{R} \times \mathbf{A}^* \\
\text{(variable)} \quad v \in \mathbf{V} & \text{(ground fact)} \quad g \in \mathbf{G} = \mathbf{R} \times \mathbf{C}^* \\
\text{(valuation)} \quad \sigma \in \mathbf{V} \to \mathbf{C} & \text{(weight)} \quad w \in \mathbb{R}+ = (0, \infty]
\end{array}
$$

$$
\begin{array}{ll}
\text{(hard constraints)} \quad H ::= \{h_1, ..., h_n\}, & h ::= \bigwedge_{i=1}^{n} t_i \Rightarrow \bigvee_{i=1}^{m} t'_i \\
\text{(soft constraints)} \quad S ::= \{s_1, ..., s_n\}, & s ::= (h, w)
\end{array}
$$

$$
\text{(weighted constraints)} \quad C ::= (H, S) \qquad \text{(input, output)} \quad P, Q \subseteq \mathbf{G}
$$

Fig. 1. Syntax of weighted EPR constraints.

VOLT provides the first setting that formally compares and clarifies the relationship between these various techniques.

We have implemented the VOLT framework and its instantiations. It allows any off-the-shelf MaxSAT solver to be used in each iteration of the lazy grounding process. We evaluate the performance of VOLT under different state-of-the-art MaxSAT solvers using a particular instantiation. Our evaluation shows that existing lazy grounding techniques can produce instances that are beyond the reach of exact MaxSAT solvers. This in turn leads these techniques to sacrifice optimality, soundness, or scalability. VOLT is only a starting point and seeks to motivate further advances in lazy grounding and MaxSAT solving.

2   VOLT: A Lazy Grounding Framework

The first step in solving large MaxSAT instances is to succinctly represent them. VOLT uses a variant of *effectively propositional logic* (EPR) [11]. Our variant operates on relations over finite domains and has an optional weight associated with each clause. Figure 1 shows the syntax of a weighted EPR formula C, which consists of a set of hard constraints and a set of soft constraints. For convenience in formulating problems, we augment C with an *input* P which defines a set of ground facts (extensional database or EDB). Its solution, *output* Q, defines a set of ground facts that are true (intensional database or IDB).

Weighted EPR formulae are grounded by instantiating the relations over all constants in their corresponding input domains. We presume a grounding procedure $[\![\cdot]\!]$ that grounds each constraint into a set of corresponding clauses. For example, $[\![h]\!] = \bigwedge_\sigma [\![h]\!]_\sigma$ grounds the hard constraint h by enumerating all possible groundings σ of variables to constants, yielding a different clause for each unique valuation to the variables in h. The ground clauses represent a MaxSAT problem which can be solved to produce a solution that satisfies all hard clauses and maximizes the sum of the weights of satisfied soft clauses.

Enumerating all possible valuations, called *full grounding*, does not scale to real-world problems. Our framework VOLT, described in Algorithm 1, uses lazy grounding to address this problem.[2] The framework is parametric in procedures

---

[2] We assume that any input P is encoded as part of the hard constraints H. For brevity, we assume that the hard constraints H are satisfiable, allowing us to elide showing UNSAT as a possible alternative to output Q.

Table 1. Instantiating lazy grounding approaches with VOLT where $\text{Active}(h,Q) = \{ \ [\![h]\!]_\sigma \ | \ (h = \bigwedge_{i=1}^n t_i \Rightarrow \bigvee_{i=1}^m t'_i) \ and \ (\exists i : [\![t_i]\!]_\sigma \in Q \vee [\![t'_i]\!]_\sigma \in Q) \}$ and $\text{Violate}(h,Q) = \{ \ [\![h]\!]_\sigma \ | \ Q \not\models [\![h]\!]_\sigma \}$.

approach	$(\phi, \psi) := \text{INIT}(H, S)$	$(\phi, \psi) := \text{GROUND}(H, S, Q)$	$\text{DONE}(\phi, \phi', \psi, \psi', w, i)$
SoftCegar [4]	$\phi := true$ $\psi := [\![S]\!]$	$\phi := \bigwedge_{h \in H} \bigwedge \text{Violate}(h,Q)$ $\psi := true$	$\phi' = true$
Cutting Plane [16,17]	$\phi := true$ $\psi := true$	$\phi := \bigwedge_{h \in H} \bigwedge \text{Violate}(h,Q)$ $\psi ::= \bigwedge_{(h,w) \in S} \bigwedge \{ (\rho, w) \ | \ \rho \in \text{Violate}(h,Q) \}$	clauses in ϕ', ψ' $\subseteq$ clauses in ϕ, ψ
Alchemy [9] Tuffy [15]	$\phi := \bigwedge_{h \in H} \bigwedge \text{Active}(h,P)$ $\psi := \bigwedge_{(h,w) \in S} \bigwedge \{ (\rho, w) \ | \ \rho \in \text{Active}(h,P) \}$	$\phi := \bigwedge_{h \in H} \bigwedge \text{Active}(h,Q)$ $\psi := \bigwedge_{(h,w) \in S} \bigwedge \{ (\rho, w) \ | \ \rho \in \text{Active}(h,Q) \}$	$i > maxIters$ $\vee \ w > target$
AbsRefine [19]	$\phi := (\bigoplus_{a \in A} a) \wedge \neg q$ $\psi := \bigwedge_{a \in A} (a, w)$	$\phi := \bigwedge \{ \ \bigvee_{i=1}^n \neg [\![t_i]\!]_\sigma \vee [\![t_0]\!]_\sigma \ | \ (\bigwedge_{i=1}^n t_i \Rightarrow t_0) \in H \wedge \forall i \in [0..n] : [\![t_i]\!]_\sigma \in G \}$ $\psi := true$ $where \ G = \text{lfp} \ \lambda G'. \ G' \cup \{ \ [\![t_0]\!]_\sigma \ | \ (\bigwedge_{i=1}^n t_i \Rightarrow t_0) \in H \wedge \forall i \in [1..n] : [\![t_i]\!]_\sigma \in (G' \cup Q) \}$	$\phi' = true$

INIT, GROUND, and DONE. Diverse lazy grounding algorithms in the literature can be derived by different instantiations of these three procedures.

In line 3, VOLT invokes the INIT procedure to compute an initial set of hard clauses ϕ and soft clauses ψ. Next, VOLT enters the loop defined in lines 5–11. In each iteration of the loop, the algorithm keeps track of the previous solution Q, and the weight w of the solution Q by calling the Weight procedure that returns the sum of the weights of the soft clauses satisfied by Q. Initially, the solution is empty with weight zero (line 4). In line 7, VOLT invokes the GROUND procedure to compute the set of hard clauses ϕ' and soft clauses ψ' to be grounded next. Typically, ϕ' and ψ' correspond to the set of hard and soft clauses violated

---

Algorithm 1. VOLT

1: **input** (H, S): Weighted constraints.
2: **output** Q: Solution (assumes $[\![H]\!]$ is satisfiable).
3: $(\phi, \psi) := \text{INIT}(H, S)$
4: $Q := \emptyset; \ w := 0; \ i := 0$
5: **loop**
6: $\quad i := i + 1$
7: $\quad (\phi', \psi') := \text{GROUND}(H, S, Q)$
8: $\quad$ **if** $\text{DONE}(\phi, \phi', \psi, \psi', w, i)$ **return** Q
9: $\quad (\phi, \psi) := (\phi \wedge \phi', \psi \wedge \psi')$
10: $\quad Q := \text{MaxSAT}(\phi, \psi)$
11: $\quad w := \text{Weight}(Q, \psi)$

---

by the previous solution Q. Next, in line 8, the algorithm checks if Q satisfies the terminating condition by invoking the DONE procedure. If not, then in line 9, both sets of grounded clauses ϕ' and ψ' are added to the corresponding sets of grounded hard clauses ϕ and grounded soft clauses ψ respectively. In line 10,

Table 2. Benchmark program characteristics.

	brief description	# classes	# methods	bytecode (KB)	source (KLOC)
antlr	parser/translator generator	350	2,370	186	119
luindex	document indexing and search tool	619	3,732	235	170
lusearch	text indexing and search tool	640	3,923	250	178
avrora	microcontroller simulator/analyzer	1,544	6,247	325	178
xalan	XSLT processor to transform XML	903	6,053	354	285

this updated set ϕ of hard clauses and set ψ of soft clauses are fed to the MaxSAT procedure to produce a new solution Q and its corresponding weight w.

Instantiations. Table 1 shows various lazy grounding algorithms from the literature as instantiations of the VOLT framework. SoftCegar [4] grounds all the soft clauses upfront but lazily grounds the hard clauses. In each iteration, this approach grounds all the hard clauses violated by the current solution Q. Note that the Violate procedure takes as input a hard constraint h and a MaxSAT solution Q, and returns all grounded instances of h that are violated by Q. The algorithm terminates when no further hard clauses are violated.

Cutting Plane Inference (CPI) [16,17], on the other hand, is lazier than SoftCegar and grounds no clauses upfront. In each iteration, both, hard and soft constraints are checked for violations, and any violated clauses are grounded. The algorithm terminates when no new constraints are violated.

A common approach, used in statistical relational learning tools like Alchemy [9] and Tuffy [15], relies on the observation that most ground facts are false in the final solution, and thereby most clauses are trivially true (since most clauses are Horn in these applications). An active ground fact is one that has a value of true. In each iteration, the clauses grounded are such that they contain at least one active fact as per the current solution. Initially, only the input facts P are considered active. This approach terminates after a fixed number of iterations or after the weight of the satisfied clauses is greater than a target weight.

Finally, the AbsRefine approach tackles a central problem in program analysis of efficiently finding a program abstraction that keeps only information relevant for proving properties of interest. In particular, this approach uses the counterexample-guided abstraction refinement (CEGAR) method [5] to efficiently find a suitable abstraction to prove a particular program property when the program analysis is expressed in Datalog. For such analyses, a set of hard Horn constraints expresses the analysis rules. A set of input ground facts A expresses the space of abstractions, with each ground fact in A representing a unique abstraction of cost w. The query q is a unique ground fact and proving the query implies having q as false in the final solution Q. The problem is to then find a solution with the lowest cost abstraction such that the query fact does not hold and all the analysis rules are satisfied. To lazily solve this problem, AbsRefine initially grounds hard constraints to ensure that in the final solution, the query fact q is false and only a single abstraction is true. Also, soft constraints specifying the abstraction costs are grounded upfront. Next, in the GROUND procedure, AbsRefine grounds not only the hard clauses violated by the current solution, but uses the Horn nature of the constraints to ground additional clauses that would be necessarily grounded in future iterations. Specifically, it calls a Datalog solver, with the Horn constraints and the current solution Q as input, to compute the corresponding least fixed point (lfp) solution G. Any clause which has all of its ground facts in set G is added to the set ϕ' of hard clauses to be grounded. This approach terminates when no further hard clauses are grounded.

Implementation. We have implemented the VOLT framework in Java. To compute the set of clauses to be grounded when the hard constraints are in the form of Horn clauses, as in [19], we use bddbddb [18], a Datalog solver. To compute Violate, the grounded constraints that are violated by a solution, we follow existing techniques [15,16] and use SQL queries implemented using PostgreSQL.

3 Empirical Evaluation

We evaluate VOLT by instantiating it with the AbsRefine approach for the problem of finding suitable abstractions for proving safety of downcasts in five Java benchmark programs. A safe downcast is one that cannot fail because the object to which it is applied is guaranteed to be a subtype of the target type. Our experiments were done using a Linux server with 64GB RAM and 3.0GHz CPUs.

Table 2 shows statistics of the five Java programs (antlr, lusearch, luindex, avrora, xalan) from the DaCapo suite [3], each comprising 119–285 thousand lines of code. Note that these are fairly large real-world programs and allow us to study the limits of VOLT's scalability with existing MaxSAT solvers.

We use complete weighted partial MaxSAT solvers that were available from the top performers in Random, Crafted and Industrial categories of the 9th MaxSAT Evaluation [1]. In particular, the solvers we use are CCLS2akms [10,12], Eva500a [14], MaxHS [6], wmifumax [7], MSCG [8,13] and WPM-2014-co [2].

Table 3 summarizes the results of running VOLT with the different MaxSAT solvers on our benchmarks. The 'total time' column shows the total running time of VOLT. A '-' indicates an incomplete run either because the underlying MaxSAT solver crashed or timed out (ran for >18000 seconds) on a particular instance. The next column '# iterations' provides the number of iterations needed by the lazy VOLT algorithm. In cases where VOLT did not terminate, this indicates the iteration in which the MaxSAT solver failed. The 'avg solver time' column provides the average time spent by the MaxSAT solver in solving an instance. It does not include the time spent by the solver on a failed run. The 'ground clauses' column provides the distinct number of clauses grounded by VOLT in the process of solving the weighted constraints. In other words, it indicates the size of the problem fed to the MaxSAT solver in the final iteration of the VOLT algorithm. The 'total clauses' column reports the theoretical upper bound for the number of ground clauses if all the constraints were grounded naively.

The evaluation results indicate that the MaxSAT instances generated by VOLT are many orders of magnitude smaller than the full MaxSAT instance. It is clear from these numbers that any approach attempting to tackle problems of this

Table 3. Results of VOLT on program analysis benchmarks. Highlighted rows indicate cases where the MaxSAT solver used finishes successfully in all iterations.

benchmark	solver	total time (min)	# iterations	avg solver time (secs)	grounded clauses ($\times 10^6$)	total clauses
antlr	CCLS2akms	–	1	–	7.8	8.5×10^{35}
	Eva500a	124	15	64.8	10.4	
	MaxHS	117	14	71.2	10.1	
	wmifumax	109	14	44.4	10.3	
	MSCG	–	5	22.2	7.9	
	WPM-2014-co	115	14	40.3	10.3	
lusearch	CCLS2akms	–	1	–	4.6	1×10^{37}
	Eva500a	127	14	78.6	14.7	
	MaxHS	144	14	123.1	19.1	
	wmifumax	119	15	51.2	10.2	
	MSCG	–	6	17	7.5	
	WPM-2014-co	196	14	332.7	16	
luindex	CCLS2akms	–	1	–	5.2	4.5×10^{36}
	Eva500a	172	23	45.2	5.9	
	MaxHS	161	22	52.5	5.9	
	wmifumax	169	23	34.1	6.9	
	MSCG	–	6	17.8	9	
	WPM-2014-co	216	21	226.3	5.7	
avrora	CCLS2akms	–	1	–	7	4×10^{37}
	Eva500a	–	4	80.2	17.6	
	MaxHS	–	13	136.7	15.5	
	wmifumax	–	13	115.1	9.1	
	MSCG	–	5	31.6	16.9	
	WPM-2014-co	–	12	2135.9	14.8	
xalan	CCLS2akms	–	1	–	10	3.8×10^{39}
	Eva500a	–	5	96.6	19.2	
	MaxHS	–	18	571.6	> 4290	
	wmifumax	–	14	78.7	42.9	
	MSCG	–	5	47.6	19.7	
	WPM-2014-co	–	12	505.7	44.3	

scale needs to employ lazy techniques for solving such instances. On the other hand, we also observe that many of the solvers are unable to solve these relatively smaller instances generated by VOLT. For example, VOLT does not terminate using any of the solvers for avrora and xalan.

The lack of scalability of existing solvers on the larger MaxSAT instances from our evaluation suggests the need for further research in both, lazy grounding approaches as well as MaxSAT solvers. A possible next step is to make lazy grounding more demand-driven. This is motivated by the fact many applications including ours are only concerned with the value of a particular variable instead of the entire MaxSAT solution. We intend to make the MaxSAT instances generated in our evaluation publicly available to facilitate future research.

4   Conclusion

Emerging problems in fields like statistical relational learning and program analysis are being cast as very large MaxSAT instances. Researchers in these areas have developed approaches that lazily ground weighted EPR formulae to solve such instances. We have presented a framework VOLT that captures the essence of lazy grounding techniques in the literature. VOLT not only allows to formally compare and clarify the relationship between diverse lazy grounding techniques but also enables to empirically evaluate different MaxSAT solvers. We hope that VOLT will stimulate further advances in lazy grounding and MaxSAT solving.

Acknowledgments. We thank the anonymous referees for helpful feedback. This work was supported by DARPA contract #FA8750-15-2-0009 and by NSF awards #1253867 and #1526270.

References

1. http://www.maxsat.udl.cat/14/index.html
2. http://web.udl.es/usuaris/q4374304/
3. Blackburn, S.M., Garner, R., Hoffman, C., Khan, A.M., McKinley, K.S., Bentzur, R., Diwan, A., Feinberg, D., Frampton, D., Guyer, S.Z., Hirzel, M., Hosking, A., Jump, M., Lee, H., Moss, J.E.B., Phansalkar, A., Stefanović, D., VanDrunen, T., von Dincklage, D., Wiedermann, B.: The DaCapo benchmarks: Java benchmarking development and analysis. In: OOPSLA (2006)
4. Chaganty, A., Lal, A., Nori, A.V., Rajamani, S.K.: Combining relational learning with SMT solvers using CEGAR. In: Sharygina, N., Veith, H. (eds.) CAV 2013. LNCS, vol. 8044, pp. 447–462. Springer, Heidelberg (2013)
5. Clarke, E., Grumberg, O., Jha, S., Lu, Y., Veith, H.: Counterexample-guided abstraction refinement for symbolic model checking. JACM **50**(5) (2003)
6. Davies, J., Bacchus, F.: Postponing optimization to speed up MAXSAT solving. In: CP (2013)
7. Janota, M.: MiFuMax – a literate MaxSAT solver (2013)
8. Marques-Silva, J., Ignatiev, A., Morgado, A.: MSCG - Maximum Satisfiability: a Core-Guided approach (2014)
9. Kok, S., Sumner, M., Richardson, M., Singla, P., Poon, H., Lowd, D., Domingos, P.: The alchemy system for statistical relational AI. Tech. rep., Department of Computer Science and Engineering, University of Washington, Seattle, WA (2007). http://alchemy.cs.washington.edu
10. Kügel, A.: Improved exact solver for the weighted MAX-SAT problem. In: POS 2010. Pragmatics of SAT (2010)
11. Lewis, H.R.: Complexity results for classes of quantificational formulas. J. Comput. Syst. Sci. **21**(3), 317–353 (1980)
12. Luo, C., Cai, S., Wu, W., Jie, Z., Su, K.: CCLS: an efficient local search algorithm for weighted maximum satisfiability. IEEE Trans. Computers **64**(7), 1830–1843 (2015)
13. Morgado, A., Dodaro, C., Marques-Silva, J.: Core-guided MaxSAT with soft cardinality constraints. In: O'Sullivan, B. (ed.) CP 2014. LNCS, vol. 8656, pp. 564–573. Springer, Heidelberg (2014)
14. Narodytska, N., Bacchus, F.: Maximum satisfiability using core-guided MaxSAT resolution. In: AAAI (2014)
15. Niu, F., Ré, C., Doan, A., Shavlik, J.W.: Tuffy: scaling up statistical inference in markov logic networks using an RDBMS. In: VLDB (2011)

16. Noessner, J., Niepert, M., Stuckenschmidt, H.: RockIt: Exploiting parallelism and symmetry for MAP inference in statistical relational models. In: AAAI (2013)
17. Riedel, S.: Improving the accuracy and efficiency of MAP inference for markov logic. In: UAI (2008)
18. Whaley, J., Lam, M.: Cloning-based context-sensitive pointer alias analysis using binary decision diagrams. In: PLDI (2004)
19. Zhang, X., Mangal, R., Grigore, R., Naik, M., Yang, H.: On abstraction refinement for program analyses in datalog. In: PLDI (2014)

Between SAT and UNSAT: The Fundamental Difference in CDCL SAT

Chanseok Oh[(✉)]

New York University, New York, NY 10012, USA
chanseok@cs.nyu.edu

Abstract. The way CDCL SAT solvers find a satisfying assignment is very different from the way they prove unsatisfiability. We propose an explanation to the difference by identifying direct connections to the workings of some of the most important elements in CDCL solvers: the effects of restarts and VSIDS, and the roles of learned clauses. We give a wide range of concrete evidence that highlights the varying effects and roles of these elements. As a result, this paper also sheds a new light on the internal workings of CDCL. Based on our reasoning on the difference in solver behaviors, we present several ideas for optimizing SAT solvers for either SAT or UNSAT instances. We then show that we can achieve improvements on both SAT and UNSAT at the same time by judiciously exploiting the difference. We have implemented a hybrid idea mixing two different restart strategies on top of our new solver COMiniSatPS and observed substantial performance improvement.

1 Introduction

Annual SAT Competitions have always been very competitive, but particularly, the recent SAT Competitions in the application domain have become extremely intense, showing the clear indication that modern solvers have reached a state of saturated performance. A difference of solving one or two more problem instances may completely shuffle the ranks of the top-performing solvers. For example, the number of solved instances by the top 13 solvers in the application SAT+UNSAT track in 2014 ranges between 221 and 231, and between 98 and 110 by the top 20 solvers in the SAT track. Particularly notable is that MiniSat [13] hack solvers, despite their simplicity and the legacy of the base solver, are as good as any top performing solvers. The top 13 and 20 solvers above include, respectively, two and three MiniSat hack solvers, and one of them (minisat_blbd [11]) was actually the winner in the SAT track[1]. Also in the SAT+UNSAT track, MiniSat_HACK_999ED [25] solved just four less problems than bronze-awardee Riss BlackBox [2]. Moreover, in Configuration SAT Solver Challenge 2014, MiniSat_HACK_999ED was a close runner-up to the winner (Lingeling [8]) in the industrial track and was the top solver when using default parameters [20].

---

[1] As an important note, minisat_blbd uses many hand-picked magic constants overly tuned for the competition benchmarks and selects them in a controversial way.

© Springer International Publishing Switzerland 2015
M. Heule and S. Weaver (Eds.): SAT 2015, LNCS 9340, pp. 307–323, 2015.
DOI: 10.1007/978-3-319-24318-4_23

This trend extends to previous years, e.g, in 2013, MiniSat hack solvers took the 4th place in the application SAT+UNSAT track, and the 1st and the 3rd in the SAT track[2]. In 2011, the top two solvers in the application track were MiniSat hacks.

These results suggest that today's solvers, including MiniSat hacks, are increasingly showing similar and saturated performance. We have witnessed impressive advancements in SAT research since the inception of CDCL [29], but the speed and degree of improvements is declining. Now we seem to have faced one of the ceilings that calls for a breakthrough. However, the empirical and NP-complete nature of practical SAT research offers opportunities to anyone in this field for making a breakthrough at any stage. Although what we report in this paper is never close to providing such opportunities leading to a breakthrough, we believe that it has enough potential to push the ceiling further up.

The central theme of this paper is the well-known fact that the way CDCL SAT solvers find a satisfying assignment is very different from the way they prove unsatisfiability in practice [12]. Although the fact itself is very well known, it is not well understood how and why they work differently and what can be done accordingly to realize improvements. In this paper, we give partial explanations to these questions from certain aspects. Understanding the reasons for the difference will not only be interesting from the theoretical perspective in explaining the actual workings of CDCL but allow us to leverage the difference in an effective way to bring further improvements. As a proof, we implemented simple techniques based on our reasoning on the difference in our new solver COMiniSatPS. The main contributions of this paper is summarized as follows:

1. Reasoning on the SAT/UNSAT Difference. The SAT community is well aware that CDCL solvers work differently between SAT and UNSAT, but today's solvers are not leveraging this difference to the fullest degree. This is not an irony, because how and why they are different has not been explained much. We will provide our explanations to the reasons for the difference. We will support our claims with a wide range of evidence, and the main evidence is the varying roles and effects of learned clauses, restarts, and the VSIDS heuristic [23]. The evidence will give fresh insights on the workings of these elements in CDCL.

2. Promoting Attention to the Difference. Historically, effects of a new technique have not been analyzed separately on SAT and UNSAT in many occasions, or only superficially if done. Likewise, in the 2014 Competition, every solver used the same binary in both SAT and UNSAT tracks[3], except ROKK [36]. Moreover, SAT-Race 2015 will not have an independent SAT or UNSAT track. We call for more attention from the SAT community to this issue of neglecting the SAT/UNSAT difference. Particularly, we strongly suggest that techniques and solvers be evaluated separately on SAT and UNSAT whenever possible.

---

[2] Assuming that the authors of these solvers indicated participation in the main track.

[3] Some solvers disabled certain complex simplifications for UNSAT, but it was only to be able to generate verifiable proofs. Moreover, the SAT-focused ROKK was also limited to adjusting several parameters.

3. Performance Improvements. First, we will come to understand how to make solvers stronger on SAT at the expense of making it weaker on UNSAT (and vice versa). Ultimately, we will highlight the potential of exploiting the SAT/UNSAT difference for achieving improvements on both SAT and UNSAT by presenting the latest version of COMiniSatPS and its results.

4. Uncovering Potential Values of Neglected Techniques. Our supporting evidence includes our explanations to the effectiveness (and ineffectiveness) of the Luby-series [22] restarts. Although still used by some solvers, the Luby strategy has largely been replaced with much more rapid restarts (e.g, Glucose-style restarts [6]) in modern solvers. This is because rapid restarts are shown to be vastly superior in a universal sense. However, we will show that Luby is superior to rapid restarts if restricted to satisfiable instances. With this observation, we further raise a concern that techniques of the past can be overshadowed and discarded too easily in favor of new ones, e.g, when the SAT community neglects the SAT/UNSAT difference. Like in the Luby case, revisiting past and current research with the difference in mind may reveal new insights. In the same vein, we will uncover some interesting ideas hidden in the results of the past SAT Competitions in the course of our discussion.

2   Background

It is assumed that readers are familiar with basic CDCL knowledge. After we introduce COMiniSatPS, we briefly cover VSIDS and a few strategies for restarts and learned clause management in CDCL that appear in this paper.

2.1   COMiniSatPS

COMiniSatPS[4] is our new solver designed to exploit the SAT/UNSAT difference. It is officially a successor to the award-winning[5] solver SWDiA5BY [25]. SWDiA5BY in turn merely implements on top of Glucose a tiny hack of another award-winning[6] MiniSat hack solver MiniSat_HACK_999ED [25]. We will actually use empirical data generated from COMiniSatPS to highlight the SAT/UNSAT difference. However, we assure readers that COMiniSatPS is a simple MiniSat (practically, Glucose) derivation, and all of its essence will be covered eventually in the course of discussion.

2.2   VSIDS Branching Heuristic

VSIDS [23] is a branching heuristic to choose a decision variable for searching. The heuristic favors variables that are more active in terms of being involved

---

[4] Source is available at http://www.cs.nyu.edu/~chanseok/cominisatps.
[5] Three medals in SAT Competition 2014.
[6] Collectively three medals in SAT Competition 2014 and Configurable SAT Solver Challenge 2014.

in recent conflict analyses. The activity scores however slowly decay over time, which naturally penalizes variables that have been inactive for a long time. It has long been the standard heuristic adopted by almost all modern solvers, with small variation at best.

2.3 Learned Clause Management

MiniSat's Clause Activity Scheme. MiniSat removes (roughly) half of the entire learned clauses based on their activities periodically. The notion of activity is same as in VSIDS and thus dynamic: clauses involved in recent conflict analyses are awarded with bumped scores, and the scores decay over time. The size of the clause database is capped to follow a geometric progression by periodic reduction. The base of the progression is determined by the size of an input problem (retaining more clauses for large problems).

Glucose's LBD Scheme. Instead of using clause activities, it uses LBD [4] to prioritize which clauses to remove. In short, LBD is a number of different decision levels of variables in a clause (hence never greater than the clause size), and low LBD is favored. Unlike MiniSat, the LBD value is mostly static and determined at the time of clause creation (can only decrease occasionally). Another critical difference from MiniSat is its aggressive tendency to maintain a very compact database with short intervals between reductions.

2.4 Restart Strategies

MiniSat's Luby-series Restarts. MiniSat's default restart strategy that was once a standard. The intervals between restarts (in terms of conflicts) are fixed to follow the Luby sequence [22], each multiplied by 100: 1, 1, 2, 1, 1, 2, 4, 1, 1, 2, 1, 1, 2, 4, 8, ... The sequence is known to be log optimal when the runtime distribution of a problem is unknown in the theoretical sense [22].

Glucose's Dynamic and Rapid Restarts. Restarts are dynamic [6] in that it initiates a restart when the solver appears to learn clauses with higher LBD than the global average. This typically results in (relatively) much more rapid restarts. Later versions of Glucose added a method to skip restarts when the solver seems likely to have got close to a satisfying assignment (precisely speaking, when a lot of variables are suddenly and unusually assigned) [6].

Some of other restart strategies are worth mentioning: Lingeling's agility and saturation [9], and progressive saving based quality measure (PSM) [3].

3   The SAT/UNSAT Difference of CDCL in Practice

It is well known that the way a CDCL SAT solver finds a satisfying assignment is very different from the way it finds a refutation proof for unsatisfiability. From the complexity theory point of view, showing that a Boolean formula is satisfiable is normally believed to reside in a different complexity class than proving that it is unsatisfiable. There always exists a polynomial-length witness for any satisfiable formula, but it is generally believed that not every unsatisfiable formula can have a short proof (the question of NP = co-NP). It is well known that the proof system of a broad class of CDCL is as powerful as general resolution [27], and that the resolution-based proofs for certain problems (e.g, pigeon-hole) are exponential in size [18].

In this section, we discuss in detail the varying degrees of roles and effects of learned clauses, restarts and the VSIDS heuristic between SAT and UNSAT in an attempt to understand the nature of the SAT/UNSAT difference in CDCL.

3.1 Roles of Learned Clauses

Background. Recently, learned clause management has been an active topic of research. Solvers maintain a huge number of learned clauses by periodically removing them (typically halving) to contain the fast growth rate of the clause database. The followings are some of the main goals of this periodic reduction:

- We want to accumulate clauses, since learning more lemmas is advantageous for diverse reasons.
- However, we need to periodically forget some that seem less helpful, since keeping too many clauses severely penalizes propagation efficiency.
- Finally, we need to make the database grow over time, e.g, to avoid repetitive learning [3] or to ensure making progress (and for completeness too).

Each issue has its own ground for consideration based on some commonly held assumptions about learned clauses. At the root of such assumptions is often the view that learned clauses in CDCL are the most important asset we learn during solving. Many believe that keeping learned clauses is essential to avoid repetition or to ensure making progress. Some believe that it would always help if we could predict and keep clauses that will be used frequently in future propagations or conflicts. Even though we lack clear understanding about the roles of learned clauses at this stage, most of such assumptions seem too obvious not to accept. However, we will see later that some of such traditional beliefs do not hold firm ground or justify much consideration in practice. We had already questioned the validity of such beliefs and submitted a prototype SWDiA5BY to the 2014 Competition to challenge them. The results suggest that, in an ultimate sense, clauses with LBD >5 are largely meaningless and that repetitive learning is either infrequent or negligible. We will further show that, for satisfiable instances, even LBD greater than 1 or 2 are not so helpful and VSIDS scores may be a more important asset than learned clauses. Before that, we begin with a short survey that hints the different roles of learned clauses between SAT and UNSAT.

A Short Survey. Fig. 1 summarizes a short survey of running Glucose 2.3 (participant of SAT Competition 2014[7]) on 135 benchmark problems from 2013 and

---

[7] The authors of Glucose specified the version as 3.0 in the Competition, but the code is precisely Glucose 2.3. This is not a mistake, since, for sequential SAT solving, there is no-real difference between 2.3, 3.0, and 4.0.

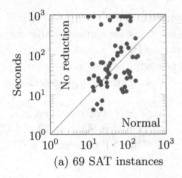 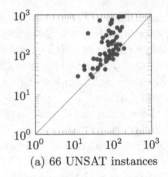

(a) 69 SAT instances   (a) 66 UNSAT instances

Fig. 1. Comparison of runtimes (secs): with and without database reduction

2014 Competitions with a short timeout of 900 seconds on Intel Core i5-4460S @ 2.90GHz and 12GB RAM. We selected *easy* problems that Glucose solved roughly between 15 and 200 seconds according to the competition data (excluding some that are too big). Almost all other solvers solved them very efficiently too. The figures compare runtimes of original Glucose with a variant that never removes learned clauses (unless satisfied). Expectations with the variant on these easy problems could be that it would still solve them efficiently, or sometimes more efficiently as Glucose may be removing clauses too aggressively. However, for SAT instances, we observe large variation. We are often lucky to find a model much faster, but sometimes the solver becomes completely lost to take significantly more time (9 timed out). In contrast, the result is stable and robust in the UNSAT case. It is rare to take less time, and even if it does, the gain is negligible. The overall variation is by far smaller too (3 timed out). In fact, this kind of difference in solver stability between SAT and UNSAT has been known to researchers [30]. The reason becomes more clear if we look at another metric.

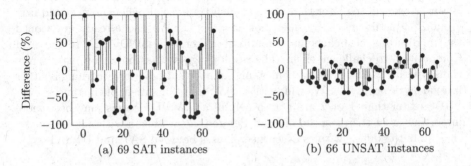

(a) 69 SAT instances   (b) 66 UNSAT instances

Fig. 2. Difference in no. conflicts (%): with and without database reduction

Fig. 2 plots the difference (%-increase) of the number of conflicts required to solve a problem before and after disabling clause removal. For SAT instances, the difference is unstable and substantial in almost all cases without clear trend. The graph is capped at 100%, and the differences of several hundreds % are common (up to 1200%). In contrast, in the UNSAT case, we see the trend of moderately reduced conflicts for most cases. The overall variation is comparably small too. We conjecture that this trend and high stability in UNSAT is because the solver has to derive an empty clause (i.e, an UNSAT proof) by successive resolutions based on existing clauses. That is, ignoring slowdown of unit propagation, learned clauses are certainly useful to accumulate to prove UNSAT, and keeping every clause would generally bring substantial improvement for UNSAT under this assumption. However, this does not apply for SAT, and it may adversely make the solver very unstable as observed. Therefore, for SAT, keeping non-essential clauses might be disadvantageous rather than just being useless. However, this short survey is too primitive to draw such a firm conclusion, so we will now present compelling evidence. The evidence will show that learned clauses play surprisingly insignificant roles, particularly on SAT instances.

Table 1. Solved instances with 600 (300 SAT/300 UNSAT) competition benchmarks

		SWDiA5BY	COMiniSatPS							C	Lingeling		Glucose
Core LBD cut		5	0	1	2	3	4	5	6	3	ayv	aqw	
2013	SAT	109	120	131	129	129	124	129	128	133	122	119	104
	UNSAT	123	47	88	113	116	120	126	122	132	107	112	112
	Total	232	167	219	242	245	244	255	250	265	229	231	216
2011	SAT	88	88	93	91	92	90	90	95	94	86	88	85
	UNSAT	104	73	92	97	97	101	101	99	102	94	93	108
	Total	192	161	185	188	189	191	191	194	196	180	181	193

Varying Roles of Learned Clauses. Table 1 shows results of running COMin-iSatPS with different *core LBD cuts* [25] on the 2013 and 2011 Competition benchmarks with timeouts of 4,200 and 1,500 seconds, respectively, on the same machine as before. Results of Lingeling ayv (2014 Competition winner), SWDiA5BY (runner-up), Lingeling aqw (2013 winner), and Glucose (2014) are also included. The solver named C is a refined COMiniSatPS covered in Sec. 4.

As a minimum base, all the COMiniSatPS settings in the table manage up to 30,000 learned clauses by MiniSat's clause activities (i.e, *no LBD*) and employ hybrid restarts (Sec. 3.2). This is precisely the 0-LBD cut setting. The maximum limit of 30,000 clauses is indeed very low as solvers routinely learn thousands of clauses per second. Note that this 0-LBD cut can solve implausibly many problems. Particularly for SAT, it is comparable with Lingeling. We even observed, for certain satisfiable benchmarks (e.g, 001-010.cnf from 2013), this "unreasonable" setting is exceptionally effective and almost optimal. However, this setting is very poor on UNSAT and particularly disastrous on the 2013 benchmarks. Next, the 1-LBD cut keeps *forever* learned clauses (apart from those 30,000) that ever attained LBD 1. Note that LBD 1 is observable only when dynamically updating LBD during conflict analysis, in which case we learn a unit clause and backtrack to the top decision level. Note the dramatic improvement on UNSAT, whose trend continues to the next LBD cut of 2. Certainly, learned clauses are far more required for UNSAT than for SAT. This makes sense when considering how resolution derives a refutation proof. In contrast, learned clauses play a far less significant role for SAT, and what is probably more important could be the evolution of VSIDS scores and variable phases [26]. In some sense, this working is rather similar to local search algorithms being able to find a satisfying assignment by evolving the current assignment set with phase flips. In this sense, changes to the VSIDS heuristic might be a key for future improvement for SAT.

The table also shows that not-critically-low LBD is barely useful from a practical sense. SWDiA5BY already proved it openly in the 2014 Competition. The table fortifies this view by showing that an increase of the LBD cut after 1 does not help on SAT, and helps rather marginally on UNSAT. Our pessimistic hypothesis is that as SAT being NP-complete, we can only derive an easy UNSAT proof in general (i.e, only using very low LBD) for easy (e.g, industrial) problems. In this sense, LBD seems to be a great static measure for how much a clause would help in composing an easy proof. In fact, we tested using a clause size of 12 instead of LBD 5 as a core cut limit (i.e., keeping forever clauses of size ≤12). We chose the size 12 based on previous work [21]. Table 2 compares the results of MiniSat_HACK_999ED using the clause size of 12 and using LBD 5 (original MiniSat_HACK_999ED) as a core cut. We used the 2013 Competition benchmarks with a timeout of 5,000 seconds on another machine with Intel Core 2 Duo E8400 @ 3.00GHz and 4G RAM. Note the much degraded performance on UNSAT. This is in contrast with the considerably better result on SAT. This shows one way of improving performance on SAT at the expense of having degraded performance on UNSAT.

However, it is critical to understand that completely ignoring high-LBD or large clauses will not simply work in CDCL. We still need to keep recent or active clauses around for a while to efficiently drive search by conflicts. This is why COMiniSatPS manage up to 30,000 clauses, and MiniSat's clause activity scheme seems to be a great choice for this purpose.

Table 2. LBD vs. clause size as the core cut with MiniSat_HACK_999ED

Core cut	LBD 5	Size 12
SAT	100	107
UNSAT	103	92

Glucose's success comes with its continuous evolution of increased aggressiveness in clause database reduction [4]. We will explain the efficiency of Glucose's aggressive clause removal in our terms. We realized that only low-LBD clauses are left after each database reduction, whose trend is only reinforced with increased aggressiveness. Fig. 3 is a typical graph of average LBD of the entire learned clauses that confirms this behavior. The average increases over time

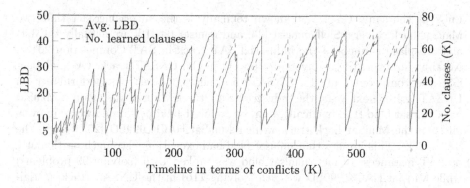

Fig. 3. Glucose running on minxor128.cnf from SAT Competition 2014

locally and globally. However, reducing the database always drops the average sharply to about 5, which accounts for the great efficiency of Glucose. Remember that, once a clause have attained a sufficiently low LBD, it tends to remain in the database in a stable manner, often forever if the LBD is critically low. The lower the LBD of a clause is, the higher is its chance of being fixed into the database. This is because low-LBD clauses have far more chances to be used and hence updated with a lower LBD. We can also infer that, considering the sharp LBD drop after each reduction which halves the database, most learned clauses are generated with a much higher LBD. Therefore, from our perspective, most clauses are largely useless, and Glucose has been very good at striking the right balance to accumulate more and more critically-low-LBD clauses while constantly truncating a large body of useless clauses by aggressive reduction. The reduction intervals also seem to be fine-tuned so that each reduction does not remove critically-low-LBD clauses, or only a small portion if any.

3.2   Effects of Restarts

Restarts in CDCL are crucial and very effective to counter the heavy-tailed phenomena [15] in search space exploration. In this section, we will discuss varying effects of restarts on SAT and UNSAT in relation to the frequency of restarts.

The Value of Luby Restarts. The Luby restart strategy was once a standard in the past after shown to be empirically superior to other existing schemes [19]. Recently, however, the huge success and continued innovations of Glucose have popularized the trend of dynamic and rapid restarts. The result is the currently dominant state of rapid restarts in recent solvers. This is not surprising since, e.g, Glucose's restart strategy is decisively superior to Luby. However, we recently became aware that Luby outperforms Glucose-style restarts in certain benchmarks, particularly on satisfiable industrial instances.

The value of Luby is highlighted by many pieces of evidence, and we will list some interesting ones shortly. Such evidence also reveals the weakness of

Luby at the same time. Not known to many is the surprising fact that two MiniSat hack solvers, SINNminisat [35] and minisat_bit [10] are actually the 1st- and 3rd-place winners of the industrial SAT track in SAT Competition 2013[8]. Notably, however, these solvers did poorly in the UNSAT track. Original MiniSat also competed (only for UNSAT) and showed disastrous performance on UNSAT. Later next year, the industrial SAT track winner was again a MiniSat hack minisat_blbd [11]. Ironically, however, minisat_blbd performed worse in overall than the MiniSat hack track winner MiniSat_HACK_999ED [25] (5th in the SAT track). This implies that minisat_blbd had exceptional strength particularly on SAT instances. In fact, minisat_blbd was ranked 13th (solving 99 problems) while MiniSat_HACK_999ED was 4th (solving 116) in the UNSAT track. (Original MiniSat did not compete in 2014.) Then, the organizers of Competition 2011 already reported in the past that there were many good MiniSat hack solvers for application SAT, including the top two Contrasat [14] and CIR_minisat [31]: six out of the top 10 solvers were MiniSat hacks [5]. One common property of all those MiniSat hacks is the Luby restarts. To be qualified as a hack, most of the hack solvers were not able to change MiniSat's Luby strategy. This ironically made the hack solvers excel in the SAT track (but perform poorly on UNSAT).

There exist many other examples of Luby's strength in recent competitions. One good example is satUZK [16] in 2013. The solver won a bronze medal in the SAT track while ranked 23rd in the SAT+UNSAT track. Notably, satUZK abandoned Luby to use the Glucose-style restarts in the following year.

There also have been hybrid restart strategies using Luby. It is well known that a portfolio-based parallel approach is very effective [1]. As such, there always have been attempts to diversify search characteristics in sequential solvers too, e.g, by changing various major parameters dynamically or taking a hybrid setting. Solvers combining different restart strategies have been around for years, and the first appearances in a competitive event known to us are SINN [32], TENN [33] and ZENN [34] (all from the same authors) in SAT Challenge 2012. These solvers periodically switch between Luby and (relatively) much more rapid restarts (e.g, Glucose restarts). It is interesting that *this hybrid approach is equally very good on both SAT and UNSAT*. SINN took 2nd, ZENN 3rd, TENN 8th in the (single-engine) application track, and ZENN 3rd in the combinatorial track. ZENN in 2013 was officially 2nd in the SAT track and 3rd in the SAT+UNSAT. ROKK [36] in 2014 using the same approach was also very successful. Solving one more problem than MiniSat_HACK_999ED, ROKK was ranked 7th in the SAT+UNSAT track. Here, knowing that MiniSat_HACK_999ED was ranked 11th, we again verify the performance saturation of today's solvers.

Varying Effects of Restarts on SAT and UNSAT. Glucose has constantly shown its particular strength on industrial UNSAT since its first release in 2009. Relatively, however, it has been much weaker on SAT. In the application UNSAT tracks, it was ranked 1st in 2009, 2nd in 2011, 1st in 2013, and 2nd in 2014. In contrast, in the SAT tracks, it was ranked 8th in 2009, 10th in 2011, 12th

---

[8] They did not win medals as they only indicated participation in the hack track.

in 2013, and 14th in 2014. The authors were clearly aware of this weakness to employ the clever measure of blocking restarts to compensate for this weakness on SAT [6]. This restart blocking has brought substantial improvement on SAT, but the recent competitions still show Glucose's weakness on SAT. This situation is complete opposite to the Luby-employing solvers that we have seen previously.

Table 3. Luby vs. Glucose restarts with MiniSat_HACK_999ED

	No. solved		Avg CPU time	
	Luby	Glucose-style	Luby	Glucose-style
SAT	119	100	356.7	405.1
UNSAT	85	107	1102.4	675.8

To verify this, we modified MiniSat_HACK_999ED to use Luby and compared the result with original MiniSat_HACK_999ED that faithfully implements Glucose restarts (Table 3). We used the 2013 Competition benchmarks on the Intel Core 2 Duo machine with a timeout of 5,000 seconds as before. The rightmost two columns compare the average CPU time only for the problems that *both solvers* were able to solve. It is clear that Luby is vastly inferior to Glucose restarts on UNSAT in terms of both CPU time and the number of solved instances. In contrast, Luby is shown to be very powerful on SAT, which explains the good results of Luby-employing solvers in the 2013 Competition. However, we caution the reader that the huge win over Glucose restarts comes from one benchmark series (001-010.cnf, 30 instances, all SAT). If we exclude those benchmarks, Luby solves 107 instances. (No difference with Glucose restarts since it solved none of the instances.) Therefore, even if we ignore the said benchmarks entirely, Luby is still superior to Glucose restarts on SAT. This makes a stark contrast in that Luby is significantly bad on UNSAT.

It is worth mentioning more about the benchmarks 001-010.cnf above. Most of the solvers in the 2013 Competition used Glucose-style or similarly rapid restarts. According to the competition data, all of them could solve about two out of 30 instances from this benchmark series. In contrast, MiniSat hacks, satUZK, and ZENN utilizing Luby could solve usually more than half of them. It is in fact these benchmarks that gave the latter solvers a great advantage compensating for the poor performance on UNSAT in the Competition.

In this context, the prominent difference between Luby and Glucose is the frequency of restarts. Although Luby was considered "rapid" restarts at the time of its introduction, nowadays it is very infrequent compared to Glucose restarts in general. Notably, Luby has a distinctive feature that it occasionally guarantees an extended periods of no restarts. As such, we conjecture that rapid[9] restarts generally help deriving a refutation proof (e.g, by lowering the average size of clauses [28]), while remaining in the current branch in the search space increases

---

[9] We mean being rapid in today's sense. For example, "aggressive" or "frequent" in [7] is now seen infrequent.

the chance of reaching a model. This view is not really new in that Glucose blocks restarts to compensate for the weakness on SAT [6] in recent versions. In fact, the authors of Glucose added this blocking feature after noting the use of Luby by the top-performing MiniSat hacks in the 2011 Competition [5]. However, it is not uncommon that there exist largely different views that, e.g, future solvers will evolve towards ultra rapid restarts [17].

To prove our theory, we designed and tested a very crude hybrid restart strategy: alternating between a *no-restart phase* and a Glucose *restart phase*. The basic idea is to force extended periods of no restarts periodically. We allocated twice more time to the Glucose phase than to the no-restart. One alternating cycle starts with 300 conflicts (100 for no-restart and 200 for Glucose), and the length of the following cycle increases by 10% (i.e, 330 conflicts). The global and local LBD averages used for Glucose's dynamic restarts are computed and preserved only throughout the Glucose phases, because clauses learned in the no-restart phase will show completely different characteristics. The restart blocking in the Glucose phase is disabled given that we have the no-restart phase. We will discuss the result later after explaining all other changes we add to this strategy.

One important lesson in this section is that many studies on restarts (e.g, [17], [31], [24]) have been carried out without considering the SAT/UNSAT difference. This has contributed to the currently dominant state of rapid restarts in recent solvers that quickly replaced Luby, even though we now see that slow Luby is superior to rapid restarts on SAT.

3.3   VSIDS and Variable Decay Factor

We observed that Luby having occasional and extended periods of no restarts make a solver stronger on SAT (while making it weaker on UNSAT) in general. Our natural deduction for the reason has been that long periods of no restarts increase chances of reaching the bottom of the search space (i.e, a model) by giving sufficient time to the conflict-driven search before giving up with too frequent restarts. From this perspective, we hypothesized that making search more stable and steady may have positive effect on SAT. We tested our hypothesis by focusing on making changes to VSIDS to alter the stability in search. The focus on VSIDS was also a good starting point based on our conjecture that VSIDS may be a more influential factor than learned clauses on SAT.

The VSIDS scores of variables "decay" over time. In MiniSat, the decay rate is controlled by a parameter whose default value is 0.95^{10}. A lower value implies more dynamic and reactive nature in decision variable selection as activity scores of old variables decay fast. That is, a lower factor makes recently active variables more influential, while a higher factor leads to higher stability in search space exploration. Using the value of 0.95 is almost a standard. The recent versions of Glucose (and SWDiA5BY) initially start with a lower factor of 0.8, but the factor increases and eventually reaches 0.95.

---

[10] "Decaying" in MiniSat is different from original Chaff [23]: it is simulated by bumping scores with a value that is continuously increased by the ratio of 1/0.95 per conflict.

Table 4. Different variable decay factors with MiniSat_HACK_999ED

Decay factors in each phase	NR 0.95	G 0.95	NR 0.999	G 0.6	NR 0.999	G 0.85	NR 0.999	G 0.95
SAT	110		111		117		114	
UNSAT	107		95		99		107	

Changing the decay factor has a profound effect on solver performance. Our preliminary research showed that using a factor of 0.999 in the no restart phase slightly increases strength on SAT. Table 4 compares the number of solved instances using different decay factors for each restart phase (2013 Competition benchmarks on the Intel Core 2 Duo machine and the timeout of 5,000 seconds as before). NR and G in the table refer to, respectively, the no restart phase and the Glucose phase. The hybrid strategy was implemented on top of MiniSat_HACK_999ED. The first solver using the default decay factor of 0.95 for both phases simply implements the two alternating restart phases. The other three solvers switch decay factors for respective phases (and thus disable Glucose's feature of initially starting with the value of 0.8). Note that we tested only 0.999 for the no restart phase. The reason is that 0.999 was our first choice that showed immediate improvement in our preliminary research. Having little computing resource, we did not try other values for the no restart phase. From these observations, we decided to use the factor of 0.999 in the no restart phase. For the Glucose phase, however, we retained the default value of 0.95, since this value appears to be an optimal value for the moment.

Ultimately, we implemented an elaborated approach of maintaining two separate sets of VSIDS scores (hence two separate priority queues for variable selection), each used exclusively for one restart phase. Our motivation was to reduce interference on VSIDS scores between SAT and UNSAT. In either phase, VSIDS scores in both sets are bumped and decayed together as usual, but with different decay factors of 0.999 and 0.95 for each respective set. This scheme seems to give more robust outcomes and work better than simply switching the decay factors (also better on the 2014 Competition benchmarks).

In fact, a very similar idea already appeared once in one of the past competitions (but not in the literature). As mentioned, the signature feature of ZENN is the search diversification with a hybrid approach. The authors of ZENN seemed to have tried many interesting hybridization ideas. One of the ideas is to use different decay factors (0.99 and 0.8) in a way similar to ours: switching the factors between two different restart strategies. However, they abandoned this idea in their new solvers in the following year. Because there exists no publication, it is not clear why they decided to implement and later abandon the idea. Like Luby, it would be interesting to revisit this idea in more depth.

The results of implementing the simple hybrid restart strategy from the previous section together with the alternating decay factors are already presented in Table 1, in the columns under COMiniSatPS using 7 different LBD cuts (0 to 6). Since SWDiA5BY uses the LBD cut of 5, it is best to compare the

5-LBD-cut version against SWDiA5BY. Note that because of the said instances 001-010.cnf in the 2013 benchmarks, the table shows heavily biased results of dramatic improvement on 2013 SAT for all LBD cuts. For 2011 SAT, it shows a slightly better result, but the degree of improvement is very marginal. For 2013 UNSAT, the LBD-cut-5 version has a slightly better result than SWDiA5BY, which is not the case with 2011 UNSAT. In fact, we report that the overall strength on UNSAT is slightly reduced, particularly in terms of CPU time. This overhead on UNSAT is not really a surprise, since the entire runtime is divided into two different restart phases. Considering that only two third of the runtime is spent for the Glucose phase, this slight degradation on UNSAT is actually encouraging.

This hybrid strategy is still very primitive, and we will show one possible way of achieving further improvement on both SAT and UNSAT.

4   Refining the Hybrid Strategy

Now that we gained better understanding on the varying degrees of effects and roles of learned clauses, restarts, and VSIDS between SAT and UNSAT, we attempted refining the hybrid strategy with the aim of achieving further improvement for both SAT and UNSAT. Before we explain our refinement, we report that we tested several other ideas and verified that there exist many ways to improve performance on SAT while having negative impact on UNSAT (and vice versa), which we omit in this paper due to page limit.

We have seen that clauses of LBD >5 are barely useful, in a practical and global sense. For UNSAT, it is the low-LBD clauses that play a central role of establishing a foundation that provides sufficient lemmas (clauses) necessary for constructing an easy UNSAT proof. However, giving a little bit more room for clauses having slightly higher but sufficiently low LBD may be advantageous in deriving a proof too. On the other hand, even LBD >1 does not seem to help much for SAT. There could be a compromise that satisfies both SAT and UNSAT in this context. We designed and tried an idea of adding a mid-tier in the clause database, in addition to the existing *core* and *local* tiers [25]. The idea is to lower the core LBD cut to have a more compact database mainly for SAT, while the mid-tier accommodates recently used clauses of higher LBD between 4 and 6 for UNSAT. The mid-tier functions as a buffer and staging area in that clauses may stay *as long as but only if* they have been involved in recent conflict analyses. The mid-tier is checked for reduction at every 10,000 conflicts, and clauses not used in the past 30,000 conflicts are demoted to the local tier. There are still a few subtleties in the actual implementation details, but this workings of the mid-tier is the essence of the refinement. To recap, (1) we bring down the core LBD cut to 3 for increased efficiency on SAT; while (2) we retain recently used clauses of LBD up to 6 in addition to local clauses in the hope that those mid-tier clauses can be used efficiently as bridging elements for an UNSAT proof. The result is presented in Table 1 as Solver C. It shows substantial improvement both on SAT and UNSAT with the 2013 benchmarks, although the improvement is only marginal on the 2011 benchmarks with the short timeout.

In fact, this idea of keeping clauses that were touched in recent conflicts was inspired by ROKK [36]. ROKK showed remarkable performance in the 2014 Competition (7th in SAT+UNSAT and 6th in SAT). The solver uses a hybrid strategy, and its learned clause management is very peculiar. Basically, the solver reduces the database at every 10,000 conflicts (i.e, high tendency towards shrinking to 5,000 clauses over time), while protecting recently used clauses in a similar (but much more complex) way to ours.

An interesting observation is the completely different characteristics of this mid-tier exhibited in each restart phase. For the no-restart phase, the size of the tier decreases quickly over time. Literals per learned clause (in an overall sense) tend to increase quickly too. The general implication is that, when remaining in the current search space without restarts, new learned clauses are used mostly locally and rarely get reused. However, the situation is opposite in the Glucose-style restart phase, although the size of the tier does not grow much anyway due to the low limit of 30,000 conflicts to be considered recent.

5   Conclusion

CDCL SAT solvers prove satisfiability and unsatisfiability in very different ways. We proposed an explanation to the difference and presented a wide range of interesting evidence that supports our explanation. In the course, we provided additional insights on the roles and effects of learned clauses, restarts and VSIDS, and particularly, their varying degrees of effects between SAT and UNSAT. We uncovered virtues of past and hidden techniques including Luby and hybridization. We showed that there exist ways to make solvers stronger on SAT at the expense of making it weaker on UNSAT (and vice versa), e.g, by suppressing rapid restarts. We also suggested a possible way to improve performance on both SAT and UNSAT by judiciously exploiting the SAT/UNSAT difference. However, the hybrid strategy of COMiniSatPS is very primitive, and the current state is far from maintaining an optimal balance between SAT and UNSAT. Moreover, there may exist largely different ways that better exploit the SAT/UNSAT difference. We also want to make a note that revisiting previous research with the difference in mind may shed more light on the internal workings of CDCL.

Lastly, we emphasize that all the arguments in this paper applies only to the problems from the industrial domain. Particularly, it may not work as explained for the hand-crafted problems. It is well known that industrial problems are very different from hand-crafted ones although they share some similarities. It will be interesting to find out in which ways they are different, which might give new insights on the CDCL workings. Similarly, the structure of one benchmark series can be very different from the structure of another. One strategy cannot be universally good to every series and may be good for only a certain set of benchmark families.

Acknowledgments. This work was supported by the National Science Foundation under grant CCF-1350574. The author thanks (in alphabetical order) Laurent Simon, Mate Soos, Thomas Wies, and all the reviewers for helping to improve the paper.

References

1. Aigner, M., Biere, A., Kirsch, C.M., Niemetz, A., Preiner, M.: Analysis of portfolio-style parallel SAT solving on current multi-core architectures. In: POS (2013)
2. Alfonso, E.M., Manthey, N.: Riss 4.27 BlackBox. In: SAT-COMP (2014)
3. Audemard, G., Lagniez, J.-M., Mazure, B., Saïs, L.: On freezing and reactivating learnt clauses. In: Sakallah, K.A., Simon, L. (eds.) SAT 2011. LNCS, vol. 6695, pp. 188–200. Springer, Heidelberg (2011)
4. Audemard, G., Simon, L.: Predicting learnt clauses quality in modern SAT solvers. In: IJCAI (2009)
5. Audemard, G., Simon, L.: Glucose 2.1: aggressive but reactive clause database management, dynamic restarts. In: POS (2012)
6. Audemard, G., Simon, L.: Refining restarts strategies for SAT and UNSAT. In: Milano, M. (ed.) CP 2012. LNCS, vol. 7514, pp. 118–126. Springer, Heidelberg (2012)
7. Biere, A.: Adaptive restart strategies for conflict driven SAT solvers. In: Kleine Büning, H., Zhao, X. (eds.) SAT 2008. LNCS, vol. 4996, pp. 28–33. Springer, Heidelberg (2008)
8. Biere, A.: Lingeling, plingeling and treengeling entering the SAT competition 2013. In: SAT-COMP (2013)
9. Biere, A.: Yet another local search solver and lingeling and friends entering the SAT competition 2014. In: SAT-COMP (2014)
10. Chen, J.: Solvers with a bit-encoding phase selection policy and a decision-depth-sensitive restart policy. In: SAT-COMP (2013)
11. Chen, J.: Minisat_blbd. In: SAT-COMP (2014)
12. Dubois, O., Andre, P., Boufkhad, Y., Carlier, J.: SAT versus UNSAT. In: DIMACS Cliques, Coloring and Satisfiability (1996)
13. Eén, N., Sörensson, N.: An extensible SAT-solver. In: Giunchiglia, E., Tacchella, A. (eds.) SAT 2003. LNCS, vol. 2919, pp. 502–518. Springer, Heidelberg (2004)
14. Gelder, A.V.: Contrasat - A contrarian SAT solver. JSAT (2012)
15. Gomes, C.P., Selman, B., Crato, N., Kautz, H.A.: Heavy-tailed phenomena in satisfiability and constraint satisfaction problems. J. Autom. Reasoning (2000)
16. van der Grinten, A., Wotzlaw, A., Speckenmeyer, E., Porschen, S.: SATUZK: solver description. In: SAT-COMP (2013)
17. Haim, S., Heule, M.: Towards ultra rapid restarts. CoRR (2014)
18. Haken, A.: The intractability of resolution. Theor. Comput. Sci. (1985)
19. Huang, J.: The effect of restarts on the efficiency of clause learning. In: IJCAI (2007)
20. Hutter, F., Lindauer, M., Balint, A., Bayless, S., Hoos, H., Leyton-Brown, K.: The Configurable SAT Solver Challenge (CSSC). Under review at AIJ; preprint available on arXiv: (2015). http://arxiv.org/abs/1505.01221
21. Jabbour, S., Lonlac, J., Sais, L., Salhi, Y.: Revisiting the learned clauses database reduction strategies. CoRR (2014)
22. Luby, M., Sinclair, A., Zuckerman, D.: Optimal speedup of las vegas algorithms. Inf. Process. Lett. (1993)
23. Moskewicz, M.W., Madigan, C.F., Zhao, Y., Zhang, L., Malik, S.: Chaff: engineering an efficient SAT solver. In: DAC (2001)
24. Nossum, V.: SAT-based preimage attacks on SHA-1. Master's thesis, University of Oslo (2012)

25. Oh, C.: MiniSat_HACK_999ED, MiniSat_HACK_1430ED, and SWDiA5BY. In: SAT-COMP (2014)
26. Pipatsrisawat, K., Darwiche, A.: A lightweight component caching scheme for satisfiability solvers. In: Marques-Silva, J., Sakallah, K.A. (eds.) SAT 2007. LNCS, vol. 4501, pp. 294–299. Springer, Heidelberg (2007)
27. Pipatsrisawat, K., Darwiche, A.: On the power of clause-learning SAT solvers as resolution engines. Artif. Intell. (2011)
28. Ryvchin, V., Strichman, O.: Local restarts. In: Kleine Büning, H., Zhao, X. (eds.) SAT 2008. LNCS, vol. 4996, pp. 271–276. Springer, Heidelberg (2008)
29. Silva, J.P.M., Sakallah, K.A.: GRASP: A search algorithm for propositional satisfiability. IEEE Trans. Computers (1999)
30. Simon, L.: Post mortem analysis of SAT solver proofs. In: POS (2014)
31. Sonobe, T., Inaba, M.: Counter implication restart for parallel SAT solvers. In: Hamadi, Y., Schoenauer, M. (eds.) LION 2012. LNCS, vol. 7219, pp. 485–490. Springer, Heidelberg (2012)
32. Yasumoto, T.: SINN. In: SC (2012)
33. Yasumoto, T.: TENN. In: SC (2012)
34. Yasumoto, T.: ZENN. In: SC (2012)
35. Yasumoto, T., Okugawa, T.: SINNminisat. In: SAT-COMP (2013)
36. Yasumoto, T., Okugawa, T.: ROKK. In: SAT-COMP (2014)

Efficient MUS Enumeration of Horn Formulae with Applications to Axiom Pinpointing

M. Fareed Arif[1], Carlos Mencía[1]([⊠]), and Joao Marques-Silva[1,2]

[1] CASL, University College Dublin, Dublin, Ireland
farif@ucdconnect.ie, {carlos.mencia,jpms}@ucd.ie
[2] INESC-ID, IST, ULisboa, Lisbon, Portugal

Abstract. The enumeration of minimal unsatisfiable subsets (MUSes) finds a growing number of practical applications, that includes a wide range of diagnosis problems. As a concrete example, the problem of axiom pinpointing in the $\mathcal{EL}$ family of description logics (DLs) can be modeled as the enumeration of the group-MUSes of Horn formulae. In turn, axiom pinpointing for the $\mathcal{EL}$ family of DLs finds important applications, such as debugging medical ontologies, of which SNOMED CT is the best known example. The main contribution of this paper is to develop an efficient group-MUS enumerator for Horn formulae, HGMUS, that finds immediate application in axiom pinpointing for the $\mathcal{EL}$ family of DLs. In the process of developing HGMUS, the paper also identifies performance bottlenecks of existing solutions. The new algorithm is shown to outperform all alternative approaches when the problem domain targeted by group-MUS enumeration of Horn formulae is axiom pinpointing for the $\mathcal{EL}$ family of DLs, with a representative suite of examples taken from different medical ontologies.

1 Introduction

Description Logics (DLs) are well-known knowledge representation formalisms [4]. DLs find a wide range of applications in computer science, including the semantic web and representation of ontologies, but also in medical bioinformatics.

Given an ontology (that consists of a set of axioms) and a subsumption relation entailed by the ontology, *axiom pinpointing* is the problem of finding minimal axiom sets (MinAs), equivalently minimal sub-ontologies, each one entailing the given subsumption relation [48]. So, each MinA represents a minimal explanation or justification for the subsumption relation. Example applications of axiom pinpointing include context-based reasoning, error-tolerant reasoning [32], and ontology debugging and revision [26,49]. Axiom pinpointing for different description logics (DLs) has been studied extensively for more than a decade, with related work in the mid 90s [1,3,5–8,25,31,33,37,39,40,42,48–51,53].

The $\mathcal{EL}$ family of DLs is well-known for being tractable (i.e. polynomial-time decidable). Despite being inexpressive, the $\mathcal{EL}$ family of DLs, concretely by using the more expressive, and still tractable, $\mathcal{EL}^+$, has been used for representing ontologies in the medical sciences, including the well-known SNOMED CT

© Springer International Publishing Switzerland 2015
M. Heule and S. Weaver (Eds.): SAT 2015, LNCS 9340, pp. 324–342, 2015.
DOI: 10.1007/978-3-319-24318-4_24

ontology [55]. Work on axiom pinpointing for the $\mathcal{EL}$ family of DLs can be traced to 2006, namely the CEL tool [5]. Later, in 2009, the use of SAT was proposed for axiom pinpointing in the $\mathcal{EL}$ family of DLs [50,51,56], concretely for the more expressive DL $\mathcal{EL}^+$. This seminal work proposed a propositional Horn encoding that can be exponentially smaller than earlier work [5,7,8]. Moreover, the use of SAT for axiom pinpointing for the $\mathcal{EL}$ family of DLs, named EL$^+$SAT [50,51,56], was shown to consistently outperform earlier work, concretely CEL [5]. Recent work [1] builds on these propositional encodings, but exploits the relationship between axiom pinpointing and enumeration of minimal unsatisfiable subsets (MUSes) [30], achieving conclusive performance gains over earlier work.

Nevertheless, this recent work has a number of potential drawbacks that will be analyzed later in the paper.

The relationship between axiom pinpointing and MUS enumeration was also studied elsewhere [33]. Instead of exploiting hitting set dualization, this alternative approach exploits the enumeration of implicants [33].

The main contribution of this paper is to develop an efficient group-MUS enumerator for Horn formulae, referred to as HGMUS, that finds immediate application in axiom pinpointing for the $\mathcal{EL}$ family of DLs. In the process of developing HGMUS, the paper also identifies performance bottlenecks of existing solutions, in particular EL$^+$SAT [50,51]. The new group-MUS enumerator for Horn formulae builds on the large body of recent work on problem solving with SAT oracles. This includes, among others, MUS extraction [12], MCS extraction and enumeration [34], and partial MUS enumeration [28,29,44]. HGMUS also exploits earlier work on solving Horn propositional formulae [17,38], and develops novel algorithms for MUS extraction in propositional Horn formulae. The experimental results, using well-known problem instances, demonstrate conclusive performance improvements over all other existing approaches, in most cases by several orders of magnitude.

The paper is organized as follows. Section 2 introduces the notation and definitions used throughout the paper. Section 3 reviews recent work on MUS enumeration, which serves as the basis for HGMUS. Afterwards, the new group-MUS enumerator HGMUS is described in Section 4. Section 5 compares HGMUS with existing alternatives. Experimental results on well-known problem instances from axiom pinpointing for the $\mathcal{EL}$ family of DLs are analyzed in Section 6. The paper concludes in Section 7.

2   Preliminaries

Standard definitions of propositional logic are assumed [13]. This paper considers Boolean formulae in Conjunctive Normal Form (CNF). A CNF formula $\mathcal{F}$ is defined over a set of Boolean variables $V(\mathcal{F}) = \{x_1, ..., x_n\}$ as a conjunction of clauses $(c_1 \wedge ... \wedge c_m)$. A clause c is a disjunction of literals $(l_1 \vee ... \vee l_k)$ and a literal l is either a variable x or its negation $\neg x$. We refer to the set of literals appearing in $\mathcal{F}$ as $L(\mathcal{F})$. Formulae can also be represented as sets of clauses, and clauses as sets of literals.

A truth assignment, or interpretation, is a mapping $\mu : V(\mathcal{F}) \rightarrow \{0, 1\}$. If all the variables in $V(\mathcal{F})$ are assigned a truth value, μ is referred to as a *complete* assignment. Interpretations can also be seen as conjunctions or sets of literals. Truth valuations are lifted to clauses and formulae as follows: μ satisfies a clause c if it contains at least one of its literals. Given a formula $\mathcal{F}$, μ satisfies $\mathcal{F}$ (written $\mu \vDash \mathcal{F}$) if it satisfies all its clauses, being μ referred to as a *model* of $\mathcal{F}$.

Given two formulae $\mathcal{F}$ and $\mathcal{G}$, $\mathcal{F}$ entails $\mathcal{G}$ (written $\mathcal{F} \vDash \mathcal{G}$) iff all the models of $\mathcal{F}$ are also models of $\mathcal{G}$. $\mathcal{F}$ and $\mathcal{G}$ are equivalent (written $\mathcal{F} \equiv \mathcal{G}$) iff $\mathcal{F} \vDash \mathcal{G}$ and $\mathcal{G} \vDash \mathcal{F}$.

A formula $\mathcal{F}$ is satisfiable ($\mathcal{F} \nvDash \bot$) if there exists a model for it. Otherwise it is unsatisfiable ($\mathcal{F} \vDash \bot$). SAT is the decision problem of determining the satisfiability of a propositional formula. This problem is in general NP-*complete* [15].

Some applications require computing certain types of models. In this paper, we will make use of maximal models, i.e. models such that a set-wise maximal subset of the variables are assigned value 1:

Definition 1 (MxM). *Let $\mathcal{F}$ be a satisfiable propositional formula, $\mu \vDash \mathcal{F}$ a model of $\mathcal{F}$ and $P \subseteq V(\mathcal{F})$ the set of variables appearing in μ with positive polarity. μ is a maximal model (MxM) of $\mathcal{F}$ iff $\mathcal{F} \cup P \nvDash \bot$ and for all $v \in V(\mathcal{F}) \setminus P$, $\mathcal{F} \cup P \cup \{v\} \vDash \bot$.*

Herein, we will denote a maximal model by P, i.e. the set of its positive literals.

Horn formulae constitute an important subclass of propositional logic. These are composed of Horn clauses, which have at most one positive literal. Satisfiability of Horn formulae is decidable in polynomial time [17,23,38].

Given an unsatisfiable formula $\mathcal{F}$, the following subsets represent different notions regarding (set-wise) minimal unsatisfiability and maximal satisfiability [30,34]:

Definition 2 (MUS). $\mathcal{M} \subseteq \mathcal{F}$ *is a* Minimally Unsatisfiable Subset *(MUS) of $\mathcal{F}$ iff $\mathcal{M}$ is unsatisfiable and $\forall c \in \mathcal{M}, \mathcal{M} \setminus \{c\}$ is satisfiable.*

Definition 3 (MCS). $\mathcal{C} \subseteq \mathcal{F}$ *is a* Minimal Correction Subset *(MCS) iff $\mathcal{F} \setminus \mathcal{C}$ is satisfiable and $\forall c \in \mathcal{C}, \mathcal{F} \setminus (\mathcal{C} \setminus \{c\})$ is unsatisfiable.*

Definition 4 (MSS). $\mathcal{S} \subseteq \mathcal{F}$ *is a* Maximal Satisfiable Subset *(MSS) iff $\mathcal{S}$ is satisfiable and $\forall c \in \mathcal{F} \setminus \mathcal{S}, \mathcal{S} \cup \{c\}$ is unsatisfiable.*

An MSS is the complement of an MCS. MUSes and MCSes are closely related by the well-known hitting set duality [10,14,46,54]: Every MCS (MUS) is an irreducible hitting set of all MUSes (MCSes) of $\mathcal{F}$. In the worst case, there can be an exponential number of MUSes and MCSes [30,41]. Besides, MCSes are related to the MaxSAT problem, which consists in finding an assignment satisfying as many clauses as possible. The smallest MCS (largest MSS) represents an optimal solution to MaxSAT.

Motivated by several applications, MUSes and related concepts have been extended to CNF formulae where clauses are partitioned into disjoint sets called groups [30].

Definition 5 (Group-Oriented MUS). *Given an explicitly partitioned unsatisfiable CNF formula $\mathcal{F} = \mathcal{G}_0 \cup ... \cup \mathcal{G}_k$, a group-oriented MUS (or group-MUS) of $\mathcal{F}$ is a set of groups $\mathcal{G} \subseteq \{\mathcal{G}_1, ..., \mathcal{G}_k\}$, such that $\mathcal{G}_0 \cup \mathcal{G}$ is unsatisfiable, and for every $\mathcal{G}_i \in \mathcal{G}$, $\mathcal{G}_0 \cup (\mathcal{G} \setminus \mathcal{G}_i)$ is satisfiable.*

Note the special role $\mathcal{G}_0$ (*group-0*); this group consists of *background* clauses that are included in every group-MUS. Because of $\mathcal{G}_0$ a group-MUS, as opposed to MUS, can be empty. Nevertheless, in this paper we assume that $\mathcal{G}_0$ is satisfiable.

Equivalently, the related concepts of group-MCS and group-MSS can be defined in the same way. We omit these definitions here due to lack of space. In the case of MaxSAT, the use of groups is investigated in detail in [22].

3 MUS Enumeration in Horn Formulae

Enumeration of MUSes has been the subject of research that can be traced to the seminal work of Reiter [46]. A well-known family of algorithms uses (explicit) minimal hitting set dualization [10,14,30]. The organization of these algorithms can be summarized as follows. First compute all the MCSes of a CNF formula. Second, MUSes are obtained by computing the minimal hitting sets of the set of MCSes. The main drawback of explicit minimal hitting set dualization is that, if the number of MCSes is exponentially large, these approaches will be unable to compute MUSes, even if the total number of MUSes is small. As a result, recent work considered what can be described as implicit minimal hitting set dualization [28,29,44]. In these approaches (namely EMUS [44] and MARCO [29] MUS enumerators), either an MUS or an MCS is computed at each step of the algorithm, with the guarantee that one or more MUSes will be computed at the outset. In some settings, implicit minimal hitting set dualization is the only solution for finding some MUSes of a CNF formula. As pointed out in this recent work, implicit minimal hitting set dualization aims to complement, but not replace, the explicit dualization alternative, and in some settings where enumeration of MCSes is feasible, the latter may be the preferred option [29,44].

Algorithm 1 shows the EMUS enumeration algorithm [44], also used in the most recent version of MARCO [29]. It relies on a two-solver approach aimed at enumerating the MUSes/MCSes of an unsatisfiable formula $\mathcal{F}$. On the one hand, a formula $\mathcal{Q}$ is used to enumerate subsets of $\mathcal{F}$. This formula is defined over a set of variables $I = \{p_i \mid c_i \in \mathcal{F}\}$, each one of them associated with one clause $c_i \in \mathcal{F}$. Iteratively until $\mathcal{Q}$ becomes unsatisfiable, EMUS computes a maximal model P of $\mathcal{Q}$ and tests the satisfiability of the corresponding subformula $\mathcal{F}' \subseteq \mathcal{F}$. If it is satisfiable, $\mathcal{F}'$ represents an MSS of $\mathcal{F}$, and the clause $I \setminus P$ is added to $\mathcal{Q}$, preventing the algorithm from generating any subset of the MSS (superset of the MCS) again. Otherwise, if $\mathcal{F}'$ is unsatisfiable, it is reduced to an MUS $\mathcal{M}$, which is blocked adding to $\mathcal{Q}$ a clause made of the variables in I associated with $\mathcal{M}$ with negative polarity. This way, no superset of $\mathcal{M}$ will be generated. Algorithm 1 is guaranteed to find all MUSes and MCSes of $\mathcal{F}$, in a number of iterations that corresponds to the sum of the number of MUSes and MCSes.

This paper considers the problem of enumerating the group-MUSes of an unsatisfiable Horn formula. As highlighted earlier, and as discussed later in the

---

Algorithm 1. EMUS [44] / MARCO [29]

Input: $\mathcal{F}$ a CNF formula
Output: Reports the set of MUSes of $\mathcal{F}$

```
1  I ← {p_i | c_i ∈ F}                          // Variable p_i picks clause c_i
2  Q ← ∅
3  while true do
4      (st, P) ← MaximalModel(Q)
5      if not st then return
6      F' ← {c_i | p_i ∈ P}                      // Pick selected clauses
7      if not SAT(F') then
8          M ← ComputeMUS(F')
9          ReportMUS(M)
10         b ← {¬p_i | c_i ∈ M}                  // Negative clause blocking the MUS
11     else
12         b ← {p_i | p_i ∈ I \ P}               // Positive clause blocking the MCS
13     Q ← Q ∪ {b}
```

---

paper, enumeration of the group-MUSes of unsatisfiable Horn formulae finds important applications in axiom pinpointing for the $\mathcal{EL}$ family of DLs, including $\mathcal{EL}^+$. It should be observed that the difference between the enumeration of plain MUSes of Horn formulae and the enumeration of group-MUSes is significant. First, enumeration of group-MUSes of Horn formulae cannot be achieved in total polynomial time, unless P = NP. This is an immediate consequence from the fact that axiom pinpointing for the $\mathcal{EL}$ family of DLs cannot be achieved in total polynomial time, unless P = NP [7], and that axiom pinpointing for the $\mathcal{EL}$ family of DLs can be reduced in polynomial time to group-MUS enumeration of Horn formulae [1]. Second, enumeration of MUSes of Horn formulae can be achieved in total polynomial time (actually with polynomial delay) [43].

Given the above, a possible approach for enumerating group-MUSes of Horn formulae is to use an existing solution, either based on explicit or implicit minimal hitting set dualization. For example, the use of explicit minimal hitting dualization was recently proposed in EL2MCS [1]. Alternatively, either EMUS [44] or the different versions of MARCO [28,29] could be used, as also pointed out in [33].

This paper opts instead to exploit the implicit minimal hitting set dualization approach [28,29,44], but develops a solution that is specific to the problem formulation. This solution is described in the next section.

4   Algorithm for Group-MUS Enumeration in Horn Formulae

This section describes HGMUS, a novel and efficient group-MUS enumerator for Horn formulae based on implicit minimal hitting set dualization. In this section, $\mathcal{H}$ denotes the group of clauses $\mathcal{G}_0$, i.e. the background clauses. Moreover, $\mathcal{I}$ denotes the set of (individual) groups of clauses, with $\mathcal{I} = \{\mathcal{G}_1, \ldots, \mathcal{G}_k\}$. So, the unsatisfiable group-Horn formula corresponds to $\mathcal{F} = \mathcal{H} \cup \mathcal{I}$. Also, in this

---

Algorithm 2. Computation of Maximal Models

    Input: $\mathcal{Q}$ a CNF formula
    Output: (st, P): with st a Boolean and P an MxM (if it exists)

1 $(P, U, B) \leftarrow (\{\{x\} \mid \neg x \notin L(\mathcal{Q})\}, \{\{x\} \mid \neg x \in L(\mathcal{Q})\}, \emptyset)$
2 $(st, P, U) \leftarrow \texttt{InitialAssignment}(\mathcal{Q} \cup P)$
3 **if not** st **then return** $(\text{false}, \emptyset)$
4 **while** $U \neq \emptyset$ **do**
5     $l \leftarrow \texttt{SelectLiteral}(U)$
6     $(st, \mu) = \texttt{SAT}(\mathcal{Q} \cup P \cup B \cup \{l\})$
7   **if** st **then** $(P, U) \leftarrow \texttt{UpdateSatClauses}(\mu, P, U)$
8   **else** $(U, B) \leftarrow (U \setminus \{l\}, B \cup \{\neg l\})$
9 **return** (true, P) // P is an MxM of $\mathcal{Q}$

---

section, the formula $\mathcal{Q}$ shown in Algorithm 1 is defined on a set of variables associated to the groups in $\mathcal{I}$. For the problem instances considered later in the paper (obtained from axiom pinpointing for the $\mathcal{EL}$ family of DLs), each group of clauses contains a single unit clause. However, the algorithm would work for arbitrary groups of clauses.

4.1 Organization

The high-level organization of HGMUS mimics that of EMUS/MARCO (see Algorithm 1), with a few essential differences. First, the satisfiability testing step (because it operates on Horn formulae) uses the dedicated linear time algorithm LTUR [38]. LTUR can be viewed as one-sided unit propagation, since only variables assigned value 1 are propagated. Moreover, the simplicity of LTUR enables very efficient implementations, that use adjacency lists for representing clauses instead of the now more commonly used watched literals. Second, the problem formulation motivates using a dedicated MUS extraction algorithm, which is shown to be more effective in this concrete case than other well-known approaches [12]. Third, we also highlight important aspects of the EMUS/MARCO implicit minimal hitting set dualization approach, which we claim have been overlooked in earlier work [51, 56].

4.2 Computing Maximal Models

The use of maximal models for computing either MCSes of a formula or a set of clauses that contain an MUS was proposed in earlier work [44], which exploited SAT with preferences for computing maximal models [20, 47]. The use of SAT with preferences for computing maximal models is also exploited in related work [50, 51].

Computing maximal models of a formula $\mathcal{Q}$ can be reduced to the problem of extracting an MSS of a formula $\mathcal{Q}'$ [34], where the clauses of $\mathcal{Q}$ are hard and, for each variable $x_i \in V(\mathcal{Q})$, it includes a unit soft clause $c_i \equiv \{x_i\}$. Also, recent work [9, 21, 34, 36] has shown that state-of-the-art MCS/MSS computation approaches outperform SAT with preferences. HGMUS uses a dedicated algorithm based on the LinearSearch MCS extraction algorithm [34], due to its good performance in MCS enumeration. Since all soft clauses are unit, it can also be related with the novel Literal-Based eXtractor algorithm [36]. Shown in Algorithm 2, it relies on making successive calls to a SAT solver. It maintains three sets of literals: P, an under-approximation of an MxM (i.e. positive literals s.t. $Q \cup P \nvDash \perp$), B, with negative literals $\neg l$ such that $Q \cup P \cup \{l\} \vDash \perp$ (i.e. *backbone literals*), and U, with the remaining set of positive literals to be tested. Initially, P and U are initialized from a model $\mu \vDash Q$, P (U) including the literals appearing with positive (negative) polarity in μ. Then, iteratively, it tries to extend P with a new literal $l \in U$, by testing the satisfiability of $Q \cup P \cup B \cup \{l\}$. If it is satisfiable, all the literals in U satisfied by the model (including l) are moved to P. Otherwise, l is removed from U and $\neg l$ is added to B. This algorithm has a query complexity of $\mathcal{O}(|V(Q)|)$.

Algorithm 2 integrates a new technique, which consists in pre-initializing P with the pure positive literals appearing in Q and U with the remaining ones (line 1), and then requiring the literals of P to be satisfied by the initial assignment (line 2). It can be easily proved that these pure literals are included in all MxMs of Q, so a number of calls to the SAT solver could be avoided. Moreover, the SAT solver will never branch on these variables, easing the decision problems. This technique is expected to be effective in HGMUS. Note that, in this context, Q is made of two types of clauses: positive clauses blocking MCSes of the Horn formula, and negative clauses blocking MUSes. So, with this technique, the computation of MxMs is restricted to the variables representing groups appearing in some MUS of the Horn formula.[1]

4.3   Adding Blocking Clauses

One important aspect of HGMUS are the blocking clauses created and added to the formula Q (see Algorithm 1). These follow what was first proposed in EMUS [44] and MARCO [28,29]. For each MUS, the blocking clause consists of a set of negative literals, requiring at least one of the clauses in the MUS *not* to be included in future selected sets of clauses. For each MCS, the blocking clause consists of a set of positive literals, requiring at least one of the clauses in the MCS to be included in future selected sets of clauses. The way MCSes are handled is essential to prevent that MCS and sets containing the same MCS to be selected again. Although conceptually simple, it can be shown that existing approaches may not guarantee that supersets of MCSes (or subsets of the MSSes) are not selected. As argued later, this is the case with EL$^+$SAT [51,56].

4.4   Deciding Satisfiability of Horn Formulae

It is well-known that Horn formulae can be decided in linear time [17,23,38]. HGMUS implements the LTUR algorithm [38]. There are important reasons for this choice. First, LTUR is expected to be more efficient than plain unit propagation, since only variables assigned value 1 need to be propagated. Second, most implementations of unit propagation in CDCL SAT solvers (i.e. that use watched literals) are not guaranteed to run in linear time [19]; this is for example the case with *all* implementations of Minisat [18] and its variants, for which unit

---

[1] SATPin [33] also exploits this insight of *relevant* variables, but not in the contexts of MxMs.

Algorithm 3. Insertion-based [16] MUS extraction using LTUR [38]

Input: $\mathcal{H}$, denotes the $\mathcal{G}_0$ clauses; $\mathcal{I}$, denotes the set of (individual) group clauses
Output: $\mathcal{M}$, denotes the computed MUS

```
1  (M, c_r) ← (H, 0)
2  LTUR_prop(M, M)                         // Start by propagating G_0 clauses
3  while true do
4  |   if c_r > 0 then
5  |   |   M ← M ∪ {c_r}                    // Add transition clause c_r to M
6  |   |   if not LTUR_prop(M, {c_r}) then
7  |   |   |   LTUR_undo(M, M)
8  |   |   |   return M \ H                 // Remove G_0 clauses from computed MUS
9  |   S ← ∅
10 |   while true do
11 |   |   c_r ← SelectRemoveClause(I)      // Target transition clause
12 |   |   S ← S ∪ {c_r}
13 |   |   if not LTUR_prop(M ∪ S, {c_r}) then
14 |   |   |   I ← S \ {c_r}                // Update working set of groups
15 |   |   |   LTUR_undo(M, S)
16 |   |   |   break                        // c_r represents a transition clause
```

---

propagation runs in worst-case quadratic time. As a result, using an off-the-shelf SAT solver and exploiting only unit propagation (as is done for example in earlier work [33,50,51]) is unlikely to be the most efficient solution. Besides the advantages listed above, the use of a linear time algorithm for deciding the satisfiability of Horn formulae turns out to be instrumental for MUS extraction, as shown in the next section. In order to use LTUR for MUS extraction, an incremental version has been implemented, which allows for the incremental addition of clauses to the formula and incremental identification of variables assigned value 1. Clearly, the amortized run time of LTUR, after adding $m = |\mathcal{F}|$ clauses, is $\mathcal{O}(\|\mathcal{F}\|)$, with $\|\mathcal{F}\|$ the number of literals appearing in $\mathcal{F}$.

4.5   MUS Extraction in Horn Formulae

For arbitrary CNF formulae, a number of approaches exist for MUS extraction, with the most commonly used one being the deletion-based approach [11,12], but other alternatives include the QuickXplain algorithm [24] and the more recent Progression algorithm [35]. It is also well-known and generally accepted that, due to its query complexity, the insertion-based algorithm [16] for MUS extraction is in practice not competitive with existing alternatives [12].

Somewhat surprisingly, this is not the case with Horn formulae when (an incremental implementation of) the LTUR algorithm is used. A modified insertion-based MUS extraction algorithm that exploits LTUR is shown in Algorithm 3. LTUR_prop propagates the consequences of adding some new set of clauses, given some existing incremental context. LTUR_undo unpropagates the consequences of adding some set of clauses (in order), given some existing incremental context. The organization of the algorithm mimics the standard insertion-based MUS extraction algorithm [16], but the use of the incremental LTUR yields run time complexity that improves over other approaches. Consider the operation of the standard insertion-based algorithm [16], in which clauses are iteratively added to the working formula. When the formula becomes unsatisfiable, a *transition clause* [12] has been identified, which is then added to the MUS being constructed. The well-known query complexity of the insertion-based algorithm is $\mathcal{O}(m \times k)$ where m is the number of clauses and k is the size of a largest MUS. Now consider that the incremental LTUR algorithm is used. To find the first transition clause, the amortized run time is $\mathcal{O}(||\mathcal{F}||)$. Clearly, this holds true for *any* transition clause, and so the run time of MUS extraction with the LTUR algorithm becomes $\mathcal{O}(|\mathcal{M}| \times ||\mathcal{F}||)$, where $\mathcal{M} \subseteq \mathcal{I}$ is a largest MUS. Algorithm 3 highlights the main differences with respect to a standard insertion-based MUS extraction algorithm. In contrast, observe that for a deletion-based algorithm the run time complexity will be $\mathcal{O}(|\mathcal{I}| \times ||\mathcal{F}||)$. In situations where the sizes of MUSes are much smaller than the number of groups in $\mathcal{I}$, this difference can be significant. As a result, when extracting MUSes from Horn formulae, and when using a polynomial time incremental decision procedure, an insertion-based algorithm should be used instead of other more commonly used alternatives.

5   Comparison with Existing Alternatives

This section compares HGMUS with the group MUS enumerators used in EL$^+$SAT [50,51], EL2MCS [1] and SATPin [33]. An experimental comparison with these and other methods for axiom pinpointing for the $\mathcal{EL}$ family of DLs is presented in Section 6.

5.1   EL$^+$SAT

The best known SAT-based approach for axiom pinpointing is EL$^+$SAT [50, 51,56]. EL$^+$SAT is composed of two main phases. The first phase compiles the axiom pinpointing problem to a Horn formula. The second phase enumerates the so-called MinAs, and corresponds to group-MUS enumeration for this Horn formula [1]. Although existing references emphasize the enumeration of MinAs (MUSes) using an AllSAT approach (itself inspired by an AllSMT approach [27]), the connection with MUS enumeration is immediate [1]. More importantly, EL$^+$SAT shares a number of similarities with implicit minimal hitting set dualization, but also crucial differences, which we now analyze.

Similar to EMUS, EL$^+$SAT selects subformulae of an unsatisfiable Horn formula. This is achieved with a SAT solver that always assigns variables value 1 when branching [51]. This corresponds to solving SAT with preferences [20,47], and so it corresponds to computing a maximal model, inasmuch the same way as EMUS operates.

In EL$^+$SAT, the approach for deciding the satisfiability of Horn subformulae is based on running the unit propagation engine of a CDCL SAT solver. As explained earlier, this can be inefficient when compared with the dedicated LTUR algorithm for Horn formulae [38]. Moreover, in EL$^+$SAT, MUSes are extracted with what can be viewed as a deletion-based algorithm [11,12]. Although more efficient alternatives are suggested, none is as asymptotically as efficient as the dedicated algorithm proposed in Section 4.5.

Finally, the most important drawback is the blocking of sets of clauses that do not contain an MUS/MinA. In our setting of implicit minimal hitting set dualization, this represents one MCS. The approach used in EL$^+$SAT consists of creating a blocking clause solely based on the decision variables (which are *always* assigned value 1) [51,56][2]. This means that MUSes (or MinAs) and MCSes/MSSes are blocked the same way. Thus, the learned clauses, although blocking one MCS (and corresponding MSS), do *not* block supersets of MCSes (and the corresponding subsets of the MSSes). This can result in exponentially more iterations than necessary, and explains in part the poor performance of EL$^+$SAT in practice. It should be further observed that this drawback becomes easier to spot once the problem is described as MUS enumeration by implicit minimal hitting set dualization.

5.2 EL2MCS

EL2MCS [1] implements explicit minimal hitting set dualization. In a first phase the MCS enumerator CAMUS2 [34] is used (the original CAMUS cannot be used because the formula has groups). This is achieved by iterated MaxSAT enumeration. In a second phase the MUS enumerator CAMUS [30] is used. The differences to HGMUS are clear, in that EL2MCS uses explicit minimal hitting set dualization and HGMUS uses implicit minimal hitting set dualization. Thus, there are (possibly many) instances for which EL2MCS will be unable to compute MUSes, because it will be unable to enumerate all MCSes, and this will not be the case with HGMUS. Another potential drawback of EL2MCS is that it uses a MaxSAT solver for MCS enumeration, although there are better alternatives [34]. Nevertheless, EL2MCS outperforms other existing approaches [5,31,33,50,51]. As shown later, the HGMUS approach proposed in this paper is the only one that consistently outperforms EL2MCS.

5.3 SATPin

SATPin [33] represents a recent SAT-based alternative for axiom pinpointing for the $\mathcal{EL}$ family of DLs, that focuses on optimizing the low-level implementation details of the CDCL SAT solver, including the use of incremental SAT solving. As indicated above, HGMUS opts to revisit instead the LTUR [38] algorithm from the late 80s, since it is guaranteed to run in linear time for Horn formulae, and can be implemented with small overhead. The SATPin approach is presented in terms of iteratively computing implicants. Some aspects of the organization of SATPin can be related with those of EL$^+$SAT, namely the procedure for extracting MUSes/MinAs. Although the actual enumeration of candidate sets is not detailed in [33], the description of SATPin suggests the use of model enumeration with some essential pruning techniques.

---

[2] The clause learning mechanism used in EL$^+$SAT is detailed in [51], page 17, first paragraph.

6   Experimental Results

This section evaluates group-MUS enumerators for Horn formulae obtained from axiom pinpointing problems for the $\mathcal{EL}$ family of DLs, particularly applied to medical ontologies. A set of standard benchmarks is considered. These have been used in earlier work, e.g. [1,5,31,33,50].

Since all experiments consist of converting axiom pinpointing problems into group-MUS enumeration problems, the tool that uses HGMUS[3] as its back-end is named EL2MUS. Thus, in this section, the results for EL2MUS illustrate the performance of the group-MUS enumerator described in this paper.

6.1   Experimental Setup

Each considered instance represents the problem of explaining a particular subsumption relation (query) entailed in a medical ontology. Four medical ontologies[4] are considered: GALEN [45], GENE [2], NCI [52] and SNOMED CT [55]. For GALEN, we consider two variants: FULL-GALEN and NOT-GALEN. The most important ontology is SNOMED CT and, due to its huge size, it also produces the hardest axiom pinpointing instances. For each ontology (including the GALEN variants) 100 queries are considered; 50 random (expected to be easier) and 50 sorted (expected to have a large number of minimal explanations) queries. So, there are 500 queries in total.

Given an ontology, the encoding proposed in [50,51] produces a Horn formula that represents the reasoning steps taken in the deduction of all the subsumption relations entailed by the ontology. In this formula, every variable represents a subsumption relation between two concepts. As a result, the encoding also produces a set of variables corresponding to the original axioms of the ontology, which may be responsible for any subsumption relation. Explaining a given subsumption relation (query) can be then transformed into a group-MUS enumeration problem where the original Horn formula and a unit clause with the negated query forms group-0 and each original axiom constitutes a group containing only a unit clause. Noticeably, any general Horn group-MUS problem can be converted to this particular format.

Two different experiments were considered by applying two different simplification techniques to the problem instances, both of which were proposed in [51]. The first one uses the Cone-Of-Influence (COI) reduction. These are reduced instances in both the size of the Horn formula and the number of axioms, but are still quite large. Similar techniques are exploited in related work [5,31,33]. The second one considers the more effective reduction technique (which we refer to as x2), consisting in applying the COI technique, re-encoding the Horn formula into a reduced ontology, and encoding this ontology again into a Horn formula. This results in small Horn formulae, which will be useful to evaluate the algorithms when there are a large number of MUSes/MCSes.

---

[3] HGMUS is available at http://logos.ucd.ie/web/doku.php?id=hgmus.

[4] GENE, GALEN and NCI ontologies are freely available at http://lat.inf.tu-dresden. de/~meng/toyont.html. The SNOMED CT ontology was requested from IHTSDO under a nondisclosure license agreement.

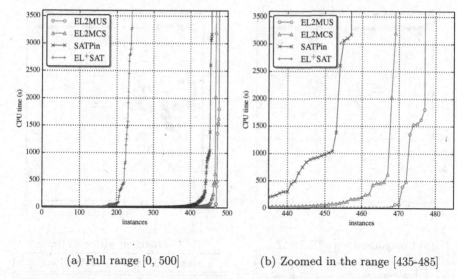

(a) Full range [0, 500]   (b) Zoomed in the range [435-485]

Fig. 1. Cactus plots comparing EL⁺SAT, SATPin, EL2MCS and EL2MUS on the COI instances

The experiments compare EL2MUS to different algorithms, namely EL⁺SAT [50,51], CEL [5], JUST [31], EL2MCS [1] and SATPin [33]. EL⁺SAT [51] has been shown to outperform CEL [5], whereas SATPin [33] has been shown to outperform the MUS enumerator MARCO [29].

The comparison with CEL and JUST imposes a number of constraints. First, CEL only computes 10 MinAs, so all comparisons with CEL only consider reporting the first 10 MinAs/MUSes. Also, CEL uses a simplification technique similar to COI, so CEL is considered in the first experiments. Second, JUST operates on selected subsets of $\mathcal{EL}^+$, i.e. the description logic used in most medical ontologies. As a result, all comparisons with JUST consider solely the problem instances for which JUST can compute correct results. JUST accepts the simplified x2 ontologies, so it is considered in the second experiments. The comparison with these tools is presented at the end of the section.

EL2MUS interfaces the SAT solver Minisat 2.2 [18] for computing maximal models. All the experiments were performed on a Linux cluster (2 GHz) and the algorithms were given a time limit of 3600s and a memory limit of 4 GB[5].

6.2 COI Instances

Figure 1 summarizes the results for EL⁺SAT, EL2MCS, SATPin and EL2MUS. As can be observed, EL2MCS has a slight performance advantage over SATPin, and EL2MUS terminates for more instances than any of the other tools. Figure 2 shows scatter plots comparing the different tools. As can be concluded, and with a few outliers, the performance of EL2MUS exceeds the performance of any of the other tools by at least one order of magnitude (and often by more). Figure 2d

---

[5] Only a sample of the results can be presented in this section due to space restrictions. Additional results are available at http://logos.ucd.ie/web/doku.php?id=hgmus.

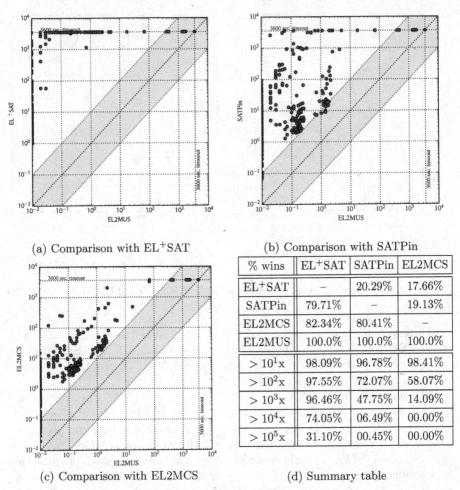

(a) Comparison with EL⁺SAT

(b) Comparison with SATPin

(c) Comparison with EL2MCS

(d) Summary table

% wins	EL⁺SAT	SATPin	EL2MCS
EL⁺SAT	–	20.29%	17.66%
SATPin	79.71%	–	19.13%
EL2MCS	82.34%	80.41%	–
EL2MUS	100.0%	100.0%	100.0%
$> 10^1$x	98.09%	96.78%	98.41%
$> 10^2$x	97.55%	72.07%	58.07%
$> 10^3$x	96.46%	47.75%	14.09%
$> 10^4$x	74.05%	06.49%	00.00%
$> 10^5$x	31.10%	00.45%	00.00%

Fig. 2. Scatter plots for COI instances

summarizes the results in the scatter plots, where the percentages shown are computed for problem instances for which at least one of the tools takes more than 0.001s. As can be observed, EL2MUS outperforms any of the other tools in all of the problem instances and, for many cases, with two or more orders of magnitude improvement.

6.3 x2 Instances

The x2 instances are significantly simpler than the COI instances. Thus, whereas the COI instances can serve to assess the scalability of each approach, the x2 instances highlight the expected performance in representative settings. Figure 3a summarizes the performance of the tools EL⁺SAT, SATPin, EL2MCS and EL2MUS. Due to its poor performance, EL⁺SAT does not show in the plot (it terminates on 317 instances). Moreover, and as before in terms of terminated instances, EL2MUS exhibits an observable performance edge.

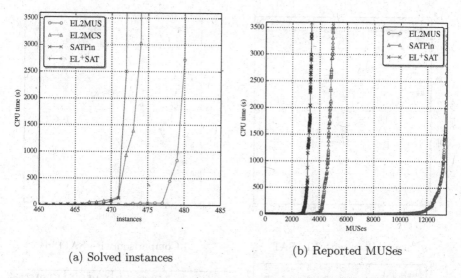

(a) Solved instances

(b) Reported MUSes

Fig. 3. Cactus plots comparing EL$^+$SAT, SATPin, EL2MCS and EL2MUS on the x2 instances

A pairwise comparison between the different tools is summarized in Figure 4. Although not as impressive as for the COI instances, EL2MUS still consistently outperforms all other tools. Figure 4d summarizes the results, where as before the percentages shown are computed for problem instances for which at least one of the tools takes more than 0.001s. Observe that, for these easier instances, SATPin becomes competitive with EL2MUS. Nevertheless, for instances taking more than 0.1s, EL2MUS outperforms SATPin on 100% of the instances. Thus, the 67.69% shown in the table result from instances for which both SATPin and EL2MUS take at most 0.04s. The summary table also lists the number of computed MUSes for the 19 instances for which EL2MUS does *not* terminate (all of the other tools also do not terminate for these 19 instances). EL2MUS computes 9948 MUSes in total. As can be observed from the table, the other tools lag behind, and compute significantly fewer MUSes. Also, as noted earlier in the paper, the main issue with EL2MCS is demonstrated with these results; for these 19 instances, EL2MCS is unable to compute any MUSes. The comparison with the other tools, EL$^+$SAT and SATPin, reveals that EL2MUS computes respectively in excess of a factor of 10 and of 5 more MUSes.

EL2MUS not only terminates on more instances than any other approach and computes more MUSes for the unsolved instances; it also reports the sequences of MUSes much faster. Figure 3b shows, for each computed MUS over the whole set of instances, the time each MUS was reported. This figure compares EL$^+$SAT, SATPin and EL2MUS, as these are the only methods able to report MUSes from the beginning. The results confirm that EL2MUS is able to find many more MUSes in less time than the alternatives.

These experimental results suggest that, not only is EL2MUS the best performing axiom pinpointing tool, on both the COI and x2 problem instances, but

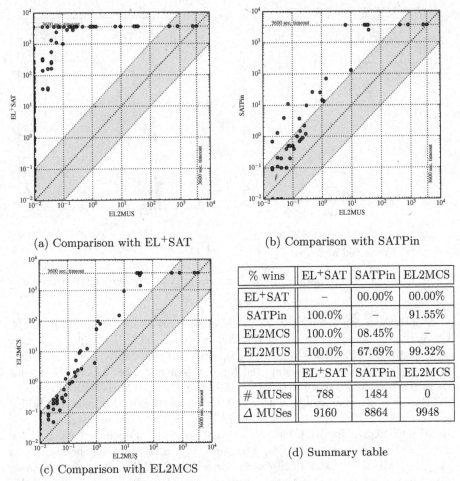

(a) Comparison with EL⁺SAT

(b) Comparison with SATPin

(c) Comparison with EL2MCS

% wins	EL⁺SAT	SATPin	EL2MCS
EL⁺SAT	–	00.00%	00.00%
SATPin	100.0%	–	91.55%
EL2MCS	100.0%	08.45%	–
EL2MUS	100.0%	67.69%	99.32%

	EL⁺SAT	SATPin	EL2MCS
# MUSes	788	1484	0
Δ MUSes	9160	8864	9948

(d) Summary table

Fig. 4. Scatter plots for x2 instances

it is also the one that is expected to scale better for more challenging problem instances, given the results on the COI instances.

6.4 Assessment of Non SAT-Based Axiom Pinpointing Tools

Figure 5 shows scatter plots comparing EL2MUS with CEL [5] and JUST [31], respectively for the COI and x2 instances[6]. As indicated earlier, CEL only computes 10 MinAs, and so the run times shown are for computing the first 10 MinA/MUSes. As can be observed, the performance edge of EL2MUS is clear, with the performance gap exceeding 1 order of magnitude almost without exception. Moreover, JUST [31] is a recent state of the art axiom pinpointing tool for the less expressive $\mathcal{ELH}$ DL. Thus, not all subsumption relations can be represented and analyzed. The results shown are for the subsumption relations for

---

[6] Due to lack of space the other scatter plots are not shown, but the conclusions are the same.

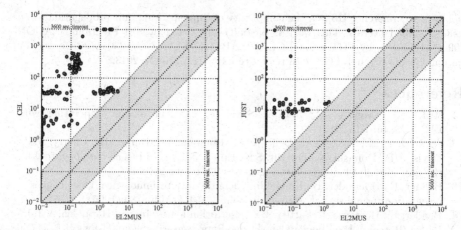

(a) EL2MUS vs. CEL on the COI instances   (b) EL2MUS vs. JUST on the x2 instances

Fig. 5. Comparison of EL2MUS with CEL and with JUST

which JUST gives the correct results. In total, 382 instances could be considered and are shown in the plot. As before, the performance edge of EL2MUS is clear, with the performance gap exceeding 1 order of magnitude without exception. In this case, since the x2 instances are in general much simpler, the performance gap is even more significant.

7   Conclusions and Future Work

Enumeration of group MUS for Horn formulae finds important applications, including axiom pinpointing for the $\mathcal{EL}$ family of DLs. Since the $\mathcal{EL}$ family of DLs is widely used for representing medical ontologies, namely with $\mathcal{EL}^+$, enumeration of group MUSes for Horn formulae represents a promising and strategic application of SAT technology. This includes, among others, SAT solvers, MCS extractors and enumerators, and MUS extractors and enumerators. This paper develops a highly optimized group MUS enumerator for Horn formulae, which is shown to extensively outperform any other existing approach. Performance gains are almost without exception at least one order of magnitude, and most often significantly more than that. More importantly, the experimental results demonstrate that SAT-based approaches are by far the most effective approaches for axiom pinpointing for the $\mathcal{EL}$ family of DLs. When compared with other non SAT-based approaches, the performance gains are also conclusive.

Future work will exploit integration of additional recent work on SAT-based problem solving, e.g. in MCS enumeration and MUS enumeration, to further improve performance of axiom pinpointing.

Acknowledgments. We thank the authors of EL$^+$SAT, R. Sebastiani and M. Vescovi, for authorizing the use of the most recent, yet unpublished, version of their work [51]. We thank the authors of SATPin [33], N. Manthey and R. Peñaloza, for bringing SATPin to our attention, and for allowing us to use their tool. We thank A. Biere for pointing out reference [19], on the complexity of implementing unit propagation when using watched literals. This work is partially supported by SFI PI grant BEACON (09/IN.1/I2618), by FCT grant POLARIS (PTDC/EIA-CCO/123051/2010), and by national funds through FCT with reference UID/CEC/50021/2013.

References

1. Arif, M.F., Mencía, C., Marques-Silva, J.: Efficient axiom pinpointing with EL2MCS. In: KI (2015)
2. Ashburner, M., Ball, C.A., Blake, J.A., Botstein, D., Butler, H., Cherry, J.M., Davis, A.P., Dolinski, K., Dwight, S.S., Eppig, J.T., et al.: Gene ontology: tool for the unification of biology. Nature genetics **25**(1), 25–29 (2000)
3. Baader, F., Hollunder, B.: Embedding defaults into terminological knowledge representation formalisms. J. Autom. Reasoning **14**(1), 149–180 (1995)
4. Baader, F., Horrocks, I., Sattler, U.; Description logics. In: van Harmelen, V.L.F., Porter, B. (eds.), Handbook of Knowledge Representation, Foundations of Artificial Intelligence, chapter 3, pp. 135–179. Elsevier (2008)
5. Baader, F., Lutz, C., Suntisrivaraporn, B.: CEL — a polynomial-time reasoner for life science ontologies. In: Furbach, U., Shankar, N. (eds.) IJCAR 2006. LNCS (LNAI), vol. 4130, pp. 287–291. Springer, Heidelberg (2006)
6. Baader, F., Peñaloza, R.: Axiom pinpointing in general tableaux. J. Log. Comput. **20**(1), 5–34 (2010)
7. Baader, F., Peñaloza, R., Suntisrivaraporn, B.: Pinpointing in the description logic $\mathcal{EL}^+$. In: KI, pp. 52–67 (2007)
8. Baader, F., Suntisrivaraporn, B.: Debugging SNOMED CT using axiom pinpointing in the description logic $\mathcal{EL}^+$. In: KR-MED (2008)
9. Bacchus, F., Davies, J., Tsimpoukelli, M., Katsirelos, G.: Relaxation search: a simple way of managing optional clauses. In: AAAI, pp. 835–841 (2014)
10. Bailey, J., Stuckey, P.J.: Discovery of minimal unsatisfiable subsets of constraints using hitting set dualization. In: Hermenegildo, M.V., Cabeza, D. (eds.) PADL 2004. LNCS, vol. 3350, pp. 174–186. Springer, Heidelberg (2005)
11. Bakker, R.R., Dikker, F., Tempelman, F., Wognum, P.M.: Diagnosing and solving over-determined constraint satisfaction problems. In: IJCAI, pp. 276–281 (1993)
12. Belov, A., Lynce, I., Marques-Silva, J.: Towards efficient MUS extraction. AI Commun. **25**(2), 97–116 (2012)
13. Biere, A., Heule, M., van Maaren, H., Walsh, T. (eds.): Handbook of Satisfiability, Frontiers in Artificial Intelligence and Applications, vol. 185. IOS Press (2009)
14. Birnbaum, E., Lozinskii, E.L.: Consistent subsets of inconsistent systems: structure and behaviour. J. Exp. Theor. Artif. Intell. **15**(1), 25–46 (2003)
15. Cook, S.A.: The complexity of theorem-proving procedures. In: STOC, pp. 151–158 (1971)
16. de Siqueira, N.J.L., Puget, J.-F.: Explanation-based generalisation of failures. In: ECAI, pp. 339–344 (1988)
17. Dowling, W.F., Gallier, J.H.: Linear-time algorithms for testing the satisfiability of propositional Horn formulae. J. Log. Program. **1**(3), 267–284 (1984)
18. Eén, N., Sörensson, N.: An extensible SAT-solver. In: Giunchiglia, E., Tacchella, A. (eds.) SAT 2003. LNCS, vol. 2919, pp. 502–518. Springer, Heidelberg (2004)
19. Gent, I.: Optimal implementation of watched literals and more general techniques. Journal of Artificial Intelligence Research **48**, 231–252 (2013)
20. Giunchiglia, E., Maratea, M.: Solving optimization problems with DLL. In: ECAI, pp. 377–381 (2006)

21. Grégoire, É., Lagniez, J., Mazure, B.: An experimentally efficient method for (MSS, CoMSS) partitioning. In: AAAI, pp. 2666–2673 (2014)
22. Heras, F., Morgado, A., Marques-Silva, J.: MaxSAT-based encodings for group MaxSAT. AI Commun. **28**(2), 195–214 (2015)
23. Itai, A., Makowsky, J.A.: Unification as a complexity measure for logic programming. J. Log. Program. **4**(2), 105–117 (1987)
24. Junker, U.: QuickXplain: preferred explanations and relaxations for overconstrained problems. In: AAAI, pp. 167–172 (2004)
25. Kalyanpur, A., Parsia, B., Horridge, M., Sirin, E.: Finding all justifications of OWL DL entailments. In: Aberer, K., Choi, K.-S., Noy, N., Allemang, D., Lee, K.-I., Nixon, L.J.B., Golbeck, J., Mika, P., Maynard, D., Mizoguchi, R., Schreiber, G., Cudré-Mauroux, P. (eds.) ASWC 2007 and ISWC 2007. LNCS, vol. 4825, pp. 267–280. Springer, Heidelberg (2007)
26. Kalyanpur, A., Parsia, B., Sirin, E., Cuenca-Grau, B.: Repairing unsatisfiable concepts in OWL ontologies. In: Sure, Y., Domingue, J. (eds.) ESWC 2006. LNCS, vol. 4011, pp. 170–184. Springer, Heidelberg (2006)
27. Lahiri, S.K., Nieuwenhuis, R., Oliveras, A.: SMT techniques for fast predicate abstraction. In: Ball, T., Jones, R.B. (eds.) CAV 2006. LNCS, vol. 4144, pp. 424–437. Springer, Heidelberg (2006)
28. Liffiton, M.H., Malik, A.: Enumerating infeasibility: finding multiple MUSes quickly. In: Gomes, C., Sellmann, M. (eds.) CPAIOR 2013. LNCS, vol. 7874, pp. 160–175. Springer, Heidelberg (2013)
29. Liffiton, M.H., Previti, A., Malik, A., Marques-Silva, J.: Fast, flexible MUS enumeration. Constraints (2015). http://link.springer.com/article/10.1007/s10601-015-9183-0
30. Liffiton, M.H., Sakallah, K.A.: Algorithms for computing minimal unsatisfiable subsets of constraints. J. Autom. Reasoning **40**(1), 1–33 (2008)
31. Ludwig, M.: Just: a tool for computing justifications w.r.t. ELH ontologies. In: ORE (2014)
32. Ludwig, M., Peñaloza, R.: Error-tolerant reasoning in the description logic $\mathcal{EL}$. In: Fermé, E., Leite, J. (eds.) JELIA 2014. LNCS, vol. 8761, pp. 107–121. Springer, Heidelberg (2014)
33. Manthey, N., Peñaloza, R.: Exploiting SAT technology for axiom pinpointing. Technical Report LTCS 15–05, Chair of Automata Theory, Institute of Theoretical Computer Science, Technische Universität Dresden, April 2015. https://ddll.inf.tu-dresden.de/web/Techreport3010
34. Marques-Silva, J., Heras, F., Janota, M., Previti, A., Belov, A.: On computing minimal correction subsets. In: IJCAI, pp. 615–622 (2013)
35. Marques-Silva, J., Janota, M., Belov, A.: Minimal sets over monotone predicates in boolean formulae. In: Sharygina, N., Veith, H. (eds.) CAV 2013. LNCS, vol. 8044, pp. 592–607. Springer, Heidelberg (2013)
36. Mencía, C., Previti, A., Marques-Silva, J.: Literal-based MCS extraction. In: IJCAI, pp. 1973–1979 (2015)
37. Meyer, T.A., Lee, K., Booth, R., Pan, J.Z.: Finding maximally satisfiable terminologies for the description logic $\mathcal{EL}^+$. In: AAAI, pp. 269–274 (2006)
38. Minoux, M.: LTUR: A simplified linear-time unit resolution algorithm for Horn formulae and computer implementation. Inf. Process. Lett. **29**(1), 1–12 (1988)
39. Moodley, K., Meyer, T., Varzinczak, I.J.: Root justifications for ontology repair. In: Rudolph, S., Gutierrez, C. (eds.) RR 2011. LNCS, vol. 6902, pp. 275–280. Springer, Heidelberg (2011)

40. Nguyen, H.H., Alechina, N., Logan, B.: Axiom pinpointing using an assumption-based truth maintenance system. In: DL (2012)
41. O'Sullivan, B., Papadopoulos, A., Faltings, B., Pu, P.: Representative explanations for over-constrained problems. In: AAAI, pp. 323–328 (2007)
42. Parsia, B., Sirin, E., Kalyanpur, A.; Debugging OWL ontologies. In: WWW, pp. 633–640 (2005)
43. Peñaloza, R., Sertkaya, B.: On the complexity of axiom pinpointing in the EL family of description logics. In: KR (2010)
44. Previti, A., Marques-Silva, J.: Partial MUS enumeration. In: AAAI, pp. 818–825 (2013)
45. Rector, A.L., Horrocks, I.R.: Experience building a large, re-usable medical ontology using a description logic with transitivity and concept inclusions. In: Workshop on Ontological Engineering, pp. 414–418 (1997)
46. Reiter, R.: A theory of diagnosis from first principles. Artif. Intell. **32**(1), 57–95 (1987)
47. Rosa, E.D., Giunchiglia, E.: Combining approaches for solving satisfiability problems with qualitative preferences. AI Commun. **26**(4), 395–408 (2013)
48. Schlobach, S., Cornet, R.: Non-standard reasoning services for the debugging of description logic terminologies. In: IJCAI, pp. 355–362 (2003)
49. Schlobach, S., Huang, Z., Cornet, R., van Harmelen, F.: Debugging incoherent terminologies. J. Autom. Reasoning **39**(3), 317–349 (2007)
50. Sebastiani, R., Vescovi, M.: Axiom pinpointing in lightweight description logics via horn-SAT encoding and conflict analysis. In: Schmidt, R.A. (ed.) CADE-22. LNCS, vol. 5663, pp. 84–99. Springer, Heidelberg (2009)
51. Sebastiani, R., Vescovi, M.: Axiom pinpointing in large $\mathcal{EL}^+$ ontologies via SAT and SMT techniques. Technical Report DISI-15-010, DISI, University of Trento, Italy, April 2015. Under Journal Submission. http://disi.unitn.it/rseba/elsat/elsat_techrep.pdf
52. Sioutos, N., de Coronado, S., Haber, M.W., Hartel, F.W., Shaiu, W., Wright, L.W.: NCI thesaurus: A semantic model integrating cancer-related clinical and molecular information. Journal of Biomedical Informatics **40**(1), 30–43 (2007)
53. Sirin, E., Parsia, B., Grau, B.C., Kalyanpur, A., Katz, Y.: Pellet: A practical OWL-DL reasoner. J. Web Sem. **5**(2), 51–53 (2007)
54. Slaney, J.: Set-theoretic duality: a fundamental feature of combinatorial optimisation. In: ECAI, pp. 843–848 (2014)
55. Spackman, K.A., Campbell, K.E., Côté, R.A.: SNOMED RT: a reference terminology for health care. In: AMIA (1997)
56. Vescovi, M.: Exploiting SAT and SMT Techniques for Automated Reasoning and Ontology Manipulation in Description Logics. Ph.D. thesis, University of Trento (2011)

QELL: QBF Reasoning with Extended Clause Learning and Levelized SAT Solving

Kuan-Hua Tu, Tzu-Chien Hsu, and Jie-Hong R. Jiang[✉]

Department of Electrical Engineering /
Graduate Institute of Electronics Engineering,
National Taiwan University, Taipei 10617, Taiwan
jhjiang@ntu.edu.tw

Abstract. Quantified Boolean satisfiability (QSAT) is natural formulation of many decision problems and yet awaits further breakthroughs to reach the maturity enabling industrial applications. Recent advancements on quantified Boolean formula (QBF) proof systems sharpen our understanding of their proof complexities and shed light on solver improvement. Particularly QBF solving based on formula expansion has been theoretically and practically demonstrated to be more powerful than non-expansion based solving. However recursive expansion suffers from exponential formula explosion and has to be carefully managed. In this paper, we propose a QBF solver using levelized SAT solving in the flavor of formula expansion. New learning techniques based on circuit structure reconstruction, complete and incomplete ALLSAT learning, core expansion, bounded recursion, and other methods are devised to control formula growth. Experimental results on application benchmarks show that our prototype implementation is comparable with state-of-the-art solvers and outperforms other solvers in certain instances.

1 Introduction

Quantified Boolean satisfiability (QSAT) is a natural formulation of various decision problems such as verification [7], planning [19], synthesis [12], and other computer science applications. Quantified Boolean formulas (QBFs) extend propositional logic and permit variables being existentially or universally quantified. This quantification capability gives QBFs distinct power to compactly encode logical constraints with an exponential reduction. There have been extensive research efforts to develop efficient QBF solvers and preprocessors, e.g., [1,2,6,9–11,15,16,21,24,25]. State-of-the-art QBF solvers employ techniques such as conflict/solution-driven learning [10], long-distance resolution [25], formula expansion [1], duality recovery [9,15,24], among many others. Due to its intrinsic hardness (PSPACE-complete complexity in contrast to the NP-completeness of SAT), QSAT has yet to reach the maturity to spark popular industrial adoptions and awaits further breakthroughs.

This work was supported in part by the Ministry of Science and Technology of Taiwan under grants MOST 103-2221-E-002-273 and 104-2628-E-002-013-MY3.

© Springer International Publishing Switzerland 2015
M. Heule and S. Weaver (Eds.): SAT 2015, LNCS 9340, pp. 343–359, 2015.
DOI: 10.1007/978-3-319-24318-4_25

Among QBF solvers, DEPQBF [16] and RAREQS [11] are the representatives of resolution-based and expansion-based solvers, respectively. The former adopts the standard QDPLL-style reasoning with resolution-based conflict/solution learning; the latter employs the counterexample guided abstraction refinement (CEGAR) paradigm for expansion-based learning. The two solvers exhibit different solving characteristics on conquering different application instances. Recent progress in QBF proof complexities makes clear the relative power between resolution-based and expansion-based systems [3,5]. Essentially expansion-based QBF solving may potentially yield proofs of size exponentially shorter than nonexpansion-based solving [3]. Therefore formula expansion can be a key technique for efficient QBF solving. Another line of research was studied in OoQ [9] to bridge the duality gap between solution learning and conflict learning in resolution-based solving. Circuit information [17,22] is recovered from the QBF under evaluation to facilitate solution learning. However, it is unclear how to close this duality gap in expansion-based solving.

In this paper, we consider formula expansion-based learning, in contrast to the resolution-based conflict/solution-driven learning [10], in a QDPLL-style reasoning. A QBF evaluation framework using levelized SAT solving [21] is proposed for integration with formula expansion-based learning. Unlike the recursive QBF solving proposed by RAREQS, our search space exploration is non-recursive, and allows tighter integration with SAT solving and better control managing formula expansion for both solution and conflict learnings than RAREQS. For solution learning we present a (complete or incomplete) ALLSAT based expansion; for conflict learning we present a localized method for recursive expansion. Moreover, we show how circuit information can be incorporated in solution learning simplification. Experimental results show that our prototyped solver is comparable to other state-of-the-art solvers (including RAREQS, DEPQBF, and OoQ). We expect to achieve further improvement by implementation optimization.

2   Preliminaries

Given a set $X = \{x_1, \ldots, x_k\}$ of Boolean variables (with domain values $\{0, 1\}$ or $\{\text{FALSE}, \text{TRUE}\}$), the set of valuation of X is denoted as $[\![X]\!]$. A Boolean variable x may appear in a propositional formula in the form of a *positive literal* (x) or a *negative literal* ($\overline{x}$ or $\neg x$). We denote the variable corresponding to a literal l by $var(l)$. A *clause* (resp. *cube*) is a disjunction (resp. conjunction) of literals. We denote the empty clause (the clause without any literals) as $\bot$ and denote the empty cube as $\top$. A *conjunctive normal form* (CNF) formula consists of a conjunction of clauses, and a *disjunction normal form* (DNF) formula consists of a disjunction of cubes. In the sequel, we alternatively represent a CNF (DNF) formula in terms of a set of clauses (cubes) and represent a clause (cube) in terms of a set of literals. We use the Boolean connectives $\neg, \wedge, \vee, \rightarrow, \leftrightarrow$ with their standard interpretations.

A *quantified Boolean formula* (QBF) Φ over variables $X = X_1 \cup \cdots \cup X_n$ in the *prenex conjunctive normal form* (PCNF) can be expressed as

$$Q_1 X_1 \cdots Q_n X_n . \phi, \qquad (1)$$

where the *prefix* $\Phi_{\mathrm{pfx}} = Q_1 X_1 \cdots Q_n X_n$ consists of quantifiers $Q_i \in \{\exists, \forall\}$ and variable sets X_i with $X_i \cap X_j = \emptyset$ for $i \neq j$, and the *matrix* $\Phi_{\mathrm{mtx}} = \phi$ is a quantifier-free CNF formula over variables X. In the sequel, we assume that a QBF is in the PCNF form Eq. (1) and let $X_i \subseteq X_\exists$ or $X_i \subseteq X_\forall$ be maximal in that $Q_i \neq Q_{i+1}$ for $i = 1, \ldots, n-1$. Moreover, we assume that a QBF is *closed*, that is, all variables are quantified.

The set X of variables of Φ can be partitioned into *existential variables* $X_\exists = \{x_i \in X \mid Q_i = \exists\}$ and *universal variables* $X_\forall = \{x_i \in X \mid Q_i = \forall\}$. A literal l is called an *existential literal* and a *universal literal* if $var(l)$ is in $X_\exists$ and $X_\forall$, respectively. Given a QBF over variables X, the *quantification level* of variable $x \in X$, denoted $lev(x)$, is defined to be the number of quantifier alternations between the quantifiers $\exists$ and $\forall$ from left (outer) to right (inner) plus 1. The same level definition extends to a literal l, i.e., $lev(l) = lev(var(l))$.

Without loss of generality, we assume the inner most quantifier Q_n to be existential because quantifying out a universal variable x at the inner most quantification level from a CNF formula φ is equivalent to simply removing the literals x and $\neg x$ in φ.

An assignment $\alpha : X \rightarrow \{0, 1\}$ on variables X is a mapping that assigns each variable in X to $\{0, 1\}$. We alternatively represent the mappings $\alpha(x) \mapsto 0$ and $\alpha(x) \mapsto 1$ for $x \in X$ as literals $\overline{x}$ and x, respectively. Therefore we consider an assignment α as a set of literals (denoting a cube). A Boolean formula ϕ over a set X of variables subject to some truth assignment $\alpha : X' \rightarrow \{0, 1\}$ on variables $X' \subseteq X$ is denoted as $\phi \mid_\alpha$ to mean the formula of ϕ induced under α. Similarly, given a QBF Φ its *induced* QBF with respect to the assignment α, denoted as $\Phi \mid_\alpha$, is defined to be the QBF with the prefix same as Φ_{pfx} except for the removal of the quantifications on variables in α and with the matrix being $\Phi_{\mathrm{mtx}} \mid_\alpha$.

A QBF is true (resp. false) if and only if it has a Skolem-function model (Herbrand-function countermodel) [4]. A model (resp. counter-model) of a QBF Φ can be presented as a tree. In the tree, each leaf node is labelled with 0 (FALSE) or 1 (TRUE); each non-leaf node u is labelled with a variable x such that u has two child nodes if $x \in X_\forall$ (resp. $x \in X_\exists$) and has one child node if $x \in X_\exists$ (resp. $x \in X_\forall$); each edge (u, v) corresponds to a 0- or 1-assignment to the variable x of u. A leaf node corresponds to either a constant 1 or constant 0. Also for each path from the root node to a leaf node in the tree, the labels must respect the prefix order of Φ. Collecting labels of nodes from the root to any leaf node can get a partial assignment α such that $\Phi_{\mathrm{mtx}} \mid_\alpha = 1$ (resp. $\Phi_{\mathrm{mtx}} \mid_\alpha = 0$).

Given a clause set D over variables X, we say that a variable $x \in X$ is *defined* under D if for any assignment α on $X \backslash \{x\} \neq \emptyset$ there is exactly one assignment β to x such that the CNF formula of D is satisfied under the combined assignment (α, β). We say that D forms a *definition* of x, call x the *defined variable*, and call $X \backslash \{x\}$ the *support variables* of x in D, denoted as $sup(x)$. A set of definitions forms a *circuit* if each defined variable is distinct. We represent a circuit by a directed graph $G(V, E)$, where each vertex $u \in V$ represents a variable and each edge $(u, v) \in E \subseteq V \times V$ indicates $u \in sup(v)$ (we do not distinguish a vertex and its represented variable). A circuit can be *cyclic* or *acyclic* depending on whether or not its graph representation is cyclic. A *feedback vertex set* of a cyclic graph is a subset of vertices such that their removal from the graph makes the induced graph acyclic. For a circuit, we alternatively call the defined and non-defined variables as the *output* and *input variables*, respectively. Note that for an acyclic circuit any assignment to all the input variables uniquely determines the values of the output variables.

When an acyclic circuit $\mathcal{C}$ appears in the context of a QBF Φ, we say that a variable x is *defined at the ith quantification level*, denoted $dlev(x) = i$, if $lev(x) = i$ for x being a non-defined variable in $\mathcal{C}$ or $\max_{y \in sup(x)}\{dlev(y)\} = i$ for x being a defined variable in $\mathcal{C}$.

3   Algorithmic Flow

Our proposed procedure for QBF evaluation is sketched in Figure 1. The algorithm takes as input a QBF $\Phi = Q_1 X_1, \ldots, Q_n X_n.\varphi$, for $Q_1 = Q_n = \exists$ and $X_i \neq \emptyset$ for $i \geq 2$ as we should assume in the sequel, and returns as output the truth or falsity of Φ along with the assignment to variables X_1. The algorithm maintains two CNF formulas $\sigma_\exists$ and $\sigma_\forall$, which are used to exclude already confirmed conflict and solution assignments to existential and universal variables, respectively, from future search. (Note that, although the interpretation of quantifiers in $\sigma_\forall$ should be inverted, i.e., $Q_i = \exists$ and $\forall$ should be interpreted as $\forall$ and $\exists$, respectively, in the sequel we always refer to the quantifiers of the original prefix to avoid confusion.) In Line 1, $\sigma_\exists$ and $\sigma_\forall$ are initialized to φ and TRUE, respectively. In Line 2, the current assignments $\alpha[1]$ to X_1, ..., $\alpha[n]$ to X_n, whose collection is denoted $\boldsymbol{\alpha}$, are initialized to $\emptyset$. Moreover, assignments $\iota_\exists$ and $\iota_\forall$ denote the sets of literals implied in $\sigma_\exists$ and $\sigma_\forall$ with respect to the assignment $\boldsymbol{\alpha}$. They are initialized to $\emptyset$ in Line 2. In Line 3, the current quantification level ℓ is initialized to 2 if $X_1 = \emptyset$ and to 1 otherwise.

After the above initialization, QBF evaluation process repeats in Lines 4-31. In Line 5, $\iota_\exists$ and $\iota_\forall$ are obtained through procedure *GetImplication*. Depending on the quantifier type of the current level, SAT solving is performed to search a proper assignment on X_ℓ in Line 7 on $\sigma_\exists$ with respect to assignments $\boldsymbol{\alpha}$ and $\iota_\forall$ if $Q_\ell = \exists$, and in Line 23 on $\sigma_\forall$ with respect to assignments $\boldsymbol{\alpha}$ and $\iota_\exists$ if $Q_\ell = \forall$. For $Q_\ell = \exists$, if the assignment for X_ℓ cannot be found, then either in Line 10 declare the QBF Φ false if $\ell = 1$ or in Lines 11-13 strengthen $\sigma_\exists$ by procedure *ConflictLearn* and backtrack to quantification level $\ell - 2$. On the contrary, if the assignment for X_ℓ is found, then in Line 16 declare Φ false if $\ell = n = 1$, in Lines 17-19 strengthen $\sigma_\forall$ by procedure *SolutionLearn* and backtrack to quantification level $\ell - 1$ if $\ell = n \neq 1$, or in Line 21 continue to quantification level $\ell + 1$ if $\ell \neq n$. On the other hand, for $Q_\ell = \forall$, if the assignment for X_ℓ cannot be found, then either in Line 26 declare the QBF Φ

```
QELL
    input: QBF Φ = Q₁X₁,...,QₙXₙ.φ with Q₁ = Qₙ = ∃ and Xᵢ ≠ ∅ for i ≥ 2
    output: truth or falsity of Φ, and 1st level assignment
    begin
01      σ∃ := φ; σ∀ := TRUE;
02      α[1], ...,α[n] := ∅; ι∃ := ∅; ι∀ := ∅;
03      ℓ := (X₁ = ∅)? 2 : 1;
04      while (TRUE)
05          (ι∃, ι∀, σ∃, σ∀) := GetImplication(σ∃, σ∀, α);
06          if (Qℓ = ∃)
07              (res, αtmp) := SatSolve(σ∃, (α, ι∀));
08              α[ℓ] := Project(αtmp, Xℓ);
09              if (res = FALSE)
10                  if (ℓ = 1) return (FALSE, ∅);
11                  σ∃ := ConflictLearn(Φ, σ∃, ℓ, α);
12                  α[ℓ], α[ℓ − 1] := ∅;
13                  ℓ := ℓ − 2;
14              else
15                  if (ℓ = n)
16                      if (n = 1) return (TRUE, α[1]);
17                      σ∀ := SolutionLearn(Φ, σ∀, ℓ, α);
18                      α[ℓ] := ∅;
19                      ℓ := ℓ − 1;
20                  else
21                      ℓ := ℓ + 1;
22          else //Qℓ = ∀
23              (res, αtmp) := SatSolve(σ∀, (α, ι∃));
24              α[ℓ] := Project(αtmp, Xℓ);
25              if (res = FALSE)
26                  if (ℓ = 2) return (TRUE, α[1]);
27                  σ∀ := SolutionLearn(Φ, σ∀, ℓ, α);
28                  α[ℓ], α[ℓ − 1] := ∅;
29                  ℓ := ℓ − 2;
30              else
31                  ℓ := ℓ + 1;
    end
```

Fig. 1. Algorithm: QELL with regular backtrack

true if $\ell = 2$ or in Lines 27-29 strengthen $\sigma_\forall$ by procedure *SolutionLearn* and backtrack to quantification level $\ell - 2$. On the contrary, if the assignment for X_ℓ is found, then in Line 31 continue to quantification level $\ell + 1$.

The procedure *GetImplication* of Figure 1 primarily collects the existential literals $\iota_\exists$ being implied in $\sigma_\exists$ and the universal literals $\iota_\forall$ implied in $\sigma_\forall$ due to assignment α. In addition, the procedure may strengthen $\sigma_\exists$ and $\sigma_\forall$ under the following two circumstances. When a universal literal is implied in $\sigma_\exists$ under assignment α, this implication indicates the QBF Φ is false under α and the cause of the implication can be learned to strengthen $\sigma_\exists$ in prevention of wasteful search. Similarly, when an existential literal is implied in $\sigma_\forall$ under assignment α, this implication indicates Φ is true under α and the cause of the implication can be learned to strengthen $\sigma_\forall$. The learning can be performed in different ways as to be discussed in Section 4.

Notice that the implication processes of $\sigma_\exists$ and $\sigma_\forall$ in procedure *GetImplication* are mutual. That is, the (universal) literals of $\sigma_\forall$ are incorporated with α for assignment in $\sigma_\exists$ and the (existential) literals of $\sigma_\exists$ are incorporated with α for assignment in $\sigma_\forall$ to generate further implications and/or trigger additional learning.

```
QELL
   input: QBF Φ = Q₁X₁,...,QₙXₙ.φ with Q₁ = Qₙ = ∃ and Xᵢ ≠ ∅ for i ≥ 2
   output: truth or falsity of Φ, and 1st level assignment
   begin
01    σ∃ := φ; σ∀ := TRUE;
02    α[1], ...,α[n] := ∅; ι∃ := ∅; ι∀ := ∅;
03    ℓ := (X₁ = ∅)? 2 : 1;
04    while (TRUE)
05      (ι∃, ι∀, σ∃, σ∀, ℓ, α) := GetImplication(σ∃, σ∀, α);
06      if (Qₗ = ∃)
07        (res, αₜₘₚ) := SatSolve(σ∃, (α, ι∀));
08        α[ℓ] := Project(αₜₘₚ, Xₗ);
09        if (res = FALSE)
10          if (ℓ = 1) return (FALSE, ∅);
11          σ∃ := ConflictLearn(Φ, σ∃, ℓ, α);
12          (ℓ, α) := UpdateLevel(Φ, σ∃, σ∀, ℓ, α);
13        else
14          if (ℓ = n)
15            if (n = 1) return (TRUE, α[1]);
16            σ∀ := SolutionLearn(Φ, σ∀, ℓ, α);
17            (ℓ, α) := UpdateLevel(Φ, σ∃, σ∀, ℓ, α);
18          else
19            ℓ := ℓ + 1;
20      else //Qₗ = ∀
21        (res, αₜₘₚ) := SatSolve(σ∀, (α, ι∃));
22        α[ℓ] := Project(αₜₘₚ, Xₗ);
23        if (res = FALSE)
24          if (ℓ = 2) return (TRUE, α[1]);
25          σ∀ := SolutionLearn(Φ, σ∀, ℓ, α);
26          (ℓ, α) := UpdateLevel(Φ, σ∃, σ∀, ℓ, α);
27        else
28          ℓ := ℓ + 1;
   end
```

Fig. 2. Algorithm: QELL with aggressive backtrack

The procedure *SatSolve* of Figure 1 returns the (un)satisfiability of a CNF formula under some *unit assumptions* [8]. If the formula is satisfiable, it also returns a satisfying assignment. Otherwise, an unsatisfiable set of assumptions can be derived.

For solution learning, there are two cases. For the case due to $\sigma_\exists$ being satisfied under assignments $\alpha[1],...,\alpha[\ell]$ with $\ell = n$ (Line 17), the assignment $\alpha[n]$ can be seen as a model to existential variables X_n with respect to assignments $\alpha[1],...,\alpha[n-1]$. The backtrack should return to level $\ell - 1$ and $\sigma_\forall$ is strengthened. For the other case due to $\sigma_\forall$ being unsatisfiable under assignments $\alpha[1],...,\alpha[\ell]$ (Line 27), the QBF $\Phi|_{\alpha[1],...,\alpha[\ell]}$ induced by assignments $\alpha[1],...,\alpha[\ell]$ is true, and so is $\Phi|_{\alpha[1],...,\alpha[\ell-2]}$. The backtrack should return to level $\ell - 2$ and $\sigma_\forall$ is strengthened. On the other hand, for conflict learning, due to $\sigma_\exists$ being unsatisfiable under assignments $\alpha[1],...,\alpha[\ell]$ (Line 11), the QBF $\Phi|_{\alpha[1],...,\alpha[\ell]}$ induced by assignments $\alpha[1],...,\alpha[\ell]$ is false, and so is $\Phi|_{\alpha[1],...,\alpha[\ell-2]}$. The backtrack should return to level $\ell - 2$ and $\sigma_\exists$ is strengthened.

Note that there may be new variables being introduced during learning, but the number of quantification levels remains unchanged. Note also that *universal reduction* [14] can be applied on $\sigma_\exists$ and *existential reduction* [10] on $\sigma_\forall$ (by "existential" reduction, recall our reference to the quantifiers of the original prefix).

The backtracks of the algorithm in Figure 1 can be improved to admit a greater leap of backtrack as the procedure outlined in Figure 2, where procedures *GetImplication* and *UpdateLevel* compute proper levels for backtrack. The backtrack level can be determined through analyzing conflicting unit assumptions on assignment to $\sigma_\exists$ and $\sigma_\forall$ using incremental SAT solving. Specifically, if the set of conflicting assumptions is empty or contains only universal variables, then the algorithm backtracks to quantification level 1 and concludes the falsity of the QBF. Otherwise, the backtrack level equals the maximal quantification level of all the existential variables in the assumptions. Moreover, assignment and implication propagation can be performed simultaneously on $\sigma_\exists$ and $\sigma_\forall$ (similar to the implication of DEPQBF).

4   Learning

Learning is a vital process in QBF evaluation. Below we elaborate the learning techniques employed in our proposed algorithm.

4.1   Solution Learning by Levelized Blocking

We exploit the following proposition for solution learning.

Proposition 1. *Given a QBF* $\Phi = Q_1 X_1, \ldots, Q_n X_n.\varphi$ *with* $Q_i = \forall$, *assume the QBF* $\Phi|_{\alpha,\beta}$ *induced under assignments* $\alpha \in [\![X_1 \cup \ldots \cup X_i]\!]$ *and* $\beta \in [\![X_{i+1}]\!]$ *is true (i.e., has a model). Let* φ_α *be the set of clauses in* φ *that are satisfied by the literals in* α. *Then* $\Phi|_{\alpha'}$ *is true for any assignment* $\alpha' \in [\![X_1 \cup \ldots \cup X_i]\!]$ *that satisfies* $\varphi_\alpha|_\beta$.

For the special case of $n = 3$ with $X_1 = \emptyset$, the proposition reduces to that of [13]. For the general case $n \geq 3$, the proposition differs from [11] in that prior work asserts that $\Phi|_{\alpha'}$ is true for $\alpha' \in [\![X_1 \cup \ldots \cup X_i]\!]$ if $(\Phi|_\beta)|_{\alpha'}$ is true. Because the number of assignments $\alpha' \in [\![X_1 \cup \ldots \cup X_i]\!]$ satisfying $\varphi_\alpha|_\beta$ is no greater than the number of assignments $\alpha' \in [\![X_1 \cup \ldots \cup X_i]\!]$ such that $\Phi|_{\alpha'}$ is true, the formula $(\Phi|_\beta)|_{\alpha'}$ is stronger than φ_α in solution learning. However it is computationally more expensive to check whether $\Phi|_{\alpha'}$ is true than to check whether $\varphi_\alpha|_{\alpha',\beta}$ is satisfiable.

By Proposition 1 once the truth of QBF $\Phi|_{\alpha,\beta}$ is established, we know that $\Phi|_{\alpha'}$ is guaranteed to be true for any assignment $\alpha' \in [\![X_1 \cup \ldots \cup X_i]\!]$ that satisfies $\varphi_\alpha|_\beta$. Therefore one can perform solution learning by blocking all assignments $\alpha' \in [\![X_1 \cup \ldots \cup X_i]\!]$ that satisfy $\varphi_\alpha|_\beta$ from future search. In this paper, we explore two approaches to implement such solution learning. One is to negate $\varphi_\alpha|_\beta$ and the other is to conjunct the negation of every satisfying solution to $\varphi_\alpha|_\beta$. For the former, since $\varphi_\alpha|_\beta$ is in CNF, fresh variables need to be introduced similar to [13] to express $\neg\varphi_\alpha|_\beta$ in CNF. For the latter, ALLSAT computation [23] can be performed to enumerate the satisfying solutions to $\varphi_\alpha|_\beta$ (by hitting set[1] generation to cover clauses $\varphi_\alpha|_\beta$), which can then be negated in CNF. Note that the ALLSAT computation does not need to be complete. It is not necessary to compute all the solutions to $\varphi_\alpha|_\beta$ because the amount of solutions being blocked only affects the strength of learning but not the correctness.

For the ALLSAT based learning, consider QBF $\Phi = Q_1X_1,\ldots,Q_nX_n.\varphi$ with $Q_i = \forall$ for $i \leq n-1$ being the quantification level under learning. Assume $\Phi|_{\alpha,\beta}$ is true under assignments $\alpha = (\alpha_1,\alpha_2)$ for $\alpha_1 \in [\![X_1 \cup \ldots \cup X_{i-1}]\!]$ and $\alpha_2 \in [\![X_i]\!]$, and $\beta \in [\![X_{i+1}]\!]$. We partition the set of clauses of φ into three subsets: The set of *outer level clauses* are those satisfied through literals in α_1; the set of *current level clauses* are those not satisfied through literals in α_1 but satisfied through literals in α_2; the set of *inner level clauses* are the rest. Let φ_α be the formula consisting of outer and current level clauses. By Proposition 1 $\Phi|_{\alpha'}$ is true for any assignment $\alpha' \in [\![X_1 \cup \ldots \cup X_i]\!]$ that satisfies $\varphi_\alpha|_\beta$. Because $Q_i = \forall$, there may be many $\alpha'_2 \in [\![X_i]\!]$ for which $\Phi|_{\alpha_1,\alpha'_2,\beta}$ are true. Therefore the set of clauses in φ that are satisfied by the literals in (α_1,α_2) and the set of clauses in φ that are satisfied by the literals in (α_1,α'_2) may overlap to some extent. Particularly, the clauses that are satisfied by the literals in α_1 are the same. Therefore, we compute only one hitting set for the outer level clauses and compute (complete or incomplete) hitting sets by ALLSAT enumeration [23] for the current level clauses. Combining the hitting set of the outer level clauses with each of the hitting sets of the current level clauses yields a (partial) assignment that satisfies $\varphi_\alpha|_\beta$. Therefore, for solution learning the solution clauses $\sigma_\forall$ of the procedures of Figures 1 and 2 can be augmented by the complements of the combined hitting set assignments.

4.2 Solution Learning in the Presence of Circuit Information

When the matrix of a QBF contains some form of circuit structures, the information may facilitate QBF evaluation as the following example illustrates.

Example 1. Consider the QBF $\forall X, \exists y, z, T.\varphi$ with the matrix

$$\varphi = (t_1 \leftrightarrow (x_1 \oplus x_2))(t_2 \leftrightarrow (t_1 \oplus x_3)) \cdots (t_{k-1} \leftrightarrow (t_{k-2} \oplus x_k))(z \leftrightarrow (t_{k-1} \oplus y)),$$

where $X = \{x_1,\ldots,x_k\}$, $T = \{t_1,\ldots,t_{k-1}\}$ and "$\oplus$" denotes Boolean XOR operation. Observe that the singleton variable set $\{t_{k-1}\}$ separates the clause set of subformula $\varphi_A = (t_1 \leftrightarrow (x_1 \oplus x_2))(t_2 \leftrightarrow (t_1 \oplus x_3)) \cdots (t_{k-1} \leftrightarrow (t_{k-2} \oplus x_k))$ and that of $\varphi_B = (z \leftrightarrow (t_{k-1} \oplus y))$. Observe also that, for every assignment to X, there exists some unique assignment to T that satisfies φ_A.

To evaluate the above QBF, an assignment, say, $(x_1,\ldots,x_k) = (0,\ldots,0)$, $(t_1,\ldots,t_{k-1}) = (0,\ldots,0)$, $y = 0$, and $z = 0$, satisfying φ can be found. By the above two observations, one can conclude that any assignment $\alpha \in [\![X]\!]$ that makes t_{k-1} false will make the QBF induced under α, i.e., $\exists y, z.\varphi_B|_{\neg t_{k-1}}$, true.

---

[1] Given a CNF formula φ over variables X and an assignment $\alpha \in [\![X]\!]$, the *hitting set* of φ with respect to α is a partial assignment $\beta \subseteq \alpha$ such that each clause in φ contains at least one literal in β.

Therefore, in the next trial one only needs to search for another assignment to X that makes t_{k-1} true. Suppose that a new assignment, say, $(x_1, \ldots, x_{k-1}, x_k) = (0, \ldots, 0, 1)$, $(t_1, \ldots, t_{k-2}, t_{k-1}) = (0, \ldots, 0, 1)$, $y = 0$, and $z = 1$ is found. One can further conclude that any assignment $\alpha \in [\![X]\!]$ that makes t_{k-1} true will make the QBF induced under α, i.e., $\exists y, z.\varphi_B|_{t_{k-1}}$, true. Consequently the truth of the QBF can be concluded by just two SAT solving trials.

Without any formula preprocessing, state-of-the-art solvers, such as RAREQS and DEPQBF, may take exponential time to solve the above QBF. For RAREQS, since each refinement step only blocks two assignments to $(x_1, \ldots, x_k)$, which are consistent with the current assignment to $(t_1, \ldots, t_{k-1})$, in total $O(2^k)$ refinement steps are required to determine the truth of the QBF. Notice that, although the above formula can be directly solved by preprocessing using block clause elimination [2], sophistication can be imposed on the formula to prevent preprocessing taking any effect while keeping RAREQS and DEPQBF inefficient. This example illustrates the potential usefulness of circuit information.

Notice that when circuit information is used to reconstruct partial duality [9], the information about quantification levels of variables has to be taken into account. Let $\mathcal{C}$ be the circuit constructed from a QBF Φ. Then $\mathcal{C}$ can be used for solution learning if

1. all defined variables are existential,
2. $\max_{y \in sup(x)}\{dlev(y)\} \leq lev(x)$ for each defined variable x, and
3. $\mathcal{C}$ is acyclic.

In contrast to the three conditions used in [9] for definition extraction, the above three conditions are used for examining a circuit's legitimacy for solution learning. Prior work [9] extracts definitions with respect to a strict quantification order of the prefix (even for variables of the same quantification level) and thus may miss definitions that only present in a different order. In contract to [9], we first find all possible definition candidates without imposing any variable orders and then construct a circuit by adding definitions from the candidates. Therefore, our method may construct circuits not obtainable previously.

If the constructed circuit $\mathcal{C}$ of a QBF is cyclic, we rewrite the QBF with respect to a feedback vertex set of $\mathcal{C}$ to break all cycles while maintaining the equisatisfiability between the original and modified QBFs. Because a variable in a cycle is both a support variable and a defined variable, we instantiate each variable x in the feedback vertex set by introducing a fresh new variable x' such that the roles of support and defined variables are separated. In addition, the equivalence constraint $(x \leftrightarrow x')$ is added to the matrix to assert equisatisfiability. The following example illustrates the QBF rewriting.

Example 2. Consider the QBF

$$\forall u \exists a, b.(a \leftrightarrow (b \wedge u))(b \leftrightarrow (a \wedge u)).$$

A cyclic circuit $(a \leftrightarrow (b \wedge u))(b \leftrightarrow (a \wedge u))$ can be extracted. To make it acyclic, we modify the original matrix clauses. A fresh new variable b' is inserted to the quantification block of level $lev(b)$, and the equivalence relation $(b \leftrightarrow b')$ is added to the original matrix clauses. The QBF after rewriting becomes

$$\forall u \exists a, b, b'.(a \leftrightarrow (b \wedge u))(b' \leftrightarrow (a \wedge u))(b \leftrightarrow b'),$$

and the clause set of the constructed circuit $\mathcal{C}$ is $(a \leftrightarrow (b \wedge u))(b' \leftrightarrow (a \wedge u))$.

Given a QBF $\Phi = Q_1 X_1, \ldots, Q_n X_n.\varphi$ and its constructed circuit $\mathcal{C}$, we obtain an initial set of cubes for QBF solving as follows. Let $\varphi_{\mathcal{C}} \subseteq \varphi$ be the circuit clauses of $\mathcal{C}$. We generate a new CNF formula $\varphi_{\mathcal{C}}^+$ by replacing all of the defined variables V in $\varphi_{\mathcal{C}}$ with their respective fresh variables. Let V^+ be the set of the introduced fresh variables. Let Φ_{pfx}^+ be the new quantification prefix $Q_1 X_1, \ldots, Q_n X_n, \forall V^+$. Then let $\sigma_\forall$ be $\varphi_{\mathcal{C}}^+$ in Line 1 of the algorithms in Figures 1 and 2 as the initial cubes.

Example 3. Continue Example 2. New variables a_u and b'_u are introduced for variables a and b', respectively. The new quantification prefix, the initial $\sigma_\exists$, and the initial $\sigma_\forall$ are

$$\Phi_{\text{pfx}}^+ = \forall u \exists a, b, b' \forall a_u, b'_u,$$
$$\sigma_\exists = (a \leftrightarrow (b \wedge u))(b' \leftrightarrow (a \wedge u))(b \leftrightarrow b'),$$
$$\sigma_\forall = \varphi_{\mathcal{C}}^+ = (a_u \leftrightarrow (b \wedge u))(b'_u \leftrightarrow (a_u \wedge u)).$$

The following proposition shows the soundness of adding $\varphi_{\mathcal{C}}^+$ as initial cubes.

Proposition 2 ([9]). *Given a QBF $\Phi = Q_1 X_1, \ldots, Q_n X_n.\varphi$ and its constructed circuit $\mathcal{C}$, let $\Phi^+ = Q_1 X_1, \ldots, Q_n X_n, \forall V^+.(\varphi \vee \neg \varphi_{\mathcal{C}}^+)$, where V^+ is the set of the fresh variables corresponding to the defined variables V in $\varphi_{\mathcal{C}}$. Then Φ and Φ^+ are equisatisfiable.*

In addition, circuit information gives the flexibility to change the quantification levels of defined variables as stated in the following proposition.

Proposition 3. *Given a QBF $\Phi = Q_1 X_1, \ldots, Q_n X_n.\varphi$ and its constructed circuit $\mathcal{C}$, let v be a defined variable with $dlev(v) = i$ for $i \leq n$ and $lev(v) = j$ for $i \leq j \leq n$. Let v be repositioned in Φ_{pfx} to any quantification level k for $i \leq k \leq n$ and $Q_k = \exists$; let Φ'_{pfx} be the new prefix. Then the QBFs Φ and $\Phi'_{\text{pfx}}.\varphi$ are equisatisfiable.*

In our algorithm, solution learning takes place in Line 23 of Figure 1 and in Line 21 of Figure 2 when $res = false$, i.e., $\sigma_\forall$ is unsatisfiable under the current assignment α. To perform solution learning, we obtain the *learning level* with respect to α by analyzing conflicting unit assumptions on α in incremental SAT solving of $\sigma_\forall$. The *learning level* of α is $q = max_{l \in L}\{lev(l)\}$ for $L = \{l \in \alpha| var(l) \in X_\forall\}$. Note that $Q_q = \forall$ and $1 \leq q \leq n$.

Given a QBF $\Phi = Q_1 X_1, \ldots, Q_n X_n.\varphi$ with its constructed circuit $\mathcal{C}$, to perform solution learning with i being the learning level we define the circuit

C_i at the ith quantification level to be the induced sub-circuit of C whose input variables are of quantification levels less than or equal to i. Assume that $\Phi|_\alpha$ is true under assignment $\alpha \in [\![X_1 \cup \ldots \cup X_i]\!]$. We partition the clauses of φ into three subsets: Let φ_1 be the set of clauses of C_i; let φ_2 be the set of clauses that are not in φ_1 and are satisfied by some literal l where l is in α or $var(l)$ is an output variable of C_i. Let φ_3 be the rest of the clauses of φ. For solution learning, we generate a hitting set α' for φ_2, where α' only contains literals whose variables are in $X_1 \cup \ldots \cup X_i \cup D_i$ for D_i being the set of the output variables of C_i. Let α'^+ be derived from α' with the output variables of C_i being replaced by their corresponding variables in V^+. Then $\sigma_\forall$ can be augmented with $\neg\alpha'^+$ for solution learning.

The following proposition states that solution learning by adding $\neg\alpha'^+$ to $\sigma_\forall$ in our algorithm is sound by showing that $\Phi|_\mu$ is true for any assignment μ blocked by $(\varphi_C^+ \wedge \neg\alpha'^+)$, i.e., $(\varphi_C^+ \wedge \neg\alpha'^+)|_\mu$ is unsatisfiable. Note that in our algorithm (in Line 23 of Figure 1 or Line 21 of Figure 2), the satisfiability of $(\varphi_C^+ \wedge \neg\alpha'^+)|_\mu$ is checked only if the current quantification level ℓ is universal.

Proposition 4. *Given a QBF $\Phi = Q_1 X_1, \ldots, Q_n X_n . \varphi$ and its constructed circuit C, let the current quantification level $\ell = k$ with $Q_k = \forall$. For any assignment $\mu \in [\![X_1 \cup \ldots \cup X_k]\!]$, the QBF $\Phi|_\mu$ is true if $(\varphi_C^+ \wedge \neg\alpha'^+)|_\mu$ is false for any learned hitting set α'^+ as specified above.*

In our algorithm, we actually check the satisfiability of $\sigma_\forall|_\mu$, which contains φ_C^+ and all previously learned solutions. In the following, we prove the soundness of solution learning with circuit information by showing that for any assignment μ blocked by $\sigma_\forall$, QBF $\Phi|_\mu$ is true.

Lemma 1. *Given two cubes μ_1 and μ_2, let cube μ_3 be the resolvent[2] of μ_1 and μ_2 if it exists. If $\Phi|_{\mu_1}$ and $\Phi|_{\mu_2}$ are both true, then $\Phi|_{\mu_3}$ is true.*

Proposition 5. *Let the current level be $\ell = k$ with $Q_k = \forall$ in our algorithm. For any assignment $\mu \in [\![X_1 \cup \cdots \cup X_k]\!]$ if $\sigma_\forall|_\mu$ is false, then $\Phi|_\mu$ is true.*

The following proposition asserts that by adding $\neg\alpha'^+$ to $\sigma_\forall$, the current assignment α will be blocked by $\sigma_\forall$.

Proposition 6. *Let α be the current assignment, α' be the hitting set as specified above, and i be the learning level when α'^+ is learned. Then, $(\varphi_C^+ \wedge \neg\alpha'^+)|_\alpha$ is false.*

As a consequence, our solution learning is complete.

4.3 Conflict Learning

We exploit the following lemma for conflict detection (in Line 7 of Fig 1 or Fig 2).

---

[2] Given two cubes $c_1 = c_1' \wedge l$ and $c_2 = c_2' \wedge \neg l$, if $c_1 = c_1' \wedge l$, $c_2 = c_2' \wedge \neg l$ for some literal l, and $\{x | x \in c_1' \wedge \neg x \in c_2'\}$ is empty, then the *resolvent* of c_1 and c_2 exists and is defined as $\{x | x \in c_1' \vee x \in c_2'\}$.

Lemma 2. *Given a QBF $\Phi = Q_1 X_1, \ldots, Q_n X_n.\varphi$ with $Q_i = \forall$ for some $i \leq n-1$, assume $\varphi|_{\alpha,\beta}$ is unsatisfiable under assignments $\alpha \in [\![X_1 \cup \ldots \cup X_{i-1}]\!]$ and $\beta \in [\![X_i]\!]$. Then QBF $\Phi|_\alpha$ is false.*

If $\varphi|_{\alpha,\beta}$ is unsatisfiable under assignments $\alpha \in [\![X_1 \cup \ldots \cup X_{i-1}]\!]$ and $\beta \in [\![X_i]\!]$, our algorithm can detect and do conflict learning when the current quantification level $\ell = i + 1$. In contrast, RAREQS does not use SAT solving to check the satisfiability of $\varphi|_{\alpha,\beta}$. It may keep deciding the unassigned variables at the currently unassigned outermost quantification level and can possibly use SAT solving only when the innermost quantification level is under decision. Moreover, RAREQS can only backtrack at most two quantification levels. Therefore, RAREQS may potentially take more time to conclude the unsatisfiability of $\Phi|_{\alpha,\beta}$ than our algorithm.

Furthermore, when our algorithm detects that $\varphi|_{\alpha,\beta}$ is unsatisfiable with the current quantification level $\ell = i + 1$, the QBF is *extended* such that α will falsify the new QBF matrix by procedure *ConflictLearn* based on Proposition 7. The following definitions are used in Proposition 7. For a CNF formula φ unsatisfiable under an assignment α, the *unsat core* ψ of φ with respect to α is referred to as a (minimal) subset of the clauses in φ and ψ is unsatisfiable under α. Also given a formula φ, we denote the new formula of φ with variables Z in φ being substituted with Z' as $\varphi[Z \leftarrow Z']$.

Proposition 7. *Given a QBF $\Phi = Q_1 X_1, \ldots, Q_n X_n.\varphi$ with $Q_i = \forall$ for some $i \leq n-1$, assume $\varphi|_{\alpha,\beta}$ is unsatisfiable under assignments $\alpha \in [\![X_1 \cup \ldots \cup X_{i-1}]\!]$ and $\beta \in [\![X_i]\!]$. Let ψ be an unsat core of φ with respect to the assignment α, β and let $\psi' = \psi|_\beta[Z \leftarrow Z']$, where Z includes all variables inner to X_i and Z' are fresh new variables for Z's substitution. Then formula $(\varphi \wedge \psi')$ is unsatisfiable under α.*

Continuing Proposition 7, the following proposition asserts the equisatisfiability between an QBF and its extended QBF under conflict learning.

Proposition 8. *Let $\Phi' = Q_1 X_1, \ldots, Q_{i-1} X_{i-1} \cup Z', Q_i X_i, \ldots, Q_n X_n.(\varphi \wedge \psi')$ be extended from $\Phi = Q_1 X_1, \ldots, Q_n X_n.\varphi$ with $Q_i = \forall$ through conflict learning under (α, β) for $\alpha \in [\![X_1 \cup \ldots \cup X_{i-1}]\!]$ and $\beta \in [\![X_i]\!]$. Then Φ' is false if and only if Φ is false.*

The following proposition states the completeness of our conflict learning.

Proposition 9. *Given a false QBF $\Phi = Q_1 X_1, \ldots, Q_n X_n.\varphi$, our algorithm will terminate and generate an extended QBF Φ' of Φ such that Φ'_{mtx} is unsatisfiable.*

In the worst case, the number of clauses in the expended QBF may double for each formula expansion. Nevertheless to reduce memory consumption, we can delete learned clauses by the following proposition.

Proposition 10. *Assume that a conflict learning occurs at quantification level i and our procedure backtracks to level i' for some $i' < i$. If another conflict or solution learning occurs later at level j and our procedure first backtracks to level*

j' for some $j' < i'$, then the learned clauses generated in the conflict learning at level i can be removed without affecting soundness and completeness of our algorithm.

5 Related Work

RAREQS is an expansion-based solver that employs an abstraction-refinement scheme for recursive QBF game solving. An abstract game between the existential and universal players is iteratively refined until the winner is determined at the first quantification level. The abstract game consists of a collection of QBFs and searching a winning move amounts to solving the abstract game QBFs recursively. This recursion may incur more QBFs being generated and thus substantial computation overhead. To alleviate this deficiency, RAREQS avoids maintaining all the learned QBFs by setting a quantification level limit and recompute QBFs when needed.

In contrast to RAREQS, our method also takes a QBF as a game, but we simply use non-quantified formulas rather than QBFs to block unsuccessful earlier moves. By this way, we avoid recomputing a winning move from solving QBFs. With an activity analysis heuristic, we manage to remove inactive learned formulas reduce memory consumption. In addition, we incorporate circuit information to facilitate solution learning.

DEPQBF is a resolution-based solver that implements a QDPLL algorithm with conflict and solution driven learnings. It assigns variables one at a time according to the prefix structure. It maintains two managers, one for clauses and the other for cubes, while both clauses and cubes are combined for joint implication propagation. For conflict learning (of both clause and cube managers), a UIP-based backtracking is applied. For solution learning, a hitting set of literals with respect to the original set of clauses (i.e., the matrix of the underlying QBF) is derived as a cube for solution blocking. OoQ extends DEPQBF by detecting circuit information from the matrix as initial cubes for solution learning. Also the identified circuit clauses are excluded from the original set of clauses in hitting set computation.

In contrast to the conflict and solution learnings of DEPQBF, ours are mainly expansion-based. We assign multiple variables in a quantification block at a time. Our hitting set computation is localized with respect to a subset of the original clauses (excluding circuit clauses) depending on the current quantification level. This localization helps strengthen solution learning. However, our current implementation does not combine clauses and cubes for joint implication propagation and may be improved in the future. In contrast to OoQ, we generalize definition detection to allow cyclic circuits and make simplification using circuit information workable under our expansion-based solving. However, our current implementation does not consider partial definition [17] and may be improved.

Table 1. Statistics of solved instances.

Track	total	solved by some	RAREQS	DEPQBF	OOQ	QELL
QBFLib	276	146	78	88	103	106
Preprocessing	243	139	104	99	73	85
Application	735	483	403	273	246	426
total	1254	768	585	460	422	617

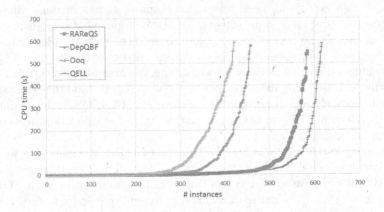

Fig. 3. Performance comparison of QBF solvers (# solved instances vs. runtime).

6   Experimental Results

Our proposed algorithm, named QELL, was implemented in the C++ language using MINISAT [8]. The experiments were conducted on a Linux machine with a Xeon 2.53 GHz CPU and 48 GB RAM. The set of benchmark instances of QBF Gallery 2014 [18] were taken for experiments. It consists of three tracks: *QBFLib, preprocessing,* and *application.* We only compared solvers closest to ours, including RAREQS, DEPQBF, and OOQ under their default settings. For QELL, circuit detection was enabled by default, an upper bound of 100 clauses per ALLSAT enumeration of solution learning was set, and aggressive backtrack were used. The benchmarks were taken as they are without being further preprocessed. A CPU time limit of 600 seconds is set for each instance.

Table 1 shows the statistics. Columns 2-7 show for each track the total number of instances, the number of instances solved by any of the solvers, the number of instances solved by RAREQS, DEPQBF, OOQ, and QELL, respectively. Figure 3 plots the number of solved instances from all three tracks vs. the CPU time, by which the instances are sorted in an ascending order for each solver. Overall QELL solves more instances than the other solvers.

We note that the formulas of the first track mostly contain circuit information, which makes OOQ and QELL effective. In particular, QELL outperforms other solvers in families nusmv.tcas and vis.prodcell. For families s298, s713, s820, and cmu, QELL and OOQ both outperform RAREQS and DEPQBF. In contrast, DEPQBF and OOQ perform much better than the other two solvers

Table 2. QELL vs. another solver in uniquely and commonly solved instances.

Track	Description	RAREQS	DEPQBF	OOQ
QBFLib	# instances only solved by competing solver	6	30	32
	# instances only solved by QELL	34	48	35
	# instances solved by both solvers	72	58	71
Preprocessing	# instances only solved by competing solver	23	31	21
	# instances only solved by QELL	4	17	33
	# instances solved by both solvers	81	68	52
Application	# instances only solved by competing solver	28	37	8
	# instances only solved by QELL	51	190	188
	# instances solved by both solvers	375	236	234
total	# instances only solved by competing solver	57	98	61
	# instances only solved by QELL	89	255	256
	# instances solved by both solvers	528	362	357

Table 3. QELL with and without circuit information.

Track	QELL w/ ckt	QELL w/o ckt
QBFLib	106	72
Preprocessing	85	84
Application	426	421
Total solved	617	577

for families `umbrella`, `w4-umbrella`, `biu.mv.xl`, and `core1108`. For the second track, only a few instances have circuit information detected by OOQ and QELL. (OOQ detected more than QELL in these instances due to its capability of recovering from partial circuit definitions.) Circuit information does not help OOQ and QELL, and RAREQS and DEPQBF tend to be more effective in this track. For the third track, RAREQS and QELL outperforms DEPQBF and OOQ. Moreover, QELL was able to solve more instances than RAREQS.

Table 2 compares QELL against every other solver in terms of the numbers of uniquely and commonly solved instances. The numbers for the first track reveal that the solving behavior of QELL is closer to RAREQS (due to expansion-based solving) and OOQ (due to exploitation of circuit information) than to DEPQBF. For the second and third tracks, without much circuit information QELL is closer to RAREQS than the other two solvers. We note that there are many cases that can only be solved by DEPQBF but not QELL. It suggests the potential benefit to combine resolution-based learning in QELL.

To see the effect of circuit information in QELL, Table 3 shows the number of instances solved by QELL with and without circuit information. QELL benefits from circuit information for instances of the first track, the only track that QELL can reconstruct circuit information for almost all instances. As expected, circuit information is in general helpful (though not always).

7   Conclusion and Future Work

We have proposed a new approach to expansion-based QBF solving. For solution learning, we have adopted ALLSAT computation to avoid recursive expansion.

For conflict learning, we have used localized recursive expansion to reduce formula growth. In addition, we have exploited circuit information for solution learning as an integral part of our levelized SAT solving scheme. Experiments have demonstrated the feasibility of our methods. For future work, we plan to explore different implementation choices for optimization.

References

1. Benedetti, M.: Evaluating QBFs via symbolic skolemization. In: Baader, F., Voronkov, A. (eds.) LPAR 2004. LNCS (LNAI), vol. 3452, pp. 285–300. Springer, Heidelberg (2005)
2. Biere, A., Lonsing, F., Seidl, M.: Blocked clause elimination for QBF. In: Bjørner, N., Sofronie-Stokkermans, V. (eds.) CADE 2011. LNCS, vol. 6803, pp. 101–115. Springer, Heidelberg (2011)
3. Beyersdorff, O., Chew, L., Janota, M.: Proof complexity of resolution-base QBF calculi. In: Proc. Symposium on Theoretical Aspects of Computer Science (STACS), pp. 76–89 (2015)
4. Balabanov, V., Jiang, J.-H.R.: Unified QBF Certification and its Applications. Formal Methods in System Design 41(1), 45–65 (2012)
5. Balabanov, V., Widl, M., Jiang, J.-H.R.: QBF resolution systems and their proof complexities. In: Sinz, C., Egly, U. (eds.) SAT 2014. LNCS, vol. 8561, pp. 154–169. Springer, Heidelberg (2014)
6. Cadoli, M., Schaerf, M., Giovanardi, A., Giovanardi, M.: An Algorithm to Evaluate Quantified Boolean Formulae and Its Experimental Evaluation. Journal of Automated Reasoning 28(2), 101–142 (2002)
7. Dershowitz, N., Hanna, Z., Katz, J.: Bounded model checking with QBF. In: Bacchus, F., Walsh, T. (eds.) SAT 2005. LNCS, vol. 3569, pp. 408–414. Springer, Heidelberg (2005)
8. Eén, N., Sörensson, N.: An extensible SAT-solver. In: Giunchiglia, E., Tacchella, A. (eds.) SAT 2003. LNCS, vol. 2919, pp. 502–518. Springer, Heidelberg (2004)
9. Goultiaeva, A., Bacchus, F.: Recovering and utilizing partial duality in QBF. In: Järvisalo, M., Van Gelder, A. (eds.) SAT 2013. LNCS, vol. 7962, pp. 83–99. Springer, Heidelberg (2013)
10. Giunchiglia, E., Narizzano, M., Tacchella, A.: Clause-Term Resolution and Learning in Quantified Boolean Logic Satisfiability. Artificial Intelligence Research 26, 371–416 (2006)
11. Janota, M., Klieber, W., Marques-Silva, J., Clarke, E.: Solving QBF with counterexample guided refinement. In: Cimatti, A., Sebastiani, R. (eds.) SAT 2012. LNCS, vol. 7317, pp. 114–128. Springer, Heidelberg (2012)
12. Jiang, J.-H.R., Lin, H.-P., Hung, W.-L.: Interpolating functions from large boolean relations. In: Proc. Int'l Conf. on Computer-Aided Design (ICCAD), pp. 779–784 (2009)
13. Janota, M., Marques-Silva, J.: Abstraction-based algorithm for 2QBF. In: Sakallah, K.A., Simon, L. (eds.) SAT 2011. LNCS, vol. 6695, pp. 230–244. Springer, Heidelberg (2011)
14. Büning, H.K., Karpinski, M., Flögel, A.: Resolution for Quantified Boolean Formulas. Information and Computation 117(1), 12–18 (1995)
15. Klieber, W., Sapra, S., Gao, S., Clarke, E.: A non-prenex, non-clausal QBF solver with game-state learning. In: Strichman, O., Szeider, S. (eds.) SAT 2010. LNCS, vol. 6175, pp. 128–142. Springer, Heidelberg (2010)

16. Lonsing, F., Biere, A.: DepQBF: A Dependency Aware QBF Solver (system description). Journal on Satisfiability, Boolean Modeling and Computation **7**, 71–76 (2010)

17. Plaisted, D., Greenbaum, S.: A structure-preserving clause form translation. J. Symbolic Computation **2**, 293–304 (1986)

18. QBF Gallery (2014). http://qbf.satisfiability.org/gallery/

19. Rintanen, J.: Asymptotically optimal encodings of conformant planning in QBF. In: National Conference on Artificial Intelligence (AAAI), pp. 1045–1050 (2007)

20. The Quantified Boolean Formulas Satisfiability Library. http://www.qbflib.org/

21. Samulowitz, H., Bacchus, F.: Using SAT in QBF. In: van Beek, P. (ed.) CP 2005. LNCS, vol. 3709, pp. 578–592. Springer, Heidelberg (2005)

22. Tseitin, G.: On the complexity of derivation in propositional calculus. In: Studies in Constructive Mathematics and Mathematical Logic, pp. 466–483 (1970)

23. Yu, Y., Subramanyan, P., Tsiskaridze, N., Malik, S.: All-SAT using minimal blocking clauses. In: Proc. Int'l Conf. on VLSI Design, pp. 86–91 (2014)

24. Zhang, L.: Solving QBF by combining conjunctive and disjunctive normal forms. In: Proc. National Conference on Artificial Intelligence (AAAI), pp. 143–150 (2006)

25. Zhang, L., Malik, S.: Conflict driven learning in a quantified boolean satisfiability solver. In: Proc. Int'l Conf. on Computer-Aided Design (ICCAD), pp. 442–449 (2002)

SMT-RAT: An Open Source C++ Toolbox for Strategic and Parallel SMT Solving

Florian Corzilius[✉], Gereon Kremer, Sebastian Junges,
Stefan Schupp, and Erika Ábrahám

RWTH Aachen University, Aachen, Germany
corzilius@cs.rwth-aachen.de

Abstract. During the last decade, popular SMT solvers have been extended step-by-step with a wide range of decision procedures for different theories. Some SMT solvers also support the user-defined tuning and combination of such procedures, typically via command-line options. However, configuring solvers this way is a tedious task with restricted options.

In this paper we present our modular and extensible C++ library SMT-RAT, which offers numerous parameterized procedure modules for different logics. These modules can be configured and combined into an SMT solver using a comprehensible whilst powerful strategy, which can be specified via a graphical user interface. This makes it easier to construct a solver which is tuned for a specific set of problem instances. Compared to a previous version, we have extended our library with a number of new modules and support for parallelization in strategies. An additional contribution is our thread-safe and generic C++ library CArL, offering efficient data structures and basic operations for real arithmetic, which can be used for the fast implementation of new theory-solving procedures.

1 Introduction

The *satisfiability problem* (SAT) poses the question whether a given propositional formula has a solution. *Satisfiability-modulo-theories* (SMT) tackles its natural extension, where we allow *theory constraints* in place of propositions. Lazy SMT solving [33] uses a *SAT solver* to find solutions of the *Boolean skeleton* of an *SMT formula* and invokes dedicated *theory solvers* to check the consistency in the underlying theory. Whereas full lazy approaches search for a complete Boolean solution before invoking theory solvers, *less lazy* techniques consult them more frequently. This cooperation highly benefits from an *SMT-compliant* theory solver, which (1) works *incrementally*, i.e., it should be able to exploit results from previous consistency checks; (2) it can *backtrack* according to the SAT solving; (3) for inconsistent constraint sets, it should be able to find an *infeasible subset* as explanation.

Most activities in the area of SMT solving focus on theories such as bit vectors (BV), uninterpreted functions (UF) or linear arithmetic over the reals (LRA)

© Springer International Publishing Switzerland 2015
M. Heule and S. Weaver (Eds.): SAT 2015, LNCS 9340, pp. 360–368, 2015.
DOI: 10.1007/978-3-319-24318-4_26

and integers (LIA) resulting in the SMT solvers, e. g., CVC4 [3], MathSAT5 [8], Yices2 [15] or OpenSMT2 [6]. However, less activity can be observed for SMT solvers for (the existential fragment of) *non-linear real arithmetic* (NRA): besides some incomplete solvers like MiniSmt [38] and iSAT3 [17,32], we are only aware of one SMT solver Z3 [24,28] that is *complete* for NRA. Even fewer SMT solvers are available for (the existential fragment of) *non-linear integer arithmetic* (NIA), which is undecidable in general. To the best of our knowledge, only Z3 and the SMT solving spin-off of Aprove [9] can tackle this theory.

One of the most widely used decision procedures for NRA is the *cylindrical algebraic decomposition* (CAD) method [10]. Other well-known methods use, e.g., *Gröbner bases* (GB) [35] or the *realization of sign conditions* [4]. Also some incomplete methods based on, e.g., *interval constraint propagation* (ICP) [17] or the *virtual substitution* (VS) [37] can handle significant fragments. However, the exponential worst-case complexity of solving NRA formulas [22,36] makes it challenging to develop practically feasible solutions. Embedding the above NRA decision procedures in SMT solvers as theory solvers is a promising symbiosis. Highly efficient SAT solvers can handle the Boolean problem structure and learn from previous (SAT and theory) conflicts. The expensive theory consistency checks then only concern conjunctions of theory constraints.

Available implementations of the above decision procedures are seldom available as *libraries*, and even if they are, they are not SMT compliant. Thus, for an SMT embedding, these mathematically complex decision procedures had to be adapted and extended before an SMT-compliant implementation could be realized. For the implementation, an *efficient library for basic computations with polynomials* was needed, which, if we want to have the door open for parallelization, must be additionally *thread-safe*. Furthermore, on a given problem instance there might be significant differences in the running times of different theory solvers. Therefore, we aim at their *strategic combination* [29] to increase usability.

We have developed the C++ library SMT-RAT containing a variety of *modules* implementing SMT-compliant solving procedures. The modular design of SMT-RAT facilitates an easy extension by further solving procedures. Modules share a common interface allowing their combination according to a user-defined strategy resulting in an SMT solver. Currently, SMT-RAT can solve problems of (the quantifier-free fragments of) LRA, LIA, NRA and NIA. Compared to the previous version of SMT-RAT [12], (1) we have extended and optimized the VS module, the GB module (can now handle inequalities and simplify formulas), and the CAD module (can now handle arbitrary instead of only univariate polynomials); (2) we have implemented a Simplex module [15], an ICP module [18], a module embedding a SAT solver and a module simplifying polynomial constraints using non-trivial factorization and sum-of-squares decomposition; (3) we have implemented a general branch-and-bound method for finding integer solutions with NRA modules, where the splitting decisions are lifted to the SAT level; (4) we have extended SMT-RAT to support strategies, which compose procedures such that they run in parallel on multiple cores and implemented an easy-to-grasp graphical user interface for the construction of such a strategy; (5) we have extended the module interfaces to support lemma exchange and lightweight invocation, where it is allowed to avoid hard obstacles during solving at the price of possibly not finding a conclusive answer.

2 System Architecture

2.1 Data Structures and Basic Procedures: CArL [26]

The current version of SMT-RAT integrates custom-designed data structures for SMT formulas and basic functions to manipulate them, bundled in the library CArL, which has also been successfully used in the tool Prophesy [13].

While there exist C++ libraries for the manipulation of polynomials such as CoCoA[1] and GiNaC [5], these libraries share some common deficits. First of all, they lack customization possibilities and are usually tied to one fixed *representation of numbers*. Secondly, the libraries are often not flexible when it comes to manipulation of *variable* (and polynomial) *orderings*, which is essential for efficient implementations of a CAD or a GB procedure. Thirdly, the libraries are usually not thread-safe, which precludes the design of parallel solvers.

In CArL, the data structure for *SMT formulas* is a directed acyclic graph, with Boolean operators as inner nodes and Boolean variables or theory constraints, e.g., polynomial inequalities, as leafs. Essential simplifications and normalizations [14] are applied by default and identical formulas are stored only once.

Polynomials are represented by default as a sum of terms. We mark leading and constant terms and sort all terms only on demand. The data structure is templated in several ways. Amongst others, we can use rational numbers, native numbers, intervals and polynomials as coefficients. Furthermore, we can use different orderings and store additional information with the polynomials with minimal overhead by utilizing policy templates. Besides, CArL supports univariate representations of multivariate polynomials, which is essential for, i. a., the CAD.

Variables are represented by bit vectors, encoding their identity, their domain and their rank (for support of fast custom-ordering of variables). Additional information is stored in a central pool. For the *representation of rational numbers*, we support gmp [20] (thread-safe) and cln [21] (faster single-threaded). Algebraic numbers are represented by the interval-isolated root of a univariate polynomial. Intervals in CArL are an extension of boost intervals also allowing open bounds.

Besides standard arithmetic operations, CArL includes the required procedures for CAD, including Sturm sequences and root isolation, and a variant of the Buchberger algorithm to compute Gröbner bases. The implemented methods are specifically tailored towards SMT compliance.

2.2 Interfaces and Strategic Compositions of Procedures: SMT-RAT [34]

Based on CArL's data structures and basic functions, a rich set of SMT-compliant implementations of NRA/NIA procedures is provided by SMT-RAT. Each procedure

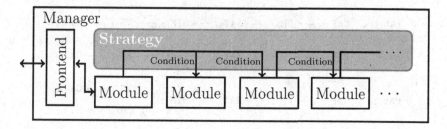

Fig. 1. A snapshot of an SMT-RAT composition of an SMT solver.

is encapsulated in a *module*, which fixes a common interface. Modules can be composed to a solver according to a user-defined *strategy*. The *manager* class provides the API, including the parsing of an SMT-LIBv2 input file, and a manager instance maintains the allocation of *solving tasks* to modules according to the strategy. An overview is given in Figure 1.

Modules. Each module m has an initially empty *set of received formulas* $C_{rcv}(m)$. We can manipulate $C_{rcv}(m)$ by adding (removing) formulas φ to (from) it with `add(`φ`)` (`remove(`φ`)`). The main function of a module is `check(bool full)`, which either decides whether the conjunction of the received formulas in $C_{rcv}(m)$ is satisfiable or not, returning `sat` or `unsat`, respectively, or returns `unknown`. If the function's argument `full` is set to `false`, the underlying procedure of m is allowed to omit hard obstacles during solving at the cost of returning `unknown` in more cases. Usually, $C_{rcv}(m)$ is only slightly changed between two consecutive `check` calls, hence, the solver's performance can be significantly improved if a module works incrementally and supports backtracking. In case m determines the unsatisfiability of $C_{rcv}(m)$, it can return an *infeasible subset* $C_{inf}(m) \subseteq C_{rcv}(m)$. Moreover, a module can specify *lemmas*, which are valid formulas. They encapsulate information which can be extracted from a module's internal state and propagated among other modules. Furthermore, a module itself can ask other modules for the satisfiability of its *set of passed formulas* denoted by $C_{pas}(m)$, if it invokes the procedure `runBackends(bool full)` (controlled by the manager). It thereby delegates work to modules that may be more suitable for the (sub-) problems in $C_{pas}(m)$.

Strategy. SMT-RAT supports user-defined *strategies* for the composition of modules. A graphical user interface can be used to specify strategies as directed trees $T := (V, E)$ with a set V of modules as nodes and the transitions $E \subseteq V \times \Omega \times \Sigma \times V$, with Ω being a set of *conditions* and Σ being a set of *priority values*. A condition is an arbitrary Boolean combination of formula properties, such as propositions about the Boolean structure of the formula, e.g., whether it is in conjunctive normal form (CNF), about the constraints, e.g., whether it contains equations, or about the polynomials, e.g., whether they are linear. Furthermore, each edge carries a unique priority value from $\Sigma = \{1, \ldots, |E|\}$.

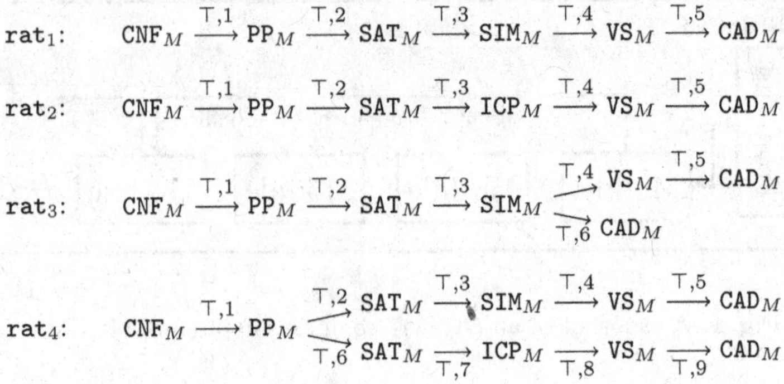

$$\mathbf{rat_1}: \quad \mathrm{CNF}_M \xrightarrow{\mathsf{T},1} \mathrm{PP}_M \xrightarrow{\mathsf{T},2} \mathrm{SAT}_M \xrightarrow{\mathsf{T},3} \mathrm{SIM}_M \xrightarrow{\mathsf{T},4} \mathrm{VS}_M \xrightarrow{\mathsf{T},5} \mathrm{CAD}_M$$

$$\mathbf{rat_2}: \quad \mathrm{CNF}_M \xrightarrow{\mathsf{T},1} \mathrm{PP}_M \xrightarrow{\mathsf{T},2} \mathrm{SAT}_M \xrightarrow{\mathsf{T},3} \mathrm{ICP}_M \xrightarrow{\mathsf{T},4} \mathrm{VS}_M \xrightarrow{\mathsf{T},5} \mathrm{CAD}_M$$

$$\mathbf{rat_3}: \quad \mathrm{CNF}_M \xrightarrow{\mathsf{T},1} \mathrm{PP}_M \xrightarrow{\mathsf{T},2} \mathrm{SAT}_M \xrightarrow{\mathsf{T},3} \mathrm{SIM}_M \begin{array}{c} \xrightarrow{\mathsf{T},4} \mathrm{VS}_M \xrightarrow{\mathsf{T},5} \mathrm{CAD}_M \\ \xrightarrow{\mathsf{T},6} \mathrm{CAD}_M \end{array}$$

$$\mathbf{rat_4}: \quad \mathrm{CNF}_M \xrightarrow{\mathsf{T},1} \mathrm{PP}_M \begin{array}{c} \xrightarrow{\mathsf{T},2} \mathrm{SAT}_M \xrightarrow{\mathsf{T},3} \mathrm{SIM}_M \xrightarrow{\mathsf{T},4} \mathrm{VS}_M \xrightarrow{\mathsf{T},5} \mathrm{CAD}_M \\ \xrightarrow{\mathsf{T},6} \mathrm{SAT}_M \xrightarrow{\mathsf{T},7} \mathrm{ICP}_M \xrightarrow{\mathsf{T},8} \mathrm{VS}_M \xrightarrow{\mathsf{T},9} \mathrm{CAD}_M \end{array}$$

Fig. 2. Example strategies with SMT-RAT ($\mathsf{T} \mathrel{\hat{=}}$ no condition).

Manager. The *manager* holds the strategy $T = (V, E)$ and the SMT solver's input formula C_{input}. Initially, the manager calls the method check of the module m_r, being the root of T, with $C_{rcv}(m_r) = C_{input}$. Whenever a module $m \in V$ calls runBackends, the manager adds a *solving task* (σ, m, m') to its priority queue Q of solving tasks (ordered by the priority value), if there exists an edge $(m, \omega, \sigma, m') \in E$ such that ω holds for $C_{pas}(m)$. If a processor p on the machine on which SMT-RAT is executed is available, the first solving task of Q is assigned to p and popped from Q. The manager thereby starts check of m' with $C_{rcv}(m') = C_{pas}(m)$ and passes the result (including infeasible subsets and lemmas) back to m, which can now benefit in its solving and reasoning process from this shared information. Note that a strategy-based composition of modules works incrementally and supports backtracking not just within one module but as a whole. Therefore, each module m stores the subsets of $C_{rcv}(m)$, which form the reasons for a passed formula being added. In order to exploit the incrementality of the modules, all backends executed in parallel terminate in a consistent state (instead of being killed), if one of them finds an answer.

Procedures implemented as modules. Usually, a SAT solver forms the heart of an SMT solver. In SMT-RAT, the module SAT_M abstracts $C_{rcv}(\mathrm{SAT}_M)$ to propositional logic and uses the efficient SAT solver minisat [16] to find a satisfying solution for the Boolean abstraction. It invokes runBackends where $C_{pas}(\mathrm{SAT}_M)$ contains the constraints abstracted by the assigned Boolean variables in a less-lazy fashion [33]. The module SIM_M implements the Simplex method equipped with branch-and-bound and cutting-plane procedures as presented in [15]. We apply it on the linear constraints of any conjunction of NRA/NIA constraints. For a conjunction of nonlinear constraints SMT-RAT provides the modules GB_M, VS_M and CAD_M, implementing GB [25], VS [11] and CAD [27] procedures, respectively. Moreover, the module ICP_M uses ICP similar as presented in [18], lifting splitting decisions and contraction lemmas to a preceding SAT_M and harnessing other modules for nonlinear conjunctions of constraints as backends. The

Table 1. Results in seconds (timeout = 200s) obtained on a 2.1 GHz AMD.

Benchmark (#examples)	Z3		rat$_1$		rat$_2$		rat$_3$		rat$_4$	
	solved	time	solved	time	solved	time	solved	time	solved	time
HONG (20)	50.0%	72.8	15.0%	< 1.0	100.0%	< 1.0	15.0%	< 1.0	100.0%	< 1.0
- sat	0	0.0	0	0.0	0	0.0	0	0.0	0	0.0
- unsat	10	72.8	3	< 1.0	20	< 1.0	3	< 1.0	20	< 1.0
KISSING (45)	68.9%	1155.9	17.8%	50.2	35.6%	375.9	28.9%	26.5	28.9%	54.4
- sat	31	1155.9	8	50.2	16	375.9	13	26.5	13	54.4
- unsat	0	0.0	0	0.0	0	0.0	0	0.0	0	0.0
METITARSKI (7713)	99.9%	370.5	92.7%	4964.3	92.8%	4658.3	93.2%	3974.8	95.6%	3109.4
- sat	5025	133.7	4766	2180.8	4740	2951.1	4802	1803.8	4815	2290.4
- unsat	2684	236.8	2385	2783.4	2418	1706.2	2388	2170.9	2560	819.0
KEYMAERA (421)	99.8%	11.5	97.6%	26.0	96.9%	17.0	96.4%	74.7	98.1%	25.3
- sat	0	0.0	0	0.0	0	0.0	0	0.0	0	0.0
- unsat	420	11.5	411	26.0	408	17.0	406	74.7	413	25.3
WITNESS (99)	21.2%	107.1	72.7%	2110.9	64.6%	332.2	21.2%	10.9	75.8%	937.9
- sat	4	75.3	55	2110.6	47	331.9	4	9.8	58	937.6
- unsat	17	31.8	17	< 1.0	17	< 1.0	17	1.1	17	< 1.0
APROVE (8829)	94.0%	12011.6	79.5%	5077.8	80.3%	6128.4	76.6%	10645	80.0%	3886.3
- sat	8014	11090.9	6965	5038.7	7038	5695.5	6698	10181.3	7009	3782.3
- unsat	284	920.7	50	39.1	56	432.9	68	463.6	58	104.0
CALYPTO (177)	98.9%	11.6	83.6%	123.3	78.0%	323.9	37.3%	402.1	85.3%	308.3
- sat	79	7.5	64	46.5	59	236.5	21	304.5	67	224.7
- unsat	96	4.1	84	76.7	79	87.4	45	97.7	84	83.6

module CNF$_M$ invokes `runBackends` on $C_{pas}(\text{CNF}_M)$ being a formula in CNF which is satisfiability-equivalent to $C_{rcv}(\text{CNF}_M)$. The module PP$_M$ performs some preprocessing based on factorizations and sum-of-square decompositions of polynomials.

3 Experimental Results and Future Work

We evaluated the four strategies specified in Figure 2 on the five NRA benchmark sets HONG [23], KISSING (both crafted and dimension dependent), METITARSKI [2], KEYMAERA [30], WITNESS [31] (generated by theorem proving, counterexample-guided synthesis and formal verification, respectively) and the two NIA benchmark sets APROVE [19] and CALYPTO [7] (generated by automated termination analysis and sequential equivalence checking, respectively). The first two strategies, rat$_1$ and rat$_2$, are sequential, using a nested combination of Simplex/ ICP, VS and CAD. The third strategy rat$_3$ extends the first one by applying CAD in parallel to the nested combination of VS and CAD. The last strategy rat$_4$ basically runs the first two strategies in parallel.

Table 1 shows the experimental results, which compare the four SMT-RAT strategies with the currently fastest SMT solver for these theories, Z3, showing that SMT-RAT is already competitive. We ran Z3 sequentially and in parallel and took the best of both real-time performances for each instance. The column "solved" shows the number of solved instances and the column "time" states the accumulated solving time not including timeouts. On WITNESS, SMT-RAT performs even better than Z3, as it benefits from the algebraic procedures being tuned for small variable domains as occurring in these examples. It also performs better on HONG, where it highly profits from the ICP module. Even though

rat_4 is the best SMT-RAT strategy overall, we observed that both parallel strategies perform worse than expected, which is due to CAD_M currently not always being able to terminate quickly with a consistent state when called in parallel. We want to extend SMT-RAT with further modules based on linearization, bit-blasting and further preprocessing. More experimental results can be found on our website [34].

References

1. Abbott, J., Bigatti, A.M.: CoCoALib: a C++ library for computations in commutative algebra.. and beyond. In: Fukuda, K., Hoeven, J., Joswig, M., Takayama, N. (eds.) ICMS 2010. LNCS, vol. 6327, pp. 73–76. Springer, Heidelberg (2010)
2. Akbarpour, B., Paulson, L.C.: Metitarski: An automatic theorem prover for real-valued special functions. Journal of Automated Reasoning 44(3), 175–205 (2010)
3. Barrett, C., Conway, C.L., Deters, M., Hadarean, L., Jovanović, D., King, T., Reynolds, A., Tinelli, C.: CVC4. In: Gopalakrishnan, G., Qadeer, S. (eds.) CAV 2011. LNCS, vol. 6806, pp. 171–177. Springer, Heidelberg (2011)
4. Basu, S., Pollack, R., Roy, M.: Algorithms in Real Algebraic Geometry. Springer (2010)
5. Bauer, C., Frink, A., Kreckel, R.: Introduction to the GiNaC framework for symbolic computation within the C++ programming language. Journal of Symbolic Computation 33(1), 1–12 (2002)
6. Bruttomesso, R., Pek, E., Sharygina, N., Tsitovich, A.: The openSMT solver. In: Esparza, J., Majumdar, R. (eds.) TACAS 2010. LNCS, vol. 6015, pp. 150–153. Springer, Heidelberg (2010)
7. Chauhan, P., Goyal, D., Hasteer, G., Mathur, A., Sharma, N.: Non-cycle-accurate sequential equivalence checking. In: Proc. of DAC 2009, pp. 460–465. ACM Press (2009)
8. Cimatti, A., Griggio, A., Schaafsma, B.J., Sebastiani, R.: The mathSAT5 SMT solver. In: Piterman, N., Smolka, S.A. (eds.) TACAS 2013 (ETAPS 2013). LNCS, vol. 7795, pp. 93–107. Springer, Heidelberg (2013)
9. Codish, M., Fekete, Y., Fuhs, C., Giesl, J., Waldmann, J.: Exotic semi-ring constraints. In: Proc. of SMT 2012. EPiC Series, vol. 20, pp. 88–97. EasyChair (2013)
10. Collins, G.E.: Quantifier elimination for real closed fields by cylindrical algebraic decompostion. In: Brakhage, H. (ed.) Automata Theory and Formal Languages. LNCS, vol. 33, pp. 134–183. Springer, Heidelberg (1975)
11. Corzilius, F., Ábrahám, E.: Virtual substitution for SMT-solving. In: Owe, O., Steffen, M., Telle, J.A. (eds.) FCT 2011. LNCS, vol. 6914, pp. 360–371. Springer, Heidelberg (2011)
12. Corzilius, F., Loup, U., Junges, S., Ábrahám, E.: SMT-RAT: an SMT-compliant nonlinear real arithmetic toolbox. In: Cimatti, A., Sebastiani, R. (eds.) SAT 2012. LNCS, vol. 7317, pp. 442–448. Springer, Heidelberg (2012)
13. Dehnert, C., Junges, S., Jansen, N., Corzilius, F., Volk, M., Bruintjes, H., Katoen, J.P., Ábrahám, E.: PROPhESY: a PRObabilistic ParamEter SYthesis tool. In: Proc. of CAV 2015. LNCS, vol. 9207. Springer (2015)
14. Dolzmann, A., Sturm, T.: Simplification of quantifier-free formulas over ordered fields. Journal of Symbolic Computation 24, 209–231 (1995)
15. Dutertre, B., de Moura, L.: A fast linear-arithmetic solver for DPLL(T). In: Ball, T., Jones, R.B. (eds.) CAV 2006. LNCS, vol. 4144, pp. 81–94. Springer, Heidelberg (2006)

16. Eén, N., Sörensson, N.: An extensible SAT-solver. In: Giunchiglia, E., Tacchella, A. (eds.) SAT 2003. LNCS, vol. 2919, pp. 502–518. Springer, Heidelberg (2004)

17. Fränzle, M., et al.: Efficient solving of large non-linear arithmetic constraint systems with complex Boolean structure. Journal on Satisfiability, Boolean Modeling and Computation 1(3–4), 209–236 (2007)

18. Gao, S., Ganai, M.K., Ivancic, F., Gupta, A., Sankaranarayanan, S., Clarke, E.M.: Integrating ICP and LRA solvers for deciding nonlinear real arithmetic problems. In: Proc. of FMCAD 2010, pp. 81–89. IEEE (2010)

19. Giesl, J., Brockschmidt, M., Emmes, F., Frohn, F., Fuhs, C., Otto, C., Plücker, M., Schneider-Kamp, P., Ströder, T., Swiderski, S., Thiemann, R.: Proving termination of programs automatically with AProVE. In: Demri, S., Kapur, D., Weidenbach, C. (eds.) IJCAR 2014. LNCS, vol. 8562, pp. 184–191. Springer, Heidelberg (2014)

20. Granlund, T.: the GMP development team: GNU MP: The GNU Multiple Precision Arithmetic Library. http://gmplib.org

21. Haible, B., Kreckel, R.B.: CLN: Class Library for Numbers. http://www.ginac.de/CLN

22. Heintz, J., Roy, M.F., Solerno, P.: On the theoretical and practical complexity of the existential theory of reals. The Computer Journal 36(5), 427–431 (1993)

23. Hong, H.: Comparison of several decision algorithms for the existential theory of the reals. Tech. Rep. 91–41, Research Institute for Symbolic Computation, Johannes Kepler University Linz (1991)

24. Jovanović, D., de Moura, L.: Solving non-linear arithmetic. In: Gramlich, B., Miller, D., Sattler, U. (eds.) IJCAR 2012. LNCS, vol. 7364, pp. 339–354. Springer, Heidelberg (2012)

25. Junges, S., Loup, U., Corzilius, F., Ábrahám, E.: On Gröbner bases in the context of satisfiability-modulo-theories solving over the real numbers. In: Muntean, T., Poulakis, D., Rolland, R. (eds.) CAI 2013. LNCS, vol. 8080, pp. 186–198. Springer, Heidelberg (2013)

26. Kremer, G., Corzilius, F., Junges, S., Schupp, S., Ábrahám, E.: CArL: Computer ARithmetic and Logic Library. https://github.com/smtrat/carl

27. Loup, U., Scheibler, K., Corzilius, F., Ábrahám, E., Becker, B.: A symbiosis of interval constraint propagation and cylindrical algebraic decomposition. In: Bonacina, M.P. (ed.) CADE 2013. LNCS, vol. 7898, pp. 193–207. Springer, Heidelberg (2013)

28. de Moura, L., Bjørner, N.S.: Z3: an efficient SMT solver. In: Ramakrishnan, C.R., Rehof, J. (eds.) TACAS 2008. LNCS, vol. 4963, pp. 337–340. Springer, Heidelberg (2008)

29. de Moura, L., Passmore, G.O.: The strategy challenge in SMT solving. In: Bonacina, M.P., Stickel, M.E. (eds.) Automated Reasoning and Mathematics. LNCS, vol. 7788, pp. 15–44. Springer, Heidelberg (2013)

30. Platzer, A., Quesel, J.-D., Rümmer, P.: Real world verification. In: Schmidt, R.A. (ed.) CADE-22. LNCS, vol. 5663, pp. 485–501. Springer, Heidelberg (2009)

31. Ravanbakhsh, H., Sankaranarayanan, S.: Counterexample guided synthesis of switched controllers for reach-while-stay properties. CoRR abs/1505.01180 (2015). http://arxiv.org/abs/1505.01180

32. Scheibler, K., Kupferschmid, S., Becker, B.: Recent improvements in the SMT solver iSAT. In: Proc. of MBMV 2013, pp. 231–241. Institut für Angewandte Mikroelektronik und Datentechnik, Fakultät für Informatik und Elektrotechnik, Universität Rostock (2013)

33. Sebastiani, R.: Lazy satisfiability modulo theories. Journal on Satisfiability, Boolean Modeling and Computation 3, 141–224 (2007)

34. https://github.com/smtrat/smtrat/wiki
35. Weispfenning, V.: A new approach to quantifier elimination for real algebra. In: Quantifier Elimination and Cylindrical Algebraic Decomposition. Texts and Monographs in Symbolic Computation, pp. 376–392. Springer (1998)
36. Weispfenning, V.: The complexity of linear problems in fields. Journal of Symbolic Computation 5(1–2), 3–27 (1988)
37. Weispfenning, V.: Quantifier elimination for real algebra - the quadratic case and beyond. Appl. Algebra Eng. Commun. Comput. 8(2), 85–101 (1997)
38. Zankl, H., Middeldorp, A.: Satisfiability of non-linear (ir)rational arithmetic. In: Clarke, E.M., Voronkov, A. (eds.) LPAR-16 2010. LNCS, vol. 6355, pp. 481–500. Springer, Heidelberg (2010)

Search-Space Partitioning
for Parallelizing SMT Solvers

Antti E.J. Hyvärinen$^{(\boxtimes)}$, Matteo Marescotti, and Natasha Sharygina

Faculty of Informatics, University of Lugano,
Via Giuseppe Buffi 13, 6904 Lugano, Switzerland
antti.hyvaerinen@gmail.com

Abstract. This paper studies how parallel computing can be used to reduce the time required to solve instances of the Satisfiability Modulo Theories problem (SMT). We address the problem in two orthogonal ways: (i) by distributing the computation using algorithm portfolios, search space partitioning techniques, and their combinations; and (ii) by studying the effect of partitioning heuristics, and in particular the lookahead heuristic, to the efficiency of the partitioning. We implemented the approaches in the OpenSMT2 solver and experimented with the QF_UF theory on a computing cloud. The results show a consistent speed-up on hard instances with up to an order of magnitude run time reduction and more instances being solved within the timeout compared to the sequential implementation.

1 Introduction

The *Satisfiability Modulo Theories* problem [9,27] (SMT) is the problem of determining whether a propositional formula is satisfiable, given that some of the Boolean variables have an interpretation as equalities or inequalities in a background theory. The problem has recently gained importance as a modeling approach for a vast range of application domains from software model checking (see, for instance, [1,12]) to optimization [4,21,26,30] due to its expressiveness and the efficient implementations. One of the features that make the SMT framework inviting for domain specialists is its flexibility in admitting a wide range of theories. The theory of quantifier-free uninterpreted functions with equalities (QF_UF) [9] is one of the most fundamental and applicable background theories, being widely used in combination with other theories (for a list of SMT theories see http://www.smt-lib.org/), and for instance as an abstraction to obtain more efficient decision procedure implementations [6]. The computational cost of solving SMT instances can be very high, given that already determining the propositional satisfiability is an NP-complete problem and the introduction of background theories can only make the problem harder. Nevertheless there has been relatively little research on how parallel computing can be used to speed up the solving of SMT instances (see the related work section for a survey).

This work addresses the challenge of parallelization. We introduce an abstract parallel algorithmic framework for SMT called the *parallelization tree*. The

© Springer International Publishing Switzerland 2015
M. Heule and S. Weaver (Eds.): SAT 2015, LNCS 9340, pp. 369–386, 2015.
DOI: 10.1007/978-3-319-24318-4_27

framework allows combining two important approaches for parallelization: algorithm portfolios and the divide-and-conquer approach. The key idea of the framework is that both solving and partitioning the search space can be done with a portfolio. The approach is applicable to all SMT solvers based on the DPLL(T) paradigm independent of the used theories, and our experiments address the central QF_UF theory. We show experimentally, both with a parallel solver implemented in a cloud computing environment and with an experimentation in a more controlled environment, that several instantiations of the parallelization tree framework are very efficient in solving SMT instances. We are able to solve more instances within a given timeout, observe sometimes an order of magnitude speed-ups, and are competitive with optimized SMT solvers on hard instances.

Most SMT solvers, including [3,6,7,10,24], consist of a SAT solver that searches for a satisfying assignment for a problem instance represented as a set of clauses, and theory solvers that check whether the assignment is consistent with respect to the theory in question. The SAT solver finds a satisfying assignment for its clause set, and the theory solvers check the consistency of the set of Boolean variables that have an interpretation in the theories. The found inconsistencies are communicated to the SAT solver as clauses that the solver adds as refinements to its clause set. The process terminates once the SAT solver has found a satisfying assignment consistent with the theories, or when the theory solvers have provided enough clauses for the SAT solver to determine unsatisfiability.

One of the challenges in parallelizing SMT solvers using the divide-and-conquer approach is that the clause set of the SAT solver does not initially contain the full information on the SMT instance unlike in SAT solving. As a result the approaches for parallelizing SAT solvers are not directly applicable to SMT solving. Our approach addresses this challenge in two ways. For constructing partitions we develop versions of the lookahead and the VSIDS heuristic [23] that are both made aware of the theory solver. The parallelization tree approach, on the other hand, is used to increase the probability of quick solving through the use of portfolios both for solving instances and constructing partitions. To the best of our knowledge the applications of the parallelization tree framework, partitioning, and lookahead in SMT with QF_UF are all new.

Related Work. The portfolio approach for parallel SMT solving is studied in [31] for problems from the QF_IDL logic. The system, implemented in the Z3 SMT solver, provides an efficient clause-sharing strategy for the workers and concludes that the best results are obtained with a random portfolio similar to ours. In this work we use instead the QF_UF logic, study different types of parallelization approaches, and scale the solver to more CPUs. A divide-and-conquer approach for the QF_BV logic is studied in [29]. The procedure tries to solve the formula with the Boolector SMT solver within a given timeout. If no result is obtained within the timeout the formula is divided into two partitions using a heuristic based on lookahead and the search is continued in parallel on the resulting partitions. The procedure terminates when Boolector returns a satisfiable result for one formula, or all formulas are shown unsatisfiable. We also study the lookahead heuristic for constructing partitions but concentrate on the QF_UF logic and provide a more general parallelization algorithm. A portfolio-style parallelization approach for the QF_ABV logic is presented in [28]. While the work uses several SMT solvers as the portfolio, in our work we concentrate on implementing all the parallelization approaches inside a single solver.

There is a substantial body of recent work on parallel SAT solving, and for instance [22] gives a good overview of the recent advances. A number of future challenges in parallel SAT solving in particular and in parallel constraint programming in general is given in [13]. Our work discusses in part the challenge of combining the portfolio style search and the divide-and-conquer approach, a topic we believe to be orthogonal to the ones presented in [13]. More recently [2] presents an approach based on search space partitioning, both with and without clause sharing. A theoretically oriented study in [20] suggests that parallelizing a SAT solver using a portfolio might in some cases be inherently difficult because of the resolution structure. We seem to observe a similar barrier with SMT solvers, and show experimentally that the divide-and-conquer approach, when used in combination with a portfolio, seems to overcome this problem. The lookahead heuristic [15] has been previously used in SAT solving and proved particularly effective in constructing partitions [14,18]. In this work we extend this line of work and use for the first time lookahead for parallelizing SMT solvers with QF_UF.

2   Preliminaries

Given a finite set of Boolean variables $B = \{x_1, \ldots, x_n\}$, a *clause* is a set of *literals*, that is, positive and negative Boolean variables $x, \neg x, x \in B$. A *propositional formula in conjunctive normal form* (CNF) is a conjunction of clauses. In the context of this work an *SMT formula F* is a propositional formula given in CNF where the variables of a subset $B_T \subseteq B$ have an interpretation as equalities over terms of a theory T. If $x \in B_T$, the literal $\neg x$ is interpreted as the corresponding disequality in the theory T. In this work we will consider the quantifier free theory of uninterpreted functions with equalities (QF_UF).

A *truth assignment* $\sigma \subseteq \{x, \neg x \mid x \in B\}$ is a set of literals such that for no $x \in B$ both $x \in \sigma$ and $\neg x \in \sigma$. A truth assignment is *total* if for all $x \in B$ either $x \in \sigma$ or $\neg x \in \sigma$. A clause c is *propositionally satisfied* if $\sigma \cap c \neq \emptyset$ and a CNF formula F is propositionally satisfied if all its clauses are propositionally satisfied. The formula F is *satisfiable* if there is an assignment σ satisfying propositionally F, and the set of equalities and disequalities imposed by the assignment σ on the equalities in B_T interpreted in the theory T is consistent. A literal l is *implied* under σ in F if there is a clause $c \in F$ such that $l \in c, l \notin \sigma$ and for all other $l' \in c, l' \neq l$ it holds that $\neg l' \in \sigma$.[1] A truth assignment σ is *conflicting* if for some variable x either both x and $\neg x$ are implied or for some $l \in \sigma$, $\neg l$ is implied.

An *SMT solver* consists of a SAT solver and a theory solver that communicate by exchanging equalities and negations of equalities from the set B_T as clauses.

---

[1] We follow the convention that $\neg \neg x = x$ for $x \in B$.

The SAT solver starts with an initial set of clauses F. The solver communicates periodically non-conflicting assignments to a theory solver. Upon receiving an assignment σ, the theory solver interprets the set $\sigma_T = \sigma \cap \{x, \neg x \mid x \in B_T\}$ as equalities and disequalities and determines whether σ_T is inconsistent or consistent with the theory T. In case of consistency the theory solver may provide the SAT solver with a set of *theory-implied literals* $l \notin \sigma$. If the theory solver agrees on the consistency of a total truth assignment σ, the assignment σ is returned as a proof of satisfiability for F. If the theory solver finds theory-implied literals these are communicated to the SAT solver as clauses that imply the theory-implied literals. For any theory-inconsistent truth assignment the theory solver will communicate a set of *theory clauses* consisting of variables in B_T and expressing the reason for the inconsistency of the assignment. During the search the SAT solver can find *learned clauses*, that is, clauses that the SAT solver has derived using its current clause database with conflict analysis based on resolution. A learned clause c_l has the property that if F is the current set of clauses, then any truth assignment σ propositionally satisfying F also propositionally satisfies c_l. In contrast a theory clause learned right after communicating the satisfying truth assignment σ for F is not propositionally satisfied by σ. Finally, the *unit propagation closure* $U(F, \sigma)$ is the smallest set of literals containing σ that is closed under a rule that includes to σ all implied and theory-implied literals.

3   Parallelization Approaches for SMT

Heuristics for guiding the search on a boolean structure play an important role in both SAT and SMT solvers. As a natural consequence of the computational difficulty of the SMT problem heuristics are inaccurate and small changes can result in significant differences in run times. This phenomenon can be used to obtain speed-up in a parallel setting using a *portfolio* of algorithms. The main challenge in parallelizing SMT solvers this way is that portfolio-style solving seems to hit a scalability limit where adding more CPUs does not provide more speed-up [18–20]. The scalability problem of the portfolio-style solving can be addressed by allowing the search processes to share information such as learned and theory clauses. The approach has been studied for SMT in [31] where it was shown that sharing both types of clauses helps speeding up the solver. In this work we use the simple portfolio obtained by forcing the SAT solver to make certain choices randomly, potentially against the heuristic values. This approach was found to be efficient in SAT solving [17] and was identified to be the best-performing strategy for SMT solvers in [31].

However, this paper targets also the scalability limit in an orthogonal way by using a divide-and-conquer approach where several solvers work in parallel on problem instances that are constructed by partitioning the search-space of the original instance and hence are different from each other. The solution to the original problem instance can be obtained by combining the results from the partitioned instances. This approach has an inherent problem that needs to be addressed to obtain good results: If the original problem instance is unsatisfiable, the variance in run time results in decreased performance. Instead of having to solve a single instance the solver needs to solve several instances that, despite being usually easier than the original, might still be challenging. While the variance in solver run time makes the portfolio approach efficient, it degrades the performance of the divide-and-conquer approach, effectively resulting in the solver having to wait for the "unluckiest" instance to be solved. Under certain assumptions it can be shown that for implementations based on pure divide-and-conquer it is possible to come up with a run-time behavior that results in increased run time when parallelized this way, and that a different organization of the search can help to avoid this problem [19].

We show experimentally that often the partitions constructed from the original problem are somewhat easier but not significantly so, and this results in such slowdown anomalies. However, even in cases where the instances do not get significantly easier it is possible to obtain good speed-up by using a portfolio approach on both constructing and solving the partitions. To present our approach we will formalize the idea of combining divide-and-conquer with portfolio. We introduce an abstract parallelization algorithm framework called *parallelization tree* and give five concrete instantiations of the framework. In addition to providing us with a convenient tool for discussing different parallelization algorithms the framework is also used as a tool for explaining the performance results we present in Sec. 5. We will introduce an even more practical implementation of the framework in Sec. 4.2 which uses also a load balancing schema.

In the following we first discuss certain approaches for partitioning the search-space in SMT and then describe the parallelization tree framework. We conclude with concrete examples of the framework.

3.1 Search-Space Partitioning in SMT

The basic approach for constructing partitions in SMT uses a *partitioning function*, denoted by $partf_n : F \mapsto F_1, \ldots, F_n$, to divide an SMT instance F into n partitions $F_1, \ldots, F_n$. The function satisfies the conditions that F is satisfiable if and only if $F_1 \vee \ldots \vee F_n$ is satisfiable and no two partitions $F_i, F_j, i \neq j$, share a satisfying truth assignment. We construct partitions by conjoining *partitioning constraints* $P_1, \ldots, P_n$, in general set of clauses, to F. We use two types of partitioning constraints: the ones obtained with the *scattering approach* [18] and the ones obtained with *guiding paths* [5,32].

The scattering approach. The scattering approach is a technique for partitioning an instance into arbitrary number of partitions. Each partitioning constraint P_i is obtained by heuristically selecting a set of *scattering literals* $l_1^i, \ldots, l_{k_i}^i$ and conjoining these literals with the clauses obtained by negations of the previous scattering literals. More formally this can expressed as $P_i := l_1^i \wedge \ldots \wedge l_{k_i}^i \wedge (\neg l_1^{i-1} \vee \ldots \vee \neg l_{k_{i-1}}^{i-1}) \wedge \ldots \wedge (\neg l_1^1 \vee \ldots \vee \neg l_{k_1}^1)$. The number of scattering literals k_i are selected so that the partitions constructed are approximately equally sized under the assumption that fixing a literal will reduce the search space with a constant factor. If n is the number of partitions to be generated, it can be shown that the fraction obtained by fixing the scattering literals $l_1^i, \ldots, l_{k_i}^i$ should be

$r_i = \frac{1}{n-i+1}$ of the previous instance F_{i-1} [16]. We simply assume that fixing a literal will half the search space, and this results in us choosing the number of scattering literals k_i minimizing the difference $|r_i - 2^{-k_i}|$.

The guiding paths. As an alternative to the scattering based method for constructing the partitions we use a simple variant of the *guiding path* approach [32] where a binary tree with literals as nodes represents the $n = 2^k, k \geq 1$ partitions. The root of the tree consists of the *true* literal, and the rest of the nodes are either leaves with no children or have exactly two children v and $\neg v$ where $v \in B$. Each path $true, l_1, \ldots, l_k$ from the root to a leaf l_k corresponds to a partitioning constraint $l_1 \wedge \ldots \wedge l_k$.

3.2 Combining Search Space Partitioning and Portfolio

The key idea in obtaining well-performing parallel solvers where search-space partitioning plays a role is to combine elements from both the search-space partitioning and the algorithm portfolio.

The *parallelization tree* abstract algorithmic framework provides a unified way of presenting and comparing different parallelization algorithms. The parallelization tree consists of two types of nodes: *and-nodes* and *or-nodes*. The root and the leaves of the parallelization tree are and-nodes. Each and-node is associated with an SMT instance and, with the possible exception of the root of the parallelization tree, with one or more SMT solvers. The instance at the root of the parallelization tree is satisfiable if any instance in the and-nodes is shown satisfiable. A subtree rooted at an and-node is unsatisfiable if one of its children is unsatisfiable or at least one of the solvers associated with the and-node has shown the instance unsatisfiable. A tree rooted at an or-node is unsatisfiable if every tree rooted at its children is unsatisfiable.

We use a *partitioning operator* $\mathrm{split}^k(n_1, \ldots, n_k, F)$ to construct the parallelization tree. The result of applying the operator split^k on an and-node F is a tree rooted at the and-node F having k children $o_1, \ldots, o_k$. Each child node o_i is an or-node and has as children the and-nodes $a_1^i, \ldots, a_{n_i}^i$. Finally, each and-node a_j^i is associated with the partition obtained by applying the (randomized) partitioning function $partf_{n_i}$ on the formula F.

As instances of the parallelization tree we identify five particularly interesting parallelization algorithms.

- The *plain* partitioning approach $plain(n, F)$ corresponds to the parallelization tree $\mathrm{split}^1(n, F)$ where each of the instances associated with the nodes $a_1^1, \ldots, a_n^1$ is solved with a single SMT solver.
- The *portfolio* approach $portf(k, F)$ corresponds to the parallelization tree consisting of the root associated with the instance F and using k SMT solvers to solve the instance.
- The *safe* partitioning approach $safe(n, s, F)$ corresponds to the parallelization tree $\mathrm{split}^1(n, F)$ and solving each of the instances $a_1^1, \ldots, a_n^1$ with s SMT solvers.

- The *repeated* partitioning approach $rep(n, k, F)$ corresponds to the parallelization tree $split^k(n, \ldots, n, F)$ where each instance associated with the nodes $a_1^1, \ldots, a_n^1, \ldots, a_1^k, \ldots, a_n^k$ is solved with one SMT solver.
- The *iterative* partitioning approach $iter(k, F)$ corresponds to the infinite parallelization tree where every instance associated with an and-node is being solved with a single SMT solver and every and-node associated with an instance F_a has the single or-child and and-grandchildren constructed by applying the operator $split^1(n, F_a)$.

Figure 1 illustrates the corresponding parallelization trees and the solver assignments. When clear from the context, we omit the formula F as well as the other parameters from the partitioning approach.

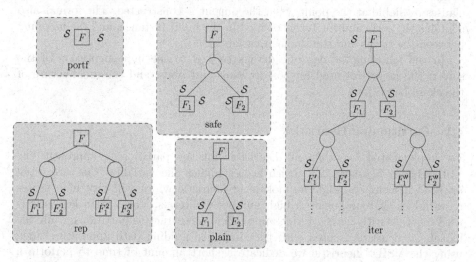

Fig. 1. Example parallelization trees (clockwise from the top left): $portf(2, F)$, $safe(2, 2, F)$, $iter(2, F)$, $plain(2, F)$, and $rep(2, 2, F)$. The and-nodes are drawn with boxes, and the or-nodes with circles. The SMT solvers are indicated with the symbol $\mathcal{S}$.

Concrete SMT instantiations of the parallelization tree include the CVC4 and Z3 SMT solvers which implement a portfolio, and PBoolector [29] which implements an iterative partitioning approach.

4   A Cloud-Based Parallel SMT Solver for QF_UF

We have implemented the approaches discussed in this work into the OpenSMT2 solver. The solver is a complete rewrite of the SMT solver OpenSMT [6] and includes all the algorithmic optimizations present in OpenSMT. However, due to improved memory management, reduced memory footprint of the critical data structures, and certain bug fixes, the current version is approximately 10% faster on the QF_UF family of benchmarks.

The SAT solver inside OpenSMT2 is based on MiniSAT 2.0 [11], a conflict-driven clause-learning SAT solver. The congruence closure algorithm implemented in OpenSMT2 employs the algorithm from [25] for communicating small reasons for unsatisfiability of equalities containing uninterpreted functions to the SAT solver.

To be able to distribute in an unambiguous way the partitioning constraints and partitions to the parallel working solvers we have implemented a format where the propositional encoding into CNF is made explicit. Due to the format we are able to do the partitioning in a more general way, using also Boolean variables created with the Tseitin transformation that are not in general part of the original problem description. The format consists of the CNF corresponding to the Tseitin encoding of the SMT instance, and the learned and the theory clauses available at the point when the output is constructed. The format also contains the sorts and the terms defined in the input instance, and the mapping between the terms and the Boolean variables.

In the following we describe details related to the implementation of the solver: the heuristics used for constructing partitions, and the architecture of the cloud-based tool.

4.1   Partitioning Heuristics

We implemented two different heuristics for the partitioning approach: the VSIDS-based heuristic [23] which scores higher the variables that are often involved in conflicts, and the lookahead heuristic which will give high scores to variables that propagate a high number of literals. The VSIDS heuristic is used together with the scattering approach for constructing partitions, while the lookahead heuristic is used with the simplified guiding path approach. When using the VSIDS heuristic we dedicate a short amount of time to perform a search on the instance so that the VSIDS heuristic gets reasonable scores for the variables.

The lookahead heuristic starts with an assignment σ and computes for each variable $x \notin \sigma$ the sizes of the sets $U(F, \sigma \cup \{x\})$ and $U(F, \sigma \cup \{\neg x\})$. The highest score is assigned to the variable that maximizes the minimum of the sizes of these two sets. As a result the heuristic favors variables that construct similar sized partitions having few variables. We have implemented a few important optimizations for the approach: if a literal l propagates n literals $l_1, \ldots, l_n$, the heuristic sets the number n for the upper bounds for all literals $l_1, \ldots, l_n$. If a variable is propagated both in the positive and in the negative polarity, and if the lower of the upper bounds is lower than the current best value, the lookahead on this variable can be safely skipped. Additionally if the literal propagation results in a conflict this means that the solver can learn an arbitrary number of learned and theory clauses and the current heuristic values are no longer valid. In this case the heuristic does not restart the lookahead from the beginning, but continues instead from the next literal. This trick is known to reduce the run time of the lookahead in practice by a linear factor.

4.2   The Client-Server Architecture

We implemented a subset of the functionality of the parallelization tree framework into the OpenSMT2 solver designed to run in a computing cloud or a cluster. The tool is available at http://verify.inf.usi.ch/parallel-opensmt2/.

The system follows the client-server architecture where the server receives a set of SMT instances and an arbitrary number of connections from clients. The server then manages the construction of partitions from the original instances and distributes the partitions to the clients for solving. The server and the clients communicate using our custom-built protocol through TCP/IP sockets, making the solution light-weight, portable, easy to modify and easy to use.

The clients are implemented as processes with two threads, one communicating with the server and the other responsible for the solving of the SMT instance. The communicating thread waits for an instance from the server, passes the instance to the solving thread and then continues to listen to further commands from the server. Currently the supported commands are initiation of a solving of a partition and termination of the solving. If the thread solves the instance before server sends the terminate command the result is communicated to the server and the process returns to the initial state waiting for a new instance from the server. The client is implemented in C++ on top of OpenSMT2, and an architectural overview is given in Fig. 2.

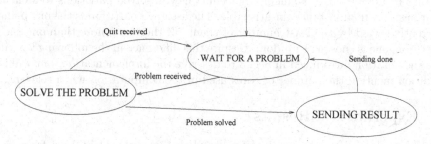

Fig. 2. The client architecture

The server is implemented as a two-threaded Python/2.7 program that calls OpenSMT2 to construct the partitions. Both threads listen to connections on separate TCP ports, the command thread for commands from the user and the worker thread for communication with the clients on solving SMT instances. The worker thread maintains a list of all connected clients and the list of partitions to be solved, and constructs the partitions using the partitioning heuristic and the parallelization tree. In addition the worker thread provides the unsolved partitions to the clients again based on the selected parallelization tree. The command thread communicates with the user. The user may send commands such as initiations of the solving of a new SMT instances, requests to terminate the current solving, or a request to print the status of current jobs. The implementation is capable of handling client failures, exits, and new connections

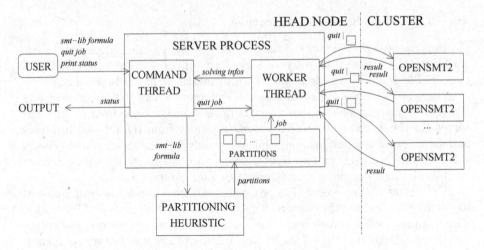

Fig. 3. The Server Architecture

seamlessly and completely automatically. The architecture of the server and the communication between the clients and the user is described in Fig. 3.

Currently the implementation has been adapted to the parallelization algorithm *safe*. The implementation differs from the algorithm discussed in Sec. 3 in that it provides load balancing by feeding new unsolved partitions to a client that has shown an instance unsatisfiable. The assigned partition is the one being currently solved by the least number of clients. If there are more than one such partition, one is chosen at random. Despite this difference in the following we will use the notation introduced in Sec. 3 to describe the implementation, but will in addition mention the number of cores used in the computation when relevant.

5   Experimental Results

This section presents the results of some of the algorithms obtained from the *parallelization tree* algorithmic framework, using the cloud-based implementation presented in Sec. 4.2 as a uniform platform for testing. For completeness we also report experimentations against other SMT solvers in Sec. 5.3. The experiment set contains of all instances from the QF_UF category of the SMT-LIB benchmark collection (http://www.smt-lib.org/) having run time longer than one minute with the default configuration of OpenSMT2. This set consists of 54 instances, 11 of which could not be solved within the 1000 seconds timeout and 4 GB memory limit. All the instances we could solve from this set turned out to be unsatisfiable, and therefore we also added randomly selected 100 easier satisfiable and unsatisfiable instances, resulting in total 254 benchmark instances. All the experiments were run on a cluster consisting of nodes with two AMD quad-core Opteron 2344 HE CPUs and each node was running at most four solver processes. All times are reported in seconds.

We show the results for the hardest instances in our benchmark set in Table 1. To the table we have selected certain approaches that illustrate the behavior of

Table 1. Instances solved with at least one of the approaches, but where the portfolio approach required over 100 seconds with 64 CPUs. All the instances are unsatisfiable.

Name	OSMT2[1]	portf(64)	rep(2,32)	safe(2,32)	plain(64)	rep(8,8)	safe(8,8)	OSMT2[64]
PEQ003_size9	437.37	299.76	336.00	232.81	431.44	286.11	248.32	**195.70**
PEQ004_size8	124.85	117.17	109.15	110.76	125.70	108.07	115.34	**15.85**
PEQ011_size8	572.12	302.12	267.11	265.91	388.54	309.68	280.54	**258.94**
PEQ012_size6	—	—	456.56	507.76	621.24	574.61	532.06	**382.44**
PEQ014_size11	737.58	**338.56**	482.68	564.22	—	—	540.05	539.28
PEQ016_size6	223.68	181.19	158.60	168.57	188.14	158.02	176.94	**20.20**
PEQ018_size7	192.58	144.96	155.99	**139.68**	182.50	207.46	218.02	168.18
PEQ020_size6	511.26	409.50	379.89	**314.26**	405.37	401.27	371.41	337.04
SEQ005_size8	174.85	159.70	144.13	144.28	132.26	148.76	131.84	**16.84**
SEQ010_size8	244.22	190.11	123.36	166.96	196.38	157.59	155.92	160.94
SEQ026_size7	890.18	708.89	731.43	794.43	671.05	774.14	725.79	**686.63**
SEQ038_size8	—	826.11	903.32	751.73	751.03	745.75	792.07	819.41
NEQ006_size6	—	—	—	—	—	—	—	**16.62**
NEQ016_size8	774.57	616.73	682.64	575.34	—	625.47	**419.60**	592.15
NEQ023_size7	—	—	—	—	—	—	—	**50.75**
NEQ032_size6	—	830.03	407.59	**373.25**	—	836.31	865.95	532.08
NEQ048_size8	476.46	430.31	421.38	**341.27**	479.38	349.94	445.24	458.11
NEQ048_size9	—	815.72	**759.96**	804.73	849.76	832.16	833.86	846.21
Total solved	12	15	16	16	13	15	16	**18**

the parallelization algorithms well. The columns OSMT2[1] and OSMT2[64] represent, respectively, the sequential run of the OpenSMT2 solver and the run of the cloud-based implementation described in Sec. 4.2, using the VSIDS scattering heuristic, the algorithm $safe(8, 8)$ and 64 CPU cores. The rest of the columns correspond to instantiations of the parallelization algorithms discussed in Sec. 3 with scattering and VSIDS heuristic. The reported times include also the time to run the partitioning. The best run time for a given instance is shown in boldface and the dashes indicate timeouts.

The results suggest that for hard instances OSMT2[64] performs very well compared to the other approaches, solving the largest number of instances and usually with a very good run time. The implementation is the fastest solver for eight instances, the runner-up being the parallelization algorithm $safe(2, 32)$ with four fastest times. The run time of the implementation is very competitive with the parallelization algorithm $safe(8, 8)$. When a formula is partitioned to many sub-instances some of them will be easy. The load-balancing present in the implementation allows the solving approach to concentrate on the hard instances.

The parallelization algorithm *plain* performs badly for our benchmark instances when compared to the other parallelization algorithms. This suggests that often the partitions are not significantly easier. Therefore constructing a large number of partitions with mutually exclusive models easily results in more work for the parallel solver. To illustrate this better we show in Fig. 4 (top left) a comparison between the algorithms $plain(64)$ and $rep(2, 32)$ using the full benchmark set. The run times include the time required for the partitioning. Here, as well as in all other similar graphs, we denote satisfiable instances by × and unsatisfiable by □, and highlight timeouts with a red color. We can see that the algorithm $plain(64)$ is almost always worse. A closer analysis reveals that when considering only the instances that both approaches could solve the algorithm $rep(2, 32)$ solves the full problem set 9 times faster than the algorithm $plain(64)$.

Finally we point out that while the parallelization algorithm $portf$ works relatively well it seems to loose in the hard instances when compared to the approaches that combine elements from portfolio and search-space partitioning. In fact if we do not consider the sequential execution and the algorithm $plain$, other algorithms perform better than our implementation of the portfolio.

In the following subsections we will study in more detail the observations made based on Table 1, consider role of the heuristic used in constructing the partitions, and finally conclude with a comparison of our implementation against some other well-known SMT solvers.

5.1   Comparing the Implementation to the Portfolio Approach

Given a fixed amount of parallel CPUs, there is an interesting tradeoff between the number of partitions constructed from an instance and the number of solvers that can be assigned for each partition. The comparison in Table 1 suggests that with this benchmark set and the VSIDS scattering heuristic good performance is obtained by constructing only two partitions and dedicating a large number of solvers for the two partitions (32 in the experiment) when using the paralleliza-tion algorithms. The situation changes when we use the parallel implementation with the load-balancing schema since usually some of the constructed partitions are much easier than others, therefore freeing up resources for solving the yet unsolved partitions. Figure 4 compares the implementation of the parallelization algorithms $safe(2, 32)$, $safe(8, 8)$, and $safe(64, 1)$ against the algorithm $portf(64)$ all using 64 cores. The results show that while the easy instances suffer from the overhead caused by the communication in the network and the time required to construct the partitions, the harder instances with run times more than 10 sec-onds usually profit from the partitioning. The positive effect of partitioning the problem into more than two parts can in particular be seen when comparing the implementations of $safe(2, 32)$ and $safe(8, 8)$. Unlike in the abstract algorithms the implementation with $safe(8, 8)$ performs clearly better than $safe(2, 32)$ solv-ing one more instance and providing a total speed-up of roughly 10% on the instances solved by both approaches. The implementation of $plain(64)$, while still not competitive, also performs significantly better as a result of the load balancing.

5.2   Comparing the Partitioning Heuristics

Since the ability of the partitioning to construct easy instances plays such a criti-cal role in the overall success of the partitioning based instances, it is interesting to study the effect of partitioning heuristics. We compare here two different types of heuristics, the VSIDS scattering and the lookahead with our guiding path implementation. We also experimented with alternations of the VSIDS scat-tering heuristics that prefer the equalities in the set B_T and the purely Boolean

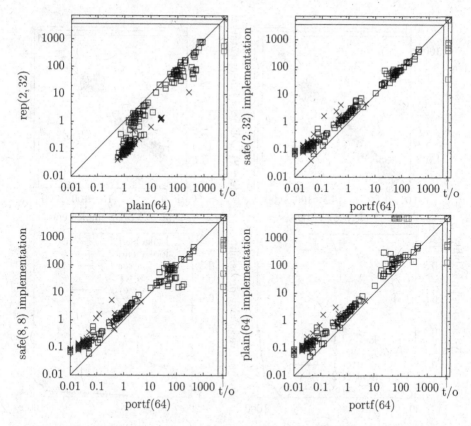

Fig. 4. The run times for the parallelization algorithms *plain*(64) and *rep*(2, 32) (*top left*). The *portf*(64) algorithm compared to the load-balancing implementations of *safe*(2, 32) (*top right*), *safe*(8, 8) (*bottom left*) and *plain*(64) (*bottom right*) on 64 cores.

variables $B \setminus B_T$. These however did not result in significant differences in our benchmark set and the results are therefore not shown.

The results for the comparison are given in Fig. 5. Excluding the time to construct the partitions, the lookahead gives a 40% reduction in the run time of the solver when using the abstract algorithms *safe*(8, 8), the average speed-up being 2. However, when the time required to construct the partitions is included, the lookahead-based heuristic looses the edge and becomes slightly worse compared to the VSIDS-based heuristic. This results mainly from the implementation of the lookahead-heuristic. The current implementation is not as optimized as the VSIDS implementation, but we believe that the heuristic can be made more efficient.

To understand the impressive efficiency of the lookahead heuristic we study closer two examples where the abstract parallelization algorithm *safe*(2, 32) performs well with the lookahead heuristic and with the VSIDS heuristic. The graphs on the bottom of Fig. 5 report the cumulative run-time distributions of

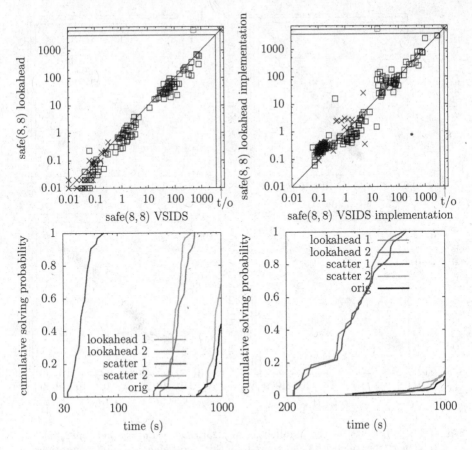

Fig. 5. The lookahead heuristic compared to the VSIDS-based scattering heuristic

the original instance and the four partitions constructed with the two heuristics. In the first example the lookahead heuristic finds a partitioning where the two partitions have a very similar run time distribution, whereas the VSIDS heuristic results in a very uneven distribution where one partition is significantly easier to solve than the other. In the second case (lower right graph in Fig. 5) the lookahead heuristic performed on a single run worse than the VSIDS heuristic. In this case both the heuristics resulted in a very uneven partitions. However it would seem that the cumulative run-time distribution of the VSIDS heuristic dominates on a wide area the distribution of the lookahead-based heuristic. Interestingly there seems to be a small probability that the lookahead-heuristic can solve the problem somewhat faster than the original problem, suggesting that the implementation with load balancing should be capable of performing well on this instance also for the lookahead heuristic.

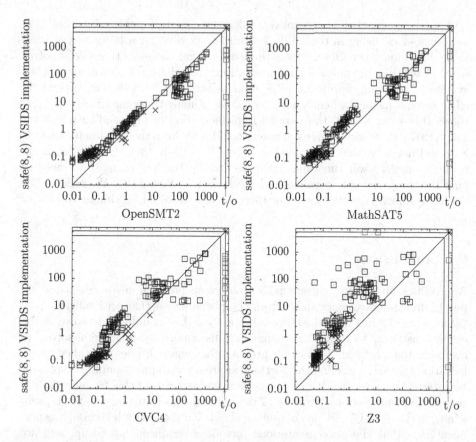

Fig. 6. Comparison of the $safe(8,8)$ implementation with the scattering heuristic against other SMT solvers.

5.3 Comparison to Other SMT Solvers

Finally we report the comparison of the parallel implementation and in particular the implementation of the parallel algorithm $safe(8,8)$ against other solvers in Fig. 6. All solvers were run with the default configurations. We first note that the implementation provides a clear speed-up against the sequential version of OpenSMT2, being 75% faster in the total run time over the benchmark set and solving six more instances within the timeout. However, the parallel implementation suffers a penalty related to the communication delays and constructing partitions when the instances are easy. The comparison against MathSAT5 [7] shows similar behavior: the parallel implementation can solve a handful of instances within the timeout that MathSAT5 could not solve. Comparing the solver against

CVC4 [3] reveals that the parallel implementation is capable of solving nine more instances, using in total 12% less wall-clock time for solving all instances in the benchmark set. Nevertheless there are several instances that CVC4 solves much faster than the parallel implementation. We believe the reason for this is a symmetry breaking simplification [8] implemented in CVC4 that is particularly effective on some of the benchmarks in our set. Finally the comparison to Z3 [24] shows that even though the parallel implementation of OpenSMT2 is not yet competitive, there are some instances we could solve from the benchmark set that Z3 could not solve and several others where it is likely that the parallelization results in much lower run times compared to Z3. For lack of space we need to omit the comparison to certain other solvers such as Yices2 [10]. We believe that due to optimizations the results of the comparison would be similar to that of Z3.

6   Conclusions

Approaches for solving unsatisfiable constraint problem instances based on purely divide-and-conquer suffer from the phenomenon that an inefficient partitioning results in several instances that are roughly as difficult to solve as the original instance. As a result it is common to use a portfolio of different solvers to overcome this problem. This paper presents the generic framework called parallelization tree for combining the portfolio approach with partitioning. We present how several parallel algorithms can be seen as instances of this framework, and provide implementations for some of the parallel algorithms for the SMT problem with the logic QF_UF in computing cloud. We show with a thorough experimentation that the implementations provide a significant speed-up, and are capable of solving several more instances within a given time-out compared to the sequential implementation. Furthermore we show that the implementations are competitive against many state-of-the-art SMT solvers.

Based on the results we are able to point out certain directions for future research. We believe that there is still work to be done in the heuristics for constructing partitions: our implementation of the lookahead heuristic is fairly straightforward and there are several techniques that can be used to improve its performance. One such technique is identifying equalities and inequalities of the variables. Also generalizing the lookahead to a portfolio in the way it was done for the VSIDS heuristic seems like a viable alternative for obtaining efficient partitionings. Finally we are interested in applying the knowledge obtained in this study to a setting where we allow the parallel-running solvers to exchange also learned and theory clauses.

Acknowledgments. We thank the anonymous reviewers for their valuable comments. This work was financially supported by SNF project number 153402.

References

1. Alberti, F., Bruttomesso, R., Ghilardi, S., Ranise, S., Sharygina, N.: SAFARI: SMT-based abstraction for arrays with interpolants. In: Madhusudan, P., Seshia, S.A. (eds.) CAV 2012. LNCS, vol. 7358, pp. 679–685. Springer, Heidelberg (2012)
2. Audemard, G., Hoessen, B., Jabbour, S., Piette, C.: Dolius: a distributed parallel SAT solving framework. In: Berre, D.L. (ed.) POS-2014. EPiC Series, vol. 27, pp. 1–11. EasyChair (2014)
3. Barrett, C., Conway, C.L., Deters, M., Hadarean, L., Jovanović, D., King, T., Reynolds, A., Tinelli, C.: CVC4. In: Gopalakrishnan, G., Qadeer, S. (eds.) CAV 2011. LNCS, vol. 6806, pp. 171–177. Springer, Heidelberg (2011)
4. Bjørner, N., Phan, A.-D., Fleckenstein, L.: νZ - an optimizing SMT solver. In: Baier, C., Tinelli, C. (eds.) TACAS 2015. LNCS, vol. 9035, pp. 194–199. Springer, Heidelberg (2015)
5. Böhm, M., Speckenmeyer, E.: A fast parallel SAT-solver: Efficient workload balancing. Annals of Mathematics and Artificial Intelligence 17(4-3), 381–400 (1996)
6. Bruttomesso, R., Pek, E., Sharygina, N., Tsitovich, A.: The OpenSMT solver. In: Esparza, J., Majumdar, R. (eds.) TACAS 2010. LNCS, vol. 6015, pp. 150–153. Springer, Heidelberg (2010)
7. Cimatti, A., Griggio, A., Schaafsma, B.J., Sebastiani, R.: The MathSAT5 SMT solver. In: Piterman, N., Smolka, S.A. (eds.) TACAS 2013 (ETAPS 2013). LNCS, vol. 7795, pp. 93–107. Springer, Heidelberg (2013)
8. Déharbe, D., Fontaine, P., Merz, S., Woltzenlogel Paleo, B.: Exploiting symmetry in SMT problems. In: Bjørner, N., Sofronie-Stokkermans, V. (eds.) CADE 2011. LNCS, vol. 6803, pp. 222–236. Springer, Heidelberg (2011)
9. Detlefs, D., Nelson, G., Saxe, J.B.: Simplify: a theorem prover for program checking. Journal of the ACM 52(3), 365–473 (2005)
10. Dutertre, B.: Yices 2.2. In: Biere, A., Bloem, R. (eds.) CAV 2014. LNCS, vol. 8559, pp. 737–744. Springer, Heidelberg (2014)
11. Eén, N., Sörensson, N.: An extensible SAT-solver. In: Giunchiglia, E., Tacchella, A. (eds.) SAT 2003. LNCS, vol. 2919, pp. 502–518. Springer, Heidelberg (2004)
12. Ghilardi, S., Ranise, S.: MCMT: a model checker modulo theories. In: Giesl, J., Hähnle, R. (eds.) IJCAR 2010. LNCS, vol. 6173, pp. 22–29. Springer, Heidelberg (2010)
13. Hamadi, Y., Wintersteiger, C.M.: Seven challenges in parallel SAT solving. AI Magazine 34(2), 99–106 (2013)
14. Heule, M.J.H., Kullmann, O., Wieringa, S., Biere, A.: Cube and conquer: guiding CDCL SAT solvers by lookaheads. In: Eder, K., Lourenço, J., Shehory, O. (eds.) HVC 2011. LNCS, vol. 7261, pp. 50–65. Springer, Heidelberg (2012)
15. Heule, M., van Maaren, H.: Look-ahead based SAT solvers. Handbook of Satisfiability, Frontiers in Artificial Intelligence and Applications 185, 155–184 (2009). IOS Press
16. Hyvärinen, A.E.J.: Grid-Based Propositional Satisfiability Solving. Ph.D. thesis, Aalto University School of Science, Aalto Print, Helsinki, Finland, November 2011
17. Hyvärinen, A.E.J., Junttila, T.A., Niemelä, I.: Incorporating clause learning in grid-based randomized SAT solving. Journal on Satisfiability Boolean Modeling and Computation 6(4), 223–244 (2009)
18. Hyvärinen, A.E.J., Junttila, T.A., Niemelä, I.: Partitioning search spaces of a randomized search. Fundamenta Informaticae 107(2-3), 289–311 (2011)

19. Hyvärinen, A.E.J., Manthey, N.: Designing scalable parallel SAT solvers. In: Cimatti, A., Sebastiani, R. (eds.) SAT 2012. LNCS, vol. 7317, pp. 214–227. Springer, Heidelberg (2012)

20. Katsirelos, G., Sabharwal, A., Samulowitz, H., Simon, L.: Resolution and parallelizability: barriers to the efficient parallelization of SAT solvers. In: desJardins, M., Littman, M.L. (eds.) Proc. AAAI 2013. AAAI Press (2013)

21. Li, Y., Albarghouthi, A., Kincaid, Z., Gurfinkel, A., Chechik, M.: Symbolic optimization with SMT solvers. In: Proc. POPL 2014, pp. 607–618. ACM (2014)

22. Martins, R., Manquinho, V.M., Lynce, I.: An overview of parallel SAT solving. Constraints 17(3), 304–347 (2012)

23. Moskewicz, M.W., Madigan, C.F., Zhao, Y., Zhang, L., Malik, S.: Chaff: engineering an efficient SAT solver. In: Proc. DAC 2001, pp. 530–535. ACM (2001)

24. de Moura, L., Bjørner, N.S.: Z3: an efficient SMT solver. In: Ramakrishnan, C.R., Rehof, J. (eds.) TACAS 2008. LNCS, vol. 4963, pp. 337–340. Springer, Heidelberg (2008)

25. Nieuwenhuis, R., Oliveras, A.: Proof-producing congruence closure. In: Giesl, J. (ed.) RTA 2005. LNCS, vol. 3467, pp. 453–468. Springer, Heidelberg (2005)

26. Nieuwenhuis, R., Oliveras, A.: On SAT modulo theories and optimization problems. In: Biere, A., Gomes, C.P. (eds.) SAT 2006. LNCS, vol. 4121, pp. 156–169. Springer, Heidelberg (2006)

27. Nieuwenhuis, R., Oliveras, A., Tinelli, C.: Solving SAT and SAT modulo theories: From an abstract Davis-Putnam-Logemann-Loveland procedure to DPLL(T). Journal of the ACM 53(6), 937–977 (2006)

28. Palikareva, H., Cadar, C.: Multi-solver support in symbolic execution. In: Sharygina, N., Veith, H. (eds.) CAV 2013. LNCS, vol. 8044, pp. 53–68. Springer, Heidelberg (2013)

29. Reisenberger, C.: PBoolector: a Parallel SMT Solver for QF_BV by Combining Bit-Blasting with Look-Ahead. Master's thesis, Johannes Kepler Univesität Linz, Linz, Austria (2014)

30. Sebastiani, R., Tomasi, S.: Optimization in SMT with $\mathcal{LA}$ ($\mathbb{Q}$) cost functions. In: Gramlich, B., Miller, D., Sattler, U. (eds.) IJCAR 2012. LNCS, vol. 7364, pp. 484–498. Springer, Heidelberg (2012)

31. Wintersteiger, C.M., Hamadi, Y., de Moura, L.: A concurrent portfolio approach to SMT solving. In: Bouajjani, A., Maler, O. (eds.) CAV 2009. LNCS, vol. 5643, pp. 715–720. Springer, Heidelberg (2009)

32. Zhang, H., Bonacina, M., Hsiang, J.: PSATO: A distributed propositional prover and its application to quasigroup problems. Journal of Symbolic Computation 21(4), 543–560 (1996). citeseer.ist.psu.edu/zhang96psato.html

A New Approach to Partial MUS Enumeration

Christian Zielke[✉] and Michael Kaufmann

Wilhelm-Schickard Institute of Computer Science, University of Tübingen,
Tübingen, Germany
{zielke,mk}@informatik.uni-tuebingen.de

Abstract. Searching for minimal explanations of infeasibility in constraint sets is a problem known for many years. Recent developments closed a gap between approaches that enumerate all minimal unsatisfiable subsets (MUSes) of an unsatisfiable formula in the Boolean domain and approaches that extract only one single MUS. These new algorithms are described as partial MUS enumerators. They offer a viable option when complete enumeration is not possible within a certain time limit.

This paper develops a novel method to identify clauses that are identical regarding their presence or absence in MUSes. With this concept we improve the performance of some of the state-of-the-art partial MUS enumerators using its already established framework. In our approach we focus mainly on determining minimal correction sets much faster to improve the MUS finding subsequently. An extensive practical analysis shows the increased performance of our extensions.

1 Introduction

Many algorithms in common applications of constraint systems cover problems of finding a satisfying assignment, commonly known as model. Applications require either a single model, a set of these or even all models for a given problem [15].

On the other hand, constraint sets without any model are target for the "infeasibility analysis" algorithms, which can be partitioned into two groups. Their tasks are a) finding a - preferably very large - part of the constraint set that is still satisfiable and b) locating the area of the constraint set where the reason for unsatisfiability lies. These two categories are known by different names: Maximal Satisfiable Subsets (MSS) and Maximum Feasibly Subset (MaxFS) for the former and Minimal(ly) Unsatisfiable Subset or Core (MUS/MUC) and Irreducible Infeasible Subsystem (IIS) for the latter. Although "Max / Min" and "SAT / UNSAT" seem to be completely opposite, they are strongly connected via a hitting set relationship [10,27].

Minimal reasons of infeasibility in linear programming [16,23] and in artificial intelligence [29] are studied since the 1980s. Finding MUSes in SAT covers a lot of applications, including debugging of relational specifications [30] or type errors [2] and model checking [9,31].

The relationship of Maximal Satisfiability and Minimal Unsatisfiability is based on the following: any satisfiable subset of an infeasible constraint set

© Springer International Publishing Switzerland 2015
M. Heule and S. Weaver (Eds.): SAT 2015, LNCS 9340, pp. 387–404, 2015.
DOI: 10.1007/978-3-319-24318-4_28

cannot completely contain any unsatisfiable subset (US) of the formula, and thus must at least exclude one constraint from every US. Searches for results of these can be guided by the results of the other. Algorithms exist that compute MUSes with the help of MSSes [2], vice versa [13] and even ones that use non-minimal USes to support MSS solution finding to finally produce MUSes [22]. The latest improvements for (partially) enumerating MUSes are based on this as well [20, 26].

We propose a novel approach to improve the partial enumeration of MUSes by using the information which clauses are very similar according to their presence or absence in MUSes. We first define basic terms and concepts (Section 2) , before describing the related work and especially the MARCO algorithm [20] (Section 3). We present the new techniques in Sections 4, 5 and an extensive practical analysis (Section 6) before concluding the work and offering some possibilities for future research (Section 7).

2 Preliminaries

Although the presented algorithm can be applied on any constraint programming problem, we focus on a special class of Boolean Satisfiability (SAT) formulae in Conjunctive Normal Form (CNF). A formula F in CNF is a conjunctive set of clauses C, each clause is a disjunctive set of literals. A literal is either a Boolean variable or its complement. These variables can be assigned $true(1)$ or $false(0)$, represented via a mapping $m : v \rightarrow \{0, 1\}$, for all variables v. If there exists at least one model - a mapping that satisfies each clause - the formula is said to be satisfiable. Otherwise, the formula F is unsatisfiable. The SAT problem describes the decision problem whether a formula is satisfiable or not [14]. In this paper we focus only on unsatisfiable instances. The following definitions are used throughout this work.

A **Minimal Unsatisfiable Subset** (MUS) is defined as follows:

Definition 1. *A subset $M \subseteq F$ is an MUS $\Leftrightarrow$ M is unsatisfiable and $\forall c \in M :$ $M \setminus \{c\}$ is satisfiable*

An MUS is essentially a set of constraints, that cannot be reduced without losing unsatisfiability. They minimize an unsatisfiable constraint set to a "core" proof of its inconsistency. They are called "unsatisfiable cores" in some work, but we use the term MUS.

A closely related concept is the one of **Minimal Correction Subsets** (MCS):

Definition 2. *A subset $M \subset F$ is an MCS $\Leftrightarrow$ F \ M is satisfiable and $\forall c \in M :$ $(F \setminus M) \cup \{c\}$ is unsatisfiable*

The removal of an MCS from the formula restores its satisfiability ("corrects" it). The minimality is again not in cardinality, but in the face that no proper subset of an MCS is a correction set itself. An MCS can also be defined as the complement of a **Maximal Satisfiable Subset** (MSS):

Definition 3. *A subset $M \subseteq F$ is an MSS $\Leftrightarrow M$ is satisfiable and $\forall c \in (F \backslash M)$: $M \cup \{c\}$ is unsatisfiable*

A very common problem regarding MSSes is finding the largest MSS. It is also well-known as the MaxSAT problem. Any MaxSAT solution is an MSS, but the converse does not necessarily hold.

Example 1: The following unsatisfiable formula F in CNF is used to explain the basics. We refer to the 6 clauses of the formula as $c_1, \ldots, c_6$.

$F = \bigwedge c_i : 1 \leq i \leq 6$	MUSes(F)	MCSes(F)	MSSes(F)
	$\{c_1, c_2, c_3\}$	$\{c_3\}$	$\{c_1, c_2, c_4, c_5, c_6\}$
$c_1 = (\overline{x_1}) \qquad c_2 = (\overline{x_2})$	$\{c_1, c_3, c_4\}$	$\{c_1, c_5\}$	$\{c_2, c_3, c_4, c_6\}$
$c_3 = (x_1 \vee x_2) \; c_4 = (x_1 \vee \overline{x_2})$	$\{c_2, c_3, c_5\}$	$\{c_2, c_4\}$	$\{c_1, c_3, c_5, c_6\}$
$c_5 = (\overline{x_1} \vee x_2) \; c_6 = (\overline{x_1} \vee \overline{x_2})$	$\{c_3, c_4, c_5, c_6\}$	$\{c_1, c_2, c_6\}$	$\{c_3, c_4, c_5\}$

The formula F has 4 MUSes and 4 MCS/MSS pairs. For simplicity we denote any MUS, MCS and MSS as a set of clauses throughout this work. Note that any MCS is a complement of an MSS and vice versa.

We use the following fact as an important connection between MUSes and MCSes heavily: the set of MUSes of a formula F and the set of MCSes of F are "hitting set duals" of one another. All MUSes of F form a set that is equivalent to the set of all irreducible hitting sets of the MCSes and analogously the set of MCSes is equivalent to all irreducible hitting sets of the MUSes. The following Theorem 1 is proven formally in [7].

Theorem 1. *Let F be an unsatisfiable formula, $MUSes(F)$ the set of all minimal unsatisfiable subsets of F and $MCSes(F)$ the set of all minimal correction sets.*

1. *$U \subset F$ is an MUS $\Leftrightarrow U$ is an irreducible hitting set of $MCSes(F)$*
2. *$C \subset F$ is an MCS $\Leftrightarrow C$ is an irreducible hitting set of $MUSes(F)$*

We recall an intuitive explanation for this from [21] here. Recall that an unsatisfiable formula F has at least one MUS M. Due to the minimality of an MUS it can be made satisfiable by simply deleting a single clause of it. Therefore, a way to make the whole formula F feasible, one has to "dispose" its MUSes by removing at least one clause from every MUS. An MCS corresponds to a set of clauses that accomplishes this: its removal restores the satisfiability of F. Thus, any MCS has to contain at least one element of every MUS of F and due to its minimality the irreducibility of the hitting set is obtained. A similar argument can be found for the fact that MUSes are irreducible hitting sets of MCSes.

Example 2: We explain the property using the example from above.

Whenever an MUS and MCS have a clause in common, an "x" denotes the fact, that the MUS and MCS hit each other. Each clause in the intersection of all MUSes infers an MCS of size one, in this example $\{c_3\}$. All other MCSes are of larger size.

	MCSes			
	$\{c_3\}$	$\{c_1, c_5\}$	$\{c_2, c_4\}$	$\{c_1, c_2, c_6\}$
$\{c_1, c_2, c_3\}$	x	x	x	x x
$\{c_1, c_3, c_4\}$	x	x		x x
$\{c_2, c_3, c_5\}$	x		x x		x
$\{c_3, c_4, c_5, c_6\}$	x		x		x		x

MUSes appears as a label on the left of the rows.

Observe that all the other MCSes hit the corresponding MUSes by their common clauses. Note that the MCS $\{c_1, c_2, c_6\}$ is the only one, where a single MUS $\{c_1, c_2, c_3\}$ is hit by more than one member of the MCS (via c_1, c_2), but neither can be removed from the MCS, because both are exclusively responsible for hitting the second and third MUS, respectively.

3   Related Work and the MARCO Algorithm

The algorithms for extracting a single MUS can be characterized as *constructive*, *destructive* or *dichotomic* [11,28]. Nearly all state-of-the-art MUS extractors [5,25] use a variant of a destructive MUS extraction algorithm, which was first proposed more than 20 years ago [3,8]. A destructive MUS extractor computes a series of reduction steps on formula F, moving into smaller unsatisfiable subsets F' until all subsets $F'' \subset F'$ are satisfiable. The development of the recursive model rotation [6] led to major improvements for MUS extraction algorithms in recent years [4].

The existing work on MUS enumeration can be divided into two main categories: a) approaches that compute MUSes directly and b) algorithms that use hitting set techniques. The direct computation of MUSes is based on an exhaustive search on the power set of subsets of the formula. The explicit enumeration uses a HS-tree data structure [18] together with pruning rules to avoid multiple satisfiability tests for a single subset. Every node in the tree corresponds to a subset S of the formula F, and every child node is labeled with a subset $S' \subset S$. In a depth-first fashion the subsets are tested for unsatisfiability. Each unsatisfiable node whose children are all found to be satisfiable is marked as an MUS. Several improvements could be made for this technique [17], but the iterative SAT-solver calls and the explicit enumeration of all possible subsets of F are a performance bottleneck [2].

Examples for the second category of algorithms are CAMUS [21] and DAA [2]. CAMUS works in two phases: the first phase computes all MCSes of a constraint set by decreasing size using MSS solutions. With every found MSS a new blocking clause is added to F' to ensure, that the same subset is not found in any further SAT-solver calls. The second phase is a hitting set approach that is completely independent on any constraint solver. It starts when all MSSes/MCSes are found.

Due to the possibility that the number of MCSes may be exponential in the size of the formula, the first phase is potentially intractable. Therefore it is not

---

Algorithm 1. MARCO
___
Input: unsatisfiable formula $F = \{c_1, \ldots, c_n\}$
Output: MCSes and MUSes of F as they are discovered
1:   $map \leftarrow$ BoolFormula($s_1, \ldots, s_n$)   ▷ s_i are selector variables
2:   **while** map is satisfiable **do**
3:      $m \leftarrow$ **getModel**(map)
4:      $seed \leftarrow \{c_i \in F : m[s_i] = True\}$ ▷ project the model to F
5:      **if** seed is satisfiable **then**
6:         $MSS \leftarrow$ ***grow***(seed,C)
7:         $MCS \leftarrow$ **complement**(MSS)
8:         **yield** MCS ▷ print the MCS without ending the algorithm
9:         $map \leftarrow map \wedge$ ***blockMCS(MCS)***
10:     **else**
11:        $MUS \leftarrow$ ***shrink***(seed,C)
12:        **yield** MUS ▷ print the MUS without ending the algorithm
13:        $map \leftarrow map \wedge$ ***blockMUS(MUS)***
14:     **end if**
15: **end while**

---

suitable in cases, where some MUSes should be found very quickly. Partial MUS enumeration resolves this problem.

The DAA (dualize and advance) algorithm is an incremental hitting set approach developed by Bailey and Stuckey [2]. It uses the same relationship of MCSes and MUSes, but computes both subsets during its execution. In every iteration a satisfiable subset is grown into an MSS. Its complement MCS is added to the set of already found MCSes. On this set of MCSes all minimal hitting sets (possible MUSes) are computed. Each MUS candidate is checked for unsatisfiability. Whenever one candidate is found to be satisfiable, it is used as a starting point for the computation of a new MSS in the next iteration. The main bottlenecks here are the computation of the hitting sets and the test whether each MUS candidate is unsatisfiable.

3.1 The MARCO Algorithm

Independently from each other Previti and Marques-Silva [26] and Liffiton and Malik [20] developed two very similar algorithms called eMUS and MARCO. Both can be seen as *partial* (or incremental) MUS enumerators, not replacing any state-of-the-art complete MUS enumerators like CAMUS [21] but providing a viable alternative for satisfiability instances where the full enumeration is computationally infeasible in limited time. Their major advantage is in reporting the first MUS as quickly as state-of-the-art MUS extractors and reporting further MUSes with a similar delay. We describe the MARCO algorithm here and use its implementation for our extensions. We refer to the pseudocode in Algorithm 1.

The novel approach to (partially) enumerate MUSes is the use of an additional SAT instance containing only the selector variables $s_1, \ldots, s_n$ (line 1) of the unsatisfiable formula $F = (c_1 \wedge \ldots \wedge c_n)$. Every model obtained from this "meta"-instance (called map) is an unexplored subset of the formula. Depending on its satisfiability, it is either used to find a new MSS (MCS) or a new MUS.

Initially the meta-instance is a tautology (true in every model), meaning that no subset has been explored. Given the fact, that the map contains the selector variables s_i that were added to every clause c_i in F each model can be projected onto F to identify an unexplored element of F's power set of possible subsets. If this subset, called "seed", is satisfiable it must be a subset of an MSS and thus can be expanded into an MSS via the **grow**-method (line 6). Likewise, if the seed is unsatisfiable, it has to be a superset of an MUS and therefore it can be used as the starting point for an MUS extraction algorithm (**shrink**-method in line 11). In either case the result is reported and used to mark a region in the map as explored. For every found MUS U and every MSS S (respectively its complement MCS C) the following clauses are added.

$$\textbf{\textit{blockMUS(U)}} : \bigvee_{i:c_i \in U} \overline{s_i} \qquad\qquad \textbf{\textit{blockMCS(C)}} : \bigvee_{i:c_i \in S} s_i$$

blockMUS(U) ensures, that at least one member of the MUS must not be present in any further seed. In other words, all proper supersets of the MUS are forbidden as new seeds. Likewise, **blockMCS(C)** ensures that at least one clause of every MCS has to be in any seed from now on. Eventually all MCSes and MUSes are enumerated and the algorithm terminates due to the map being unsatisfiable.

A major difference of the eMUS approach from Previti and Marques-Silva [26] to this MARCO algorithm is the usage of maximal models as seeds, making the **grow**-method obsolete. However, the latest MARCO versions use maximal models by default. We decided to use MARCO v1.0.1 as the base program, in which our extensions are incorporated, since it outperforms eMUS in our practical experiments (see Section 6).

Note, that the **grow**-method is used as an MSS/MCS oracle in MARCO, instead of using maximal models, any state-of-the-art MSS/MCS extractor [1,24] could be "plugged-in" to potentially speed-up the computation as well.

4   A First Extension: Determine MUS Members via Map

The map in MARCO allows us to determine the hitting set property of MUSes and MCSes. In its latest release of the MARCO algorithm a feature was added that uses top-level assignments within the map to identify clauses, that have to be present in an MUS. Top-level assignments are implications that were caused by the propositional logic of the map without any further assumptions of variables. These top-level assignments can be caused for example by unit clauses, that were added by MCSes of size one.

We decided to extend this in the following way. Given a formula F and a seed $S \subset F$. Whenever S is unsatisfiable we determine the set of positive top-level assignments of the map that are forced by adding the corresponding negative literals $\overline{s_i}$ for each clause $c_i \in F \setminus S$ as assumptions to the map. This is done via the method **getImplies(seed)**. Suppose during the execution of MARCO the

MCS $\{c_1, c_2, c_3\}$ was found and the current unsatisfiable seed obtained from the map contains s_3, but not s_1 nor s_2. Then the clause c_3 has to be present within the MUS to ensure that the hitting set property between MUSes and MCSes is preserved. The **shrink**-method would provide the result as well, but potentially has to use one SAT-solver call to determine this. By declaring some clauses as definitive members of the MUS the extraction algorithm can save unnecessary SAT-solver calls. This extension alone does not lead to an improved performance. The effort to compute the set of forced assignments is potentially larger than the savings during the **shrink**-method. This is the reason we extended the **grow**-method (see Section 5.4) to find more MCSes faster to improve the performance of the MARCO algorithm with the help of the extended **getImplies(seed)**-method.

5   Using Blocks to Speed-Up Partial MUS Enumeration

All our extensions to the state-of-the-art partial MUS enumeration approach are based on the following *block* property of clauses, which was already introduced as *generalized nodes* in an approach to speed-up Hitting-Set-Computations in Hypergraphs [19].

Definition 4 (Block property). *Given an unsatisfiable formula F. A **block** b is a set of clauses $b = \{c_x, c_y, \ldots, c_z\}$ that are always either exactly altogether present in an MUS or not:*

$$\forall M \in MUSes(F) : b \cap M = \emptyset \vee b \cap M = b$$

The blocks are clause maximal, meaning that the block b cannot be extended by any clause $c_i \in F \setminus b$ without losing the block property.

Some trivial observations derived from this definition are that every clause belongs to exactly one block and the set of blocks $\mathcal{B}$ is a partition of the unsatisfiable formula F. We denote b_0 as the block of clauses that do not belong to any MUS of F. Then $F \setminus b_0$ is the union of all MUSes.

5.1   Determine the Blocks

To obtain the set of blocks $\mathcal{B}(F)$ for an unsatisfiable formula F the straightforward approach is to enumerate all MUSes of F and use the following split routine. Initially $\mathcal{B}_0 = b_0 = F$. With no found MUSes all clauses of the formula belong to the default block of clauses. Note that M_i denotes the i-th found MUS and therefore $\mathcal{B}_i$ denotes the set of (interim) blocks that are formed by the MUSes $M_1, \ldots, M_i$. Please note that interim blocks only permit the block property for the MUSes $M_1, \ldots, M_i$, and not necessarily for the later ones. Nevertheless we drop the word "interim" from it in the remaining part of the work.

Definition 5 (Splitting blocks). *Let $\mathcal{B}_i = \{b_0, \ldots, b_x\}$ be the set of blocks for an unsatisfiable formula F, that were obtained by splitting the blocks via the MUSes $\{M_1, \ldots, M_i\}$ and $M_{i+1} \subset F$ be the next MUS that was discovered during the MUS enumeration algorithm. Then $\mathcal{B}_i$ is updated to $\mathcal{B}_{i+1}$ via Algorithm 2.*

Algorithm 2. splitblocks

Input: blocks $\mathcal{B}_i = \{b_0, \ldots, b_x\}$ and MUS $M_{i+1} \subset F = \{c_1, \ldots, c_n\}$
Output: blocks $\mathcal{B}_{i+1}$
1: $\mathcal{B}_{i+1} \leftarrow \emptyset$
2: $m \leftarrow x + 1$ ▷ new block index (b_x is last element in $\mathcal{B}_i$)
3: **for** $b_i \in \mathcal{B}_i$ **do**
4:     **if** $i == 0$ **and** $0 < |b_i \cap M_{i+1}|$ **then** ▷ clauses that were in no MUS until now
5:         $b_m \leftarrow b_i \cap M_{i+1}$ ▷ build new block b_m
6:         $b_i \leftarrow b_i \setminus b_m$ ▷ update the old block b_i
7:         $\mathcal{B}_{i+1} \leftarrow \mathcal{B}_{i+1} \cup b_i \cup b_m$ ▷ add both blocks b_i, b_m
8:         $m \leftarrow m + 1$
9:     **else if** $0 < |b_i \cap M_{i+1}| < |b_i|$ **then**   ▷ real subset
10:         $b_m \leftarrow b_i \cap M_{i+1}$ ▷ build new block b_m
11:         $b_i \leftarrow b_i \setminus b_m$ ▷ update the old block b_i
12:         $\mathcal{B}_{i+1} \leftarrow \mathcal{B}_{i+1} \cup b_i \cup b_m$ ▷ add both blocks b_i, b_m
13:         $m \leftarrow m + 1$
14:     **else**   ▷ block unchanged
15:         $\mathcal{B}_{i+1} \leftarrow \mathcal{B}_{i+1} \cup b_i$
16:     **end if**
17: **end for**

Since the blocks can only get smaller, each block $\in \mathcal{B}_k$ is an ancestor for at least one block $\in \mathcal{B}(F)$. Each ancestor block can be seen as an over-approximation of a block, that gets tighter with more MUSes found until ultimately reaching equality. Tightness is reached at latest when all MUSes were enumerated, but could be obtained earlier as well. For example, whenever $\mathcal{B}_k \neq \mathcal{B}(F)$ contains a block of size 1 that block cannot be split any further.

Example 3: The **splitblocks**-method results in the following blocks for the formula F from above and the given sequence of enumerated MUSes.

Initialization:		$\mathcal{B}_0 = \{(C_1, C_2, C_3, C_4, C_5, C_6)\}$
1st MUS:	$M_1 = \{C_1, C_2, C_3\}$	$\mathcal{B}_1 = \{(C_4, C_5, C_6), (C_1, C_2, C_3)\}$
2nd MUS:	$M_2 = \{C_1, C_3, C_4\}$	$\mathcal{B}_2 = \{(C_5, C_6), (C_2), (C_4), (C_1, C_3)\}$
3rd MUS:	$M_3 = \{C_2, C_3, C_5\}$	$\mathcal{B}_3 = \{(C_6), (C_2), (C_4), (C_1), (C_5), (C_3)\}$
4th MUS:	$M_4 = \{C_3, C_4, C_5, C_6\}$	$\mathcal{B}_4 = \mathcal{B}_3 = \mathcal{B}(F)$

Real-world instances show some important properties: The block of clauses that are present in no MUS at all (b_0) is normally the largest block and although there are a lot of blocks containing only one clause, several blocks of larger sizes are present as well.

5.2   Proving the Block Property

With the block property proven, we could save many SAT-solver calls due to the fact that whenever one clause of the block is determined to be present or absent in the current MUS within the **shrink** subroutine, all other members of the block are determined as well, without using any additional SAT-solver calls. To prove the block property we could use the available map instance. Recall that the map is used to determine already covered areas of the search space. It provides the main method with a seed from an area of the search space, that was not yet covered by the algorithm and therefore offers a new result, either an MCS or an MUS.

By adding (an over-approximation of) a block $b_i \in \mathcal{B}_k$ to the map via **blockMUS(b_i)** the map provides seeds where not the whole block b_i is present. Let c_i be the clause that was added to the map via **blockMUS(b_i)** and the seed returned from the map to be unsatisfiable. During the subsequent **shrink**-method the SAT-solver either deletes all members of b_i from the seed to find a new MUS, or at least one member of b_i is still present in the new MUS M_{k+1}. In the first case, the block b_i is not touched and thus is still valid. The algorithm could go on with the proposed block b_i added to the map. In the second case, the block b_i is divided into two new blocks $b_{i'}$ and $b_{i''}$ with $b_{i'}$ consisting of the elements of b_i that are present in the new M_{k+1}, and $b_{i''} = b_i \setminus b_{i'}$. Suppose we continue trying to prove the block property for $b_{i'}$, since b_i was proven to be an over-approximation. Adding $b_{i'}$ to the map would make b_i obsolete, since $b_{i'}$ is a subset of b_i and thus the clause $c_{i'}$ added via **blockMUS(b_i')** subsumes the clause c_i. From now on the map provides seeds where not the whole block $b_{i'}$ is present until it reaches unsatisfiability. In the end, the block property of the current block is proven, since every unsatisfiable seed that is provided by the map without the current block has to contain the whole block.

The problem of this approach to prove the block property for a block b_i is, that it finds all MUSes $\mathcal{M}$ that do not contain b_i. Thus the proven property is only available when enumerating the remaining MUSes $\mathcal{M}' = \text{MUSes(F)} \setminus \mathcal{M}$, a part of them already enumerated and used to redefine b_i by the **splitblocks**-method. Therefore, we show in the next sections how unproven blocks (that are over-approximations of blocks) are used to support the MUS and MCS detection.

5.3   Using Block Information During *shrink*-Method

The **shrink**-method can be any state-of-the-art MUS extraction algorithm. MARCO uses muser2 [5]. One major advantage and prerequisite for our extension is, that the solver is able to cope with so-called *group*-MUS instances [21].

Definition 6. *Given an explicitly partitioned unsatisfiable CNF formula $F = D \cup \bigcup_{G \in \mathcal{G}} G$ with $\mathcal{G} = \{G_1, \ldots, G_k\}$, D and G_i being disjoint sets of clauses, a **group oriented MUS** of F is a subset $\mathcal{G}'$ of $\mathcal{G}$, such that $D \cup \bigcup_{G \in \mathcal{G}'}$ is unsatisfiable, and $\forall \mathcal{G}'' \subset \mathcal{G}' : D \cup \bigcup_{G \in \mathcal{G}''}$ is satisfiable.*

D is the default group (often denoted as being group G_0) that has to be present in every MUS. It consists of the clauses that correspond to the implied variable assignments given by the map as described in Section 4.

The possibility to define partitioned groups allows us to use the block information of clauses rather straight-forward. All blocks b_i that are present in the seed get their own group $G_{n+i} = \{seed \cap b_i\}$ with n being the number of clauses in the unsatisfiable formula F. The lone exception from this rule is the block b_0 of clauses, that were not present in any MUS until now. Each of these clauses $c_i \in \{seed \cap b_0\}$ form their own group $G_i = \{c_i\}$. Due to the blocks being over-approximations, the block property is not proven for the members. Thus, executing the MUS extractor on this grouped instance does not return an MUS, but rather an over-approximation of an MUS $gM \supseteq M$ as well. We have to run the MUS extractor a second time iff $\exists G_i \in gM : |G_i| > 1$. For every $G_i \in gM$ with $|G_i| = 1$ we know that the clause representing this group has to be in M. We add that clause to G_0. Since it was found to be critical for the over-approximation gM, it has to be critical for each subset of gM, especially M, as well. A clause c is critical when the deletion of it from an unsatisfiable set of clauses U causes $U \setminus c$ being satisfiable.

For all other groups $G_j \in gM$ with $|G_j| > 1$ every clause $c_i \in G_j$ forms its own group G_i for the second call. Together with the increased G_0, that can be possibly (when $gM \subset seed$) further increased by new forced implications recognized via **getImplies(gM)** they form a new instance where every non-default group is of size one. Running the MUS extractor on this finally returns an MUS $M \subset F$.

As we have seen, using the block property may cause that two MUS extractor calls have to be used to determine a single MUS. Nevertheless the sum of SAT-solver calls in those two MUS extractor runs is usually much smaller in comparison to the normal **shrink**, when a large group G_i could be deleted from the seed within the first run.

5.4   Using Block Information to Find More MCSes

To gain additional boost of the **getImplies(seed)**-method we present a method that uses the block information to determine likely candidates for other MCSes. As we have seen in Section 4, with every **blockMCS()** a clause is added, which then can be used to infer clauses, that have to be part of an MUS.

Recall, that when two clauses c_i and c_j are present in the same block b_k, the clauses do not appear separately in any MUS. This leads to the following Lemma.

Lemma 1. *Let block b_k have at least two clauses c_i and c_j. For every MCS M with $c_i \in M$, there has to be another MCS $M' = c_j \cup (M \setminus c_i)$.*

Proof: By the hitting set property of MCSes and MUSes and the minimality of MCSes we know that there has to be at least one MUS U with $U \cap M = c_i$. If there is no such U, then M would not be minimal. It would be possible to eliminate c_i from M and not lose the hitting set property of the set of MUSes. But since c_i and c_j are in the same block b_k all MUSes that were hit by c_i are hit by c_j as well. Therefore $M' = (M \setminus c_i) \cup c_j$ is a valid MCS by the hitting set property.□

Based on this Lemma we present the following algorithm, that is called whenever a new MCS is found via the ***grow***-method (line 6 in Algorithm 1):

Algorithm 3. moreMCS

Input: blocks $\mathcal{B}_i = \{b_0, \ldots, b_x\}$ and MCS $C = \{c_1, \ldots, c_n\}, C \subset F$
Output: MCSes and MUSes of F as they are discovered
1: find block $blk(c_j) \in \mathcal{B}_i$ for every $c_j \in C$
2: **for** every possible combination $MCSc$ in $\{blk(c_1)\} \times \ldots \times \{blk(c_n)\}$ **do**
3:     $MSSc \leftarrow$ **complement**$(MCSc)$ ▷ get the MSS candidate
4:     **if** $MSSc$ is satisfiable **then**   ▷ new MCS found
5:         **yield** $MCSc$ ▷ print the MCS without ending the algorithm
6:     **else**   ▷ unsatisfiable seed for MUS extraction
7:         $MUS \leftarrow$ **shrink**$(MSSc)$ ▷ extract new MUS
8:         **yield** MUS ▷ print the MUS without ending the algorithm
9:         **splitblocks**(MUS) ▷ use MUS to split blocks
10:        find more MCSes/MUSes in combinations of split blocks   ▷ see Example 4
11:        **return**
12:     **end if**
13: **end for**

According to Lemma 1 the algorithm tests all possible combinations as long as the resulting candidate MSSes $MSSc$ are satisfiable. Whenever the algorithm detects an unsatisfiable $MSSc \subset F$, it is used as a seed to the ***shrink***-method to extract an MUS. This MUS is used to split the blocks, causing at least one of the blocks $\{blk(c_1), \ldots, blk(c_n)\}$ to be split.

Example 4: Suppose $C = \{c_1, c_4\}$ is the MCS that triggered the call of **moreMCS**, $blk(c_1) = \{c_1, c_2, c_3\}$, $blk(c_4) = \{c_4, c_5, c_6\}$. That leads to $|blk(c_1)| * |blk(c_4)| - 1 = 3 * 3 - 1 = 8$ possible new MCSes since $\{c_1, c_2\}$ has not to be tested. Suppose that $\{c_1, c_5\}, \{c_1, c_6\}$ are tested successfully as MCSes, but $\{c_2, c_4\}$ is not an MCS. We know that (at least) c_2 has to leave the block $blk(c_1)$ since it violates Lemma 1.

The algorithm ensures this due to the following observation. The seed $MSSc$ does not contain c_4, since the seed is the complement of $\{c_2, c_4\}$. Due to the hitting set property and the MCS $\{c_1, c_4\}$ at least c_1 has to be present in the new MUS, so c_2 and c_1 cannot remain in the same block after the split operation.

Suppose the new blocks after the split operation are $b_i = \{c_1, c_3\}, b_i' = \{c_2\}, b_j = \{c_4, c_5\}, b_j' = \{c_6\}$. The new possible combinations are $b_i \times b_j, b_i \times b_j', b_i' \times b_j, b_i' \times b_j'$. Please note, that in line 10 of Algorithm 3 the combinations $b_i' \times b_j, b_i' \times b_j'$ would not be tested, since no MCS was found that hits these combinations of blocks.

The implementation ensures that no subsets are tested twice during the recursion to prevent doubled results. For example, the MCS candidate $\{c_1, c_6\}$ from the combination $b_i \times b_j'$ is not tested again, but the candidate $\{c_3, c_6\}$ from the same combination has to be tested.

6   Practical Results

To evaluate the extensions to the MARCO algorithm and to compare it to the previous approaches for (partial) MUS enumeration, MARCO and eMUS, we ran all algorithms on a set of 207 instances from the Boolean Satisfiability domain. These instances were drawn from a large variety of applications, with the most prominent being hardware and software verification, product configuration and bounded model checking. The benchmark set is a subset of the MUS track of the 2011 SAT competition[1] containing only instances where at least two MUSes and one MCS are found within the time limit of one hour. This decision is based on the fact that our techniques for boosting the computation of MUSes and MCSes use the block information, which is inferred from the already enumerated MUSes. Both techniques (*shrink*, *moreMCS*) are triggered for the first time when the original MARCO algorithm found the first MCS, respectively starts to extract the second MUS from a part of the formula. Thus, we focus our analysis of the effects on the performance on these instances.

We used the latest MARCO release[2] v1.0.1 as the framework for our extension. It is written as a python script that uses the MiniSAT [12] solver for the formula F as well as the map. The *shrink*-method uses muser2 [5] as a MUS and group-MUS extraction algorithm. All experiments were run on 2.83GHz Intel Xeon CPUs with a 3600 second timeout and a 16 GB memory limit.

We use the following terminology to describe the different versions of the algorithm and its possible combinations:

- **MARCO** the original algorithm proposed by Liffiton et al. [20] (see Section 3)
- **MARCO+** more critical clauses obtained by *getImplies*-method (see Section 4)
- **MARCOs** block information used during *shrink* (see Section 5.3)
- **MARCOm** block information used to find more MCSes faster (see Section 5.4)

Thus, when mentioning MARCO+m the second and fourth option are used in parallel.

The first results (Figure 1) show that MARCO finds more MUSes than eMUS for 183 instances, 24 times eMUS reports more MUSes and both provide the same amount only twice. 44 times the number of MUSes found by MARCO is one order of magnitude higher than the number found by eMUS. In comparison to MARCO+ the results are not so clear. In that case MARCO reports more MUSes for 86 instances, in 70 out of 207 instances MARCO+ finds more MUSes and for the remaining 51 instances both versions find the same amount of MUSes. The additional effort to compute forced MUS members shown in Section 4 is not worth it when using the original MARCO algorithm without any further extensions.

---

[1] http://www.satcompetition.org/2011/
[2] http://sun.iwu.edu/~mliffito/marco/

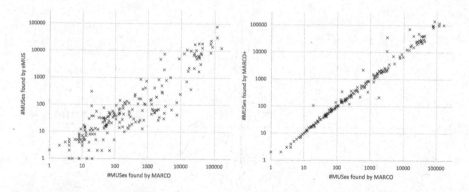

Fig. 1. Comparing MARCO to eMUS (left) and MARCO+ (right): number of MUSes found within time limits of 3600 seconds. Each point declares one out of the 207 instances

The additional use of MARCOm changes that. MARCO+ benefits from the MCSes that have been produced by MARCOm earlier. The resulting version MARCO+m reports more MUSes for 105 instances, less MUSes in 87 and for the remaining 15 instances the same amount as the original MARCO within the time limits. Nearly 94% of the found MCSes $(6,944,690$ out of $7,390,727)$ are reported by the algorithm presented in Section 5.4.

6.1 Workload Computation

The partial MUS enumerators used in this work produce two different results, MUSes and MCSes. Therefore, the evaluation and comparison of different approaches should cover both results as well.

Figure 2 shows the relative number of MUSes and MCSes found by MARCO+m and MARCO. The values on the x-axis are computed by the logarithm (to the base 2) of the fraction of MUSes found by MARCO+m and by MARCO. The y-values show the ratio of found MCSes.

Points in the positive region of the x-axis denote instances where MARCO+m found more MUSes in the same time limit. The same applies correspondingly for the y-axis and the number of MCSes. For the vast majority of 91.8 % (190 out of 207) MARCO+m reports more MCSes than MARCO.

For 50.7 % of the instances (105 out of 207) MARCO+m outperforms MARCO on both, MUSes and MCSes. The opposite holds for just 5.3 % of the instances (11 out of 207). For 10 instances the number of found MUSes and MCSes are identical. The remaining 81 instances where MARCO+m either found less MUSes, but more MCSes or vice versa, cannot simply be evaluated by the raw numbers of found MUSes/MCSes since the effort to compute one MUS, or MCS respectively, varies.

Thus, we introduce the following scoring function, called the *additional expected workload* to compare the results of two algorithms A_1 and A_2.

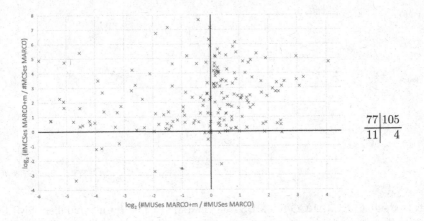

Fig. 2. $\log_2$ of the relative number of MUSes on x-axis and of MCSes on y-axis; together with the amount of points (instances) for every quadrant in the plane - 10 points are lying in the point or origin

Definition 7 (The additional expected workload). *Let Algorithm A_1 and Algorithm A_2 be two partial MUS enumerators and the respective number of found MUSes $nU(A_i)$ and MCSes $nC(A_i)$. With the time used to compute all found MUSes $tU(A_i)$ and MCSes $tC(A_i)$ for a fixed instance we define the additional expected workload of A_1 in comparison to A_2 as:*

$$wl(A_1, A_2) = nU(A_1)\frac{tU(A_1) + tU(A_2)}{nU(A_1) + nU(A_2)} + nC(A_1)\frac{tC(A_1) + tC(A_2)}{nC(A_1) + nC(A_2)} - (tU(A_1) + tC(A_1))$$

The first term of the formula describes the expected time that is needed to find the number of MUSes by algorithm A_1. It is computed via the average time both algorithms needed to compute a single MUS. The second term describes the same for the MCSes found by algorithm A_1. Subtracting the real times the algorithm A_1 spends computing MUSes and MCSes from this sum we get a positive value iff A_1 performed better than A_2 because the expected runtime is higher than the actual runtime. Note that $wl(A_1, A_2) = -wl(A_2, A_1)$.

For the aforementioned 81 instances represented by the points in the second and forth quadrant of the Cartesian plane shown in Figure 2 we get a sum of the *additional expected workload* of 3412.86 seconds for MARCO+m in comparison to MARCO. When expanding the sum to all 207 instances in the benchmark set we get the *additional expected workload* of **99911,44 seconds** with a median of 376.26 seconds and an average of 482.66 seconds. Our approach MARCO+m clearly outperforms the state-of-the-art MARCO algorithm.

To put these results in perspective: the overall runtime for MARCO+m on the 207 instances is approximately 730000 seconds, with the *additional expected workload* of 99911.44 seconds **MARCO+m performs 13.7 % better than MARCO.**

With Figure 3 we want to support the observation that with an increasing amount of known MCSes the runtime to extract one MUS can be decreased. The figure compares the runs from MARCO, MARCO+ and MARCO+m on the instance dlx2_aa[3]. This instance is a hardware verification problem and was first used in the MUS extraction track of the SAT competition 2011. On the x-axis the sequence of extracted MUSes during the enumeration is shown. The bars denote the amount of known MCSes at the moment the *shrink*-method is called. The height of the bars gives an upper bound on how many clauses of the MUS are known to be critical before the MUS extractor is started. The corresponding points on the lines represent the runtime in ms that was needed to extract the MUS in the *shrink*-method.

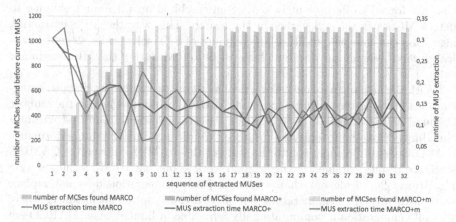

Fig. 3. Comparing MARCO+m to MARCO+ and MARCO: the difference of runtime for the MUS extraction correlates to the difference in the number of known MCSes at that time, especially for the MUSes in the range between 6 - 17

We see a significant difference in runtime in the majority of the MUS extractor calls (*shrink*-method) for the MUSes from 6 to 17. We see that the definitive MUS members found via the *getImplies*-method before the start of the MUS extractor result in a reduced runtime for MARCO+ in comparison to MARCO. Furthermore the differences in runtime from MARCO+ and MARCO+m correlate to the differences in known MCSes for the corresponding time points during the MUS enumeration algorithm. Due to the fact that at the start of an MUS extraction MARCO+m already found more MCSes, the algorithm is able to detect more forced members of the MUSes beforehand via the *getImplies*-method resulting in a significantly shorter MUS extraction time. After the 18th MUS, when all three versions have found approximately the same amount of MCSes, the runtimes of the *shrink*-methods approach each other again. The average MUS extraction time for this instance and the three MARCO versions are 154ms for MARCO, 149ms for MARCO+ and 118 ms for MARCO+m.

---

[3] Similar results can be obtained for a whole set of benchmarks.

The extensions of MARCO with activated *shrink* option (MARCOs, MARCO+s and MARCO+ms, see Section 5.3) do not result in any improvements of the performance. Thus, we do not provide any additional numbers or charts for these versions.

7   Conclusion

We presented an extension to the MARCO algorithm based on a novel **block property** for clauses. We can use this property for both interesting subset detections, the *shrink*-method for MUS extraction and for finding more MCSes whenever MARCO has found a new MCS via *grow*. The main performance boost is obtained by our extension of the already available MARCO map to infer clauses that are forced to be present in a MUS in combination with our technique to detect MCSes faster and earlier in the search. This way the MUS extraction algorithm can save potentially a large amount of SAT-solver calls, since the criticality for these clauses does not have to be tested during the MUS extractor call.

With the help of an extensive practical analysis we could show that our extensions lead to a better performance of our MARCO variant regarding the number of found MUSes, MCSes, as well as the *expected additional workload* within a time limit of one hour for every instance. We did not analyze the performance of MARCO+m in comparison to the state-of-the-art MUS enumerator CAMUS, since the results obtained in the original work about partial MUS enumeration algorithms [20, 26] do not change: partial MUS enumeration does not replace state-of-the-art MUS enumerators, but offers a viable option for instances where the full enumeration is computationally infeasible in limited time. MARCO+m offers better results than MARCO / eMUS, but does not close the gap completely.

Future research directions include exploring the possibilities to use the block information directly within the *grow*-method or any MSS/MCS oracle, as well as more engineering and extensive use of more sophisticated data-structures to boost the partial MUS enumeration further. Additionally, it would be advantageous to prove the block property without the current problems stated.

Another useful research direction can be the development of a novel quality measure of an MUS. At the moment, the number of MUSes seems to be the natural measure when assessing partial MUS enumeration techniques, but finding one MUS of superb quality with respect to the application may be better than finding many MUSes of poor quality with respect to the application.

Acknowledgments. The authors thank the anonymous reviewers for the helpful comments as well as Mark Liffiton and Ammar Malik for providing the well-documented code of MARCO.

References

1. Bacchus, F., Davies, J., Tsimpoukelli, M., Katsirelos, G.: Relaxation search: a simple way of managing optional clauses. In: AAAI 2014, pp. 835–841 (2014)

2. Bailey, J., Stuckey, P.J.: Discovery of minimal unsatisfiable subsets of constraints using hitting set dualization. In: Hermenegildo, M.V., Cabeza, D. (eds.) PADL 2004. LNCS, vol. 3350, pp. 174–186. Springer, Heidelberg (2005)
3. Bakker, R.R., Dikker, F., Tempelman, F., Wognum, P.: Diagnosing and solving over-determined constraint satisfaction problems. In: IJCAI 1993, pp. 276–281 (1993)
4. Belov, A., Lynce, I., Marques-Silva, J.: Towards efficient MUS extraction. AI Commun. **2**, 97–116 (2012)
5. Belov, A., Marques-Silva, J.: MUSer2: an efficient MUS extractor. In: JSAT 2012, pp. 123–128 (2012)
6. Belov, A., Marques-Silva, J.: Accelerating MUS extraction with recursive model rotation. In: FMCAD 2011, pp. 37–40 (2011)
7. Birnbaum, E., Lozinskii, E.L.: Consistent subsets of inconsistent systems: structure and behaviour. J. Exp. Theor. Artif. Intell. **15**(1), 25–46 (2003)
8. Chinneck, J.W., Dravnieks, E.W.: Locating Minimal Infeasible Constraint Sets in Linear Programs. INFORMS Journal on Computing **3**(2), 157–168 (1991)
9. Clarke, E., Kroning, D., Lerda, F.: A tool for checking ANSI-C programs. In: Jensen, K., Podelski, A. (eds.) TACAS 2004. LNCS, vol. 2988, pp. 168–176. Springer, Heidelberg (2004)
10. De Kleer, J., Williams, B.C.: Diagnosing multiple faults. AI **32**(1), 97–130 (1987)
11. Desrosiers, C., Galinier, P., Hertz, A., Paroz, S.: Using heuristics to find minimal unsatisfiable subformulas in satisfiability problems. J. Comb. Optim. **18**(2), 124–150 (2009)
12. Eén, N., Sörensson, N.: An extensible SAT-solver. In: Giunchiglia, E., Tacchella, A. (eds.) SAT 2003. LNCS, vol. 2919, pp. 502–518. Springer, Heidelberg (2004)
13. Fu, Z., Malik, S.: On solving the partial MAX-SAT problem. In: Biere, A., Gomes, C.P. (eds.) SAT 2006. LNCS, vol. 4121, pp. 252–265. Springer, Heidelberg (2006)
14. Garey, M.R., Johnson, D.S.: Computers and Intractability: A Guide to the Theory of NP-Completeness (1979)
15. Gomes, C.P., Sabharwal, A., Selman, B.: Model counting. In: Handbook of SAT 2009, pp. 633–654 (2009)
16. Greenberg, H.J., Murphy, F.H.: Approaches to Diagnosing Infeasible Linear Programs. INFORMS Journal on Computing **3**(3), 253–261 (1991)
17. Han, B., Lee, S.J.: Deriving minimal conflict sets by CS-trees with mark set in diagnosis from first principles. IEEE Transactions on Systems, Man, and Cybernetics, 281–286 (1999)
18. Hou, A.: A Theory of Measurement in Diagnosis from First Principles. AI **65**(2), 281–328 (1994)
19. Kavvadias, D.J., Stavropoulos, E.C.: An efficient algorithm for the transversal hypergraph generation. Journal of Graph Algorithms and Applications **9**, 239–264 (2005)
20. Liffiton, M.H., Malik, A.: Enumerating infeasibility: finding multiple MUSes quickly. In: Gomes, C., Sellmann, M. (eds.) CPAIOR 2013. LNCS, vol. 7874, pp. 160–175. Springer, Heidelberg (2013)
21. Liffiton, M.H., Sakallah, K.A.: Algorithms for Computing Minimal Unsatisfiable Subsets of Constraints. J. Autom. Reasoning **40**(1), 1–33 (2008)
22. Liffiton, M.H., Sakallah, K.A.: Generalizing core-guided Max-SAT. In: Kullmann, O. (ed.) SAT 2009. LNCS, vol. 5584, pp. 481–494. Springer, Heidelberg (2009)
23. van Loon, J.: Irreducibly inconsistent systems of linear inequalities. EJOR **8**(3), 283–288 (1981)

24. Marques-Silva, J., Heras, F., Janota, M., Previti, A., Belov, A.: On computing minimal correction subsets. In: IJCAI 2013 (2013)
25. Nadel, A., Ryvchin, V., Strichman, O.: Accelerated Deletion-based Extraction of Minimal Unsatisfiable Cores. JSAT **9**, 27–51 (2014)
26. Previti, A., Marques-Silva, J.: Partial MUS enumeration. In: AAAI 2013, pp. 818–825 (2013)
27. Reiter, R.: A Theory of Diagnosis from First Principles. Artificial Intelligence **32**(1), 57–95 (1987)
28. Silva, J.P.M.: Minimal Unsatisfiability: models, algorithms and applications (invited paper). In: ISMVL 2010, pp. 9–14 (2010)
29. de Siqueira N., J.L., Puget, J.F.: Explanation-based generalisation of failures. In: ECAI 1988, pp. 339–344 (1988)
30. Torlak, E., Chang, F.S.-H., Jackson, D.: Finding minimal unsatisfiable cores of declarative specifications. In: Cuellar, J., Sere, K. (eds.) FM 2008. LNCS, vol. 5014, pp. 326–341. Springer, Heidelberg (2008)
31. Velev, M.N.: Using rewriting rules and positive equality to formally verify wide-issue out-of-order microprocessors with a reorder buffer. In: DATE 2002, pp. 28–35 (2002)

Evaluating CDCL Variable Scoring Schemes

Armin Biere[(✉)] and Andreas Fröhlich

Johannes Kepler University, Linz, Austria
{armin.biere,andreas.froehlich}@jku.at

Abstract. The VSIDS (variable state independent decaying sum) decision heuristic invented in the context of the CDCL (conflict-driven clause learning) SAT solver Chaff, is considered crucial for achieving high efficiency of modern SAT solvers on application benchmarks. This paper proposes ACIDS (average conflict-index decision score), a variant of VSIDS. The ACIDS heuristics is compared to the original implementation of VSIDS, its popular modern implementation EVSIDS (exponential VSIDS), the VMTF (variable move-to-front) scheme, and other related decision heuristics. They all share the important principle to select those variables as decisions, which recently participated in conflicts. The main goal of the paper is to provide an empirical evaluation to serve as a starting point for trying to understand the reason for the efficiency of these decision heuristics. In our experiments, it turns out that EVSIDS, VMTF, ACIDS behave very similarly, if implemented carefully.

1 Introduction

The application track of SAT competitions [1,2] is dominated by *conflict-driven clause learning* (CDCL) [3] solvers. Beside *learning* [4], the most important feature of these solvers is the *variable state independent decaying sum* (VSIDS) decision heuristic [5], actually in its modern variant *exponential VSIDS* (EVSIDS) [6], as first implemented in the MiniSAT solver [7]. The EVSIDS heuristic allows fast selection of decision variables and adds focus to the search, but also is able to pick up long-term trends due to a "smoothing" component, as argued in [6].

On the practical side, there have been various attempts to improve on the EVSIDS scheme. These include the *variable move-to-front* (VMTF) strategy of the Siege SAT solver [8], the BerkMin strategy [9], which is focusing on recently learned clauses, and the *clause move-to-front* (CMTF) strategies of HaifaSAT [10] and PrecoSAT [11]. In this paper, we suggest another new decision heuristic, called *average conflict-index decision score* (ACIDS). Our main contribution, however, is to show that EVSIDS, VMTF, and ACIDS empirically perform equally well, if implemented carefully. Beside allowing simpler implementation, these empirical results further shed light on what EVSIDS actually

Supported by Austrian Science Fund (FWF), national research network RiSE (S11408-N23). Builds on discussions from the 2014 workshop on Theoretical Foundations of Applied SAT Solving (14w5101), hosted by Banff International Research Station, and Dagstuhl Seminar 15171 (2015), Theory and Practice of SAT Solving.

M. Heule and S. Weaver (Eds.): SAT 2015, LNCS 9340, pp. 405–422, 2015.
DOI: 10.1007/978-3-319-24318-4_29

means. They open up new directions for treating practically successful decision heuristics formally, for instance in the context of proof complexity.

Regarding alternative decision schemes, we refer to the cube-and-conquer approach [12]. It combines CDCL with classical look-ahead [13] solving, and is particularly effective for solving hard combinatorial benchmarks (in parallel). The rest of the paper will focus on decision heuristics for CDCL solving, related to VSIDS. This paper also complements recent developments which try to relate and explain VSIDS with community structure [14–16].

2   Decision Heuristics

Following the same decision order in every branch of a DPLL [17] search tree amounts to a simple *static* decision heuristic, as in ordered binary decision diagrams (BDDs) [18], which even with dynamic variable reordering are restricted to one variable order along each path from root to a leaf. The freedom of being able to pick an arbitrary variable in every node "dynamically" is generally considered an advantage of SAT over BDDs, e.g., in the context of bounded model checking [19]. These *dynamic* decision heuristics originally only took the current partial assignment in a search node into account when selecting the next decision variable. They did not consider how the search progressed to reach this point in the search space. We call this set of restricted dynamic heuristics *first-order dynamic decision heuristics*. A typical example is the dynamic literal individual sum heuristic (DLIS). It selects as next decision literal one with the largest DLIS score, which is computed as the number of still unsatisfied clauses in which a literal occurs. A well-known and often applied variant of DLIS is the Jeroslow-Wang heuristic [20], which for instance is discussed in [21], together with other related early decision heuristics, including Bohm's, MOM's, etc.

With the introduction of learning in Grasp [4], these first-order heuristics implicitly became *second-order dynamic heuristics*, since learned clauses were used in computing scores too, and they do capture the history of the search progress. An early evaluation [21] of decision heuristics, originally designed as first-order heuristics but then applied as second-order heuristics together with clause learning, showed that variants of DLIS actually perform quite well.

In principle, one has to distinguish between selecting a *decision variable* and selecting a *decision phase*, i.e., the Boolean constant to which the selected variable is assigned. However, almost all modern CDCL solvers implement *phase saving* [22], which always reassigns the decision variable to the last phase it was previously assigned. Modulo initialization, typically based on (one-sided) Jeroslow-Wang's heuristic [20], phase saving turns the decision heuristic into a variable selection heuristic. Accordingly, we focus on variable selection, which in turn will be based on selecting a variable with the highest *decision score*.

Using learned clauses for computing scores is actually quite expensive, since it requires either to traverse the whole clause data base, which is growing fast due to adding learned clauses, or requires expensive book keeping of scores during propagation of assigned variables. The latter became expensive after it was possible to reduce propagation effort through lazy clause watching techniques [5, 23], particularly since learned clauses tend to be large [24]. Thus, one of the most important observations in the seminal Chaff paper [5] was that it is possible and even beneficial to replace DLIS by an even more aggressive dynamic scoring scheme, the VSIDS (variable state independent decaying sum) scheme, which does not require to traverse the clause data base at decision variable selection, nor to use expensive full occurrence list traversal for accurate score updates.

VSIDS. The *variable state independent decaying sum* (VSIDS) of Chaff [5] maintains a *variable score* for each variable. The basic idea is that variables with large score are preferred decisions. The original VSIDS implementation in Chaff worked as follows. Variables are stored in an array used to search for a decision variable. After learning a clause, the score of its variables is incremented. Further, every 256th conflict, all variable scores are divided by 2, and the array is sorted w.r.t. decreasing score. This process is also called *variable rescoring*. Moreover, note that the order of decision variables is not changed between rescores.

The process of updating scores of variables is also referred to as *variable bumping* [7]. Note, however, that in modern solvers and also in our experiments we not only bump variables of the learned clause, but all *seen* variables occurring in antecedents used to derive the learned clause through a regular input resolution chain [25] from existing clauses.

The *decide* procedure selects the next decision variable, by searching for the first unassigned variable in the ordered array, starting at the lower end, e.g., the variable with the highest score during sorting. An essential optimization in Chaff is to cache the position of the last found decision variable with maximum score in the ordered array. This position is used as starting point for the next search. If a variable in the array with a position smaller than the cached maximum score position becomes unassigned then the maximum score position is updated to that position. During rescoring, similar updates might be necessary.

The first part of VSIDS, e.g., only incrementing scores, constitutes an approximation of dynamic DLIS. It counts occurrences of variables in clauses, ignoring whether a clause is satisfied or not, or even removed during learned clause deletions [3] (called clause database *reduction* in the following). This restricted version of VSIDS without smoothing is denoted INC (or inc in the experiments).

As an alternative to using frequent rescoring, we propose that the smoothing part of VSIDS can also be approximated by adding the *conflict-index* to the score instead of just incrementing it. The conflict-index is the total number of conflicts that occurred so far. We call this scheme SUM (or sum in our experiments).

At each conflict, a new clause is learned, except for instance if on-the-fly subsumption [26,27] is employed. This might trigger additional conflicts, through strengthening existing clauses, without learning a new clause. Our implementation does not bump variables in this case, nor does it increase the conflict-index.

EVSIDS. If variables are rescored at each conflict, a variant of VSIDS, called *normalized VSIDS* (NVSIDS) [6], is an *exponential moving average* on how often a variable occurred in antecedents of learned clauses [6]. For NVSIDS, the score s of a bumped variable is computed as $s' = f \cdot s + (1 - f)$, using a damping factor f with $0 < f < 1$. The score of other variables, which are not bumped, still have to be "rescored", e.g., $s' = f \cdot s$.

At each conflict, NVSIDS requires to update the score of *all* variables. A more efficient implementation, which we called *exponential VSIDS* (EVSIDS) in [6], was originally proposed by the authors of MiniSAT [7]. It updates only scores of (the much smaller set) of bumped variables by adding an exponential increasing score increment g^i, with i denoting the conflict-index and $g = 1/f$, thus $g > 1$. As the relative order of variables for NVSIDS and EVSIDS is identical [6], the notion of NVSIDS is only of theoretical interest (for the purpose of this paper).

Typical values for g are in the range of 1.01 to 1.2. Small values have been shown to be useful for hard satisfiable instances (like cryptographic instances). Large values are useful with very frequent restarts, particularly in combination with the *reuse-trail* technique [28]. In Glucose 2.3, even without reusing the trail, it was thus suggested to slowly decrease g over time from a large value to a small one.[1] In the (new) version of Lingeling used in our experiments, g is kept at 1.2.

Instead of rescoring variables explicitly, MiniSAT uses a priority queue, which is implemented as a binary heap. This data structure allows fast insertion and removal of variables and also updating scores, all in logarithmic time. If this priority queue was updated eagerly to contain exactly all the unassigned variables, then searching for an unassigned variable with maximal score would even be possible in constant time. However, the number of propagated variables per decision can be quite large (on average, 323 propagations per decision for 275 benchmarks in the evsids column in Tab. 2). Removing them *eagerly* is too costly.

A *lazy* alternative, as first implemented in MiniSAT [7] and now being the default implementation of modern CDCL solvers, is to remove variables with maximum score from the priority queue until the removed variable turns out to be unassigned. It is then used as the next decision variable. Note that, during backtracking, this lazy scheme still requires to insert variables back into the priority queue, as they are unassigned, in order to make sure that the priority queue contains all unassigned variables (but assigned ones are not eagerly removed).

While the original implementation of VSIDS in Chaff [5] can be considered to be lazy too, variable selection is still imprecise, since rescoring is delayed. An attempt to provide a more efficient implementation of rescoring with precise variable selection was implemented in the JeruSAT solver [29]. It still uses counters, i.e., inaccurate integer scores, but instead of using one sorted array for all variables, partitions them into doubly linked lists of variables with the same score. This allows faster insertion, removal, update, and rescoring.

Another invention in MiniSAT, particularly important for EVSIDS, is to use a precise floating-point representation instead of integers as in previous solvers. Even though we do not have separate experimental evidence in this paper, our experience suggests that using integer scores dramatically deteriorates performance compared to using floating-point scores. Even fixed-point scores (as in PrecoSAT [11]) need additional techniques like clause based decision heuristics in order to be competitive with floating-point based EVSIDS.

---

[1] Every 5000th conflict, f is increased by 0.01, starting at 0.8 until 0.95 is reached.

However, g^i usually grows very fast: Note that $1.01^{4459}, 1.2^{244} > 2^{64}$, and, more severely, $1.01^{71333}, 1.2^{3894} > 1.797 \cdot 10^{308}$ ($\approx$ maximum value in 64 bit IEEE double floating-point representation). Thus, even for EVSIDS with floating-points, the variable scores and the score increment have to be rescored occasionally, as in the VSIDS scheme. This also becomes necessary if the score of a bumped variable would overflow during an update. We will report how often this occurs and how much time is spent on rescoring in our experiments.

VMTF. Variable selection heuristics can be seen as online sorting algorithms of variable scores. This view suggests to use online algorithms with efficient amortized complexity, such as *move-to-front* (MTF) [30]. A similar motivation was given in the master thesis of Lawrence Ryan [8], which precedes MiniSAT [7] and introduced the Siege SAT solver as well as the *variable move-to-front* (VMTF, or vmtf in the experiments) strategy. As in Chaff, the restriction in Siege's VMTF bumping scheme was to only move variables in the learned clause. Actually, only a small subset of those variables, e.g., of size 8, was selected, according to [8].

The restriction in Siege to move only a small subset of variables might have been partially motivated by the cost of moving many. It is not uncommon that tens of thousands variables occur in antecedents of a learned clause, which also are rather long for some instances. In our experiments in Sect. 4, the default decision heuristic (evsids in Tab. 2) bumped on average 276 literals per learned clause of average length 105 (on 275 considered instances). Unfortunately, details on how even this restricted version of VMTF is implemented in Siege were not provided. The source code is not available either. We give details for a fast implementation of *unrestricted* VMTF in Sect. 3.

ACIDS. As further extension to the proposed SUM heuristic we want to introduce the *average conflict-index decision score* (ACIDS, or acids in our experiments). While SUM realizes a certain amount of smoothing (compared to INC) by giving a larger weight to later conflicts, this effect is rather small when compared to the exponential kind of smoothing that is applied in VSIDS and EVSIDS. However, as smoothing is conjectured to be an important part for variable score heuristics [6], the latter kind of smoothing might be preferable. We realize this as follows. In the ACIDS scheme, in the same way as for INC, SUM, VSIDS, and EVSIDS, we keep a score for each variable. Whenever a variable is bumped, its score is updated to be $s' = (s+i)/2$, with i being the conflict-index. Compared to SUM, much stronger smoothing is realized by ACIDS. In addition to giving a larger weight to later conflicts, the influence of earlier conflicts decreases exponentially in the number of times the variable is bumped.

To compare the influence of the current conflict with that of earlier ones, we can represent the score of the variable by $s = s_c + s_p$, with s_c and s_p representing the contribution of the current conflict and the previous conflicts, respectively. As before, we define i to be the current conflict-index. Further, I_p is the set of indices of all previous conflicts the variable was involved in. For SUM, $s_c = i$ and $s_p = \Sigma_{I_p} i_p$, with i_p being the elements of I_p. By definition, this will lead to $s_p > s_c$ in most cases, particularly after a certain number of conflicts occurred.

Table 1. Summary of considered variable scoring schemes, where s and s' denote current and updated variable scores, i the conflict-index, and f a damping factor with $0 < f < 1$, used in our reformulation NVSIDS of VSIDS as exponential moving average [6]. For EVSIDS, we use the inverse $g = 1/f$ of f (thus $g > 1$). For the VSIDS version implemented in Chaff, we set $h_i^m = 0.5$ if m divides i, and $h_i^m = 1$ otherwise.

| | variable score s' after i conflicts | | |
	bumped	not-bumped	
STATIC	s	s	static decision order
INC	$s + 1$	s	increment scores
SUM	$s + i$	s	sum of conflict-indices
VSIDS	$h_i^{256} \cdot s + 1$	$h_i^{256} \cdot s$	original implementation in Chaff [5]
NVSIDS	$f \cdot s + (1 - f)$	$f \cdot s$	normalized variant of VSIDS [6]
EVSIDS	$s + g^i$	s	exponential dual of NVSIDS [6,7]
ACIDS	$(s + i)/2$	s	average conflict-index decision scheme
VMTF	i	s	variable move-to-front [8]

Similarly for INC, $s_c = 1$ and $s_p = |I_p|$, which already implies $s_p > s_c$ as soon as a variable is bumped twice. However, for the ACIDS heuristic, we obviously have $s_p < s_c$ at every point in the search.

Note that, in contrast to VSIDS and NVSIDS, scores of variables that are not bumped do not change for ACIDS. This not only allows to keep track of accurate scores in each step, but also avoids (delayed) variable rescoring. Additionally, compared to EVSIDS, the scores of variables grow much slower when using the ACIDS heuristic. In particular, the score of a variable in ACIDS is bounded by the conflict-index i, instead of being exponential in the number of conflicts, as it was the case for EVSIDS. Thus, also rescoring of variables to prevent overflow does not occur in practice. Considering overall performance, our experiments in Sect. 4 show that ACIDS works as well as EVSIDS and VMTF.

Clause Based Decision Heuristics. There also is related work on using recently learned clauses in variable selection, such as the BerkMin heuristic [9], or clause-move-to-front (CMTF) strategies [10,11]. In our experience, they are inferior to variable scoring schemes as considered in this paper, and we leave it to future work for a more detailed comparison. The same applies to one-sided schemes which select literals instead of variables (without phase saving).

3   Implementation

We describe how the VMTF scheme can be implemented efficiently, as well as how these techniques can be lifted to implement a generic priority queue, which (empirically) is efficient for all the considered scoring schemes. This new implementation of a priority queue for variable selections combines ideas originally implemented in Chaff [5] and JeruSAT [29], but adds additional optimizations and works with arbitrary precise floating-point scores, in contrast to an imprecise earlier version implemented in Lingeling [31].

Variable scores play a role while (a) *bumping* variables participating in deriving a learned clause, (b) *deciding* or searching for the next decision variable, (c) *unassigning* variables during backtracking, (d) *rescoring* variable scores either for explicit smoothing in VSIDS or due to protecting scores from overflow during bumping, and (e) comparing past decisions on the trail to maximize *trail reuse* [28]. First, we explain a fast implementation for VMTF, focusing on (a)-(c). Next, we address its extension to precise scoring schemes using floating-point numbers, which in previous implementations followed the example set by MiniSAT to use a binary heap data structure. Last, we discuss (d) and (e).

3.1 Fast Queue for VMTF

According to Sect. 2, the score of a variable in VMTF is the conflict-index, e.g., the number of conflicts at the point a variable was last bumped. With this score definition, VMTF can be simulated with a binary heap. However, every bump then needs a logarithmic number of steps to "bubble-up" a bumped variable in the heap. Instead, a queue, implemented as doubly linked list which holds all variables, only requires two simple constant time operations for bumping: dequeue the variable and enqueue it back at the end of the list, which we consider as head. Even storing the score seems to be redundant.

To find the next decision variable in the queue, we could start at the end (head) of the queue and traverse it backwards until an unassigned variable is found. Unfortunately, this algorithm has quadratic accumulated complexity. For example, consider an instance with 10000 variables and a single clause containing all variables in default phase. However, we can employ the same[2] optimization as used in Chaff (see Sect. 2) and remember the variable up to which the last search proceeded until finding an unassigned variable. Since the solver will restart the next search at this variable, we call this reference *next-search*.

During backtracking, variables are unassigned and (as in Chaff) next-search potentially has to be updated to such an unassigned variable if it sits further down the queue closer to head than the next-search variable. In order to achieve this, we could use the scores of the variables for comparing queue position. However, in VMTF, variables bumped at the same conflict all get the same score, and thus simply using the score leads to violation of the following important invariant: variables right of next-search (closer to head) are assigned.

To fix this problem, we globally count enqueue operations to the queue with an *enqueue-counter* and remember with each variable the value of the enqueue-counter at the point the variable was enqueued as *enqueue-time*. Thus, the enqueue-time precisely captures the order of the elements in the queue and can be used to precisely compare the relative positions of variables in the queue. In the actual implementation, we use a 32-bit integer for the enqueue-counter,

---

[2] But in reverse order, e.g., while we prefer the variable with largest score at the end of the queue, Chaff had the variable with largest score at the first array position.

which occasionally, e.g., after billion enqueue operations, requires to reassign enqueue-times to all queue elements in a linear scan of the queue. Note that, in a dedicated queue implementation for VMTF (like queue in our experiments), the scores become redundant again, after adding enqueue-times.

3.2 Generic Queue for all Decision Heuristics

For other schemes, it is tempting to also just use a queue implemented as doubly linked list as for VMTF, maintaining both scores and enqueue-times. Every operation remains constant time except for bumping. We have to ensure that the queue is sorted w.r.t. score. However, only for VMTF, bumped variables are guaranteed to be enqueued at the end (head) of the queue, i.e., in constant time. For other scoring schemes, a linear search is required to find the right position, which risks an accumulated quadratic bumping effort. To reduce enqueue time, we propose three optimizations and two modifications to the bumping order.

The **first optimization** is inspired by bucket sort and already gives acceptable bumping times for EVSIDS. It is motivated by the following observation. For EVSIDS, rescoring to avoid floating-point overflow of scores and score increment occurs quite frequently, e.g., roughly every 2000 conflicts, as Tab. 2 suggests. Thus, the exponents of variable scores represented as floating-point numbers will tend to span the whole range of possible values[3]. So instead of a single queue, we keep a stack of queues, indexed by the exponent of the scores of variables. Variables belong to the queue of the floating-point exponent of their score. As the motivation on rescoring shows, this stack will soon grow to its maximum size for EVSIDS, but for other scoring schemes (particularly for VMTF or INC) it will only have very few elements or even just one.

Note that, since exponents can be negative, the actual index to access the stack is obtained after adding the negation of the minimum negative exponent. Furthermore, Lingeling uses its own implementation of floating-points, in order to make execution of Lingeling deterministic across different hardware, compilers, and compiler flags. These software floats have a 32 bit exponent, but we restrict exponents to 10 bits including a sign bit, by proper rescoring of large scores and truncation of small scores. MiniSAT/Glucose use 10^{100} as an upper score limit, which is only a slightly smaller maximum limit than ours $2^{512} \approx 10^{154}$, but then does not use any truncation for small scores, which means that the minimum score exponent in MiniSAT is (roughly) 2^{-10}. So Lingeling uses 9 bits for positive scores and 9 bits for negative scores, while MiniSAT uses slightly less than 9 bits for positives scores and (almost) full 10 bits for negative scores.

When searching for decisions as well as during backtracking, more specifically during unassigning variables, we additionally have to maintain the highest exponent of an unassigned variable. This follows the same idea as for next-search in a single queue and only adds constant time effort for all considered operations.

During conflict analysis, variables participating in resolutions to derive a learned clause are collected on a *seen-variables* stack, before they are bumped

---

[3] Almost 2048 values for an 11-bit exponent in IEEE representation of 64 bit doubles.

(or discarded if on-the-fly subsumption succeeds). The analysis traverses the trail of assigned variables in reverse order. Thus, there is a similarity between the order of variables on the seen-variables stack and the reverse order of assignments. However, this is not guaranteed, particularly for variables with smaller decision-level. The order of bumping these variables then follows this order too.

At a conflict, it can happen that thousands of variables with different score are bumped and end up in almost random order w.r.t score order on the seen-variables stack (or worse, in reverse order) before they are bumped. For many of these variables, even for EVSIDS, the new updated score might end up having the same exponent and all those variables have to be enqueued to the same queue. However, since their scores still differ, enqueueing them degrades to insertion-sort. There are instances where bumping leads to a time-out due to this effect.

A **first modification** to the order in which variables are bumped prevents this problem. Before actually first dequeuing a bumped variable, then updating its score, and finally enqueueing it back, we sort the seen-variables stack w.r.t. increasing score. However, a similar problem occurs if all bumped variables have the same score exponent, which also does not change during update. This is for instance almost always the case for INC. The **second modification** prevents this corner case by first dequeuing all variables on the seen-variables stack, and only then updating their score and enqueueing them back in score order.

While EVSIDS exponents of variable scores are more or less spread out, other schemes do not have this property, clearly not INC, but probably also SUM and ACIDS to a smaller extent. For these schemes, score exponents might cluster around some few values. Thus, our **second optimization** repeats the bucket sort argument w.r.t. some fixed number of highest bits of the mantissa of a variable score. For each queue (indexed by exponent), we add another cache-table (indexed by highest bits of mantissa) of references pointing to the last element in the queue with matching highest mantissa bits. This ensures that these variables referenced in the cache-table have the maximum score among variables in this queue with the same highest bits of the mantissa of their score. In our implementation, we use the highest 8 bits and thus a cache-table of size 256. This cache is only used for fast enqueue and can be ignored otherwise.

If bumping individual variables is done in the order of their scores, as suggested by the first modification above, there is a high chance that consecutively bumped variables end up in the same queue one after each other or at least close to each other. Thus, as a **third optimization**, we propose to additionally cache the *last-enqueued* variable for each (sub) queue consisting of variables with the same highest mantissa bits. In an enqueue operation, we first check whether the corresponding cache-table entry of the second optimization points to a variable with smaller (or equal) score. If this is the case, we enqueue right next to it. Otherwise, we obtain the last-enqueued variable and start searching for the proper enqueue position from there towards the end, e.g., towards larger scores. This might fail if the score of the last-enqueued variable is larger or if the last-enqueue reference is not valid, e.g., if the variable is already dequeued. We then search backwards from the cache-table reference (towards smaller scores).

Altogether, these optimizations and modifications seem to avoid the most severe worst-case corner cases. We track this by profiling relative and total decide and particularly bump time per instance. Total time summed for these over all instances are shown in Tab. 2. Further distribution plots are included in the additional material, mentioned in the results in Sect. 4.

3.3   Rescore, Reuse-Trail and Complexity

For the original array based VSIDS implementation, *rescoring* requires sorting variables. For a binary heap implementation, one would expect that the heap does not change, since rescoring does not change the relative order of variables. However, due to finite precision of scores, even when using floating-points, rescoring will make the score of some variables the same, even though they differed in score before rescoring. Moreover, scores of many variables will become zero after a few rescores (particularly in EVSIDS). In this situation, the binary heap will only remain unchanged after rescoring if the actual scores are the only mean to compare variables (and for instance the variable index is not used as a tie breaker for comparing variables with the same score). The same argument applies to our improved queue based implementation.

The reuse-trail optimization [28] is based on the following observation. After a restart, it often happens that the same decisions are taken and the trail ends up with the same assigned variables. Thus, the whole restart was useless. By comparing scores of assigned previous decisions with the score of the next decision variable before restarting, this situation can be avoided. With some effort, this technique can be lifted to our generic queue implementation. To simplify the comparison in favor of a clean experiment, the results presented in Sect 4 are without reuse-trail (except for sc14ayv, the old 2014 version of Lingeling).

While we do not have a precise complexity analysis for this new data structure, our empirical results show that it performs almost as good as a dedicated binary heap for EVSIDS (heap) and as a dedicated simplified queue for VMTF (queue). This makes our empirical comparison of decision heuristics more accurate since they all use the same implementation. This data structure should also allow to experiment with new scoring schemes without the need to implement dedicated data structures. It might also be possible to improve it further, while our binary heap implementation is close to being as fast and compact as possible.

4   Results

The variants of Lingeling used in the experiments evolved from the SAT competition 2014 version *ayv* [32] (sc14ayv)[4]. This old 2014 version of Lingeling solved the largest number of instances in the SAT+UNSAT application track. This success of Lingeling can be contributed to the rather long time limit of 5000 seconds as used in the competition. For shorter time limits, Glucose version 2.3 [33,34]

---

[4] Acronyms in sans serif font denote SAT solver versions and configurations.

(glucose-2.3) from 2013 and particularly its 2014 derivative SWDiA5BY A26 [35] (swdia5bya26) show much better performance, despite lacking many effective pre-processing and inprocessing techniques [36].

Our post competition analysis showed that this effect can be contributed to two different aspects. On the one hand, the benchmark selection scheme used in the SAT competition 2014 (and already in 2013) had a strong influence on those results. Benchmarks were selected in such a way to level out performance of solvers. The goal of the organizers was to make the competition as interesting as possible, with the unfortunate effect, however, that unique solving capabilities, such as inprocessing [36], are deemphasized. On the other hand, our analysis showed that there is indeed an algorithmic feature implemented in all the Glucose variants taking part in the competition, which on these competition benchmarks is quite effective: the Glucose restart strategy [37].

This strategy uses the glucose level of learned clauses, which is the number of different decision levels [33] in the learned clause. It compares current short term average glucose level of learned clauses with a long term average. If short term average is substantially larger than long term average (say 25%), a restart is triggered, unless a restart happened very recently (less than 50 conflicts earlier).

To derive this conclusion, we implemented all techniques used in Glucose 2.3 and SWDiA5BY A26 previously not available in Lingeling, and compared their effect on the considered SAT competition 2014 application track benchmarks. Without being able to give more details, which is also not the focus of this paper, implementing a variant of the Glucose dynamic restart scheme [37] had the largest impact and allowed us to solve a comparable number of benchmarks as the aforementioned Glucose variants even with much smaller time limits.

Beside incorporating effective techniques from Glucose and SWDiA5BY, the base line version *b7ztzu* of Lingeling (evsids), as used in this evaluation, differs from the 2014 version sc14ayv mainly in the implementation of the priority queue used for selecting decision variables as detailed in Sect. 3. In other solvers, and previously in Lingeling, the priority queue was implemented with a binary heap data structure, as pioneered by MiniSAT [7]. This change was necessary to avoid slowing down the decision selection procedure for certain decision heuristics, particularly the variable move-to-front strategy (VMTF), which does not require the overhead of a binary heap. It is also slightly faster than using a binary heap.

As Glucose (and thus SWDiA5BY) is based on MiniSAT [7] (minisat), we also include in our comparison the latest version of MiniSAT from *git-hub*, which essentially has not changed since 2011. For all these considered MiniSAT derivatives, we use the default configuration with the internal MiniSAT version of SatELite style preprocessing [38] enabled.

The experiments were performed on our benchmark cluster, consisting of 30 nodes with Intel Q9550 Core 2 Quad CPUs running at 2.83GHz and 8 GB of main memory. Each job, e.g., pair of solver (configuration) and benchmark, had exclusive access to one node and CPU, respectively. The time limit was set to 1000 seconds, which is substantially smaller than the original competition

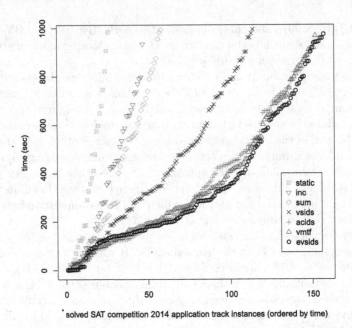

solved SAT competition 2014 application track instances (ordered by time)

Fig. 1. Lingeling with variable scoring schemes of Sect. 2 on SAT competition 2014 application track benchmarks using the generic priority queue implementation of Sect. 3.

time-out of 5000 seconds (competition hardware was further roughly 1.2 times faster). As memory limit, we used 7GB.

In this paper, we focus on the 300 instances of the SAT+UNSAT application track of the SAT competition 2014, but exclude 25 instances, which were solved by the new Lingeling base line version evsids, without producing any conflicts. Among those excluded, there are 13 satisfiable "argumentation" instances [39] submitted 2014, with name prefix "*complete...*". These excluded 13 instances have a simple solution, with all variables set to false. In contrast, if this is not detected and a more sophisticated phase initialization heuristic like Jeroslow-Wang [20] is triggered before switching to phase saving [22], they become very hard. The old SAT competition 2014 version of Lingeling sc14ayv fails to solve 7 within 1000 seconds in our set-up.

The other excluded 12 instances are unsatisfiable combinational hardware equivalence checking "miter" benchmarks [40] submitted 2013. They are solved by our base line version evsids, and all other considered new variants of Lingeling, during the first preprocessing phase, without any search. Within 1000 seconds, the three MiniSAT/Glucose variants easily solve the 13 excluded satisfiable "argumentation" instances, due to initializing the saved phase to false, but need more effort than Lingeling to solve the unsatisfiable "miters". Both glucose-2.3 and swdia5bya26 fail on benchmark *6s151*, and minisat even fails on 11 "miters" (but does solve *6s165-non*). Note that, altogether, there were 30

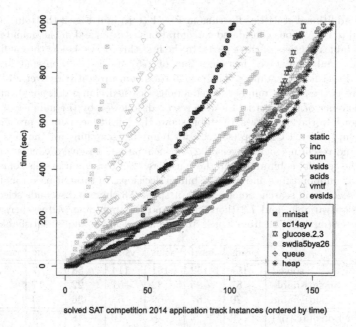

Fig. 2. Additional variants of Lingeling on SAT competition 2014 application track benchmarks as well as other state-of-the-art SAT solvers for this set-up.

"miters" in the competition. Thus, 17 "miters" remained in our subset of 275 actually compared instances, as well as 7 out of the 20 original "argumentation" benchmarks.

We describe additional specifics of the configurations used in our experiments on top of what has been explained in detail in previous sections and further summarize conclusions which can be drawn from the data provided in the tables and cactus plots. All experimental data including source code is available at http://fmv.jku.at/evalvsids/evalvsids.7z (27MB).

The main result of the paper is documented in Fig. 1. The cactus plot shows, that EVSIDS, VMTF, as well as our new ACIDS scheme, perform equally well. This is supported by the data in the upper part of Tab. 2, which corresponds to the same experiment. In the last three rows, we see that our generic priority queue is still somewhat optimized for EVSIDS and VMTF. For instance, ACIDS needs more time during bumping, which applies even more to INC and SUM.

In Fig. 2 and the lower part of Tab. 2, we compare against two variants of the new Lingeling, one using a dedicated optimized binary *heap* implementation for EVSIDS on one side, and the other one using a dedicated optimized *queue* implementation for VMTF. Both are slightly faster. Decision plus bumping time decreases. Otherwise, they show very similar behavior. We also compare against the state-of-the-art on these benchmarks, which for this small time-out of 1000 seconds, consists of SWDiA5BY A26 and also to some extent its "parent" Glucose

Table 2. Additional statistics for runs in Fig. 1 (top) and Fig. 2 (bottom). Columns correspond to the various considered configurations as discussed in the main text. Each of the two tables consists of three parts. In the first three rows, below the configuration names, the number of solved instances (out of 275) are listed, then split into unsatisfiable and satisfiable instances. The next 5 rows sum up statistics over all 275 runs. First, there is the overall number of reductions (learned clause deletions), number of restarts, number of times variables were rescored, followed by the number of conflicts and decisions. In the last 5 rows, the table shows the total time spent in pre- and inprocessing (simp), the CDCL loop (search), for bumping, searching for the next decision (decide), and rescoring (again over all 275 benchmarks). To give a concrete example, consider the "evsids" column. For all the considered 275 benchmarks, this configuration restarted 5.8 million times and used 3.7 billion decisions. In total, it used roughly 143.1 thousand seconds in search, among which it spent 2.3 thousand seconds selecting the next decision variable, and 7.8 thousand seconds for bumping. Altogether, it solved 157 instances (out of 275), from which 87 were unsatisfiable and 70 satisfiable.

	evsids	vmtf	acids	vsids	sum	inc	static
solved	157	152	151	114	58	47	26
unsatisfiable	87	85	82	51	22	17	9
satisfiable	70	67	69	63	36	30	17
reductions (1e3 #)	8	8	8	10	8	8	8
restarts (1e3 #)	5826	6000	5678	4491	2612	2387	5593
rescored (1e3 #)	253	0	0	2338	0	0	0
conflicts (1e6 #)	488	476	444	604	527	540	463
decisions (1e6 #)	3691	3581	3889	4263	2603	2567	21503
simp (1e3 sec)	29.7	30.0	29.4	32.6	34.5	34.1	31.2
search (1e3 sec)	143.1	146.4	147.9	174.9	203.9	209.7	226.7
bump (1e3 sec)	7.8	6.2	16.0	16.9	34.6	37.2	0.0
decide (1e3 sec)	2.3	2.5	2.6	2.8	1.7	1.7	12.9
rescore (1e3 sec)	0.2	0.0	0.0	2.6	0.0	0.0	0.0

	heap	queue	swd ia5by a26	glu cose 2.3	sc14 ayv	mini sat
solved	161	156	153	144	119	101
unsatisfiable	90	86	81	79	60	41
satisfiable	71	70	72	65	59	60
reductions (1e3 #)	8	8	59	10	30	—
restarts (1e3 #)	5870	6003	3210	3846	7948	1782
rescored (1e3 #)	241	0	—	—	393	—
conflicts (1e6 #)	463	474	650	728	760	1090
decisions (1e6 #)	3874	3566	5868	6818	5002	8388
simp (1e3 sec)	29.2	29.7	0.8	0.8	32.4	2.2
search (1e3 sec)	141.8	144.6	165.4	172.5	164.4	206.5
bump (1e3 sec)	3.8	4.9	—	—	3.3	—
decide (1e3 sec)	4.9	2.5	—	—	6.4	—
rescore (1e3 sec)	0.1	0.0	—	—	0.0	—

2.3. We also include MiniSAT 2.2, e.g., the "grandparent" of SWDiA5BY A26, and version *ayv* of Lingeling of the SAT Competition 2004 (sc14ayv).

5   Conclusion

In this paper, we evaluated several important CDCL decision schemes, including VSIDS [5] and the related EVSIDS [6] heuristic, which are considered to be one of the major reasons for good performance of modern SAT solvers on application benchmarks. While some reasons for the efficiency of VSIDS have been conjectured before [6], there is still a lot of ongoing research on finding good explanations for its performance, particularly related to problem structure [14–16]. Understanding VSIDS and related decision heuristics in a better way would help us to further improve performance of SAT solvers from a practical point of view, as well as open up possibilities for formal analysis in a theoretical sense.

To take a major step into that direction, we gave a detailed evaluation, comparing VSIDS and EVSIDS to several other heuristics, including static decision heuristics, a non-smoothing version of VSIDS and approximations of smoothing versions. We also proposed ACIDS, a new decision heuristic with similar properties as VSIDS, and revisited the VMTF scheme [8], which is easy to implement and also offers an alternative perspective on the meaning of the decision order of variables. We further provided a formalization of the score update as a function for each heuristic to capture its effect in a clear way.

In our experiments, it turned out that EVSIDS, VMTF, and ACIDS perform very similarly. Since efficient implementation is crucial and non-trivial for all those heuristics, we pointed out differences in underlying data structures and discussed important aspects of implementation in detail. We further provided detailed results, allowing us to analyze the effect variations in heuristics and implementations cause on the time spent in the individual steps of a search.

In addition, our results also shed new light on the performance of decision heuristics from an algorithmic point of view, as well as on many beliefs about decision heuristics that have been held previously. For instance, EVSIDS, VMTF, and ACIDS have in common that they put a very strong focus on variables that participated in the most recent conflicts. This is in contrast to heuristics, such as INC and SUM, where the occurrence in earlier conflicts also contributes significantly to the score of a decision variable throughout the whole progress of the search. While VSIDS, EVSIDS, and ACIDS implement explicit smoothing schemes to realize this kind of focus, the good performance of VMTF in our experiments shows that this is not necessarily required when directly using a more aggressive bumping strategy for recent conflict variables.

For future work, it will be interesting to analyze the contribution of the individual components in detail. Having provided a formal way of describing general scoring schemes and given several implementations of flexible data structures in a simpler way, the next steps could be motivated by theory as well as practice. For instance, combining aggressive bumping strategies in combination with particularly adapted smoothing schemes could yield even more efficient decision heuristics. Similarly, more refined functions for updating the variable scores could be beneficial as well. On the other hand, simple but yet efficient heuristics, such as VMTF, might allow us to analyze CDCL more formally, e.g., in the context of proof complexity.

References

1. Balint, A., Belov, A., Heule, M.J.H., Järvisalo, M. (eds.): Proceedings of SAT Competition 2013. Volume B-2013-1 of Department of Computer Science Series of Publications B. University of Helsinki (2013)
2. Belov, A., Heule, M.J.H., Järvisalo, M. (eds.): Proceedings of SAT Competition 2014. Volume B-2014-2 of Department of Computer Science Series of Publications B. University of Helsinki (2014)
3. Marques-Silva, J.P., Lynce, I., Malik, S.: Conflict-driven clause learning SAT solvers. [41], 131–153
4. Marques-Silva, J.P., Sakallah, K.A.: GRASP: A search algorithm for propositional satisfiability. IEEE Trans. Computers 48(5), 506–521 (1999)
5. Moskewicz, M.W., Madigan, C.F., Zhao, Y., Zhang, L., Malik, S.: Chaff: engineering an efficient SAT solver. In: Proceedings of the 38th Design Automation Conference, DAC 2001, pp. 530–535. ACM, Las Vegas, June 18–22, 2001
6. Biere, A.: Adaptive restart strategies for conflict driven SAT solvers. In: Kleine Büning, H., Zhao, X. (eds.) SAT 2008. LNCS, vol. 4996, pp. 28–33. Springer, Heidelberg (2008)
7. Eén, N., Sörensson, N.: An extensible SAT-solver. In: Giunchiglia, E., Tacchella, A. (eds.) SAT 2003. LNCS, vol. 2919, pp. 502–518. Springer, Heidelberg (2004)
8. Ryan, L.: Efficient algorithms for clause-learning SAT solvers. Master's thesis, Simon Fraser University (2004)
9. Goldberg, E.I., Novikov, Y.: Berkmin: a fast and robust sat-solver. In: 2002 Design, Automation and Test in Europe Conference and Exposition (DATE 2002), pp. 142–149. IEEE Computer Society, Paris, March 4–8, 2002
10. Gershman, R., Strichman, O.: Haifasat: A new robust SAT solver. In: Ur, S., Bin, E., Wolfsthal, Y. (eds.) Hardware and Software Verification and Testing. LNCS, vol. 3875, pp. 76–89. Springer, Heidelberg (2006)
11. Biere, A.: P{re, i}coSAT@SC 2009. In: SAT 2009 Competitive Event Booklet, pp. 42–43 (2009)
12. Heule, M.J.H., Kullmann, O., Wieringa, S., Biere, A.: Cube and conquer: guiding CDCL SAT solvers by lookaheads. In: Eder, K., Lourenço, J., Shehory, O. (eds.) HVC 2011. LNCS, vol. 7261, pp. 50–65. Springer, Heidelberg (2012)
13. Heule, M., van Maaren, H.: Look-ahead based SAT solvers. [41], 155–184
14. Ansótegui, C., Giráldez-Cru, J., Levy, J.: The community structure of SAT formulas. In: Cimatti, A., Sebastiani, R. (eds.) SAT 2012. LNCS, vol. 7317, pp. 410–423. Springer, Heidelberg (2012)
15. Newsham, Z., Ganesh, V., Fischmeister, S., Audemard, G., Simon, L.: Impact of community structure on SAT solver performance. In: Sinz, C., Egly, U. (eds.) SAT 2014. LNCS, vol. 8561, pp. 252–268. Springer, Heidelberg (2014)
16. Ansótegui, C., Bonet, M.L., Giráldez-Cru, J., Levy, J.: The fractal dimension of SAT formulas. In: Demri, S., Kapur, D., Weidenbach, C. (eds.) IJCAR 2014. LNCS, vol. 8562, pp. 107–121. Springer, Heidelberg (2014)
17. Davis, M., Logemann, G., Loveland, D.W.: A machine program for theorem-proving. Commun. ACM 5(7), 394–397 (1962)

18. Bryant, R.E.: Graph-based algorithms for boolean function manipulation. IEEE Trans. Computers **35**(8), 677–691 (1986)
19. Biere, A., Cimatti, A., Clarke, E., Zhu, Y.: Symbolic model checking without BDDs. In: Cleaveland, W.R. (ed.) TACAS 1999. LNCS, vol. 1579, p. 193. Springer, Heidelberg (1999)
20. Jeroslow, R.G., Wang, J.: Solving propositional satisfiability problems. Annals of Mathematics and Artificial Intelligence **1**(1–4), 167–187 (1990)
21. Marques-Silva, J.: The impact of branching heuristics in propositional satisfiability algorithms. In: Barahona, P., Alferes, J.J. (eds.) EPIA 1999. LNCS (LNAI), vol. 1695, pp. 62–74. Springer, Heidelberg (1999)
22. Pipatsrisawat, K., Darwiche, A.: A lightweight component caching scheme for satisfiability solvers. In: Marques-Silva, J., Sakallah, K.A. (eds.) SAT 2007. LNCS, vol. 4501, pp. 294–299. Springer, Heidelberg (2007)
23. Zhang, H.: SATO: an efficient propositional prover. In: McCune, William (ed.) CADE 1997. LNCS, vol. 1249, pp. 272–275. Springer, Heidelberg (1997)
24. Biere, A.: PicoSAT essentials. JSAT **4**(2–4), 75–97 (2008)
25. Beame, P., Kautz, H.A., Sabharwal, A.: Towards understanding and harnessing the potential of clause learning. J. Artif. Intell. Res. (JAIR) **22**, 319–351 (2004)
26. Han, H., Somenzi, F.: On-the-fly clause improvement. In: Kullmann, O. (ed.) SAT 2009. LNCS, vol. 5584, pp. 209–222. Springer, Heidelberg (2009)
27. Hamadi, Y., Jabbour, S., Sais, L.: Learning for dynamic subsumption. In: 21st IEEE International Conference on Tools with Artificial Intelligence, Newark, New Jersey, USA, ICTAI 2009, pp. 328–335. IEEE Computer Society, November 2–4, 2009
28. van der Tak, P., Ramos, A., Heule, M.J.H.: Reusing the assignment trail in CDCL solvers. JSAT **7**(4), 133–138 (2011)
29. Nadel, A.: Backtrack search algorithms for propositional logic satisfiability: Review and innovations. Master's thesis, Hebrew University (2002)
30. Sleator, D.D., Tarjan, R.E.: Amortized efficiency of list update and paging rules. Commun. ACM **28**(2), 202–208 (1985)
31. Biere, A.: Lingeling and friends entering the SAT challenge 2012. In: Balint, A., Belov, A., Diepold, D., Gerber, S., Järvisalo, M., Sinz, C. (eds.) Proceedings SAT Challenge 2012: Solver and Benchmark Descriptions. Volume B-2012-2 of Department of Computer Science Series of Publications B., University of Helsinki, pp. 33–34 (2012)
32. Biere, A.: Yet another local search solver and Lingeling and friends entering the SAT Competition 2014. [2], 39–40
33. Audemard, G., Simon, L.: Predicting learnt clauses quality in modern SAT solvers. In: Boutilier, C. (ed.) Proceedings of the 21st International Joint Conference on Artificial Intelligence, IJCAI 2009, Pasadena, California, USA, pp. 399–404, July 11–17, 2009
34. Audemard, G., Simon, L.: Glucose 2.3 in the SAT 2013 Competition. [1], 42–43
35. Oh, C.: MiniSat HACK 999ED, MiniSat HACK 1430ED and SWDiA5BY. [2], 46–47
36. Järvisalo, M., Heule, M.J.H., Biere, A.: Inprocessing rules. In: Gramlich, B., Miller, D., Sattler, U. (eds.) IJCAR 2012. LNCS, vol. 7364, pp. 355–370. Springer, Heidelberg (2012)
37. Audemard, G., Simon, L.: Refining restarts strategies for SAT and UNSAT. In: Milano, M. (ed.) CP 2012. LNCS, vol. 7514, pp. 118–126. Springer, Heidelberg (2012)

38. Eén, N., Biere, A.: Effective preprocessing in SAT through variable and clause elimination. In: Bacchus, F., Walsh, T. (eds.) SAT 2005. LNCS, vol. 3569, pp. 61–75. Springer, Heidelberg (2005)
39. Wallner, J.P.: Benchmark for complete and stable semantics for argumentation frameworks. [2], 84–85
40. Biere, A., Heule, M.J.H., Järvisalo, M., Manthey, N.: Equivalence checking of HWMCC 2012 circuits. [1], 104
41. Biere, A., Heule, M.J.H., van Maaren, H., Walsh, T. (eds.): Handbook of Satisfiability. Volume 185 of Frontiers in Artificial Intelligence and Applications. IOS Press (2009)

SAT-Based Horn Least Upper Bounds

Carlos Mencía[1]([✉]), Alessandro Previti[1], and Joao Marques-Silva[1,2]

[1] CASL, University College Dublin, Dublin, Ireland
{carlos.mencia,jpms}@ucd.ie, alessandro.previti@ucdconnect.ie
[2] INESC-ID, IST, ULisboa, Lisbon, Portugal

Abstract. Knowledge compilation and approximation finds a wide range of practical applications. One relevant task in this area is to compute the Horn least upper bound (Horn LUB) of a propositional theory F. The Horn LUB is the strongest Horn theory entailed by F. This paper studies this problem and proposes two new algorithms that rely on making successive calls to a SAT solver. The algorithms are analyzed theoretically and evaluated empirically. The results show that the proposed methods are complementary and enable computing Horn LUBs for instances with a non-negligible number of variables.

1 Introduction

Propositional logic constitutes a powerful paradigm for knowledge representation and reasoning. Its expressiveness suffices in a wealth of practical settings, and state-of-the-art SAT CDCL solvers usually allow for efficient inference in practice. Unfortunately, answering a query against a propositional knowledge base (KB) is co-NP-*complete*, so, in general there is no guarantee that a solver will be efficient. This is a clear difficulty when a KB is expected to be queried many times. In this context, knowledge compilation and approximation represents an effective alternative, with related work since the early 90s [8,9,11–15,17–19,33,34,40,47–50]. It relies on the key idea of compiling the KB into a tractable target theory in an offline step, and then using the compiled theory online in order to save time.

This paper investigates the computation of Horn least upper bounds (Horn LUBs), introduced in Selman and Kautz's seminal work [48]. Given a propositional formula F, the Horn LUB, F_{LUB}, is the strongest Horn theory entailed by F. So, given the query $F \vDash^? c$, one can check in polynomial time whether $F_{\text{LUB}} \vDash c$, knowing that, in case it is affirmative, $F \vDash c$ holds as well. In the same paper, the authors introduced the related concept of Horn greatest lower bounds (Horn GLBs), which are the weakest Horn theories entailing F. While the Boolean function represented by the Horn LUB is unique, there can be an exponential number of Horn GLBs. Horn GLBs have been studied extensively (e.g. [8,11,14,48,49]). This paper focuses solely on Horn LUBs.

A number of methods have been proposed for computing Horn LUBs. Most of them rely on resolution and restricted forms of resolution. Relevant examples include the works of Selman and Kautz [26,48,49], del Val [18,19] and Langlois, Sloan and Turán [28,29]. A different kind of algorithms, using Reduced

© Springer International Publishing Switzerland 2015
M. Heule and S. Weaver (Eds.): SAT 2015, LNCS 9340, pp. 423–433, 2015.
DOI: 10.1007/978-3-319-24318-4_30

Ordered Binary Decision Diagrams (ROBDDs), have been proposed in [45] for restricted knowledge compilation to different target theories. These algorithms were used in [46] to compare the quality of LUBs considering Horn, contra-dual Horn, Krom and affine target theories. Interestingly, a combination of Horn and dual-Horn LUBs is experimentally shown to represent the most accurate upper-approximations.

Alternatively, this paper studies the use of SAT solvers to compute Horn LUBs. SAT-based approaches constitute the current state of the art for many different function problems on Boolean formulas, such as the computation of minimal unsatisfiable subsets [1,3,5,27,32,38], minimal equivalent subformulas [4], minimal correction subsets [2,21,30,36], maximum satisfiability [22,35,39], maximum autarkies [31] or backbones [24] to mention a few.

This paper proposes two novel algorithms for computing Horn LUBs, which rely on making successive calls to a SAT oracle. These algorithms are analyzed theoretically and evaluated in an experimental study, which shows that they are complementary and are able to solve instances with a non-negligible number of variables.

The paper is organized as follows: Section 2 introduces basic definitions and notation. The proposed SAT-based algorithms for computing Horn LUBs are presented in Section 3, as well as a comparison of both approaches. Section 4 reports the results from an empirical evaluation. Finally, the paper concludes in Section 5.

2   Preliminaries

We assume familiarity with propositional logic [7] and consider Boolean formulas in Conjunctive Normal Form (CNF). A CNF formula F is defined over a set of Boolean variables $\mathsf{var}(F) = X = \{x_1, ..., x_n\}$ as a conjunction of clauses $(c_1 \wedge ... \wedge c_m)$. A clause c is a disjunction of literals $(l_1 \vee ... \vee l_k)$ and a literal l is either a variable x or its negation $\neg x$. Formulas (clauses) can be represented as sets of clauses (literals).

A truth assignment, or interpretation, is a mapping $\mu : X \rightarrow \{0, 1\}$. If all the variables in X are assigned a truth value, μ is referred to as a *complete* assignment. Interpretations can be also seen as conjunctions or sets of literals. Truth valuations are lifted to clauses and formulas as follows: μ satisfies a clause c if it contains at least one of its literals, whereas μ falsifies c if it contains the complements of all its literals. Given a formula F, μ satisfies F (written $\mu \models F$) if it satisfies all its clauses, being μ referred to as a *model* of F. Models such that a set-wise minimal set of the variables are assigned to 1 are referred to as minimal models [6].

Unless otherwise indicated, we consider that models are complete assignments. Given a model μ, its negation $\neg\mu$ is a clause including the literals in μ with complementary polarity. The same applies the other way round. For a model μ, μ^+ (resp. μ^-) denotes the set of positive (resp. negative) literals of μ. The same notation will be used for clauses (c^+, c^-). The set of models of F is denoted as $\mathcal{M}(F)$.

Given formulas F and G, F entails G (written $F \vDash G$) iff $\mathcal{M}(F) \subseteq \mathcal{M}(G)$. They are equivalent (written $F \equiv G$) iff $F \vDash G$ and $G \vDash F$.

Given F an implicant I is a conjunction/set of literals such that $I \vDash F$. An implicate c is a clause such that $F \vDash c$. Prime implicants/implicates are such that they are irreducible w.r.t set-inclusion [42]. Contradictory implicants (e.g. $x \wedge \neg x$) and tautologous implicates (e.g. $x \vee \neg x$) are excluded without loss of generality.

A formula F is satisfiable ($F \nvDash \bot$) if there exists a model for it. Otherwise it is unsatisfiable ($F \vDash \bot$). SAT is the decision problem of determining the satisfiability of a propositional formula. This problem is in general NP-*complete* [16].

Horn formulas constitute an important subclass of propositional logic. These are made of Horn clauses, i.e. clauses having at most one positive literal. Satisfiability of Horn formulas is decidable in linear time [20,23,37]. This paper focuses on computing the Horn least upper bound (Horn LUB) of a satisfiable propositional formula.

Definition 1. *(Horn LUB): Given a satisfiable formula F, a Horn least upper bound (Horn LUB) is a Horn formula H such that $F \vDash H$ and for all Horn formulas H' with $F \vDash H'$, $H \vDash H'$.*

There could be many Horn formulas equivalent to the Horn LUB. All of them represent the same Boolean function, denoted F_{LUB}, which is unique. In the worst-case, the smallest clausal representation of the Horn LUB is exponential on the size of F [49]. It is well-known that the set of all Horn prime implicates of F is equivalent to F_{LUB}.

Definition 2. *(HPI): A clause c is a Horn prime implicate (HPI) of a formula F iff c is Horn, $F \vDash c$ and for all $c' \subsetneq c$, $F \nvDash c'$.*

3 SAT-Based Algorithms

This section presents two novel algorithms for computing Horn LUBs. Both are based on making successive calls to a SAT oracle. The first one, HFLUBBER, relies on iteratively refining a upper approximation of F_{LUB} by finding new irredundant HPIs. The second one, IP-HORN, is built on a novel state-of-the-art SAT-based prime implicate compilation algorithm, which is adapted to produce HPIs. A comparison of both approaches is provided at the end of the section.

3.1 HFLUBBER: An Iterative Refinement Approach

The first approach computes a Horn LUB by iteratively discovering new HPIs that are added to a working upper-approximation F_{HPIC} (initially empty), which will eventually be equivalent to F_{LUB} and potentially much smaller than the set of all HPIs.

HFLUBBER exploits the well-known fact that any prime implicate of a Boolean formula F corresponds to a prime implicant of $\neg F$. Given an arbitrary upper-approximation F_{HPIC} of F_{LUB}, it holds that $F \vDash F_{LUB} \vDash F_{HPIC}$, i.e.,

---

Algorithm 1. HFLUBBER

---

input : F a Boolean formula
output: F_{HPIC} a Horn formula s.t. $F_{\text{HPIC}} \equiv F_{\text{LUB}}$
1 $(F_{\text{HPIC}}, B) \leftarrow (\emptyset, \emptyset)$
2 $(st, \mu) \leftarrow \text{SAT}(\neg F \wedge F_{\text{HPIC}} \wedge B)$
3 **while** st **do**
4     $c_{\text{HPI}} \leftarrow \text{FindHornPrimeImplicate}(F, \neg \mu)$
5   **if** $c_{\text{HPI}} \neq \emptyset$ **then** $F_{\text{HPIC}} \leftarrow F_{\text{HPIC}} \cup \{c_{\text{HPI}}\}$
6   **else** $B \leftarrow B \cup \{\neg \mu\}$
7     $(st, \mu) \leftarrow \text{SAT}(\neg F \wedge F_{\text{HPIC}} \wedge B)$
8 **return** F_{HPIC}

$\mathcal{M}(F) \subseteq \mathcal{M}(F_{\text{LUB}}) \subseteq \mathcal{M}(F_{\text{HPIC}})$. HFLUBBER tries to *refine* F_{HPIC} by removing some models in $\mathcal{M}(F_{\text{HPIC}})$ (not contained in $\mathcal{M}(F)$) by adding a new HPI of F to F_{HPIC}.

Its main organization is shown in Algorithm 1. At each step, it computes a model $\mu \in \mathcal{M}(F_{\text{HPIC}}) \setminus \mathcal{M}(F)$ and checks whether there exists an HPI c_{HPI} of F such that $c_{\text{HPI}} \subseteq \neg \mu$. To this aim, it calls a SAT solver on $\neg F \wedge F_{\text{HPIC}} \wedge B$, where B is a formula blocking previously found models that did not lead to finding a new HPI. Given a model $\mu \models \neg F \wedge F_{\text{HPIC}} \wedge B$, checking the existence of an HPI $c_{\text{HPI}} \subseteq \neg \mu$ can be done within a linear number of calls to the SAT solver: we only need to check maximal subsets $\neg \mu_H \subseteq \neg \mu$ containing at most one positive literal. There are at most $|(\neg \mu)^+| + 1$ options. This is illustrated in Algorithm 2. If there exists such $\neg \mu_H$, it is reduced to an HPI $c_{\text{HPI}} \subseteq \neg \mu_H$ of F and c_{HPI} is added to F_{HPIC}. Otherwise, $\neg \mu$ is added to B, preventing μ from being computed again. Reducing an implicate to a prime implicate is closely related to the task of extracting a minimal unsatisfiable subset [32], so several algorithms can be used [5,10,25,32]. We opted to use a deletion-based approach [5], which has a query complexity bounded on the size of the implicate to be minimized. This algorithm is not shown in detail due to lack of space. HFLUBBER terminates when $\neg F \wedge F_{\text{HPIC}} \wedge B \models \bot$, proving that $F_{\text{HPIC}} \equiv F_{\text{LUB}}$.

HFLUBBER exhibits an interesting property: every new HPI computed is irredundant, i.e. it is not entailed by F_{HPIC}. This is based on the following result:

Proposition 1. *Let F be a formula and an implicant $\mu \models F$. Then $\forall \mu' \subseteq \mu$, $F \not\models \neg \mu'$.*

Proof. Suppose $F \models \neg \mu'$. Then $F \wedge \mu' \models \bot$, which entails $F \wedge \mu \models \bot$. A contradiction, since μ is an implicant of F.

Corollary 1. *HFLUBBER never adds new redundant Horn prime implicates to F_{HPIC}.*

Proof. Every new discovered HPI c_{HPI} is a subset of $\neg \mu$, with $\mu \models F_{\text{HPIC}}$. So, by Proposition 1 $F_{\text{HPIC}} \not\models c_{\text{HPI}}$.

It could be the case that after adding a new HPI, some other HPIs of F_{HPIC} became redundant, so it is not guaranteed that the final F_{HPIC} will be irredundant. Anyway, this property is interesting, as every time F_{HPIC} is added a new

---

Algorithm 2. FindHornPrimeImplicate

---

input : F a Boolean formula, c clause
output: $c_{\mathrm{HPI}} \subseteq c$ an HPI of F, if it exists. Otherwise $\emptyset$

1   $(st, \mu) \leftarrow \mathrm{SAT}(F \wedge \neg(c^-))$
2   **if** $\neg st$ **then**
3      $c_{\mathrm{HPI}} \leftarrow \mathrm{ReduceImplicate}(F, c^-)$
4   **return** c_{HPI}
5   **for** $l \in c^+$ **do**
6      $(st, \mu) \leftarrow \mathrm{SAT}(F \wedge \neg(l \cup c^-))$
7   **if** $\neg st$ **then**
8         $c_{\mathrm{HPI}} \leftarrow \mathrm{ReduceImplicate}(F, l \cup c^-)$
9   **return** c_{HPI}

10 **return** $\emptyset$

---

clause, the approximation gets tighter. Correctness of the algorithm follows from the next result.

Proposition 2. *Let c_{H} be an HPI of F s.t. $F_{\mathrm{HPIC}} \not\models c_{\mathrm{H}}$. Let B be a set of clauses representing the complement of complete models $\mu_{\mathrm{NH}_i} \models \neg F$ s.t. there exists no $\neg\mu_{\mathrm{H}} \subseteq \neg\mu_{\mathrm{NH}_i}$ with $\neg\mu_{\mathrm{H}}$ an HPI of F. There exists a model $\mu \models \neg F \wedge F_{\mathrm{HPIC}} \wedge B$ s.t. $c_{\mathrm{H}} \subseteq \neg\mu$.*

Proof. First, we prove that $F_{\mathrm{HPIC}} \wedge B \not\models c_{\mathrm{H}}$. Recall that $F_{\mathrm{HPIC}} \not\models c_{\mathrm{H}}$. Suppose $F_{\mathrm{HPIC}} \wedge B \models c_{\mathrm{H}}$. Then $F_{\mathrm{HPIC}} \wedge B \wedge \neg c_{\mathrm{H}} \models \bot$, and so for all models μ' s.t. $\mu' \models F_{\mathrm{HPIC}} \wedge \neg c_{\mathrm{H}}$, there exists $\neg\mu_{\mathrm{NH}_i} \in B$ s.t. μ' falsifies $\neg\mu_{\mathrm{NH}_i}$. Equivalently, for all models μ' s.t. $\mu' \models F_{\mathrm{HPIC}} \wedge \neg c_{\mathrm{H}}$, there exists $\neg\mu_{\mathrm{NH}_i} \in B$ such that $\mu' \equiv \mu_{\mathrm{NH}_i}$, and so $\neg c_{\mathrm{H}} \subseteq \mu_{\mathrm{NH}_i}$. Hence $c_{\mathrm{H}} \subseteq \neg\mu_{\mathrm{NH}_i}$. A contradiction.

Now, let μ be a model s.t. $\mu \models F_{\mathrm{HPIC}} \wedge \neg c_{\mathrm{H}} \wedge B$. Note $\neg c_{\mathrm{H}} \subseteq \mu$ and, equivalently, $c_{\mathrm{H}} \subseteq \neg\mu$. Because of monotonicity of logical entailment, we know $\mu \models F_{\mathrm{HPIC}}$, $\mu \models \neg c_{\mathrm{H}}$ and $\mu \models B$. Also, as c_{H} is a prime implicate of F, $\neg c_{\mathrm{H}} \models \neg F$. Hence, $\mu \models \neg c_{\mathrm{H}} \models \neg F$. So, $\mu \models \neg F \wedge F_{\mathrm{HPIC}} \wedge B$ and $c_{\mathrm{H}} \subseteq \neg\mu$.

Corollary 2. *HFLUBBER is correct.*

So, HFLUBBER is guaranteed to compute a Horn formula equivalent to F_{LUB}. It is easy to see that before terminating, it needs to enumerate all the models in $\mathcal{M}(F_{\mathrm{LUB}}) \setminus \mathcal{M}(F)$, so it is expected to be more efficient when the Horn LUB is tight.

3.2   IP-HORN: Computing all Horn Prime Implicates

The second approach is based on the novel state-of-the-art *primer-b* algorithm [41] for compiling a formula F into the set of its prime implicates. This method implicitly exploits the minimal hitting set duality relationship between the set of prime implicants and prime implicates of F [44]. At each iteration it either returns a prime implicate or a prime implicant of F. Upon termination, the set of all prime implicants and a cover of the prime implicates of F are guaranteed to have been computed.

IP-HORN (shown in Algorithm 3) adapts *primer-b* in order to perform knowledge compilation restricted to the target theories of Horn LUBs. To this aim, it instruments a dual version of *primer-b* and adds a special *AtMostOneNeg* constraint. It computes the set of all HPIs. It follows a two-solver approach where an auxiliary formula Q is used to enumerate candidate sets of literals and the original formula F is used to test whether the computed candidates constitute implicants or implicates of F. Q is built using the so-called *dual-rail encoding*, necessary to compute the complete set of HPIs. Thus $\mathrm{var}(Q) \neq \mathrm{var}(F)$ (see [41] for further details). Iteratively, at each step, IP-HORN computes a minimal model μ^Q of Q (line 4), which encodes a candidate set of literals to test. The computation of μ^Q is done following an approach based on SAT with preferences [43]. Since *dual-rail encoding* is used for the formula Q, a mapping from $\mathrm{var}(Q)$ to $\mathrm{var}(F)$ is required (line 6) giving the candidate set C^F to test on F. The *AtMostOneNeg* constraint added to Q (line 2) results in C^F having at most one negative literal. By its construction, it is guaranteed that if C^F falsifies F, then its negation is an HPI of F (note that its negation has at most one positive literal). Otherwise, the candidate is an implicant, which is reduced to a prime implicant. Note that in the case of CNF formulas, computing a prime implicant from an implicant can be done in polynomial time. In both cases, either the prime implicant or the HPI are blocked (line 15) to avoid future repetitions. A proof that no HPI is repeated is a simple extension of the proofs provided in [41]. We refer to that work for further details.

While IP-HORN does not suffer from the limitation of HFLUBBER commented above, it needs to enumerate all the HPIs entailed by F and some prime implicants. If this set is large it would affect its performance, and the resulting Horn formula could be larger than necessary. However the resulting formula could be reduced by computing a minimal equivalent subformula [4].

3.3   Discussion

Both HFLUBBER and IP-HORN aim to exploit the capabilities of modern SAT solvers in the task of computing Horn LUBs. The two approaches are essentially different from each other, exhibiting different strengths and limitations, and so are expected to be well-suited for different kinds of instances.

HFLUBBER computes potentially much smaller Horn formulas than IP-HORN, but it needs to enumerate all the models of $F_{\mathrm{LUB}} \wedge \neg F$ before terminating. So, it is expected to perform well on instances where the Horn LUB is tight.

IP-HORN does not have the limitation of HFLUBBER, but computes the whole set of HPIs, resulting in larger formulas and running times if this set is very large. So, it is expected to work well when the set of HPIs is of reduced size.

Algorithm 3. IP-HORN

---

    input : Formula F
    output: F_{HPIC} a Horn formula s.t. $F_{\text{HPIC}} \equiv F_{\text{LUB}}$

1 $F_{\text{HPIC}} \leftarrow \emptyset$
2 $Q \leftarrow \{(\neg x_v \vee \neg x_{\neg v}) \,|\, v \in \mathsf{var}(F)\} \cup AtMostOneNeg$
3 **while** true **do**
4      $(\mathsf{st}, \mu^Q) \leftarrow \mathsf{MinimalModel}(Q)$
5   **if not** st **then return**
6      $C^F \leftarrow \mathsf{Map}(\mu^Q)$
7      $(\mathsf{st}, \mu^F) \leftarrow \mathsf{SAT}(C^F \wedge F)$
8   **if** st **then**   **#** $\mu^F \vDash F$; i.e. μ^F is an implicant
9         $I_n \leftarrow \mathsf{ReduceImplicant}(\mu^F, F)$
10       $b \leftarrow \{x_l \,|\, l \in I_n\}$
11   **else**   **#** $F \vDash \neg C^F$; i.e. $\neg C^F$ is an implicate
12       $c_{\text{HPI}} \leftarrow \neg C^F$
13       $F_{\text{HPIC}} \leftarrow F_{\text{HPIC}} \cup \{c_{\text{HPI}}\}$
14       $b \leftarrow \{\neg x_l \,|\, l \in c_{\text{HPI}}\}$
15    $Q \leftarrow Q \cup \{b\}$
16 **return** F_{HPIC}

An important remark is that both algorithms generate HPIs from the beginning, so this upper-approximation of F_{LUB} could be used to try to answer deduction queries.

4 Results

We evaluate the proposed algorithms[1] over a set of well-known structured satisfiable CNF formulas, taken from [24], with up to 2000 variables. Most of them have (much) more variables than the instances considered in previous works[2]. In all, there are 131 instances. All the experiments were run on a Linux cluster (2 GHz), setting a limit of 3600s and 4 GB of memory. The algorithms interface the solver Minisat.

Figure 1 shows the running times from HFLUBBER and IP-HORN. Figure 1a includes VBS (for Virtual Best Solver), which emulates a portfolio running both algorithms in parallel. It shows that IP-HORN solves some more instances than HFLUBBER (66 vs 58) by the time limit, while VBS solves more instances (70) taking less time. Nevertheless, in many cases, the results are favorable to

---

[1] Available at http://logos.ucd.ie/web/doku.php?id=hornapp.

[2] Previous works [26,28,29,46,48,49] mostly considered experiments with random 3-CNF formulas with a few tens of variables. In [26,49], the authors consider random instances with up to 200 variables, and also a structured instance with 576 variables, but their experiments are restricted to computing HPIs of size 1. In addition, [28,29] reported that previous methods (e.g. [26,48,49]) cannot scale to more than 75 variables. Finally, in [46] the authors also considered some structured instances with up to 326 variables.

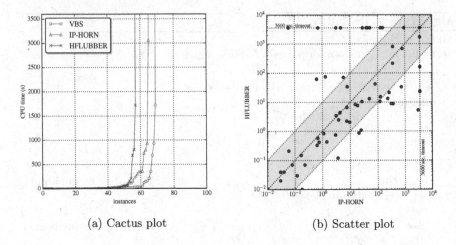

(a) Cactus plot   (b) Scatter plot

Fig. 1. Plots with results from HFLUBBER and IP-HORN.

HFLUBBER, as shown in Figure 1b. This scatter plot, reveals that the two algorithms perform indeed very differently from each other, since for several instances there are very significant differences in favor of one of the methods. This confirms that both algorithms are complementary.

Table 1 reports detailed results for some representative problem instances. For each instance, it shows the number of variables ($|X|$) and clauses ($|F|$), as well as the size of the computed Horn formulas by the time limit (#HPIs) and the time taken in seconds (T) for both algorithms. For HFLUBBER it also shows the number of clauses blocking models in B. The instances with 3600s indicate that the method timed out.

The first rows show instances where HFLUBBER performs better than IP-HORN. In these instances the difference in the number of HPIs of both approximations is remarkable. In some cases, IP-HORN computes orders of magnitude more HPIs than HFLUBBER, resulting in the former being unable to solve some instances. Also, for these instances, HFLUBBER shows that the Horn LUB is quite accurate, as $|B|$ is not very large. The second part of the table show instances where IP-HORN performs better than HFLUBBER. These instances have a small number of HPIs, so IP-HORN is able to compute the Horn LUB quickly. On the other hand, HFLUBBER is unable to terminate by the time limit due to its need of enumerating a large number of models.

In all, both methods are able to compute Horn LUBs for instances with a large number of variables, compared to what has been considered in previous works.

5   Conclusions

Computing Horn least upper bounds represents a well-known problem in knowledge compilation and approximation, which has been studied since the early 90s.

Table 1. Results from HFLUBBER and IP-HORN on some representative problem instances.

Instance	$|X|$	$|F|$	IP-HORN		HFLUBBER		
			#HPIs	T(s)	#HPIs	$|B|$	T(s)
elevator1-b6-s	1639	4437	965070	3600.0	3059	39870	**1721.3**
elevator2-b7-p	1911	5212	8921	258.4	2061	11	**21.4**
pdtvisbpb0	839	2480	2445680	3600.0	3339	4083	**161.8**
parity12-47	525	1539	685798	3063.7	1475	11	**5.3**
parity16-4	920	2712	541744	3600.0	3621	4	**22.9**
parity16-6	899	2649	715290	3600.0	2616	4	**23.8**
pdtvisbpb0	839	2480	2445680	3600.0	3339	4083	**161.8**
3blocks	283	9690	70701	**353.4**	1544	163901	805.9
dme6p1neg	1139	2651	1234	**151.1**	728	203408	3600.0
dme6ptimoneg	1067	2483	1165	**125.9**	960	230843	3600.0
dme3p1neg	622	1445	591	**4.6**	570	576686	3600.0
brpptimoneg	765	1922	728	**934,4**	588	368860	3600.0
srg5ptimoneg	267	626	986	**0.7**	322	996024	3600.0

This paper studies the use of SAT oracles for computing Horn LUBs and proposes two new algorithms, HFLUBBER and IP-HORN, that rely on making successive calls to a CDCL solver. The two approaches are analyzed theoretically and evaluated empirically. The results show that both algorithms are complementary, and enable the computation Horn LUBs for instances with a non-negligible number of variables.

Future research focuses on the development of new algorithms and techniques that mitigate the limitations of HFLUBBER and IP-HORN. Also, it would be interesting to adapt these algorithms to different target theories (e.g. k-Horn LUBs).

Acknowledgments. This work is partially supported by SFI PI grant BEACON (09/IN.1/I2618), by FCT grant POLARIS (PTDC/EIA-CCO/123051/2010), and by national funds through FCT with reference UID/CEC/50021/2013.

References

1. Audemard, G., Lagniez, J.-M., Simon, L.: Improving glucose for incremental SAT solving with assumptions: application to MUS extraction. In: Järvisalo, M., Van Gelder, A. (eds.) SAT 2013. LNCS, vol. 7962, pp. 309–317. Springer, Heidelberg (2013)

2. Bacchus, F., Davies, J., Tsimpoukelli, M., Katsirelos, G.: Relaxation search: a simple way of managing optional clauses. In: AAAI, pp. 835–841 (2014)

3. Belov, A., Heule, M.J.H., Marques-Silva, J.: MUS extraction using clausal proofs. In: Sinz, C., Egly, U. (eds.) SAT 2014. LNCS, vol. 8561, pp. 48–57. Springer, Heidelberg (2014)

4. Belov, A., Janota, M., Lynce, I., Marques-Silva, J.: Algorithms for computing minimal equivalent subformulas. Artif. Intell. **216**, 309–326 (2014)

5. Belov, A., Lynce, I., Marques-Silva, J.: Towards efficient MUS extraction. AI Commun. **25**(2), 97–116 (2012)

6. Ben-Eliyahu, R., Dechter, R.: On computing minimal models. Ann. Math. Artif. Intell. **18**(1), 3–27 (1996)
7. Biere, A., Heule, M., van Maaren, H., Walsh, T. (eds.): Handbook of Satisfiability. Frontiers in Artificial Intelligence and Applications, vol. 185. IOS Press (2009)
8. Boufkhad, Y.: Algorithms for propositional KB approximation. In: AAAI, pp. 280–285 (1998)
9. Boufkhad, Y., Grégoire, É., Marquis, P., Mazure, B., Sais, L.: Tractable cover compilations. In: IJCAI, pp. 122–127 (1997)
10. Bradley, A.R., Manna, Z.: Checking safety by inductive generalization of counterexamples to induction. In: FMCAD, pp. 173–180 (2007)
11. Cadoli, M.: Semantical and computational aspects of Horn approximations. In: IJCAI, pp. 39–45 (1993)
12. Cadoli, M., Donini, F.M.: A survey on knowledge compilation. AI Commun. **10**(3–4), 137–150 (1997)
13. Cadoli, M., Donini, F.M., Liberatore, P., Schaerf, M.: Space efficiency of propositional knowledge representation formalisms. J. Artif. Intell. Res. (JAIR) **13**, 1–31 (2000)
14. Cadoli, M., Scarcello, F.: Semantical and computational aspects of Horn approximations. Artif. Intell. **119**(1–2), 1–17 (2000)
15. Castell, T.: Computation of prime implicates and prime implicants by a variant of the Davis and Putnam procedure. In: ICTAI, pp. 428–429 (1996)
16. Cook, S.A.: The complexity of theorem-proving procedures. In: STOC, pp. 151–158 (1971)
17. Darwiche, A., Marquis, P.: A knowledge compilation map. J. Artif. Intell. Res. (JAIR) **17**, 229–264 (2002)
18. del Val, A.: An analysis of approximate knowledge compilation. In: IJCAI, pp. 830–836 (1995)
19. del Val, A.: First order LUB approximations: characterization and algorithms. Artif. Intell. **162**(1–2), 7–48 (2005)
20. Dowling, W.F., Gallier, J.H.: Linear-time algorithms for testing the satisfiability of propositional Horn formulae. J. Log. Program. **1**(3), 267–284 (1984)
21. Grégoire, É., Lagniez, J.M., Mazure, B.: An experimentally efficient method for (MSS, coMSS) partitioning. In: AAAI, pp. 2666–2673 (2014)
22. Ignatiev, A., Morgado, A., Manquinho, V.M., Lynce, I., Marques-Silva, J.: Progression in maximum satisfiability. In: ECAI, pp. 453–458 (2014)
23. Itai, A., Makowsky, J.A.: Unification as a complexity measure for logic programming. J. Log. Program. **4**(2), 105–117 (1987)
24. Janota, M., Lynce, I., Marques-Silva, J.: Algorithms for computing backbones of propositional formulae. AI Commun. **28**(2), 161–177 (2015)
25. Junker, U.: QuickXplain: preferred explanations and relaxations for overconstrained problems. In: AAAI, pp. 167–172 (2004)
26. Kautz, H.A., Selman, B.: An empirical evaluation of knowledge compilation by theory approximation. In: AAAI, pp. 155–161 (1994)
27. Lagniez, J.-M., Biere, A.: Factoring out assumptions to speed Up MUS extraction. In: Järvisalo, M., Van Gelder, A. (eds.) SAT 2013. LNCS, vol. 7962, pp. 276–292. Springer, Heidelberg (2013)
28. Langlois, M., Sloan, R.H., Turán, G.: Horn upper bounds and renaming. In: Marques-Silva, J., Sakallah, K.A. (eds.) SAT 2007. LNCS, vol. 4501, pp. 80–93. Springer, Heidelberg (2007)
29. Langlois, M., Sloan, R.H., Turán, G.: Horn upper bounds and renaming. JSAT **7**(1), 1–15 (2009)

30. Marques-Silva, J., Heras, F., Janota, F., Previti, A., Belov, A.: On computing minimal correction subsets. In: IJCAI (2013)
31. Marques-Silva, J., Ignatiev, A., Morgado, A., Manquinho, V.M., Lynce, I.: Efficient autarkies. In: ECAI, pp. 603–608 (2014)
32. Marques-Silva, J., Janota, M., Belov, A.: Minimal sets over monotone predicates in boolean formulae. In: Sharygina, N., Veith, H. (eds.) CAV 2013. LNCS, vol. 8044, pp. 592–607. Springer, Heidelberg (2013)
33. Marquis, P.: Knowledge compilation using theory prime implicates. In: IJCAI, pp. 837–845 (1995)
34. Marquis, P.: Consequence finding algorithms. In: Handbook of Defeasible Reasoning and Uncertainty Management Systems, pp. 41–145 (2000)
35. Martins, R., Manquinho, V., Lynce, I.: Open-WBO: a modular MaxSAT solver'. In: Sinz, C., Egly, U. (eds.) SAT 2014. LNCS, vol. 8561, pp. 438–445. Springer, Heidelberg (2014)
36. Mencía, C., Previti, A., Marques-Silva, J.: Literal-based MCS extraction. In: IJCAI, pp. 1973–1979 (2015)
37. Minoux, M.: LTUR: A simplified linear-time unit resolution algorithm for Horn formulae and computer implementation. Inf. Process. Lett. **29**(1), 1–12 (1988)
38. Nadel, A., Ryvchin, V., Strichman, O.: Efficient MUS extraction with resolution. In: FMCAD, pp. 197–200 (2013)
39. Narodytska, N., Bacchus, F.: Maximum satisfiability using core-guided MaxSAT resolution. In: AAAI, pp. 2717–2723 (2014)
40. Palopoli, L., Pirri, F., Pizzuti, C.: Algorithms for selective enumeration of prime implicants. Artif. Intell. **111**(1–2), 41–72 (1999)
41. Previti, A., Ignatiev, A., Morgado, A., Marques-Silva, J.: Prime compilation of non-clausal formulae. In: IJCAI, pp. 1980–1987 (2015)
42. Reiter, R.: A theory of diagnosis from first principles. Artif. Intell. **32**(1), 57–95 (1987)
43. Rosa, E.D., Giunchiglia, E., Maratea, M.: Solving satisfiability problems with preferences. Constraints **15**(4), 485–515 (2010)
44. Rymon, R.: An SE-tree-based prime implicant generation algorithm. Ann. Math. Artif. Intell. **11**(1–4), 351–366 (1994)
45. Schachte, P., Søndergaard, H.: Boolean approximation revisited. In: Miguel, I., Ruml, W. (eds.) SARA 2007. LNCS (LNAI), vol. 4612, pp. 329–343. Springer, Heidelberg (2007)
46. Schachte, P., Søndergaard, H., Whiting, L., Henshall, K.: Information loss in knowledge compilation: A comparison of boolean envelopes. Artif. Intell. **174**(9–10), 585–596 (2010)
47. Schrag, R.: Compilation for critically constrained knowledge bases. In: AAAI, pp. 510–515 (1996)
48. Selman, B., Kautz, H.A.: Knowledge compilation using Horn approximations. In: AAAI, pp. 904–909 (1991)
49. Selman, B., Kautz, H.A.: Knowledge compilation and theory approximation. J. ACM **43**(2), 193–224 (1996)
50. Simon, L., del Val, A.: Efficient consequence finding. In: IJCAI, pp. 359–370 (2001)

Author Index

Printed in the United States
By Bookmasters